FOURTH EDITION

CHEMICAL THERMODYNAMICS

Basic Theories and Methods

Dedicated to the memory of
Thomas Fraser Young
September 28, 1897–April 1, 1977
Esteemed scientist, dedicated teacher

FOURTH EDITION

CHEMICAL THERMODYNAMICS

Basic Theory and Methods

Irving M. Klotz

Northwestern University

Robert M. Rosenberg

Lawrence University

The Benjamin/Cummings Publishing Company, Inc.

Menlo Park, California • Reading, Massachusetts
Don Mills, Ontario • Wokingham, U.K. • Amsterdam • Sydney
Singapore • Tokyo • Madrid • Bogota • Santiago • San Juan

Sponsoring Editor Diane Bowen
Production Coordinator Kristina Montague
Copy Editor Carol Dondrea
Artist Ben Turner Graphics
Cover Designer Michael Rogondino

Library of Congress Cataloging in Publication Data

Klotz, Irving M. (Irving Myron), 1916–
Chemical thermodynamics.

Bibliography: p.
Includes index.
1. Thermodynamics. I. Rosenberg, Robert M.,
1926– . II. Title.
QD504.K55 1986 541.3′69 86-3303
ISBN 0-8053-5501-4

ABCDEFGHIJ-MA-89876

The Benjamin/Cummings Publishing Company, Inc.

2727 Sand Hill Road
Menlo Park, California 94025

"A theory is the more impressive the greater the simplicity of its premises is, the more different kinds of things it relates, and the more extended is its area of applicability. Therefore, the deep impression which classical thermodynamics made upon me. It is the only physical theory of universal content concerning which I am convinced that, within the framework of the applicability of its basic concepts, it will never be overthrown."

Albert Einstein,
Autobiographical Notes, page 33 in
The Library of Living Philosophers, Vol. VII;
Albert Einstein: Philosopher-Scientist,
edited by P. A. Schilpp,
Open Court Publishing Company,
La Salle, Illinois, 1973.

Preface

This revision presents the fourth version of a textbook that appeared originally thirty-five years ago. The fundamental purpose of the book remains unchanged, to present to the student the logical foundations and interrelationships of the theory of thermodynamics and to teach the student the methods by which the theoretical methods may be applied to practical problems.

In the treatment of theoretical principles, we have adopted the classical, or phenomenological, approach to thermodynamics and have excluded entirely the statistical viewpoint. This attitude has several pedagogical advantages. First, it permits the maintenance of a logical unity throughout the book. In addition, it offers an opportunity to stress the "operational" approach to abstract concepts. Furthermore, it makes some contribution toward freeing the student from a perpetual yearning for a mechanical analogue for every new concept that is introduced. Finally, and perhaps most important, it avoids the promulgation of an all-too-common point of view toward statistical thermodynamics as an appendage which can be conveniently grafted on to the body of thermodynamics. A logical development of the statistical approach should probably be based on a previous introduction to the fundamental quantum-mechanical concept of energy level and should emphasize the much broader scope of the phenomena that it can treat. An effective presentation of statistical theory should be, therefore, an independent and complementary one to phenomenological thermodynamics.

A great deal of attention is paid in this text to training the student in the application of the theory of thermodynamics to problems that are commonly encountered by the chemist, the biologist, and the geologist. The mathematical tools that are necessary for this purpose are considered in more detail than is usual. In addition, computational techniques, graphical, numerical, and analytical, are described fully and are used frequently, both in illustrative and in assigned problems. Furthermore, exercises have been designed to simulate more than in most texts the type of problem that may be encountered by the practicing scientist. Short, unrelated exercises are thus kept to a minimum, whereas series of computations or derivations, illustrating a technique or principle of general applicability, are emphasized.

We have also made a definite effort to keep this volume within limits that can be covered in a course of lectures extending over a period of twelve to fifteen weeks. Too often, a textbook that attempts to be exhaustive in its coverage merely serves to overwhelm the student. On the other hand, if a student can be guided to a sound grasp of the fundamental principles and shown how these can be applied to a few typical problems, that student will be capable of examining other special topics independently or with the aid of one of the excellent comprehensive treatises that are available.

Another feature of this book is the extensive use of subheadings in outline form to indicate the position of a given topic in the general sequence of presentation. In using this method of presentation, we have been influenced strongly by the viewpoint expressed so aptly by Poincaré: "The order in which these elements are placed is much more important than the elements themselves. If I have the feeling . . . of this order, so as to perceive at a glance the reasoning as a whole, I need no longer fear lest I forget one of the elements, for each of them will take its allotted place in the array, and that without any effort of memory on my part."[1] It is a universal experience of teachers, that students are able to retain a body of information much more effectively if they are aware of the place of the parts in the whole.

Although thermodynamics has not changed fundamentally since the first edition was published, conventions and pedagogical approaches have changed, and new applications continue to appear. The application of thermodynamics to biological and geological problems has been particularly fruitful. We have taken the opportunity, therefore, to revise our approach to some topics, to use SI units, to use the new standard state of 0.1 MPa (1 bar), and to add problems that reflect new applications. In addition, we have added a new chapter on equilibrium in gravitational and ultracentrifugal fields, a subject of both geological and biological interest.

We acknowledge helpful comments on the manuscript from Henry Bent, David Volman, O. D. Bonner, Claude Meares, John E. Bauman, and Laurence Strong. We also thank Edgar Westrum for making it clear that the title of Chapter 18 needed revision. We thank Carol Techlin for careful typing of a difficult manuscript, Cheryl Chisnell for recording index entries into the computer, and E. Virginia Hobbs of McGaffey Associates for producing the index. We are grateful to Dr. E. Richard Cohen for the tentative values of the fundamental constants in Table 2-3. A solutions manual that contains solutions to most of the problems in the text is available from the publisher.

Evanston, Illinois
I.M.K.
Appleton, Wisconsin
R.M.R.

[1]H. Poincaré, *The Foundations of Science*, translated by G. B. Halsted, Science Press, 1913.

Contents

11 The Third Law of Thermodynamics 221

12 Thermodynamics of Systems of Variable Composition 257

13 Mixtures of Gases 271

14 The Phase Rule 297

15 The Ideal Solution 313

16 Dilute Solutions of Nonelectrolytes 329

17 Activities and Standard States for Nonelectrolytes 345

18 Calculation of Partial Molar Quantities from Experimental Data: Volume and Enthalpy 367

19 Determination of Nonelectrolyte Activities 399

FOURTH EDITION

CHEMICAL THERMODYNAMICS
Basic Theories and Methods

CHAPTER 1

Introduction

1-1 ORIGINS OF CHEMICAL THERMODYNAMICS

An alert young scientist with only an elementary background in his or her field might be surprised to learn that a subject called "thermodynamics" has any relevance to chemical change or to biological and geological systems. The term *thermodynamics*, taken literally, implies a field concerned with the mechanical action produced by heat. Lord Kelvin invented the name to direct attention to the *dynamic* nature of *heat* and to contrast this perspective with previous conceptions of heat as a type of fluid.

In contrast to mechanics, electromagnetic field theory, or relativity, where the names of Newton, Maxwell, and Einstein stand out uniquely, the foundations of thermodynamics arose from the thinking of over half a dozen individuals: Carnot, Mayer, Joule, Helmholtz, Rankine, Kelvin, Clausius [1]. Each provided crucial steps leading to the grand synthesis of the two classical laws of thermodynamics.

The conceptual bottle into which was packaged eighteenth-century and early nineteenth-century views of the nature of heat was the principle of conservation of caloric. This principle is an eminently attractive basis for rationalizing simple observations such as temperature changes that occur when a cold body is placed in contact with a hot one. The cold body appears to have extracted something from the hot one. Furthermore, if both bodies are constituted of the same material, and the cold object has twice the mass of the hot one, then we observe that the increase in temperature of the former is only half the decrease in temperature of the latter. A conservation principle arises naturally. From this, the notion of the flow of a substance from the hot to the cold body appears almost intuitively, together with the concept that the total quantity of the caloric can be represented by the product of the mass multiplied by the temperature change. With these ideas in mind, Black was led to the discovery of specific heat, heat of fusion, and heat of vaporization. Such successes established the concept of caloric so solidly and persuasively that it blinded even the greatest scientists of the early nineteenth century. Thus, they missed seeing well-known facts that were common knowledge even in primitive cultures—for example, that heat can be produced by friction.

It seems clear that the earliest of the founders of thermodynamics, Carnot, accepted conservation of caloric as a basic axiom in his analysis [2] of the heat engine (although a few individuals [3] claim to see an important distinction in the contexts of Carnot's uses of "calorique" versus "chaleur").

Although Carnot's primary objective was to evaluate the mechanical efficiency of a steam engine, his analysis introduced certain broad concepts whose significance goes far beyond engineering problems. One of these concepts is the reversible process, which provides for thermodynamics the corresponding idealization that "frictionless motion" contributes to mechanics. The idea of "reversibility" has applicability much beyond ideal heat engines. Furthermore, it introduces continuity into the visualization of the process being considered; hence, it invites the introduction of the differential calculus. It was Clapeyron [4] who actually expounded Carnot's ideas in the notation of calculus and who thereby derived the vapor pressure equation associated with his name, as well as the performance characteristics of ideal engines.

Carnot also leaned strongly on the analogy between a heat engine and a hydrodynamic one (the water wheel) for, as he said,

> we can reasonably compare the motive power of heat
> with that of a head of water.

For the heat engine, one needs two temperature levels (a boiler and a condenser) that correspond to the two levels in height of a waterfall. For a waterfall, the quantity of water discharged by the wheel at the bottom level is the same as the quantity that entered originally at the top level, work being generated by the drop in gravitational level. Therefore, Carnot postulated that a corresponding thermal quantity, "calorique," was carried by the heat engine from a high temperature to a low one; the heat that entered at the upper temperature level was conserved and exited in exactly the same quantity at the lower temperature, work having been produced during the drop in temperature level. Using this postulate, he was able to answer in a general way the long-standing question of whether steam was suited uniquely for a heat engine; he did this by showing that in the ideal engine any other substance would be just as efficient. It was also from this construct that Kelvin subsequently realized that one could establish an absolute temperature scale independent of the properties of any substance.

When faced in the late 1840s with the idea of conservation of (heat plus work) proposed by Joule, Helmholtz, and Mayer, Kelvin at first rejected it (as did the Proceedings of the Royal Society when presented with one of Joule's manuscripts) because conservation of energy (work plus heat) was inconsistent with the Carnot analysis of the fall of an *unchanged* quantity of heat through an ideal thermal engine to produce work. Ultimately, however, between 1849 and 1851, Kelvin and Clausius, each reading the other's papers closely, came to recognize that Joule and Carnot could be made concordant if it was assumed that only *part* of the heat entering the Carnot engine at the high temperature was released at the lower level and that the *difference* was converted into work. Clausius was the first to express this in print. Within the next few years, Kelvin developed the mathematical expression $\Sigma Q/T = 0$ for "the second fundamental law of the dynamical theory of heat" and began to use the word *thermodynamic*, which he had actually coined earlier. Clausius's analysis [5] led him, in turn, to the mathematical formulation of $\int dQ/T \geqslant 0$ for the second law; in addition, he invented the term *entropy* (as an alternative to Kelvin's "dissipation of energy"), for, as he says,

> I hold it better to borrow terms for important magnitudes from the ancient languages so that they may be adopted unchanged in all modern languages.

Thereafter, many individuals proceeded to show that the two fundamental laws, explicitly so-called by Clausius and Kelvin, were applicable to all types of macroscopic natural phenomena and not just to heat engines. During the latter part of the nineteenth century, then, the scope of thermodynamics widened greatly. It became apparent that the same concepts that allow one to predict the maximum efficiency of a heat engine apply to other energy transformations, including transformations in chemical, biological, and geological systems in which an energy change is not obvious. For example, thermodynamic principles permit the computation of the maximum yield in the synthesis of ammonia from nitrogen and hydrogen under a variety of conditions of temperature and pressure, with important consequences to the chemical fertilizer industry. Similarly, the equilibrium distribution of sodium and potassium ions between red blood cells and blood plasma can be calculated from thermodynamic relationships. It was the observation of deviations from an equilibrium distribution that led to a search for mechanisms of active transport of these alkali metal ions across the cell membrane. Also, thermodynamic calculations of the effect of temperature and pressure on the transformation between graphite and diamond have generated hypotheses about the geological conditions under which natural diamonds can be made.

For these and other phenomena, thermal and work quantities, although controlling factors, are only of indirect interest. Accordingly, a more refined formulation of thermodynamic principles was established, particularly by J. Willard Gibbs [6], that emphasized the nature and use of a number of special energy functions to describe the state of a system. These have proved very convenient and powerful in prescribing the rules that govern chemical and physical transitions. Therefore, in a sense, the name "energetics" is more descriptive than "thermodynamics," insofar as applications to chemistry are concerned. More commonly, one affixes the adjective "chemical" to thermodynamics to indicate the change in emphasis, and to modify the literal and original meaning of thermodynamics.

1-2 OBJECTIVES OF CHEMICAL THERMODYNAMICS

In practice, the primary objective of chemical thermodynamics is to establish a criterion for determining the feasibility or spontaneity of a given physical or chemical transformation. For example, we may be interested in a criterion for determining the feasibility of a spontaneous transformation from one phase to another, such as the conversion of graphite to diamond, or the spontaneous direction of a metabolic reaction that occurs in a cell. On the basis of the first and second laws of thermodynamics, expressed in terms of Gibbs's energy functions, several additional theoretical concepts and mathematical functions have been developed that provide a powerful approach to the solution of these questions.

Once the spontaneous direction of a natural process is determined, we may wish to know how far the process will proceed before reaching equilibrium. For example, we might want to find the maximum yield of an industrial process, or the equilibrium solubility of atmospheric carbon dioxide in natural waters, or the equilibrium concentration of a group of metabolites in a cell. Thermodynamic methods provide the mathematical relations required to estimate such quantities.

Although the main objective of chemical thermodynamics is the analysis of spontaneity and equilibrium, the methods also are applicable to many other problems. For example, the study of phase equilibria, in ideal and nonideal systems, is basic to the intelligent use of the techniques of extraction, distillation, and crystallization, to metallurgical operations, and to the understanding of the species of minerals found in geological systems. Similarly, the energy changes that accompany a physical or chemical transformation, in the form of either heat or work, are of great interest, whether the transformation is the combustion of a fuel, the fission of a uranium nucleus, or the transport of a metabolite against a concentration gradient. Thermodynamic concepts and methods provide a powerful approach to the understanding of such problems.

1-3 LIMITATIONS OF CLASSICAL THERMODYNAMICS

Although descriptions of chemical change are permeated with the terms and language of molecular theory, the concepts of classical thermodynamics are independent of molecular theory; thus these concepts do not require modification as our knowledge of molecular structure changes. This feature is an advantage in a formal sense, but it is also a distinct limitation because we cannot obtain information at a molecular level from classical thermodynamics.

In contrast to molecular theory, classical thermodynamics deals only with measurable properties of matter in bulk (for example, pressure, temperature, volume, electromotive force, magnetic susceptibility, and heat capacity). It is an empirical and phenomenological science, and in this sense it resembles classical mechanics. The latter also is concerned with the behavior of macroscopic systems, with the position and the velocity of a body as a function of time, without regard to the body's molecular nature.

Statistical mechanics (or *statistical thermodynamics*) is the science that relates the properties of individual molecules and their interactions to the empirical results of classical thermodynamics. The laws of classical and quantum mechanics are applied to molecules; then, by suitable statistical averaging methods, the rules of macroscopic behavior that would be expected from an assembly of a large number of such molecules are formulated. Since classical thermodynamic results are compared to statistical averages over very large numbers of molecules, it is not surprising that fluctuation phenomena, such as Brownian motion, the "shot effect," or certain turbidity phenomena, cannot be treated by classical thermodynamics. Now we recognize that all such phenomena are expressions of local microscopic fluctuations in the behavior of a relatively few molecules that deviate randomly from the average behavior of the entire assembly. In this submicroscopic region, such random fluctuations make it impossible to assign a definite value to properties such as temperature or pressure. However, classical thermodynamics is predicated on the assumption that a definite and reproducible value always can be measured for such properties.

In addition to these formal limitations, there also are limitations of a more functional nature. Although the theory of thermodynamics provides the foundation for the solution of many chemical problems, the answers obtained generally are not definitive. Using the language of the mathematician, we might say that classical thermodynamics is capable of formulating necessary conditions but not sufficient conditions. Thus, a thermodynamic analysis may rule out a given reaction for the synthesis of some substance by indicating that such a transformation cannot proceed spontaneously under any set of available conditions. In such a case we have a definitive answer. However, if the analysis indicates that a reaction may proceed spontaneously, no statement can be made from classical thermodynamics alone indicating that it will do so in any finite time.

For example, classical thermodynamic methods predict that the maximum equilibrium yield of ammonia from nitrogen and hydrogen is obtained at low temperatures. Yet, under these optimum thermodynamic conditions, the rate of reaction is so slow that the process is not practical for industrial use. Thus, a smaller equilibrium yield at high temperature must be accepted to obtain a suitable reaction rate. However, although the thermodynamic calculations provide no assurance that an equilibrium yield will be obtained in a finite time, it was as a result of such calculations for the synthesis of ammonia that an intensive search was made for a catalyst that would allow equilibrium to be reached.

Similarly, specific catalysts called enzymes are important factors in determining what reactions occur at an appreciable rate in biological systems. For example, adenosine triphosphate is thermodynamically unstable in aqueous solution with respect to hydrolysis to adenosine diphosphate and inorganic phosphate. Yet this reaction proceeds very slowly in the absence of the specific enzyme adenosine triphosphatase. This combination of thermodynamic control of direction and enzyme control of rate makes possible the finely balanced system that is a living cell.

In the case of the graphite-to-diamond transformation, thermodynamic results predict that diamond is the stable allotrope at all temperatures below the transition temperature and that graphite is the stable allotrope at all temperatures above the transition temperature. But a practical production rate of diamond from graphite can be obtained only in a narrow temperature range just below the transition temperature, and only then at pressure sufficiently high that the transition temperature is about 2000 K.

Just as thermodynamic methods provide only a limiting value for the yield of a chemical reaction, so also do they provide only a limiting value for the work obtainable from a chemical or physical transformation. Thermodynamic functions predict the work that may be obtained if the reaction is carried out with infinite slowness, in a so-called reversible manner. However, it is impossible to specify the actual work obtained in a real or natural process in which the time interval is finite. We can state, nevertheless, that the real work will be less than the work obtainable in a reversible situation.

For example, thermodynamic calculations will provide a value for the maximum voltage of a storage battery—that is, the voltage that is obtained when no current is drawn. When current is drawn, we can predict that the voltage will be less than the maximum value, but we cannot predict how much less. Similarly, we can calculate the maximum amount of heat that can be transferred from a cold environment into a building by the expenditure of a certain amount of work in a heat pump, but the actual performance will be less satisfactory. Given a nonequilibrium distribution of ions across a cell membrane, we can calculate the *minimum* work required to maintain such a distribution. However, the actual process that occurs in the cell requires much more work than the calculated value because the process is carried out irreversibly.

Although classical thermodynamics can treat only limiting cases, such a restriction is not nearly as severe as it may seem at first glance. In many cases it is possible to approach equilibrium very closely, and the thermodynamic quantities coincide with actual values, within experimental error. In other situations thermodynamic analysis may rule out certain reactions under any conditions, and a great deal of time and effort can be saved. Even in their most constrained applications, such as limiting solutions within certain boundary values, thermodynamic methods can reduce materially the amount of experimental work necessary to yield a definitive answer to a particular problem.

REFERENCES

1. D. S. L. Cardwell, *From Watt to Clausius: The Rise of Thermodynamics in the Early Industrial Age*, Cornell University Press, Ithaca, New York, 1971.
2. S. Carnot, *Reflexions sur la puissance motrice du feu*, Bachelier, Paris, 1824. E. Mendoza, ed., *Reflections on the Motive Power of Fire by Sadi Carnot*, Dover, New York, 1960; T. S. Kuhn, *Am. J. Phys.* **23**, 91 (1955).
3. H. L. Callendar, *Proc. Phys. Soc. (London)* **23**, 153 (1911), and V. K. LaMer, *Am. J. Phys.* **23**, 95 (1955).
4. E. Clapeyron, *J. Ecole Polytech. (Paris)* **14**, 153 (1834). Mendoza, *Motive Power of Fire*.
5. R. Clausius, *Pogg. Ann. Series III* **79**, 368, 500 (1859); *Series V* **5**, 353 (1865); *Ann. Phys.* **125**, 353 (1865). Mendoza, *Motive Power of Fire*.
6. J. W. Gibbs, *Trans. Conn. Acad. Sci.* **3**, 228 (1876). See also *The Collected Works of J. Willard Gibbs*, Yale University Press, New Haven, 1928; reprinted 1957.

CHAPTER 2

Mathematical Techniques

Ordinary language is deficient in varying degrees for expressing the ideas and findings of science. An exact science must be founded on precise definitions, which are difficult to obtain by verbalization. Mathematics, however, offers a precise mode of expression. Mathematics also provides a rigorous logical procedure and a device for the development in succinct form of a long and often complicated argument. A long train of abstract thought can be condensed with full preservation of continuity into brief mathematical notation; thus, we can proceed readily with further steps in reasoning without carrying in our minds the otherwise overwhelming burden of all previous steps in the sequence.

Most branches of theoretical science can be expounded at various levels of abstraction. The most elegant and formal approach to thermodynamics, that of Caratheodory [1], depends on a familiarity with a special type of differential equation (Pfaff equation) with which the usual student of chemistry is unacquainted. However, an introductory presentation of thermodynamics follows best along historical lines of development, for which only the elementary principles of calculus are necessary. This is the approach we follow here. Nevertheless, we also discuss exact differentials and Euler's theorem, since many concepts and derivations can be presented in a more satisfying and precise manner with their use.

2-1 VARIABLES OF THERMODYNAMICS

Extensive and Intensive Quantities

In the study of thermodynamics we can distinguish between variables that are independent of the quantity of matter in a system, the *intensive variables*, and variables that depend on the quantity of matter. Of the latter group, those whose values are directly proportional to the quantity of matter are of particular interest and are simple to deal with mathematically. These are called *extensive variables*. Volume and heat capacity are typical examples of extensive variables, whereas temperature, pressure, viscosity, concentration, and molar heat capacity are examples of intensive variables.

Units and Conversion Factors

The base units of measurement under the Système International d'Unités, or SI units, are given in Table 2-1 [2].

Some SI-derived units with special names are included in Table 2-2. The standard atmosphere may be used temporarily with SI units; it is defined to be equal to 1.01325×10^5 Pa. The thermochemical calorie is no longer recommended as a unit of energy, but it is defined in terms of an SI unit as 4.184 J. The unit of volume, liter, symbol L, is now defined as 1 dm^3.

The authoritative values for physical constants and conversion factors used in thermodynamic calculations are assembled in Table 2-3 [3]. Further information about the proper use of physical quantities, units, and symbols can be found in several additional sources [4].

2-2 THEORETICAL METHODS

Partial Differentiation

Since the state of a thermodynamic system generally is a function of more than one independent variable, it is necessary to consider the mathematical

Table 2-1 Base SI Units

Quantity	Name	Symbol
Length	meter	m
Mass	kilogram	kg
Time	second	s
Electric current	ampere	A
Thermodynamic temperature[a]	kelvin	K
Amount of substance[b]	mole	mol
Luminous intensity	candela	cd

[a]The thirteenth General Conference on Weights and Measures in 1967 recommended that the kelvin, symbol K, be used both for thermodynamic temperature and for thermodynamic temperature interval, and that the unit symbols °K and deg be abandoned. The kelvin is defined as 1/(273.16) of the thermodynamic temperature of the triple point of water.

[b]The amount of substance should be expressed in units of moles, one mole being Avogadro's constant number of designated particles or groups of particles, whether these are electrons, atoms, molecules, or the reactants and products specified by a chemical equation.

Table 2-2 SI-Derived Units

Quantity	Name	Symbol	Expression in terms of other units
Force	newton	N	$m\ kg\ s^{-2}$
Pressure	pascal	Pa	$N\ m^{-2}$
	bar	bar	$N\ m^{-2}$
Energy, work, quantity of heat	joule	J	N m
Power	watt	W	$J\ s^{-1}$
Electric charge	coulomb	C	A s
Electric potential, electromotive force	volt	V	$W\ A^{-1}$, $J\ C^{-1}$
Celsius temperature	degree celsius	°C	K
Heat capacity, entropy	joule per kelvin		$J\ K^{-1}$

techniques for expressing these relationships. Many thermodynamic problems involve only two independent variables, and the extension to more variables is generally obvious, so we will limit our illustrations to functions of two variables.

Equation for the Total Differential. Let us consider a specific example—the volume of a pure substance. The molar volume is a function only of the temperature and pressure of the substance; thus, the relationship can be written in general notation as

$$v = f(P, T) \tag{2-1}$$

Table 2-3 Fundamental Constants and Conversion Factors

ice-point temperature, (0°C) = 273.15 kelvins (K)

thermodynamic calorie, cal = 4.184 joules (J)

gas constant, $R = 8.31451\ J\ K^{-1}\ mol^{-1}$ (±8.5 ppm)[a]

$= 1.98721\ cal_{th}\ K^{-1}\ mol^{-1}$

$= 0.0820578\ dm^3\ atm\ K^{-1}\ mol^{-1}$

Avogadro constant, $N_A = 6.022137 \cdot 10^{23}\ mol^{-1}$ (±1 ppm)[a]

Faraday constant, $\mathcal{F} = 96485.32\ C\ mol^{-1}$ (±0.5 ppm)[a]

elementary charge, $e = 1.6021772 \cdot 10^{-19}\ C$ (±5 ppm)[a]

Boltzmann constant, $k = 1.38066 \cdot 10^{-23}\ J/K$ (±8.5 ppm)

[a]Tentative values from the ICSU-CODATA Task Group on Fundamental Constants, E. R. Cohen, personal communication, 1986.

in which the small capital indicates a molar quantity. Utilizing the principles of calculus [5], we can write for the total differential

$$dv = \left(\frac{\partial v}{\partial P}\right)_T dP + \left(\frac{\partial v}{\partial T}\right)_P dT \tag{2-2}$$

For the special case of one mole of an ideal gas, Equation 2-1 becomes

$$v = \frac{RT}{P} = R(T)\left(\frac{1}{P}\right) \tag{2-3}$$

Since the partial derivatives are given by the expressions

$$\left(\frac{\partial v}{\partial P}\right)_T = -\frac{RT}{P^2} \tag{2-4}$$

and

$$\left(\frac{\partial v}{\partial T}\right)_P = \frac{R}{P} \tag{2-5}$$

the total differential for the special case of the ideal gas can be obtained by substituting from Equations 2-4 and 2-5 into Equation 2-2 and is given by the relationship

$$dv = -\frac{RT}{P^2} dP + \frac{R}{P} dT \tag{2-6}$$

We shall have frequent occasion to use this expression.

Conversion Formulas. Often there is no convenient experimental method for evaluating a derivative needed for the numerical solution of a problem. In this case we must convert the partial derivative to relate it to other quantities that are readily available. For example, let us convert the derivatives of the volume function discussed in the preceding section.

1. We can obtain a formula [6] by rearranging Equation 2-2 to find dv/dT and imposing the restriction that v be constant. Keeping in mind that $dv = 0$ we obtain

$$\frac{dv}{dT} = 0 = \left(\frac{\partial v}{\partial P}\right)_T \frac{dP}{dT} + \left(\frac{\partial v}{\partial T}\right)_P \tag{2-7}$$

Now if we indicate explicitly for the second factor of the first term on the right side that v is constant, and if we rearrange terms, we obtain

$$\left(\frac{\partial v}{\partial T}\right)_P = -\left(\frac{\partial v}{\partial P}\right)_T\left(\frac{\partial P}{\partial T}\right)_V \tag{2-8}$$

Thus, if we needed $(\partial v/\partial T)_P$ in some situation but had no method of direct evaluation, we could establish its value if $(\partial v/\partial P)_T$ and $(\partial P/\partial T)_V$ were available.

We can verify the validity of Equation 2-8 for an ideal gas by evaluating both sides explicitly and showing that the equality holds. The values of the partial derivatives can be determined by reference to Equation 2-3, and the following deductions can be made:

$$\frac{R}{P} = -\left(\frac{-RT}{P^2}\right)\left(\frac{R}{v}\right) = \frac{R^2T}{P^2v} = \frac{RT}{Pv}\frac{R}{P} = \frac{R}{P} \tag{2-9}$$

2. Another formula is obtained by rearranging Equation 2-2 to find dv/dP and imposing the restriction that v be constant. Thus we obtain

$$\frac{dv}{dP} = 0 = \left(\frac{\partial v}{\partial P}\right)_T + \left(\frac{\partial v}{\partial T}\right)_P\left(\frac{\partial T}{\partial P}\right)_V$$

Rearranging terms we have

$$\left(\frac{\partial v}{\partial P}\right)_T = -\left(\frac{\partial v}{\partial T}\right)_P\left(\frac{\partial T}{\partial P}\right)_V \tag{2-10}$$

3. A third formula is obtained by rearranging Equation 2-10 to

$$\left(\frac{\partial v}{\partial T}\right)_P = -\frac{(\partial v/\partial P)_T}{(\partial T/\partial P)_V} \tag{2-11}$$

and setting the right side of this expression equal to the right side of Equation 2-8:

$$-\left(\frac{\partial v}{\partial P}\right)_T\left(\frac{\partial P}{\partial T}\right)_V = -\left(\frac{\partial v}{\partial P}\right)_T\frac{1}{(\partial T/\partial P)_V} \tag{2-12}$$

Therefore

$$\left(\frac{\partial P}{\partial T}\right)_V = \frac{1}{(\partial T/\partial P)_V} \tag{2-13}$$

Thus, we see again that the derivatives can be manipulated practically as if they were fractions.

4. A fourth relationship is useful for problems in which a new dependent variable is introduced. For example, we could consider the energy, E, of a pure substance as a function of pressure and temperature:

$$E = g(P, T)$$

We then may wish to evaluate the partial derivative $(\partial V/\partial P)_E$—that is, the change of volume with change in pressure at constant energy. A suitable expression for this derivative in terms of other partial derivatives can be obtained from Equation 2-2 by finding dV/dP and explicitly adding the restriction that E is to be held constant. The result obtained is the relationship

$$\left(\frac{\partial V}{\partial P}\right)_E = \left(\frac{\partial V}{\partial P}\right)_T + \left(\frac{\partial V}{\partial T}\right)_P\left(\frac{\partial T}{\partial P}\right)_E \tag{2-14}$$

5. A fifth formula, for use in situations in which a new variable, $x(P, T)$, is to be introduced, is an example of the chain rule of differential calculus. The formula is

$$\left(\frac{\partial V}{\partial T}\right)_P = \left(\frac{\partial V}{\partial X}\right)_P\left(\frac{\partial X}{\partial T}\right)_P \tag{2-15}$$

These illustrations, based on the example of the volume function, are typical of the type of conversion that is required so frequently in thermodynamic manipulations.

Exact Differentials

Many thermodynamic relationships can be derived easily by using the properties of the exact differential. As an introduction to the characteristics of exact differentials, we shall consider the properties of certain simple functions used in connection with a gravitational field. We will use a capital D to indicate an inexact differential, as in DW, and a small d to indicate an exact differential, as in dE.

Example of the Gravitational Field. Let us compare the change in potential energy and the work done in moving a large boulder up a hill against the force of gravity. From elementary physics, we see that these two quantities, ΔE and W, differ in the following respects.

1. The change in potential energy depends only on the initial and final heights of the stone, whereas the work done (as well as the heat generated) depends on the path used. That is, the quantity of work expended if we use a pulley and tackle to raise the boulder directly will be much less than if we have to move the object up the hill by pushing it over a long, muddy, and tortuous road. However, the change in potential energy is the same for both paths, as long as they have the same starting point and the same end point.
2. There is an explicit expression for the potential energy, E, and this function can be differentiated to give dE, whereas no explicit expression leading to DW can be obtained. The function for the potential energy is a particularly simple one for the gravitational field because two of the space coordinates drop out and only the height, h, remains. That is,

$$E = \text{constant} + mgh \tag{2-16}$$

   The symbols m and g have the usual significance of mass and acceleration due to gravity, respectively.
3. A third difference between ΔE and W lies in the values obtained if one uses a cyclic path, as in moving the boulder up the hill and then back down to the initial point. For such a cyclic or closed path, the net change in potential energy is zero because the final and initial points are identical. This fact is represented by the equation

$$\oint dE = 0 \tag{2-17}$$

   in which $\oint$ denotes the integral around a closed path. However, the value of W for a complete cycle usually is not zero, and the value obtained depends on the particular cyclic path that is taken.

General Formulation. To understand the notation for exact differentials that generally is adopted, we shall rewrite Equation 2-2 to indicate explicitly that the partial derivatives are functions of the independent variables (P and T), and that the differential is a function of the independent variables and their differentials (dP and dT). That is,

$$dv(P, T, dP, dT) = M(P, T)dP + N(P, T)dT \tag{2-18}$$

in which

$$M(P, T) = \left(\frac{\partial v}{\partial P}\right)_T \tag{2-19}$$

and

$$N(P, T) = \left(\frac{\partial v}{\partial T}\right)_P \tag{2-20}$$

The notation in Equation 2-18 makes explicit the notion that, in general, dv is a function of the path chosen. More generally, for any case with two independent variables x and y, we can write

$$dL(x, y, dx, dy) = M(x, y)dx + N(x, y)dy \tag{2-21}$$

Using this expression we can summarize the characteristics of an exact differential as follows:

1. There exists a function, $f(x, y)$, such that
$$d[f(x, y)] = dL(x, y, dx, dy) \tag{2-22}$$
2. The value of the integral over any specified path—that is, the line integral [5] $\int_1^2 dL(x, y, dx, dy)$—depends only on the initial and final states and is independent of the path between them.
3. The line integral over a closed path is zero; that is,
$$\oint dL(x, y, dx, dy) = 0 \tag{2-23}$$

It is this last characteristic that is used most frequently in testing thermodynamic functions for exactness. If the differential dJ of a thermodynamic quantity J is exact, then J is called a *thermodynamic property* or a *state function*.

Reciprocity Characteristic. A common test of exactness of a differential expression, $dL(x, y, dx, dy)$, is whether the following relationship holds:

$$\left[\frac{\partial}{\partial y} M(x, y)\right]_x = \left[\frac{\partial}{\partial x} N(x, y)\right]_y \tag{2-24}$$

We can see that this relationship must be true if dL is exact, because in that case there exists a function $f(x, y)$ such that

$$df(x, y) = \left(\frac{\partial f}{\partial x}\right)_y dx + \left(\frac{\partial f}{\partial y}\right)_x dy = dL(x, y, dx, dy) \tag{2-25}$$

It follows from Equations 2-21 and 2-25 that

$$M(x, y) = \left(\frac{\partial f}{\partial x}\right)_y \tag{2-26}$$

and

$$N(x, y) = \left(\frac{\partial f}{\partial y}\right)_x \tag{2-27}$$

But, for the function $f(x, y)$, we know from the principles of calculus that

$$\frac{\partial}{\partial y}\left(\frac{\partial f}{\partial x}\right)_y = \frac{\partial^2 f}{\partial y \partial x} = \frac{\partial}{\partial x}\left(\frac{\partial f}{\partial y}\right)_x$$

Therefore if dL is exact [7]

$$\frac{\partial}{\partial y} M(x, y) = \frac{\partial}{\partial x} N(x, y) \tag{2-28}$$

To apply this criterion of exactness to a simple example, let us assume that we know only the expression for the total differential of the volume of an ideal gas (Equation 2-6) and do not know whether this differential is exact or not. Applying Equation 2-28 to Equation 2-6 we obtain

$$\frac{\partial}{\partial P}\left(\frac{R}{P}\right) = -\frac{R}{P^2} = \frac{\partial}{\partial T}\left(-\frac{RT}{P^2}\right)$$

Thus, we would know that the volume of an ideal gas is a thermodynamic property, even if we had not been aware previously of an explicit function for V.

Homogeneous Functions

In connection with the development of the thermodynamic concept of partial molar quantities, it is desirable to be familiar with a mathematical relationship known as *Euler's theorem*. Since this theorem is stated with reference to "homogeneous" functions, we will consider briefly the nature of these functions.

Definition. As a simple example, let us consider the function

$$u = ax^2 + bxy + cy^2 \tag{2-29}$$

If we replace the variables x and y by λx and λy, in which λ is a parameter, we can write

$$\begin{aligned} u^* = u(\lambda x, \lambda y) &= a(\lambda x)^2 + b(\lambda x)(\lambda y) + c(\lambda y)^2 \\ &= \lambda^2 ax^2 + \lambda^2 bxy + \lambda^2 cy^2 \\ &= \lambda^2(ax^2 + bxy + cy^2) \\ &= \lambda^2 u \end{aligned} \qquad (2\text{-}30)$$

Since the net result of multiplying each independent variable by the parameter λ merely has been to multiply the function by λ^2, the function is called *homogeneous*. Because the exponent of λ in the result is 2, the function is of the *second degree*.

Now we turn to an example of experimental significance. If we mix certain quantities of benzene and toluene, which form an ideal solution, the total volume, V, will be given by the expression

$$V = v_b n_b + v_t n_t \qquad (2\text{-}31)$$

in which n_b is the number of moles of benzene, v_b is the volume of one mole of pure benzene, n_t is the number of moles of toluene, and v_t is the volume of one mole of pure toluene. Suppose that we increase the quantity of each of the independent variables, n_b and n_t, by the same factor, say two. We know from experience that the volume of the mixture will be doubled. In terms of Equation 2-31 we also can see that if we replace n_b by λn_b and n_t by λn_t, the new volume V^* will be given by

$$\begin{aligned} V^* &= v_b \lambda n_b + v_t \lambda n_t \\ &= \lambda(v_b n_b + v_t n_t) \\ &= \lambda V \end{aligned} \qquad (2\text{-}32)$$

The volume function then is homogeneous of the first degree, since the parameter λ, which factors out, occurs to the first power. Although an ideal solution has been used in this illustration, Equation 2-31 is true of all solutions. However, for nonideal solutions the partial molar volume must be used instead of molar volumes of the pure components (see Chapter 18).

Proceeding to a general definition, we can say that a function $f(x, y, z, \ldots)$ is homogeneous of degree n if, upon replacement of each independent variable by an arbitrary parameter λ times the variable, the function is multiplied by λ^n, that is, if

$$f(\lambda x, \lambda y, \lambda z, \ldots) = \lambda^n f(x, y, z, \ldots) \qquad (2\text{-}33)$$

Euler's Theorem. The statement of the theorem can be made as follows: If $f(x, y)$ is a homogeneous function of degree n, then

$$x\left(\frac{\partial f}{\partial x}\right)_y + y\left(\frac{\partial f}{\partial y}\right)_x = nf(x, y) \tag{2-34}$$

The proof can be carried out by the following steps. First let us represent the variables λx and λy by

$$x^* = \lambda x \tag{2-35}$$

and

$$y^* = \lambda y \tag{2-36}$$

Then since $f(x, y)$ is homogeneous

$$f^* = f(x^*, y^*) = f(\lambda x, \lambda y) = \lambda^n f(x, y) \tag{2-37}$$

The total differential df^* is given by

$$df^* = \frac{\partial f^*}{\partial x^*} dx^* + \frac{\partial f^*}{\partial y^*} dy^* \tag{2-38}$$

Hence

$$\frac{df^*}{d\lambda} = \frac{\partial f^*}{\partial x^*}\frac{dx^*}{d\lambda} + \frac{\partial f^*}{\partial y^*}\frac{dy^*}{d\lambda} \tag{2-39}$$

From Equations 2-35 and 2-36

$$\frac{dx^*}{d\lambda} = x \tag{2-40}$$

and

$$\frac{dy^*}{d\lambda} = y \tag{2-41}$$

Consequently, Equation 2-39 can be rewritten as

$$\frac{df^*}{d\lambda} = \frac{\partial f^*}{\partial x^*} x + \frac{\partial f^*}{\partial y^*} y \tag{2-42}$$

Using the equalities in Equation 2-37, we can obtain

$$\frac{df^*}{d\lambda} = \frac{df(x^*, y^*)}{d\lambda} = \frac{d[\lambda^n f(x, y)]}{d\lambda} = n\lambda^{n-1} f(x, y) \tag{2-43}$$

Equating the right sides of Equations 2-42 and 2-43, we obtain

$$x\frac{\partial f^*}{\partial x^*} + y\frac{\partial f^*}{\partial y^*} = n\lambda^{n-1} f(x, y) \tag{2-44}$$

Since λ is an arbitrary parameter, Equation 2-44 must hold for any particular value. It must be true then for λ = 1. In such an instance Equation 2-44 reduces to

$$x\left(\frac{\partial f}{\partial x}\right)_y + y\left(\frac{\partial f}{\partial y}\right)_x = nf(x, y) \tag{2-34}$$

This equation is Euler's theorem.

As one example of the application of Euler's theorem, we refer again to the volume of a two-component system. Evidently the total volume is a function of the number of moles of each component:

$$V = f(n_1, n_2) \tag{2-45}$$

As we have seen previously, the volume function is known from experience to be homogeneous of the first degree; that is, if we double the number of moles of each component, we also double the total volume. Applying Euler's theorem we obtain the relationship

$$n_1\frac{\partial V}{\partial n_1} + n_2\frac{\partial V}{\partial n_2} = V \tag{2-46}$$

or,

$$V = n_1 v_1 + n_2 v_2 \tag{2-47}$$

where v_1 and v_2 are the partial molar volumes of components 1 and 2, respectively. Equation 2-47 is applicable to all solutions and is the analogue of Equation 2-31, which is applicable only to ideal solutions.

2-3 PRACTICAL TECHNIQUES

Throughout our discussions we will emphasize the application of thermodynamic methods to specific problems. Successful solutions of such problems depend on a familiarity with practical analytical and graphical techniques, as well as with the theoretical methods of mathematics. We will consider these techniques at this point, therefore, so they will be available for use when we approach specific problems.

Analytical Methods

In many cases it is possible to summarize data in terms of a convenient algebraic expression. Such an equation is desirable because it summarizes concisely a great deal of information. The data should be sufficiently precise, of course, to justify the effort of obtaining an analytical expression. We will use the method of least squares to fit a quadratic equation to a set of experimental data such as that in Table 2-4. The extension to power series with terms of higher or lower degree will be obvious. Methods of fitting other types of equations to experimental data are described in appropriate mathematical treatises [8].

We will assume that we have a series of data, such as equilibrium constants as a function of pressure, to which we wish to fit the quadratic equation

$$y = a + bx + cx^2 \tag{2-48}$$

Assuming that a quadratic equation can be used, we wish to obtain the best values of the constants a, b, and c. In the method we will use, it is assumed that all the error lies in the dependent variable y and none in the independent variable x, although the theory also has been developed for the case in which a significant error may appear in x [8].

The Method of Least Squares. With the method of least squares we obtain three independent equations that the three constants of the quadratic equation must obey. The procedure follows from the assumption that the best expression is the one for which the sum of the *squares* of the residuals is a minimum. If we define the residual for the general quadratic expression as

$$r = y - (a + bx + cx^2) \tag{2-49}$$

Table 2-4 The Chemical Equilibrium of the Ammonia Synthesis Reaction at High Temperatures and Extreme Pressures[a]

$N_2(g) + 3H_2(g) = 2NH_3(g)$

P/atm	K_P at 723 K
10	0.0000434
30	0.0000457
50	0.0000476
100	0.0000526
300	0.0000781
600	0.0001674

[a]L. J. Winchester and B. F. Dodge, *AIChE Journal*, Vol. 2, no. 4, p. 431, December 1956. Reproduced by permission of Amer. Institute of Chemical Engineers.

in which x and y refer to experimentally determined values, then we should obtain an equation for which

$$\Sigma r^2 = \text{a minimum} \tag{2-50}$$

This condition will be satisfied when the partial derivative of Σr^2 with respect to each of the constants a, b, and c, respectively, is zero. First, let us consider the partial derivative with respect to a:

$$\Sigma r^2 = [y_1 - (a + bx_1 + cx_1^2)]^2 + [y_2 - (a + bx_2 + cx_2^2)]^2 + \cdots \tag{2-51}$$

$$\begin{aligned} \left(\frac{\partial}{\partial a}\Sigma r^2\right)_{b,c} &= -2[y_1 - a - bx_1 - cx_1^2] - 2[y_2 - a - bx_2 - cx_2^2] - \cdots \\ &= 0 = -2[y_1 + y_2 + \cdots] - 2[-na] \\ &\quad -2[-bx_1 - bx_2 - \cdots] - 2[-cx_1^2 - cx_2^2 - \cdots] \end{aligned} \tag{2-52}$$

Rearrangement gives

$$\Sigma y = na + b\,\Sigma x + c\,\Sigma x^2 \tag{2-53}$$

By a similar procedure we can obtain the following expression from the partial derivative of Σr^2 with respect to the parameter b:

$$\Sigma yx = a\,\Sigma x + b\,\Sigma x^2 + c\,\Sigma x^3 \tag{2-54}$$

Similarly, the differentiation with respect to c leads to an expression that can be reduced to

$$\Sigma yx^2 = a\,\Sigma x^2 + b\,\Sigma x^3 + c\,\Sigma x^4 \tag{2-55}$$

The three simultaneous equations—2-53, 2-54, and 2-55—can be solved for the constants a, b, and c.

The expressions for a, b, and c, in terms of determinants, are

$$a = \left|\frac{\begin{matrix} \Sigma y & \Sigma x & \Sigma x^2 \\ \Sigma yx & \Sigma x^2 & \Sigma x^3 \\ \Sigma yx^2 & \Sigma x^3 & \Sigma x^4 \end{matrix}}{\begin{matrix} n & \Sigma x & \Sigma x^2 \\ \Sigma x & \Sigma x^2 & \Sigma x^3 \\ \Sigma x^2 & \Sigma x^3 & \Sigma x^4 \end{matrix}}\right| \tag{2-56}$$

$$b = \frac{\begin{vmatrix} n & \Sigma\, y & \Sigma\, x^2 \\ \Sigma\, x & \Sigma\, yx & \Sigma\, x^3 \\ \Sigma\, x^2 & \Sigma\, yx^2 & \Sigma\, x^4 \end{vmatrix}}{\begin{vmatrix} n & \Sigma\, x & \Sigma\, x^2 \\ \Sigma\, x & \Sigma\, x^2 & \Sigma\, x^3 \\ \Sigma\, x^2 & \Sigma\, x^3 & \Sigma\, x^4 \end{vmatrix}} \tag{2-57}$$

$$c = \frac{\begin{vmatrix} n & \Sigma\, x & \Sigma\, y \\ \Sigma\, x & \Sigma\, x^2 & \Sigma\, yx \\ \Sigma\, x^2 & \Sigma\, x^3 & \Sigma\, yx^2 \end{vmatrix}}{\begin{vmatrix} n & \Sigma\, x & \Sigma\, x^2 \\ \Sigma\, x & \Sigma\, x^2 & \Sigma\, x^3 \\ \Sigma\, x^2 & \Sigma\, x^3 & \Sigma\, x^4 \end{vmatrix}} \tag{2-58}$$

The algebraic expressions for the coefficients *a, b,* and *c* in terms of the summations calculated from the experimental values are

$$a = \frac{\Sigma y(\Sigma x^2\Sigma x^4 - (\Sigma x^3)^2) + \Sigma x(\Sigma x^3\Sigma yx^2 - \Sigma x^4\Sigma yx) + \Sigma x^2(\Sigma x^3\Sigma yx - \Sigma x^2\Sigma yx^2)}{n(\Sigma x^2\Sigma x^4 - (\Sigma x^3)^2) + \Sigma x(\Sigma x^3\Sigma x^2 - \Sigma x\Sigma x^4) + \Sigma x^2(\Sigma x\Sigma x^3 - (\Sigma x^2)^2)}$$

$$b = \frac{n(\Sigma yx\Sigma x^4 - \Sigma x^3\Sigma yx^2) + \Sigma y(\Sigma x^3\Sigma x^2 - \Sigma x\Sigma x^4) + \Sigma x^2(\Sigma x\Sigma yx^2 - \Sigma x^2\Sigma yx)}{n(\Sigma x^2\Sigma x^4 - (\Sigma x^3)^2) + \Sigma x(\Sigma x^3\Sigma x^2 - \Sigma x\Sigma x^4) + \Sigma x^2(\Sigma x\Sigma x^3 - (\Sigma x^2)^2)}$$

$$c = \frac{n(\Sigma x^2\Sigma yx^2 - \Sigma x^3\Sigma yx) + \Sigma x(\Sigma x^2\Sigma yx - \Sigma x\Sigma yx^2) + \Sigma y(\Sigma x\Sigma x^3 - (\Sigma x^2)^2)}{n(\Sigma x^2\Sigma x^4 - (\Sigma x^3)^2) + \Sigma x(\Sigma x^3\Sigma x^2 - \Sigma x\Sigma x^4) + \Sigma x^2(\Sigma x\Sigma x^3 - (\Sigma x^2)^2)}$$

The numerical values of *a, b,* and *c* can be found by direct substitution in the algebraic expressions if care is taken to carry an apparently excessive number of significant figures through the calculations, which involve taking small differences between large numbers. Alternatively, the determinants in Equations 2-56, 2-57, and 2-58 can be evaluated by methods described in the references, or the linear equations—2-53, 2-54, and 2-55—can be solved by matrix methods [9].

A least-squares fit to the data in Table 2-4 leads to the following equation for K_p as a function of P.

$$K_p = 4.48 \times 10^{-4} + 2.66 \times 10^{-7} P + 2.95 \times 10^{-9} P^2 \qquad (2\text{-}59)$$

The method of least squares permits us to calculate the best function of a given form for the set of data at hand, but it does not help us decide which form of analytic function to choose. Inspection of a graph of the data is helpful in such a choice. Figure 2-1 shows the data of Table 2-4, as well as the best straight line and the curve, represented by Equation 2-59, fitted to the data by the method of least squares.

Numerical and Graphical Methods

Experimental data of thermodynamic importance may be represented numerically, graphically, or in terms of an analytical equation. Often these data do not fit into a simple pattern that can be transcribed into a convenient equation. Consequently numerical and graphical techniques, particularly for differentiation and integration, are important methods of treating thermodynamic data.

Numerical Differentiation. Let us consider a set of experimental determinations of the standard potential, $\mathscr{E}^\circ$, at a series of temperatures, such as is

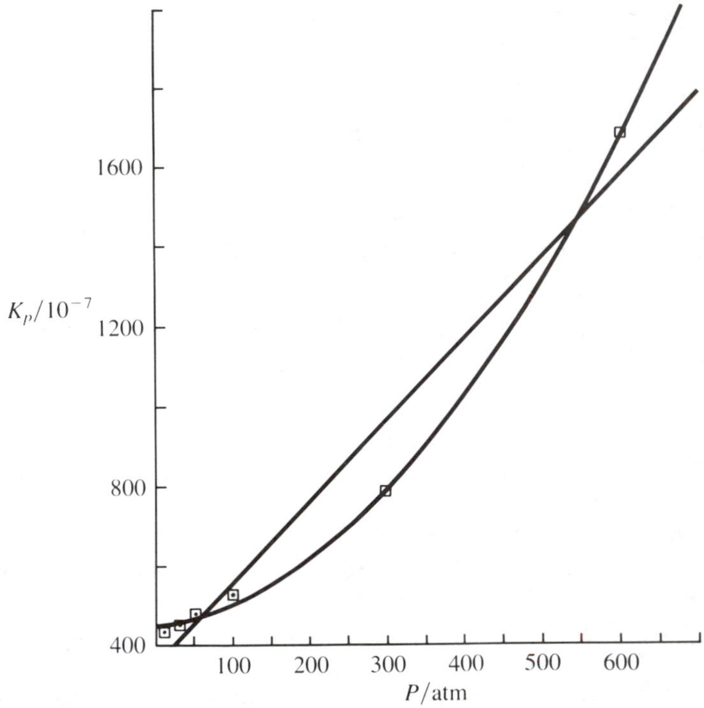

Figure 2-1. A plot of the data of Table 2-4, and the best linear and quadratic curves fitted to the data by the method of least squares.

listed in Table 2-5. A graph of these data (Figure 2-2) shows that the slope varies slowly but uniformly along the entire temperature range. For thermodynamic purposes, as in the calculation of the heat of reaction in the transformation

$$\tfrac{1}{2}H_2(g) + AgCl(s) = Ag(s) + HCl(aq)$$

it is necessary to calculate precise values of the derivative $\partial \mathscr{E}^\circ/\partial t$. A number of procedures have been described in the literature for numerical differentiation [10], but we shall use the method described by Savitzky and Golay [11].

This method is a simple extension of a procedure for filtering noise and smoothing data by the method of least squares, as opposed to a simple moving average. In the moving-average method, the average value of the dependent variable for an odd number of evenly spaced points is substituted for the value at the central point of the group. After dropping the first point of the group and adding the next point after the original group, the procedure is repeated until all data points have been treated. In the method of Savitzky and Golay, a least-squares fit to the same odd number of points is obtained, and the calculated value substituted for the central point of the group. The process is repeated, dropping the initial point and adding the next point after the original group, until all data points have been treated. The number of points used in each group and the degree of the polynomial used in the least-squares procedure

Table 2-5 Standard Potentials[a] for the Reaction

$\tfrac{1}{2}H_2(g) + AgCl(s) = Ag(s) + HCl(aq)$

$t/°C$	$\mathscr{E}^\circ$/volt
0	0.23634
5	0.23392
10	0.23126
15	0.22847
20	0.22551
25	0.22239
30	0.21912
35	0.21563
40	0.21200
45	0.20821
50	0.20437
55	0.20035
60	0.19620

[a]H. S. Harned and R. W. Ehlers, *J. Am. Chem. Soc.* **55**, 2179 (1933).

depend on the complexity of the data; we will use a five-point, quadratic polynomial fit to the data of Table 2-5.

Since the values of the independent variable are evenly spaced, the algebraic manipulations can be simplified by using an index number for each point as the independent variable. The residual square to be minimized is then

$$r^2 = [y - (a + bi + ci^2)] \tag{2-60}$$

where i varies from -2 to $+2$ in each group of five points. Since the central point of the group is that for which $i = 0$, the calculated value of y for that point is equal to a, the calculated value of the first derivative with respect to i at the central point is equal to b, and the second derivative at the central point is equal to $2c$.

In the equations for a and b following Equation 2-58, we can use i (varying from -2 to $+2$) for x, and y appears only to the first power; thus, the calculations lead to an expression for the least-square constants for each group of five points as a linear function of the five y values, and these functions are:

$$a = \frac{(-3y_{-2} + 12y_{-1} + 17y_0 + 12y_1 - 3y_2)}{35}$$

and

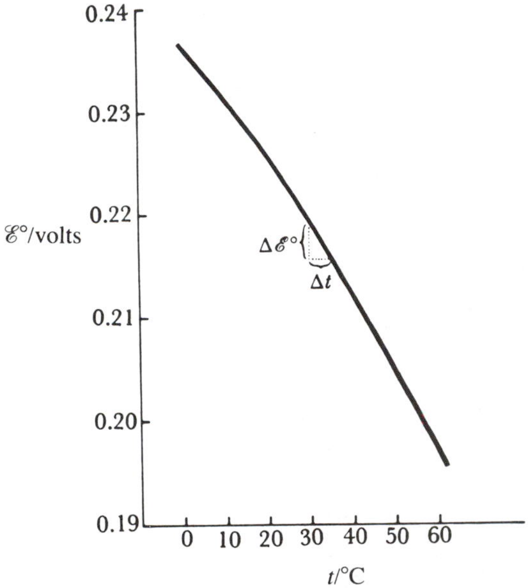

Figure 2-2. Standard electrode potentials for the reaction $\frac{1}{2}\,H_2(g) + AgCl(s) = Ag(s) + HCl(aq)$.

$$b = \frac{(-2y_{-2} - y_{-1} + y_1 + 2y_2)}{10}$$

In order to calculate the derivative with respect to x from b, we must also divide the derivative with respect to i by the real interval in the independent variable Δx.

When this procedure is applied to the data in Table 2-5, the results in Table 2-6 are obtained. Since values are obtained only for the central points of each group of points, some points are lost at each end. The procedure is clearly most useful when abscissa values are densely packed.

Numerical Integration [12]. The procedure for integration is analogous to that for differentiation. Again we will cite an example of its use in thermodynamic problems, the integration of heat capacity data. Let us consider the heat capacity data for solid n-heptane listed in Table 2-7. A graph of these data (Figure 2-3) shows a curve for which it may not be convenient to use an analytical equation. Nevertheless, in connection with determinations of certain thermodynamic functions, it may be desirable to evaluate the integral $\int_{T_1}^{T_2} C_p dT$. Therefore a numerical method is suitable.

Once again we consider small intervals of the independent variable, T, as is indicated in Figure 2-3. At the midpoint of this interval we have an average value of the heat capacity, $\ddot{C}_p$, which is indicated by the broken horizontal line in the figure. The area of the rectangle below this broken line is $\ddot{C}_p\ \Delta T$. If the interval chosen is so small that the section of the curve that has been cut is

Table 2-6 Smoothed Values of the Function and First Derivative of the Data of Table 2-5

t/°C	$\mathscr{E}°$/volt	$\mathscr{E}°$/volt (smooth)	$(d\mathscr{E}°/dt)$ / (volt/K)
0	0.23634	—	—
5	0.23392	—	—
10	0.23126	0.23127	−0.000542
15	0.22847	0.22847	−0.000576
20	0.22551	0.22551	−0.000607
25	0.22239	0.22240	−0.000641
30	0.21912	0.21911	−0.000676
35	0.21563	0.21564	−0.000710
40	0.21200	0.21199	−0.000738
45	0.20821	0.20823	−0.000764
50	0.20437	0.20436	−0.000789
55	0.20035	—	—
60	0.19620	—	—

Table 2-7 Heat Capacities[a] of Solid *n*-Heptane

T/K	C_p (J mole^{-1} K^{-1})	T/K	C_p (J mole^{-1} K^{-1})
15.14	6.276	53.18	53.56
17.52	8.828	65.25	65.65
19.74	11.422	71.86	71.30
21.80	14.238	79.18	77.53
24.00	17.205	86.56	82.97
26.68	20.648	96.20	90.29
30.44	25.430	106.25	97.15
34.34	30.836	118.55	104.98
38.43	36.531	134.28	113.60
42.96	41.92	151.11	123.60
47.87	47.53	167.38	133.72

[a]R. R. Wenner, *Thermochemical Calculations*, McGraw-Hill, New York, 1941, p. 356.

practically linear, then the area below this section of the curve is essentially the same as that of the rectangle, since the area a' is practically equal to a. Hence, it follows that the area under the curve between the limits f and g is given very closely by the sum of the areas of the rectangles taken over sufficiently short

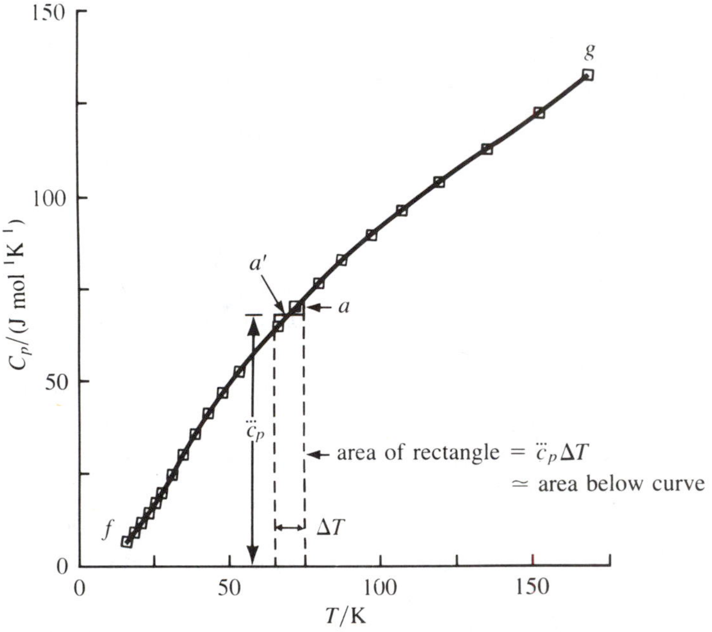

Figure 2-3. Graphical integration of heat capacity curve.

temperature intervals. Because the area under the curve corresponds to the integral $\int_f^g C_p dT$, it follows that

$$\sum_f^g \dddot{C}_p \, \Delta T \cong \int_f^g C_p dT \tag{2-61}$$

Since the first few data in Table 2-7 are given at closely successive temperatures, we can use these values to form the temperature intervals. (In many cases it is more convenient to use intervals of equal size within a given region.) For $\dddot{C}_p$ between any two temperatures, we can take the arithmetic mean between the listed experimental values. The values of ΔT and $\dddot{C}_p$ then are tabulated conveniently as in columns 3 and 4 of Table 2-8. Column 5 lists the area for the given interval. Finally, the sums of the areas of the intervals from 15.14 K are tabulated in column 6. The areas between any two of the temperatures listed in column 1 can be obtained by subtraction.

If we wish to obtain the value of the integral at some intermediate temperature not listed in Table 2-8, we can plot the values in column 6 as a function of T and read the values of the integral at the desired upper limit.

More accurate methods of numerical integration are described in the references [12].

Use of the Digital Computer. Both the numerical and analytical methods discussed in this chapter can be tedious to carry out, especially with large collections of precise data. Fortunately, the modern digital computer is ideally suited to carry out the repetitive arithmetic operations that are involved. Once a program has been written for a particular computation, whether it be numerical integration or the least-squares fitting of experimental data, it is

Table 2-8 Tabulation for Numerical Integration

1 T/K	*2* C_p	*3* ΔT	*4* $\dddot{C}_p$	*5* $\dddot{C}_p \, \Delta T$	*6* $\Sigma \dddot{C}_p \, \Delta T$
15.14	6.276				0.00
		2.38	7.552	17.99	
17.52	8.828				17.99
		2.22	10.125	22.47	
19.74	11.422				40.46
		2.06	12.832	26.44	
21.80	14.238				66.90
		2.20	15.723	34.60	
24.00	17.205				101.50

only necessary to provide a new set of data each time the computation is to be done.

Although the details of computer programming are beyond the scope of this text, the student unfamiliar with the subject is urged to consult one of the many books available [13]. Many programs designed to carry out the calculations described in this section are available commercially for use on desk-top personal computers.

Graphical Differentiation. Numerous procedures have been developed for graphical differentiation. A particularly convenient one [14], which we call the *chord-area method*, is illustrated using the same data (from Table 2-5) to which we previously applied numerical differentiation. It is clear from Figure 2-2 that if we choose a sufficiently small temperature interval, then the slope at the center of that interval will be given approximately by $\Delta\mathscr{E}^\circ/\Delta t$. In this example, with an interval of 5°C, the approximation is quite good. Then we proceed to tabulate values of $\Delta\mathscr{E}^\circ/\Delta t$ from 0°C, as illustrated in Table 2-9 for the first few data. Note that values of $\Delta\mathscr{E}^\circ$ are placed between the values of $\mathscr{E}^\circ$ to which they refer, and the temperature intervals, 5°C, are indicated between their extremities. Similarly, since $\Delta\mathscr{E}^\circ/\Delta t$ is an average value (for example, −0.000484) within a particular region (such as 0°C to 5°C), values in the fifth column also are placed between the initial and final temperatures to which they refer.

Having these *average* values of the slope, we now wish to determine the *specific* values at any given temperature. Since $\Delta\mathscr{E}^\circ/\Delta t$ is an average value, we draw it as a chord starting at the initial temperature of the interval and terminating at the final temperature. A graph of these chords over the entire temperature region from 0°C to 60°C is illustrated in Figure 2-4. To find the slope, $\partial\mathscr{E}^\circ/\partial t$, we draw a curve through these chords in such a manner that the

Table 2-9 Tabulation for Graphical Differentiation

1 t/°C	*2* $\mathscr{E}^\circ$	*3* $\Delta\mathscr{E}^\circ$	*4* Δt	*5* $\Delta\mathscr{E}^\circ/\Delta t$	*6* $\partial\mathscr{E}^\circ/\partial t$
0	0.23634				−0.000476
		−0.00242	5	−0.000484	
5	0.23392				−0.000509
		−0.00266	5	−0.000532	
10	0.23126				−0.000543
		−0.00279	5	−0.000558	
15	0.22847				−0.000576
		−0.00296	5	−0.000592	
20	0.22551				−0.000610

sum of the areas of the triangles, such as a, for which the chords form the upper sides, is equal to the sum of the areas of the triangles, such as a', for which the chords form the lower sides. This smooth curve gives $\partial\mathscr{E}^\circ/\partial t$ as a function of the temperature. Some values at several temperatures are shown in column 6 of Table 2-9; they agree quite well with those given in Table 2-6, which were obtained by numerical differentiation.

In the preceding example the chords have been taken for equal intervals, since the curve changes slope only gradually and the data are given at integral temperatures at equal intervals. In many cases, however, the intervals will not be equal, nor will they occur at whole numbers. Nevertheless, the chord-area method of differentiation can be used in substantially the same manner, although a little more care is necessary to avoid numerical errors in calculations.

Graphical Integration. The area under the curve in Figure 2-3 also can be determined graphically without recourse to numerical methods. Once the curve has been plotted, the area under the curve can be measured with a planimeter, or by cutting out the desired area, weighing the paper, and comparing the weight to that of a sample of the same paper of known area.

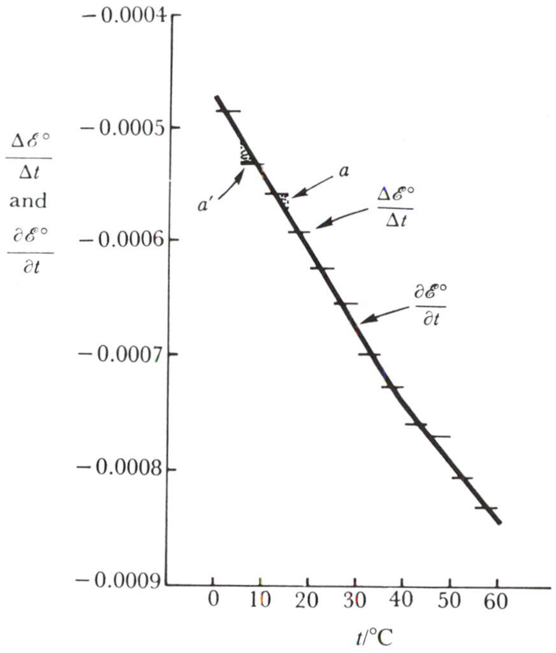

Figure 2-4. Chord-area plot of slopes of curve of Figure 2-2.

EXERCISES

1. Calculate the conversion factor for changing liter atmosphere to (a) erg, (b) joule, and (c) calorie. Calculate the conversion factor for changing atmosphere to pascal and atmosphere to bar.
2. Calculate the conversion factor for changing calorie to (a) cubic meter atmosphere and (b) volt faraday.
3. The area, a, of a rectangle can be considered to be a function of the breadth, b, and the length, l:

$$a = bl$$

The variables b and l are considered to be the independent variables; a is the dependent variable. Other possible dependent variables are the perimeter, p,

$$p = 2b + 2l$$

and the diagonal, d,

$$d = (b^2 + l^2)^{1/2}$$

a. Derive expressions for the following partial derivatives in terms of b and l, or calculate numerical answers:

$$\left(\frac{\partial a}{\partial l}\right)_b; \quad \left(\frac{\partial l}{\partial b}\right)_d; \quad \left(\frac{\partial p}{\partial l}\right)_b; \quad \left(\frac{\partial l}{\partial b}\right)_p;$$

$$\left(\frac{\partial d}{\partial b}\right)_l; \quad \left(\frac{\partial p}{\partial b}\right)_l; \quad \left(\frac{\partial a}{\partial b}\right)_l$$

b. Derive suitable conversion expressions in terms of the partial derivatives given in (a) for each of the following derivatives; then evaluate the results in terms of b and l. (Do not substitute the equation for p or d into that for a.)

$$\left(\frac{\partial a}{\partial b}\right)_d; \quad \left(\frac{\partial b}{\partial p}\right)_l; \quad \left(\frac{\partial a}{\partial b}\right)_p$$

c. Derive suitable conversion expressions in terms of the preceding partial derivatives for each of the following derivatives; then evaluate the results in terms of b and l:

$$\left(\frac{\partial p}{\partial b}\right)_d; \quad \left(\frac{\partial a}{\partial p}\right)_l; \quad \left(\frac{\partial b}{\partial p}\right)_d; \quad \left(\frac{\partial a}{\partial p}\right)_d$$

4. In a right triangle, such as that illustrated in Figure 2-5, the following relationships are valid:

$$D^2 = H^2 + B^2$$

$$P = H + B + D$$

$$A = \tfrac{1}{2}BH$$

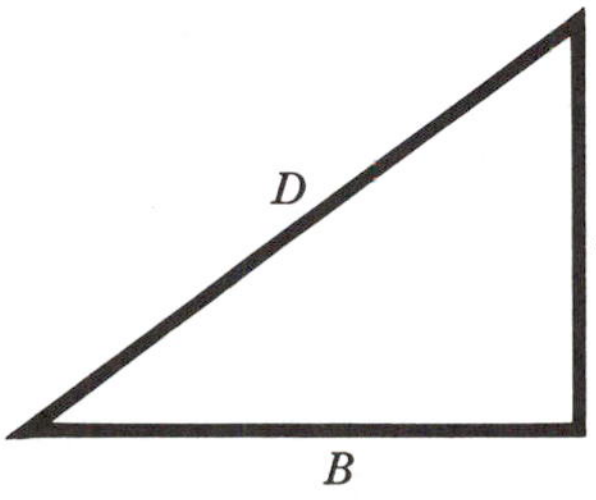

Figure 2-5. A right triangle.

a. Given the special conditions:

$$H = 1000 \text{ cm}$$

$$\left(\frac{\partial H}{\partial B}\right)_D = -0.5; \quad \left(\frac{\partial H}{\partial B}\right)_A = -2; \quad \left(\frac{\partial B}{\partial H}\right)_P = -1.309$$

compute the values of each of the following partial derivatives, using conversion relationships if necessary:

$$\left(\frac{\partial A}{\partial B}\right)_H; \quad \left(\frac{\partial A}{\partial H}\right)_B; \quad \left(\frac{\partial A}{\partial B}\right)_D; \quad \left(\frac{\partial A}{\partial H}\right)_P$$

b. Given the following different set of special conditions:

$$B = 4 \text{ cm}; \quad \left(\frac{\partial H}{\partial A}\right)_P = -0.310; \quad \left(\frac{\partial H}{\partial B}\right)_A = -2.0; \quad \left(\frac{\partial P}{\partial B}\right)_A = 2.341$$

compute the values of each of the following partial derivatives, using conversion relationships if necessary:

$$\left(\frac{\partial H}{\partial A}\right)_B; \quad \left(\frac{\partial B}{\partial A}\right)_P; \quad \left(\frac{\partial P}{\partial A}\right)_B$$

Compute A.

5. Considering E as a function of any two of the variables P, V, and T, prove that

$$\left(\frac{\partial E}{\partial T}\right)_P \left(\frac{\partial T}{\partial P}\right)_V = -\left(\frac{\partial E}{\partial V}\right)_P \left(\frac{\partial V}{\partial P}\right)_T$$

6. Using the definition $H = E + PV$ and, when necessary, obtaining conversion relationships by considering H (or E) as a function of any two of the variables P, V, and T, derive the following relationships:

a. $$\left(\frac{\partial H}{\partial T}\right)_P = \left(\frac{\partial E}{\partial T}\right)_V + \left[P + \left(\frac{\partial E}{\partial V}\right)_T\right]\left(\frac{\partial V}{\partial T}\right)_P$$

b. $$\left(\frac{\partial H}{\partial T}\right)_P = \left(\frac{\partial E}{\partial T}\right)_V + \left[V - \left(\frac{\partial H}{\partial P}\right)_T\right]\left(\frac{\partial P}{\partial T}\right)_V$$

c. $$\left(\frac{\partial E}{\partial T}\right)_V = \left(\frac{\partial H}{\partial T}\right)_P - \left[\left(\frac{\partial H}{\partial T}\right)_P\left(\frac{\partial T}{\partial P}\right)_H + V\right]\left(\frac{\partial P}{\partial T}\right)_V$$

7. By a suitable experimental arrangement it is possible to vary the total pressure, P, on a pure liquid independently of variations in the vapor pressure, p. (However, the temperature of both phases must be identical if they are in equilibrium.) For such a system, the dependence of the vapor pressure on P and T is given by

$$\left(\frac{\partial \ln p}{\partial P}\right)_T = \frac{V_1}{RT}$$

$$\left(\frac{\partial \ln p}{\partial T}\right)_P = \frac{\Delta H}{RT^2}$$

in which V_1 is the molar volume of liquid, and ΔH is the molar heat of vaporization. Prove that

$$\left(\frac{\partial P}{\partial T}\right)_p = \frac{-\Delta H}{TV_1}$$

8. The length, L, of a wire is a function of the temperature, T, and the tension, τ, on the wire. The linear expansivity, α, is defined by

$$\alpha = \frac{1}{L}\left(\frac{\partial L}{\partial T}\right)_\tau$$

and is essentially constant in a small temperature range. Likewise, the isothermal Young's modulus, Y, defined by

$$Y = \frac{L}{A}\left(\frac{\partial \tau}{\partial L}\right)_T$$

in which A is the cross-sectional area of the wire, is essentially constant in a small temperature range. Prove that

$$\left(\frac{\partial \tau}{\partial T}\right)_L = -\alpha A Y$$

9. An ideal gas in State A (Figure 2-6) is changed to State C. This transformation can be carried out by an infinite number of paths. However, only two will be considered, one along a straight line from A to C and the other from A to B to C [15].

   a. Calculate and compare the changes in volume from A to C that result from each of the two paths, AC and ABC. Proceed by integrating the differential equation

$$dV = \left(\frac{\partial V}{\partial P}\right)_T dP + \left(\frac{\partial V}{\partial T}\right)_P dT \qquad (2\text{-}2)$$

or

$$dV = -\frac{RT}{P^2}\,dP + \frac{R}{P}dT \qquad (2\text{-}6)$$

Before the integration is carried out along the path AC, use the following relationships to make the necessary substitutions:

$$\text{slope of line } AC = \frac{T_2 - T_1}{P_2 - P_1} = \frac{T - T_1}{P - P_1} \qquad (2\text{-}62)$$

Therefore

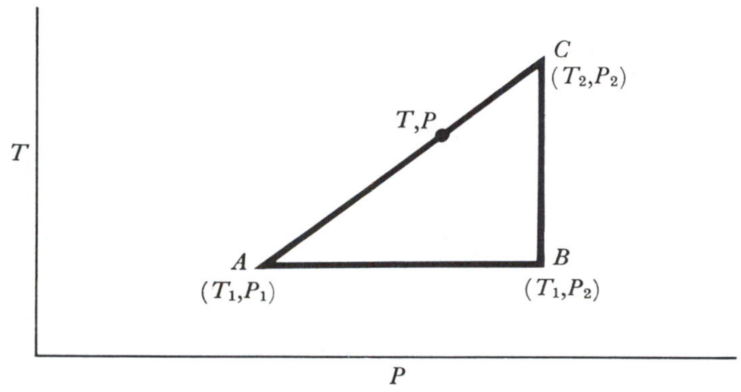

Figure 2-6. Two paths for carrying an ideal gas from State A to State C.

$$T = T_1 + \frac{T_2 - T_1}{P_2 - P_1}(P - P_1) \tag{2-63}$$

and

$$dT = \frac{T_2 - T_1}{P_2 - P_1} dP \tag{2-64}$$

Remember that T_1, T_2, P_1, and P_2 are constants in this problem.

b. Applying the reciprocity test to Equation 2-6, show that dv is an exact differential.

c. Calculate and compare the work done in going from A to C by each of the two paths. Use the relationship

$$dW = -Pdv = \frac{RT}{P} dP - RdT \tag{2-65}$$

and the substitution suggested by Equation 2-63.

d. Applying the reciprocity test to Equation 2-65 show that dW is an inexact differential.

10. For a wire the change in length, dL, can be expressed by the following differential equation:

$$dL = \frac{L}{YA} d\tau + \alpha L dT$$

in which τ is the tension, and T the temperature; A (cross-sectional area), Y, and α are essentially constant if the extension is not large.

a. Is dL an exact differential?

b. Is the differential for the work of the stretching, $dW = \tau dL$, an exact differential?

11. For an ideal gas we will show later that the entropy, s, is a function of the independent variables, volume, v, and temperature, T. The total differential, ds, is given by the equation

$$ds = (c_v/T)dT + (R/v)dv$$

in which c_v and R are constants.

a. Derive an expression for the change in volume of the gas as the temperature is changed at constant entropy—that is, for $(\partial v/\partial T)_S$. Your final answer should contain only independent variables and constants.

b. Is ds exact?

c. Derive an expression for $(\partial c_v/\partial v)_T$.

12. The compressibility, κ, and the coefficient of expansion, β, are defined by the partial derivatives:

$$\kappa = -\frac{1}{V}\left(\frac{\partial V}{\partial P}\right)_T$$

$$\beta = \frac{1}{V}\left(\frac{\partial V}{\partial T}\right)_P$$

Show that

$$\left(\frac{\partial \beta}{\partial P}\right)_T + \left(\frac{\partial \kappa}{\partial T}\right)_P = 0$$

13. For an elastic fiber such as a muscle fibril, the internal energy, E, is a function of three variables: the entropy, S, the volume, V, and the length, L. With the aid of the laws of thermodynamics it is possible to show that

$$dE = TdS - PdV + \tau dL$$

in which T is the absolute temperature, P the pressure, and τ the tension on the elastic fiber. Prove the following relationships:

$$\left(\frac{\partial E}{\partial L}\right)_{S,V} = \tau; \quad -\left(\frac{\partial \tau}{\partial V}\right)_{S,L} = \left(\frac{\partial P}{\partial L}\right)_{S,V}$$

14. The Gibbs free energy, G, is a thermodynamic property. If $(\partial G/\partial T)_P = -S$ and $(\partial G/\partial P)_T = V$, prove the following relationship:

$$\left(\frac{\partial S}{\partial P}\right)_T = -\left(\frac{\partial V}{\partial T}\right)_P$$

15. The Helmholtz free energy, A, is a thermodynamic property. If $(\partial A/\partial V)_T = -P$ and $(\partial A/\partial T)_V = -S$, prove the following relationship:

$$\left(\frac{\partial S}{\partial V}\right)_T = \left(\frac{\partial P}{\partial T}\right)_V$$

16. For a van der Waals gas

$$d\textsc{e} = C_v dT + \frac{a}{\textsc{v}^2}\,d\textsc{v}$$

in which a is independent of $\textsc{v}$.

a. Can $d\textsc{e}$ be integrated to obtain an explicit function for $\textsc{e}$?

b. Derive an expression for $(\partial C_v/\partial \textsc{v})_T$, assuming that $d\textsc{e}$ is exact.

17. Examine the following functions for homogeneity and degree of homogeneity:

a. $u = x^2y + xy^2 + 3xyz$

b. $u = \dfrac{x^3 + x^2y + y^3}{x^2 + xy + y^2}$

c. $u = (x + y)^{1/2}$

d. $u = e^{y/x}$

e. $u = \dfrac{x^2 + 3xy + 2y^3}{y^2}$

18. Complete the calculations in Table 2-9 for the graphical differentiation of the data listed in Table 2-5. Draw a graph corresponding to that of Figure 2-4, but on a larger scale for more precise readings.
19. Complete the calculations in Table 2-8 for the numerical integration of the data listed in Table 2-7. Draw a graph of $\int_0^T C_P dT$ versus temperature.

REFERENCES

1. C. Caratheodory, *Math. Ann.* **67**, 355 (1909): P. Frank, *Thermodynamics*, Brown University Press, Providence, R.I., 1945; J. T. Edsall and J. Wyman, *Biophysical Chemistry*, Vol. 1, Academic Press, New York, 1958; J. G. Kirkwood and I. Oppenheim, *Chemical Thermodynamics*, McGraw-Hill, New York, 1961.
2. NBS Special Publication 330, *The International System of Units* (SI), 1977 edition.
3. International Union of Pure and Applied Chemistry, *Manual of Symbols and Terminology for Physicochemical Quantities and Units*, Butterworth, London, 1970; ICSU-CODATA Task Group on Fundamental Constants, E. R. Cohen, chairman, CODATA Bulletin No. 11, December 1973.
4. *Quantities, Units, and Symbols*, 2d ed., a report of the Symbols Committee of the Royal Society, London, 1975; M. L. McGlashan, *Ann. Rev. Phys. Chem.* **24**, 51–76 (1973); Policy for National Bureau of Standards Usage of SI Units, *J. Chem. Educ.* **48**, 569 (1971); A. C. Norris, *J. Chem. Educ.* **48**, 797 (1971); Chemical Society Specialist Periodical Reports, *Chemical Thermodynamics*, Vol. 1, 1971.
5. W. A. Granville, P. F. Smith, and W. R. Longley, *Elements of the Differential and Integral Calculus*, Ginn, Boston, 1941; G. B. Thomas, Jr., *Calculus*, 3d ed., Addison-Wesley, Reading, Mass., 1969; J. W. Mellor, *Higher Mathematics for Students of Chemistry and Physics*, Dover, New York, 1946, pp. 68–75.
6. For a conventional rigorous derivation in terms of limits, see H. Margenau and G. M. Murphy, *The Mathematics of Physics and Chemistry*, Van Nostrand, Princeton, N.J., 1st ed., 1943, pp. 6–8; 2d ed., 1956, pp. 6–8. The approach of nonstandard analysis, however, provides a rigorous basis for the practice of treating a derivative as a ratio of infinitesimals. See A. Robinson, *Non-Standard Analysis*, North Holland, Amsterdam, 1966, pp. 79–83, pp. 269–279.
7. We have demonstrated that Equation 2-24 is a necessary condition for exactness, which is adequate for our purposes. For a proof of mathematical sufficiency also, see A. J. Rutgers, *Physical Chemistry*, Interscience, New York, 1954, p. 177, or F. T. Wall, *Chemical Thermodynamics*, 2d ed., W. H. Freeman, San Francisco, 1965, p. 425.

8. A. G. Worthing and J. Geffner, *Treatment of Experimental Data*, Wiley, New York, 1943; W. R. Steinbach and D. M. Cook, *Am. J. Phys.* **38**, 751–754 (1970); J. Mandel, *The Statistical Analysis of Experimental Data*, Interscience, New York, 1964, pp. 131–159; N. C. Barford, *Experimental Measurements: Precision, Error and Truth*, Addison-Wesley, Reading, Mass., 1967, pp. 56–71; W. E. Wentworth, *J. Chem. Ed.* **42**, 96, 162 (1965); J. S. Vandergraft, *Introduction to Numerical Computations*, Academic Press, New York, 1978, pp. 100–114.

9. B. Noble, *Numerical Methods*, Oliver and Boyd, Edinburgh, 1964, Vol. I, pp. 80–122; E. Whittaker and G. Robinson, *The Calculus of Observations*, 4th ed., Blackie and Sons, Ltd., Glasgow, 1944, pp. 71–77; Vandergraft, *Numerical Computations*, pp. 162–230.

10. Whittaker and Robinson, *Calculus of Observations*, pp. 60–68; B. Noble, *Numerical Methods*, Vol. II, Oliver and Boyd, Edinburgh, 1964, pp. 248–251; Vandergraft, *Numerical Computations*, pp. 127–137.

11. A. Savitzky and M.J.E. Golay, *Anal. Chem.* **36**, 1627–1639 (1964); C. L. Wilkins, S. P. Perone, C. E. Klopfenstein, R. C. Williams, D. E. Jones, *Digital Electronics and Laboratory Computer Experiments*, Plenum Press, New York, 1975, pp. 107–113.

12. R. Courant, *Differential and Integral Calculus*, Blackie and Sons, Ltd., London, 1959, Vol. I, pp. 342–348; F. S. Acton, *Numerical Methods That Work*, Harper and Row, New York, 1970, pp. 100–128; Whittaker and Robinson, *Calculus of Observations*, pp. 132–163; Vandergraft, *Numerical Computations*, pp. 137–161.

13. K. B. Wiberg, *Computer Programming for Chemists*, W. A. Benjamin, Menlo Park, Calif., 1965; T. L. Isenhour and P. C. Jurs, *Introduction to Computer Programming for Chemists*, Allyn and Bacon, Boston, 1972; C. L. Wilkins, C. E. Klopfenstein, T. L. Isenhour, P. C. Jurs, J. S. Evans, R. C. Williams, *Introduction to Computer Programming for Chemists—Basic Version*, Allyn and Bacon, Boston, 1974; K. Jeffrey Johnson, *Numerical Methods in Chemistry*, Dekker, New York, 1980.

14. T. R. Running, *Graphical Mathematics*, Wiley, New York, 1927, pp. 65–66.

15. Adapted from Margenau and Murphy, *Mathematics of Physics and Chemistry*, pp. 8–11.

CHAPTER 3

The First Law of Thermodynamics

Now that we have reviewed the mathematical background, we will develop the basic postulates of chemical thermodynamics upon which its theoretical framework is built. In discussing these fundamental postulates, which are essentially concise descriptions based on much experience, we will emphasize at all times their application to chemical, geological, and biological systems. However, first we must define a few of the basic concepts of thermodynamics.

3-1 DEFINITIONS

Critical studies of the logical foundations [1] of physical theory have emphasized the care that is necessary in defining fundamental concepts if contradictions between theory and observation are to be avoided. Our ultimate objective is clarity and precision in the description of the operations involved in measuring or recognizing the concepts. First let us consider a very simple example—a circle. At a primitive stage we might define a circle by the statement, "A circle is round." Such a definition would be adequate for children in the early grades of elementary school, but it could lead to long and fruitless arguments as to whether particular closed curves are circles. A much more satisfactory and refined definition is "a group of points in a plane, all of which are the same distance from an interior reference point called the center." This definition describes the operations that need to be carried out to generate a circle or to recognize one. The development of mature scientific insight involves, in part, the recognition that an early "intuitive understanding" at the primitive level often is not sound and sometimes may lead to contradictory conclusions from two apparently consistent sets of postulates and observations.

The operational approach to the definition of fundamental concepts in science has been emphasized by Mach, Poincaré, and Einstein and has been expressed in a very clear form by Bridgman [2]. (Operational definitions had been used implicitly much earlier than the twentieth century. Boyle, for example, defined a chemical element in terms of the experiments by which it might be recognized, in order to avoid the futile discussions of his predecessors, who identified elements with qualities or properties.) In this approach a concept is defined in terms of a set of experimental or mental operations used to measure or to recognize the quantity: "The concept is synonymous with the corresponding set of operations" (Bridgman). An operational definition frequently may fail to satisfy us that we know what the concept "really is." The question of scientific "reality" has been explored by many scientists and philosophers and is one that every student should examine. However, in the operational approach we are not concerned with whether our definition has told us what the concept "really is"; what we need to know is how to measure it. The operational approach has been stated succinctly by Poincaré in the course of a discussion of the concept of force:

When we say force is the cause of motion we talk metaphysics, and this definition, if we were content with it, would be absolutely sterile. For a definition to be of any use, it must teach us to *measure* force. Moreover that suffices; it is not at all necessary that it teach what force is in itself nor whether it is the cause or the effect of motion.

The power of the operational approach became strikingly evident in Einstein's theory of special relativity, with its analysis of the meaning of presumably absolute, intuitive concepts such as time or space. Newton defined absolute time as: "Absolute, True, and Mathematical Time, of itself, and from its own nature flows equably without regard to anything external." The difficulty with a definition of this type, based on properties or attributes, is that we have no assurance that anything of the given description actually exists in nature. Thus Newton's definition of time implies that it would be clear and meaningful to speak of two events in widely separated places (for example, the flaring up of two novae) as occurring simultaneously; presumably each event occurs at the same point on the time scale, which flows equably without regard to external events or of the activities of the individuals making the observations. In contrast, in relativity theory time is defined by a description of specific manipulations with clocks, light signals, and measuring rods. It turns out that events that are simultaneous for one observer will occur at different moments if viewed by another observer moving at a different velocity. Which observer is "correct"? In practice this question is meaningless. Both are correct. In fact, there is no operational meaning to "absolute simultaneity," despite its intuitive reasonableness. All operations by which time is measured are relative ones. Thus the term *absolute time* becomes meaningless.

Relativity theory, with its rigorous operational definitions of time and space, led to many unexpected results that are quite contrary to common experience. One result was that the measured length of a body depends on the speed with which the body moves with respect to the observer. These new theorems from relativity theory removed apparent contradictions that had perplexed physicists in their measurements of the speed of light, and they also allowed prediction of a variety of new phenomena that since have been verified abundantly.

Thus, physical scientists have become increasingly aware of the need to define concepts in terms of operations instead of relying on intuitive feelings of a priori recognition. To avoid possible pitfalls in thermodynamic applications, it is desirable that all thermal and energy concepts likewise be approached with an operational attitude.

Before approaching these thermodynamic concepts, we need to agree on the meaning of certain more primitive terms that will occur often in our analyses. We shall assume without analysis that the term "body" as an identifiable, definite thing has an obvious meaning. When we carry out experiments on or make observations of a body in order to characterize it, we obtain information that we call the *properties* of the body. Similarly, we shall speak of the properties of a *system*, which is any region of the universe, large or small, that is being considered in our analysis. Regions outside the boundaries of the system constitute the *surroundings*. A system is said to be in a certain state when all of its properties have specified values. The values of these properties are called *variables of state*. Generally, only a few of the properties of a system in a given state can be expressed as independent variables. Relationships between dependent variables of state and independent variables of state are specified by *equations of state*. If one or more of the properties of a system are found to be different at two different times, then in this time interval a process has taken place and a *change of state* has occurred. Occasionally, we also will speak of a *closed system*, by which we mean a system that mass neither enters nor leaves. Obviously then, an *open system* is one that mass may enter or leave. An *adiabatic system* is a closed system in which, if it is in thermal equilibrium, no change in state can be produced except by movement of its boundaries. (If electrically charged bodies are present in the system, this definition of "adiabatic" is inadequate, but such situations will not be considered here.) Such a system also is described as *thermally insulated* from the surroundings. With these primitive notions we will proceed to an analysis of thermodynamic concepts.

Temperature

The earliest concept of temperature undoubtedly was physiological, that is, based on the sensations of heat and cold. Such an approach necessarily is very crude, both in precision and accuracy. In time, people observed that the same temperature changes that produced physiological responses in themselves also produced changes in the measurable properties of matter. Among these properties are the volume of a liquid, the electrical resistance of a metal, the resonance frequency of a quartz crystal, and the volume of a gas at constant pressure.

Each of these properties can provide the basis for an operational definition of a temperature scale. For example, the Celsius temperature, θ, is defined by Equation 3-1:

$$\theta = \frac{X_\theta - X_0}{X_{100} - X_0}(100) \tag{3-1}$$

in which X_θ is the value of the property at temperature θ, X_0 is the value of the property at the temperature of a mixture of ice and water at equilibrium under a pressure of 1 atm (1.0135×10^5 Pa), and X_{100} is the value of the property at the temperature of an equilibrium mixture of water and steam under a pressure of 1 atm.

Unfortunately, when temperatures other than 0°C and 100°C are measured, the value obtained depends on the property used to measure it and, for the same property, depends on the substance whose property is measured. However, when the product of the pressure and the molar volume of a gas A, a quantity that changes in a monotonic fashion with changes of temperature, is measured at a series of pressures, and these values are extrapolated to a limit at zero pressure, the limit is the same for all gases at a given temperature. The generality of this result lends confidence that the idea of using the properties of gases at the limit of zero pressure as a fundamental basis for a temperature scale is a valid one. It also is found that the pressure–volume product extrapolates to zero at a sufficiently low temperature. This behavior leads to a definition of an absolute ideal gas temperature scale, denoted by T.

Following an international convention [3], if we take the temperature of the triple point of water as a reference temperature (T_0) and assign it the value of 273.16 K on the absolute scale, then any other temperature T is defined by the equation

$$\frac{T}{T_0} = \frac{\lim_{P\to 0}(PV)_T}{\lim_{P\to 0}(PV)_{T_0}} = \frac{T}{273.16\ \mathrm{K}} \tag{3-2}$$

Difficulties arise if the thermometer is exposed to certain types of radiation. However, calculations indicate that under normal circumstances these radiational fields raise the temperature by only about 10^{-12}°C, a quantity that is not detectable even with the most sensitive present-day instruments [4]. Similarly, we shall neglect relativistic corrections that arise at high velocities, for we do not encounter such situations in ordinary thermodynamic problems.

The word "thermodynamics" implies a relationship between thermal properties, such as temperature, and the dynamic properties described by classical mechanics. Therefore we shall consider next the dynamic concepts of work and energy and relate them to the properties of thermodynamic systems.

Work

For our purposes it is sufficient to point out that work is defined as the product of a force by a displacement. Assuming that force and displacement can be given suitable operational significance, the term "work" also will share this characteristic. The measurement of the displacement involves experimental determinations of a distance, which can be carried out, in principle, with a measuring rod. The concept of force is somewhat more complicated. It undoubtedly originated from the muscular sensation of resistance to external objects. A quantitative measurement is obtained readily with an elastic body, such as a spring, whose deformation can be utilized as a measure of the force. However, this definition of force is limited to static systems; for systems that are being accelerated, further refinements must be considered. Since these would take us too far from our main course, we merely make reference to Bridgman's critical analysis [5]. Nevertheless, for the definition of force even in the static situation, as well as for the definition of displacement, it should be emphasized that precision measurements require a number of precautions, particularly against changes in temperature. In dealing with these concepts we generally assume implicitly that such sources of error have been recognized and accounted for.

A body or system may do work (on its outside environment) or may have work done upon it. Therefore we must agree upon a sign convention for work, W. We will follow the convention that W represents the work done on the system by its surroundings. Thus a negative numerical value for W signifies that the system has done work on the surroundings; a positive value signifies that work has been done on the system by some agency in the surroundings.

In Figure 3-1, F' is the magnitude of the force exerted by the surroundings on the boundary of the system, F is the magnitude of the force exerted by the

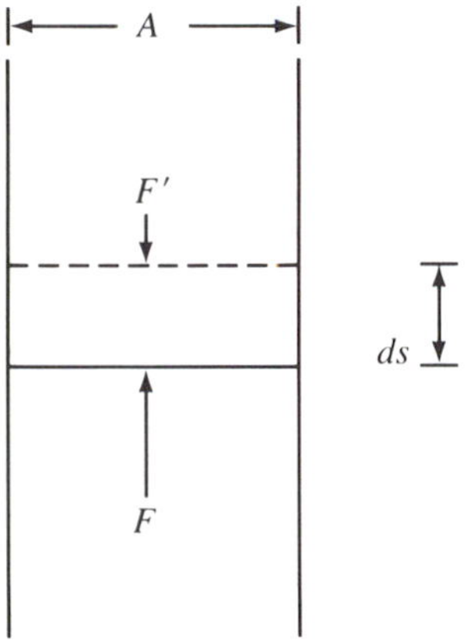

Figure 3-1. Element of work.

system on the boundary, and *ds* is the magnitude of the element of displacement of the boundary. The sign of the work calculated is consistent with the chosen convention if the infinitesimal element of work is defined as

$$DW = -F'ds \qquad (3\text{-}3)$$

The definition states explicitly that the amount of work done is determined by the *external force*, not the *internal force*.

In a noninfinitesimal change of state, the work done is obtained by integrating Equation 3-3 from a lower limit, the initial state, State 1, to an upper limit, the final state, State 2. That is,

$$W = \int_{\text{State 1}}^{\text{State 2}} -F'ds \qquad (3\text{-}4)$$

However, in order to carry out the integration, the path over which the change of state occurs must be known; that is, F' must be known as a function of the displacement s.

For example, if the bottom of Figure 3-1 represents a cylinder of gas that exerts a force F on the walls, then F' represents the external force on the piston in the cylinder (for example, that due to a weight). Since $F' = P'A$, where P' is the external pressure and A is the area of the piston, we can write Equation 3-4 as

$$\begin{aligned} W &= -\int_{\text{State 1}}^{\text{State 2}} P'Ads \\ &= -\int_{V_1}^{V_2} P'dV \end{aligned} \qquad (3\text{-}5)$$

If the change of state is carried out reversibly, so that P' is not significantly different from P throughout the change of state, then Equation 3-5 can be changed to

$$W = -\int_{V_1}^{V_2} PdV \qquad (3\text{-}6)$$

If work is done on the system, the gas is compressed, V_2 is less than V_1, and W is positive, consistent with the chosen convention. If work is done by the system, the gas expands, V_2 is greater than V_1, and W is negative.

Let us also consider the work done when a spring changes in length by an amount dL. If the original length of the spring is greater than its rest length, the spring exerts a force, in this case called a tension, τ, and the surroundings must do work on the spring if the length is to be increased further. Hence, keeping the sign convention for W in mind, we write, for a reversible change of length,

$$DW - \tau dL \qquad (3\text{-}7)$$

and W is positive if L_2, the final length, is greater than L_1, the initial length. If the original length of the spring is shorter than its rest length, it exerts a force tending to lengthen itself, and

$$dW = +\tau dL \tag{3-8}$$

since the work must be negative if L_2 is greater than L_1.

Similarly, if we focus attention on the work of raising a body a certain height, dh, against the force of gravity, we write

$$DW = mgdh \tag{3-9}$$

in which m is the mass of the body and g is the gravitational acceleration. In this case, some force other than the gravitational force does work on the body in raising it in the gravitational field, and W is positive if h_2 is greater than h_1. If the body falls in the gravitational field, it can do work, and W is negative.

From these examples equations for other types of work should follow naturally.

3-2 THE FIRST LAW OF THERMODYNAMICS

Energy

In its modern form the first law of thermodynamics has its empirical basis in a series of experiments conducted by Joule between 1843 and 1848. He did work on an adiabatic system in a variety of ways, using the rotation of a paddle wheel, the passage of an electric current, friction, and the compression of a gas. He concluded that a given amount of work done on the system, no matter how it was done, produced the same change of state, as measured by the change in the temperature of the system. Thus, to produce a given change of state, the *adiabatic work* required is *independent of the path* by which the change is achieved.

Therefore it seems appropriate to define a quantity, the energy E, whose value is characteristic of the state of a system. The difference between the value of E in one state and that in another state is equal to the adiabatic work required to bring about the change of state. That is,

$$\Delta E = E_2 - E_1 = W_{\text{adiabatic}} \tag{3-10}$$

The sign is consistent with our convention for work; if the system does work on the environment the energy of the system must decrease, and vice versa.

Heat

It is well known that the changes of state brought about by adiabatic work also can be brought about without doing work. A change can be achieved without expenditure of work by placing a system in contact with another system at a different temperature through rigid, nonadiabatic (or thermally conducting) walls. It is this exchange of energy, which is a result of a temperature difference, that we call *heat* (Q). Since the energy difference has been determined in previous experiments by measuring the adiabatic work, the amount of heat transferred can be measured by the change in energy in a system on which no work is done. That is,

$$\begin{aligned} Q &= \Delta E(\text{no work}) \\ &= E_2 - E_1(\text{no work}) \end{aligned} \tag{3-11}$$

The positive sign in Equation 3-11 expresses the convention that Q is positive when heat is transferred from the surroundings to the system. Such a transfer results in an increase in the energy of the system.

General Form of the First Law

Let us consider the general case in which a change of state is brought about both by work and by transfer of heat. It is convenient to think of the system as being in thermal contact with a body that acts as a heat bath (one that transfers heat but does no work) as well as in mechanical, but not thermal, contact with another portion of the environment, so that work can be done.

The value of ΔE in the change of state can be determined from the initial and final states of the system, as well as from a comparison to previous experiments that used only adiabatic work. The work, W, can be calculated from the mechanical changes in the environment (for example, from the change in position of a weight). The value of Q is determined from the change of state of the heat bath, also previously calibrated by experiments with adiabatic work.

Although the change in state of the heat bath, hence the value of Q, usually is determined by measuring a change in temperature, this is a matter of convenience and custom. For a pure substance, the state of a system is determined by specifying the values of two intensive variable. For a heat bath whose volume (and density) is fixed, the temperature is a convenient second variable. A measurement of the pressure, viscosity, or surface tension would determine the state of the system equally as well. This point is important to the logic of our development since a later definition of a temperature scale is based on heat measurements. To avoid circularity, the measurement of heat must be independent of the measurement of temperature.

Although the proportions of the energy change contributed by heat and by work can vary from one extreme to the other, the sum of the heat effect and the work effect is always equal to ΔE for a given change of state, as indicated in Equation 3-12:

$$\Delta E = Q + W \tag{3-12}$$

That is, Q and W can vary for a given change of state, depending on how the change is carried out, but the quantity $(Q + W)$ is always equal to ΔE, which depends only on the initial and final states.*

The energy E is called a *state function*. It is appropriate to use a quantity ΔE equal to the difference between the values of E in two different states. The quantities Q and W are not expressed as differences because their values depend on the path taken from one state to another.

For an infinitesimal change of state the analog of Equation 3-12 is

$$dE = DQ + DW \tag{3-13}$$

in which the difference in notation indicates that dE is a differential of a function, an *exact differential*, whereas DQ and DW are *inexact differentials*. Another way to express the difference is to say that the integral of dE around a closed path is equal to zero:

$$\oint dE = 0 \tag{3-14}$$

whereas the integrals of DQ and DW around a closed path can have any value, depending on the path taken.

It also should be pointed out that the very definition of the energy concept precludes the possibility of determining absolute values thermodynamically; that is, we have defined only a method of measuring *changes* in internal energy. In this regard there is a significant difference between the character of the thermodynamic property energy and that of a property such as volume. We can specify an unambiguous value for the volume of a system in a particular state. However, it is appropriate to speak of "energy" only with reference to a transition from one configuration of a system to another, that is, only with reference to an energy change. There can be no measurement of the energy of a given state. Energy differences can be measured and a value can be assigned to a particular state in terms of an arbitrary reference state. To speak of energy in an absolute sense is only a way of expressing the observation that certain aspects of a transition depend uniquely on characteristics of the initial and final states.

Equations 3-10, 3-11, and 3-12 (repeated here), together with the statement that E is a state function, constitute a complete statement of the first law. In effect, Equations 3-10 and 3-11 are operational definitions of ΔE and Q, respectively, whereas Equation 3-12 is an empirical statement of the relationship among the quantities ΔE, Q, and W:

$$\Delta E = E_2 - E_1 = W_{\text{adiabatic}} \tag{3-10}$$

*Some readers may prefer Joule's statement [6]: "... it is manifestly absurd to suppose that the power with which God has endowed matter can be destroyed any more than that they can be created by man's agency...."

$$Q = \Delta E(\text{no work}) = E_2 - E_1(\text{no work}) \tag{3-11}$$

$$\Delta E = Q + W \tag{3-12}$$

EXERCISES

1. If the temperature of 1 mL of air at 1 atm and 0°C is raised to 100°C, the volume becomes 1.3671 mL. Calculate the value of absolute zero for a thermometer using air. Compare your result with the ice-point temperature in Table 2-3.
2. Let Figure 3-1 represent a plane surface (such as a soap film between wires) that is being expanded in the direction indicated. Show that the work of reversible expansion is given by the expression

$$W = \int \gamma dA$$

in which γ is the force per unit length and A is the surface area.
3. Let τ represent the tension of a wire of length L, A represent its cross-sectional area, and Y represent the isothermal Young's modulus (see Exercise 8, p. 34). For a wire, L, A, and Y are practically constant as the tension is increased. Show that the work for an isothermal increase in tension of a wire is given by

$$W = \frac{L}{2AY}(\tau^2_{\text{final}} - \tau^2_{\text{initial}})$$

4. From Equation 3-13, show that DQ is an exact differential for a process at constant pressure in which only PV work is done.
5. If the atmosphere were isothermal, the pressure would vary with height according to the equation

$$P = P_0 \exp(-Mgh/RT)$$

where P_0 (1.01325×10^5 Pa) is the pressure at sea level, M is the average molar mass of air (considered to be 80 mol % N_2 and 20 mol% O_2), g is the acceleration due to gravity, h is the height, R is the gas constant, and T is the ideal gas temperature. If a rigid balloon containing 100×10^3 dm^3 of helium and carrying a load of 100 kg is allowed to float in an atmosphere at 298 K, at what height will it come to rest? What will be its potential energy?

REFERENCES

1. H. Poincaré, *Science and Hypothesis*, reprinted by Dover, New York, 1952; P. W. Bridgman, *The Logic of Modern Physics*, Macmillan, New York, 1927; Bridgman, *The Nature of Thermodynamics*, Harvard University Press, Cambridge, Mass., 1941.
2. Bridgman, *Logic of Modern Physics*.

3. International Union of Pure and Applied Chemistry, *Manual of Symbols and Terminology for Physiochemical Quantities and Units*, Butterworth, London, 1970.
4. Bridgman, *Nature of Thermodynamics*, p. 16.
5. Bridgman, *Logic of Modern Physics*, pp. 102–108.
6. J. P. Joule, *Scientific Papers*, Vol. I, London, 1963, pp. 269.

CHAPTER 4

Enthalpy and Heat Capacity

In the preceding chapter we defined a new function, the internal energy, E, and noted that it is a thermodynamic property; that is, dE is exact. Since Q was defined in Equation 3-11 as equal to ΔE when no work is done, the heat exchanged in a constant-volume process in which only PdV work is done is also independent of the path. For example, in a given chemical reaction carried out in a closed vessel of fixed volume, the heat absorbed (or evolved) depends only on the nature and condition of the initial reactants and of the final products; it does not depend on the mechanism by which the reaction occurs. Therefore if a catalyst speeds up the reaction by changing the mechanism, it does not affect the heat quantity accompanying the reaction.

Most chemical reactions are carried out at constant (atmospheric) pressure. It is of interest to know whether the heat absorbed in a constant-pressure reaction depends on the path—that is, on the method by which the reaction is carried out—or whether it too is a function only of the initial and final states. If the latter were true it would be possible to tabulate heat quantities for given chemical reactions and to use known values to calculate heats for new reactions that can be expressed as sums of known reactions.

Actually, this question was answered on empirical grounds long before thermodynamics was established on a sound basis. In courses in elementary chemistry students become familiar with Hess's law of constant heat summation, enunciated in 1840. Hess pointed out that the heat absorbed (or evolved) in a given chemical reaction is the same whether the process occurs in one step or in several steps. Thus, to cite a familiar example, the heat of formation [1] of CO_2 from its elements is the same if the process is the single step

$$\text{C(graphite)} + \text{O}_2\text{(gas)} = \text{CO}_2\text{(gas)}, \quad Q_{298\text{ K}} = -393.509 \text{ kJ mol}^{-1}$$

or the series of steps

$$\begin{array}{r} C(\text{graphite}) + \tfrac{1}{2}O_2(\text{gas}) = CO(\text{gas}),\ Q_{298\ K} = -110.525\ \text{kJ mol}^{-1} \\ CO(\text{gas}) + \tfrac{1}{2}O_2(\text{gas}) = CO_2(\text{gas}),\ Q_{298\ K} = -282.984\ \text{kJ mol}^{-1} \\ \hline C(\text{graphite}) + O_2(\text{gas}) = CO_2(\text{gas}),\ Q_{298\ K} = -393.509\ \text{kJ mol}^{-1} \end{array}$$

It is difficult to measure the heat of combustion of graphite to carbon monoxide because carbon dioxide always is produced as well. Thus, from Hess's law it is possible to calculate the heat of combustion of graphite to carbon monoxide from the more easily measurable heats of combustion of graphite and carbon monoxide to carbon dioxide.

We could introduce Hess's generalization into thermodynamics as another empirical law, similar to the first law. However, a firm theoretical framework depends on a minimum of empirical postulates. The power of thermodynamics lies in the fact that it leads to so many predictions, and requires only two or three basic assumptions. Hess's law need not be among these postulates since it can be derived directly from the first law of thermodynamics, perhaps most conveniently by using a new thermodynamic function—enthalpy.

4-1 ENTHALPY

Definition

The thermodynamic quantity *enthalpy* is defined in terms of thermodynamic variables that have been described already:

$$H = E + PV \tag{4-1}$$

From the definition it is evident that H, the enthalpy, is a thermodynamic property since it is defined by an explicit function. All the quantities on the right side of Equation 4-1, E, P, and V, are properties of the state of a system; consequently so is H.

It is also evident from the definition (Equation 4-1) that absolute values of H are unknown because absolute values of E cannot be obtained from classical thermodynamics alone. Therefore, from an operational point of view, it is possible only to consider *changes* in enthalpy, ΔH. Such changes can be defined readily by the expression

$$\Delta H = \Delta E + \Delta(PV) \tag{4-2}$$

Relationship between Q and ΔH

Having defined the thermodynamic property enthalpy, we will proceed to investigate the conditions under which it becomes equal to the heat that accompanies a process. Differentiating Equation 4-1 we obtain

$$dH = dE + PdV + VdP \tag{4-3}$$

From the first law of thermodynamics we can introduce

$$dE = DQ + DW \tag{3-13}$$

and obtain

$$dH = DQ + DW + PdV + VdP \tag{4-4}$$

Equation 4-4 is of general validity. Now let us consider a set of restrictions that are realized in many chemical reactions: (1) constant pressure and (2) no work other than mechanical (against the atmosphere). Under these conditions Equation 4-4 can be simplified because

$$DW = -PdV$$

and

$$dP = 0$$

Hence

$$dH_P = DQ_P = dQ_P \tag{4-5}$$

and

$$\Delta H_P = Q_P \tag{4-6}$$

in which the subscript emphasizes the restriction of constant pressure during the process. Equation 4-6 is valid *only* if no nonmechanical work is being done. Under these conditions dQ is an exact differential. In other words, for chemical reactions carried out at constant pressure (for example, at atmospheric pressure) in the usual laboratory or large-scale vessels, the heat absorbed depends only on the nature and conditions of the initial reactants and of the final products. Thus, it does not matter if a given substance is formed in one step or in many steps. As long as the starting and final materials are the same, and as long as the processes are carried out at constant pressure and with no nonmechanical work, the net Q's will be the same. Thus Hess's law is a consequence of the first law of thermodynamics.

Relationship between Q_V and Q_P

We have just proved that ΔH equals Q_P for a reaction at constant pressure. Although most calorimetric work is carried out at a constant pressure, some reactions must be observed in a closed vessel, that is, at constant volume. In such a closed system the heat quantity that is measured is Q_V. For further chemical calculations it frequently is necessary to know Q_P. Therefore it is highly desirable to derive some expression that relates these two heat quantities.

We will use the relationship implied in Equation 3-11,

$$\Delta E_V = Q_V \tag{4-7}$$

and Equation 4-6,

$$\Delta H_P = Q_P$$

together with Equation 4-2 restricted to a constant-pressure process:

$$\Delta H_P = \Delta E_P + \Delta(PV) \tag{4-8}$$

Generally, ΔE_P is not significantly different from ΔE_V. In fact, for ideal gases at a fixed temperature (as we will see in Chapter 6), E is independent of the volume or pressure. Hence, as a rule,

$$\Delta E_P \cong \Delta E_V = Q_V \tag{4-9}$$

Substituting Equations 4-6 and 4-9 into 4-8 we obtain

$$Q_P = Q_V + \Delta(PV) \tag{4-10}$$

(Q_P and Q_V refer to different changes of state.)

In reactions involving only liquids and solids the $\Delta(PV)$ term usually is negligible in comparison with Q, hence the difference between Q_P and Q_V is slight. However, in reactions involving gases $\Delta(PV)$ may be significant because the changes in volume may be large. Generally, this term can be estimated with sufficient accuracy by using the equation of state for ideal gases,

$$PV = nRT \tag{4-11}$$

in which n represents the number of moles of a particular gas. If the chemical reaction is represented by the expression

$$aA(\mathrm{g}) + bB(\mathrm{g}) + \cdots = lL(\mathrm{g}) + mM(\mathrm{g}) + \cdots \tag{4-12}$$

in which a, b, l, and m indicate the number of moles of each gas, then, since an isothermal change is being considered,

$$Q_P = Q_V + \Delta(PV) = Q_V + (n_L RT + n_M RT + \cdots - n_A RT - n_B RT - \cdots)$$

or

$$Q_P = Q_V + (\Delta n)RT \tag{4-13}$$

The symbol Δn refers to the increase in number of moles *of gases only*. In common reactions ΔnRT contributes a few kilojoules to the difference $Q_P - Q_V$ (see Exercise 3, p. 65).

4-2 HEAT CAPACITY

We introduced the enthalpy function particularly because of its usefulness as a measure of the heat that accompanies chemical reactions at constant pressure. We will find it convenient also to have a function to relate heat quantities to temperature changes, either at constant pressure or at constant volume. For this purpose we will consider a new quantity, the heat capacity.

Definition

Fundamental Statement. The heat absorbed by a body (not at a transition temperature) is proportional to the change in temperature:

$$Q = C(T_2 - T_1) \tag{4-14}$$

The proportionality constant, C, is called the *heat capacity** and is proportional to the mass of the substance undergoing the temperature change. Hence the heat capacity per gram is called the *specific heat*, and that for one mole of material is called the *molar heat capacity*.

The value of C,

$$C = \frac{Q}{T_2 - T_1} = \frac{Q}{\Delta T} \tag{4-15}$$

itself may depend on the temperature. Thus, for a rigorous definition of heat capacity, we must consider an infinitesimally small temperature interval. Consequently we define the heat capacity by the expression

$$C = \frac{DQ}{dT} \tag{4-16}$$

in which the D in DQ emphasizes the inexactness of the heat quantity. Since DQ is inexact, C has no unique value but depends on the path or conditions under which heat is supplied. We can place certain restrictions on Equation 4-16, such as constant pressure or constant volume. For these situations, we can modify Equation 4-16 to the following expressions:

$$C_p = \left(\frac{DQ}{\partial T}\right)_P \tag{4-17}$$

and

$$C_v = \left(\frac{DQ}{\partial T}\right)_V \tag{4-18}$$

Derived Relationships. Equations 4-16, 4-17, and 4-18 are fundamental definitions. From these and previous thermodynamic principles, new relationships can be derived that are very useful in further work.

*The term *heat capacity* is a historical remnant of the time when it was thought that heat is stored in an object; we now consider that thermal energy is contained in an object and that heat is energy being transferred due to a difference in temperature.

If we have a substance that is absorbing heat at a constant pressure, it is evident that the restrictions placed on Equation 4-5 are being fulfilled; that is,

$$DQ_P = dH_P$$

Substitution of Equation 4-5 into Equation 4-17 leads to the important expression

$$C_p = \left(\frac{\partial H}{\partial T}\right)_P \tag{4-19}$$

Since dH is exact, C_p has a definite value for a particular substance in a specified state.

Similarly, if we have a substance that is absorbing heat at constant volume, the restrictions placed on Equation 4-7 are being fulfilled, hence

$$DQ_V = dE_V \tag{4-20}$$

Substitution of Equation 4-20 into Equation 4-18 leads to an additional basic relationship,

$$C_v = \left(\frac{\partial E}{\partial T}\right)_V \tag{4-21}$$

Some Relationships between C_p and C_v

From the considerations of the preceding section there is no immediately apparent connection between the two heat capacities, C_p and C_v. We can illustrate the power of thermodynamic methods by developing several such relationships without any assumptions beyond the first law of thermodynamics and the definitions that have been made already. Following are three of these relationships.

1. Starting with the derived relationship for C_p (Equation 4-19),

$$C_p = \left(\frac{\partial H}{\partial T}\right)_P$$

we introduce the definition of H:

$$C_p = \left[\frac{\partial(E + PV)}{\partial T}\right]_P = \left(\frac{\partial E}{\partial T}\right)_P + P\left(\frac{\partial V}{\partial T}\right)_P \tag{4-22}$$

The partial derivative $(\partial E/\partial T)_P$ is not C_v, but if it could be expanded into some relationship with $(\partial E/\partial T)_V$, we would have succeeded in introducing C_v into Equation 4-22. The necessary relationship can be derived by considering the internal energy, E, as a function of T and V and setting up the total differential:

$$dE = \left(\frac{\partial E}{\partial T}\right)_V dT + \left(\frac{\partial E}{\partial V}\right)_T dV \tag{4-23}$$

Dividing by dT and imposing the condition of constant pressure, we obtain

$$\left(\frac{\partial E}{\partial T}\right)_P = \left(\frac{\partial E}{\partial T}\right)_V + \left(\frac{\partial E}{\partial V}\right)_T \left(\frac{\partial V}{\partial T}\right)_P \tag{4-24}$$

Substituting Equation 4-24 into Equation 4-22 and factoring out the partial derivative $(\partial V/\partial T)_P$, we obtain the desired expression:

$$C_p = \left(\frac{\partial E}{\partial T}\right)_V + \left[P + \left(\frac{\partial E}{\partial V}\right)_T\right]\left(\frac{\partial V}{\partial T}\right)_P = C_v + \left[P + \left(\frac{\partial E}{\partial V}\right)_T\right]\left(\frac{\partial V}{\partial T}\right)_P \tag{4-25}$$

This expression will be of considerable value when we consider special cases for which values or equations for the partial derivatives $(\partial E/\partial V)_T$ and $(\partial V/\partial T)_P$ are available.

2. A second relationship can be derived by substituting for the second term in Equation 4-25 so that the partial derivative $(\partial E/\partial V)_T$ is replaced by one containing H. Using the fundamental relationship between E and H we obtain

$$\left(\frac{\partial E}{\partial V}\right)_T = \left[\frac{\partial(H - PV)}{\partial V}\right]_T = \left(\frac{\partial H}{\partial V}\right)_T - P - V\left(\frac{\partial P}{\partial V}\right)_T \tag{4-26}$$

This expression can be placed in the bracketed factor in Equation 4-25 to give

$$C_p = C_v + \left[P + \left(\frac{\partial H}{\partial V}\right)_T - P - V\left(\frac{\partial P}{\partial V}\right)_T\right]\left(\frac{\partial V}{\partial T}\right)_P$$

which can be reduced to

$$C_p = C_v + \left[\left(\frac{\partial H}{\partial P}\right)_T \left(\frac{\partial P}{\partial V}\right)_T - V\left(\frac{\partial P}{\partial V}\right)_T\right]\left(\frac{\partial V}{\partial T}\right)_P$$

$$= C_v + \left[\left(\frac{\partial H}{\partial P}\right)_T - V\right]\left(\frac{\partial P}{\partial V}\right)_T\left(\frac{\partial V}{\partial T}\right)_P \tag{4-27}$$

Reference to Equation 2-8 will show that

$$\left(\frac{\partial P}{\partial V}\right)_T\left(\frac{\partial V}{\partial T}\right)_P = -\left(\frac{\partial P}{\partial T}\right)_V \tag{4-28}$$

The insertion of this expression into Equation 4-27 gives the relationship

$$C_p = C_v + \left[V - \left(\frac{\partial H}{\partial P}\right)_T\right]\left(\frac{\partial P}{\partial T}\right)_V \tag{4-29}$$

3. A third relationship between C_p and C_v can be obtained by several operations on the second term within the bracket in Equation 4-29. If we consider H as a function of T and P, then

$$dH = \left(\frac{\partial H}{\partial T}\right)_P dT + \left(\frac{\partial H}{\partial P}\right)_T dP \tag{4-30}$$

If H is held constant $dH = 0$, and

$$0 = \left(\frac{\partial H}{\partial T}\right)_P\left(\frac{\partial T}{\partial P}\right)_H + \left(\frac{\partial H}{\partial P}\right)_T \tag{4-31}$$

Using Equation 4-31 to substitute for $(\partial H/\partial P)_T$ in Equation 4-29 we obtain

$$C_p = C_v + \left[V + \left(\frac{\partial H}{\partial T}\right)_P\left(\frac{\partial T}{\partial P}\right)_H\right]\left(\frac{\partial P}{\partial T}\right)_V \tag{4-32}$$

or

$$C_p = C_v + \left[V + C_p\left(\frac{\partial T}{\partial P}\right)_H\right]\left(\frac{\partial P}{\partial T}\right)_V \tag{4-33}$$

If C_p is factored out we also can obtain

$$C_p = \frac{C_v + V(\partial P/\partial T)_V}{1 - (\partial T/\partial P)_H(\partial P/\partial T)_V} \tag{4-34}$$

Several other general relationships between C_p and C_v are obtainable by procedures similar to those just outlined.

Heat Capacities of Gases

From classical thermodynamics alone it is impossible to predict numerical values for heat capacities; hence these quantities must be determined calorimetrically. With the aid of statistical mechanics, however, it is possible to determine heat capacities from spectroscopic data, instead of from direct calorimetric measurements. Even with spectroscopic information, however, it is convenient to correlate data over a range of temperatures in terms of empirical equations [2]. The following expressions usually have been used for the molar heat capacity:

$$c_p = a + bT + cT^2 + dT^3 \tag{4-35}$$

and

$$c_p = a + bT + \frac{c'}{T^2} \tag{4-36}$$

The results of several critical surveys by H. M. Spencer and his collaborators are summarized in Table 4-1 and are illustrated in Figure 4-1. An exhaustive list of sources of tabulated thermochemical data can be found in Volume 1 of *Chemical Thermodynamics, A Specialist Periodical Report* [4].

Heat Capacities of Solids

Early in the nineteenth century, Dulong and Petit observed that the molar heat capacity of a solid element generally is near 6 cal mol^{-1} K^{-1} (25 J mol^{-1} K^{-1}). Subsequent investigation showed that c_v (or c_p) varies markedly with the temperature, in the fashion indicated by Figure 4-2. However, the upper limiting value of about 25 J mol^{-1} K^{-1} is approached by the heavier elements at room temperature.

It is impossible to predict values of heat capacities for solids by purely thermodynamic reasoning. However, the problem of the solid state has received much consideration from an extrathermodynamic view, and several important expressions for the heat capacity have been derived. For our purposes it will be sufficient to consider only the Debye equation, and in particular its limiting form at very low temperatures:

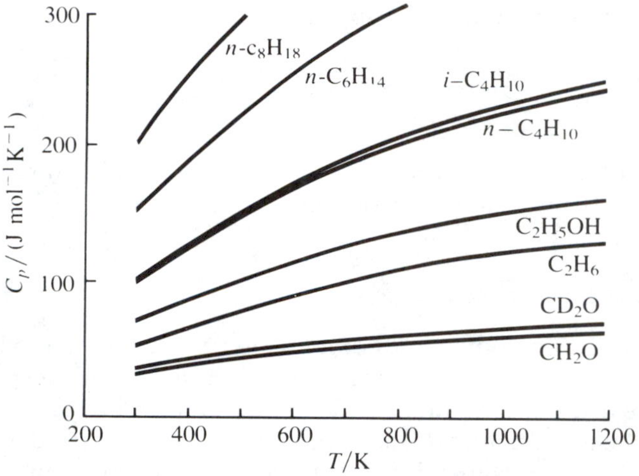

Figure 4-1. Variation of molar heat capacity with temperature for some organic compounds. The differences observed between isotopic species and the way in which heat capacity depends on molecular size and structure can be described thermodynamically, but must be explained by the methods of quantum statistical mechanics.

$$C_v = \frac{12\pi^4}{5} R \frac{T^3}{\theta^3} = 1943.8 \frac{T^3}{\theta^3} \text{ J mol}^{-1} \text{ K}^{-1} \tag{4-37}$$

The symbol θ is called the characteristic temperature and can be calculated from an experimental determination of the heat capacity at a low temperature. This equation has been very useful in the extrapolation of measured heat capacities [5] down to 0 K, particularly in connection with calculations of entropies from the third law of thermodynamics. Strictly speaking, the Debye

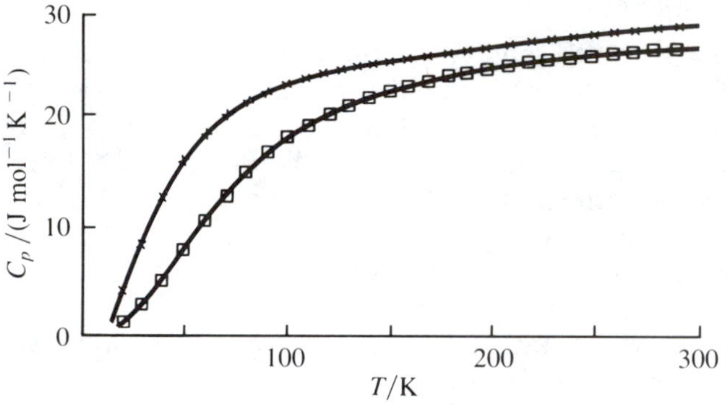

Figure 4-2. Molar heat capacities of solid sodium (x) and palladium (□). [Taken from data of G. L. Pickard and F. E. Simon, *Proc. Phys. Soc.* **61**, 1 (1948).]

Table 4-1 Molar Heat Capacities at Constant Pressure[a,b,c]

Substance	Temperature Range/K	$a/(\mathrm{J\ mol^{-1}\ K^{-1}})$	$b/(10^{-3}\ \mathrm{J\ mol^{-1}\ K^{-2}})$	$c/(10^{-7}\ \mathrm{J\ mol^{-1}\ K^{-3}})$	$d/(10^{-9}\ \mathrm{J\ mol^{-1}\ K^{-4}})$
H_2(g)	300–1500	29.066	−0.8364	20.116	
O_2(g)	300–1500	25.723	12.979	−38.62	
N_2(g)	300–1500	27.296	5.230	−0.04	
Cl_2(g)	300–1500	31.6958	10.1437	−40.376	
Br_2(g)	300–1500	35.2409	4.0748	−14.874	
H_2O(g)	300–1500	30.359	9.615	11.84	
CO_2(g)	300–1500	21.556	63.697	−405.05	9.678
CO(g)	300–1500	26.861	6.966	−8.20	
CNCl(g)	250–1000	47.296	10.213	−4.849*	
HCl(g)	300–1500	28.1663	1.8096	15.468	
SO_2(g)	300–1800	49.768	4.556	−11.054*	
SO_3(g)	300–1200	15.075	151.921	−1206.16	36.187
CH_4(g)	300–1500	17.451	60.458	11.17	−7.205
C_2H_6(g)	300–1500	5.351	177.669	−687.01	8.514
C_3H_8(g)	300–1500	−4.799	307.311	−1601.59	32.748
n-C_4H_{10}(g)	300–1500	0.469	385.38	−1988.8	39.966
n-C_5H_{12}(g)	300–1500	1.443	476.474	−2504.08	51.237
Benzene (g)	300–1500	−33.899	471.87	−2983.44	70.835
Pyridine (g)	290–1000	−12.619	368.5439	−1617.74	
Carbon (graphite)	300–1500	−5.293	58.609	−432.25	11.510

[a]H. M. Spencer and J. L. Justice, *J. Am. Chem. Soc.* **56**, 2311 (1934); H. M. Spencer and G. N. Flannagan, *J. Am. Chem. Soc.* **64**, 2511 (1942); H. M. Spencer, *J. Am. Chem. Soc.* **67**, 1859 (1945); H. M. Spencer, *Ind. Eng. Chem.* **40**, 2152 (1948). Data on numerous other substances are given in these original papers. See also K. K. Kelley, *U. S. Bur. Mines Bull.* **476** (1949) and **584** (1960).

[b]The constants are those defined in Equation 4-35, except for those with superscript *, which represent the constant $c'/(\mathrm{J\ mol^{-1}\ K})$ of Equation 4-36.

[c]Spencer and his colleagues expressed their results in terms of a defined calorie equal to 4.1833 international joules. When the international joule was replaced by the absolute joule, now called simply the joule, the defined calorie retained the same value, only now it is equal to 4.184 joules [3].

equation was derived only for an isotropic elementary substance. Nevertheless, it is applicable to most compounds, particularly in the region close to absolute zero.

Heat Capacities of Liquids

No adequate theoretical treatment has been developed that might serve as a guide in interpreting and correlating data on the heat capacities of liquids. However, it has been observed that the molar heat capacity of a pure liquid generally is near that of the solid, so if measurements are not available we may assume that c_v is 25 J mol^{-1} K^{-1}. However, the heat capacities of solutions cannot be predicted reliably from the corresponding properties of the components. Empirical methods of treating solutions will be considered in later chapters.

Integration of Heat Capacity Equations

Later, in connection with calculations of enthalpies and entropies of substances, it will become necessary to integrate the values of the heat capacity over a temperature range. Therefore we shall develop the methods for carrying out the integration at this point.

Analytical method. If the heat capacity can be expressed as an analytic function of the temperature, the integration can be done analytically. For example, if c_p can be expressed as a power series,

$$c_p = a + bT + cT^2 + \cdots \tag{4-38}$$

in which *a, b*, and *c* are constants at a fixed pressure, and if terms higher than the second degree are neglected in Equation 4-38, the integration of Equation 4-19 leads to the expression

$$dH = \int c_p \, dT = \int (a + bT + cT^2)dT \tag{4-39}$$

Carrying out an indefinite integration we obtain

$$H - H_I = aT + \frac{b}{2}T^2 + \frac{c}{3}T^3 \tag{4-40}$$

in which H_I is the constant of integration. Because *a, b,* and *c* are usually defined over a limited range of temperatures, as in Table 4-1, H_I is *not* equal to the enthalpy at $T = 0$ K. Also, since it is only possible to calculate differences in enthalpy, H_I can be evaluated only in terms of differences in enthalpy.

Graphical or Numerical Method. If the dependence of c_p on temperature is too complicated to be expressed by a simple function as, for example, in solids at low temperatures, then a graphical or numerical method of integration can be used. A typical example was worked out partially in Chapter 2 (Figure 2-2). As we pointed out there, if we take a small enough temperature interval ΔT, the product $\dddot{C}_p\ \Delta T$, in which $\dddot{C}_p$ is the average value of the molar heat capacity in that region, approximates the area under the C_p curve in that temperature interval. Consequently, the integral can be approximated to any desired degree by summation of the areas:

$$\Sigma\, \dddot{C}_p\ \Delta T \simeq \int C_p\, dT \tag{4-41}$$

The numerical method is applied extensively in integrations of heat capacity data to obtain third-law entropies for use in predicting the feasibility of chemical reactions (Chapter 11).

EXERCISES

1. Molar heat capacities of solid n-heptane are listed in Table 2-7.
   a. Calculate $H_{182.5\text{ K}} - H_{15.14\text{ K}}$ by numerical integration.
   b. Calculate $H_{15.14\text{ K}} - H_{0\text{ K}}$ by graphical integration of an extrapolation of the heat capacity curve to $C_p = 0$ at 0 K.
   c. If the heat of fusion at 182.5 K is 14,033 J mol^{-1}, calculate ΔH for the transformation:

   n-heptane(solid, 0 K) = n-heptane(liquid, 182.5 K)

2. a. Derive the Debye equation for C_v for n-heptane. Use the given value of C_p at 15.14 K (see Exercise 1 above) as a sufficiently close approximation to C_v.
   b. Calculate $H_{15.14\text{ K}} - H_{0\text{ K}}$ by integrating the Debye equation for n-heptane. Using this analytical value in place of the graphical value for the same temperature range, and using the numerical results above 15.14 K, determine $H_{182.5\text{ K}} - H_{0\text{ K}}$. Compare the answer with the result obtained by graphical and numerical methods in Exercise 1 above.

3. Calculate the differences between Q_P and Q_V in the following reactions:
   a. $H_2(g) + \frac{1}{2}O_2(g) = H_2O(l)$ at 25°C
   b. Ethyl acetate + water = ethyl alcohol + acetic acid, at 25°C
   c. Haber synthesis of ammonia, at 400°C
   d. $C_6H_{12}O_6 + 6O_2(g) = 6CO_2(g) + 6H_2O$, at 25°C
   e. α quartz = β quartz, at 846 K

4. Prove the following relationships, using only definitions and mathematical principles:

   a. $\left(\frac{\partial E}{\partial V}\right)_P = C_p\left(\frac{\partial T}{\partial V}\right)_P - P$; b. $\left(\frac{\partial E}{\partial P}\right)_V = C_v\left(\frac{\partial T}{\partial P}\right)_V$

5. Using the equations in Exercise 4, calculate $(\partial E/\partial V)_P$ and $(\partial E/\partial P)_V$ for the special case of 1 mol of an ideal gas, for which $PV = RT$.

6. Suggest a substance, and the conditions for that substance, such that $C_p = C_v$, that is, for which the second term on the right side of Equation 4-25 equals zero.

REFERENCES

1. "The NBS Tables of Chemical Thermodynamic Properties," *J. Phys. Chem. Ref. Data* **11**, 1982, Supplement No. 2.
2. A method has been described [B. L. Crawford, Jr., and R. G. Parr, *J. Chem. Phys.* **16**, 233 (1948)] by means of which one may proceed directly from spectroscopic data to an empirical equation of the form of Equation 4-35.
3. F. D. Rossini, "Units and Physical Constants," in *Combustion Calorimetry*, S. Sunner and M. Manson, Eds., Pergamon Press, Oxford, 1979, pp. 1–8.
4. E. F. G. Herington, "Thermodynamic Quantities, Thermodynamic Data, and Their Uses," in *Chemical Thermodynamics, A Specialist Periodical Report*, Vol. 1, M. L. McGlashan, Ed., The Chemical Society, London, 1973.
5. Deviations from the T^3 law and their significance have been discussed by K. Clusius and L. Schachinger, *Z. Naturforsch* **2a**, 90 (1947), and by G. L. Pickard and F. E. Simon, *Proc. Phys. Soc.* **61**, 1 (1948).

CHAPTER 5

Enthalpy of Reaction

In Chapter 4 we introduced a new function, enthalpy, and found among its properties a correspondence with the heat of reaction at constant pressure (when the only work is due to a volume change against that pressure). Most chemical reactions are carried out in a vessel exposed to the atmosphere and under conditions such that no work other than that against the atmosphere is produced. For this reason, and because ΔH is independent of the path of a reaction and is measured by the heat transferred in the reaction, the enthalpy change is a useful quantity.

Enthalpy changes also give pertinent information for several other problems in chemical thermodynamics. For a long time it was thought that the sign of ΔH, that is, whether heat is absorbed or evolved, is a criterion of the spontaneity of a reaction. When this misconception was cleared up, it was evident that this criterion is still useful within certain limitations. If the ΔH values are large enough, their signs still can be used as the basis for a first guess regarding the feasibility of a reaction.

In more rigorous applications enthalpy changes must be known in addition to other quantities. Also, ΔH values are required to establish the magnitude of the temperature dependence of equilibrium constants. For all these reasons, it is desirable to have tables of ΔH values available, so that the enthalpies of various transformations can be calculated readily. In many of these calculations we make use of Hess's law, now firmly established on the basis of the first law of thermodynamics, and we use known values of ΔH to calculate enthalpies for new reactions that can be expressed as sums of known reactions.

5-1 DEFINITIONS AND CONVENTIONS

The enthalpy of a reaction depends on the states of the substances involved. Thus in the formation of water,

$$H_2(g) + \tfrac{1}{2}O_2(g) = H_2O \tag{5-1}$$

if the H_2O produced is a liquid, ΔH will differ from that observed if the H_2O is a vapor. Similarly, the enthalpy of the reaction depends on the pressure of any gases involved. Furthermore, for reactions involving solids, such as sulfur, ΔH depends on which crystalline form (for example, rhombic or monoclinic) participates in the reaction. For these reasons, tabulated values of ΔH refer to reactions with the reactants and products in specified standard states.

Some Standard States

The states that have been agreed upon as reference states in tabulating enthalpies of reaction are summarized in Table 5-1. Other standard states may be adopted in special problems. When no state is specified, it can be assumed to be that listed in Table 5-1.

Enthalpy of Formation

The tables of enthalpies of reaction generally list the enthalpies of formation of various compounds in their standard states from the elements in their

Table 5-1 Standards and Conventions for Enthalpies of Reaction[a]

Standard state of solid	The most stable form at 0.1 MPa (1 bar) pressure and the specified temperature (unless otherwise specified)[b]
Standard state of liquid	The most stable form at 0.1 MPa (1 bar) pressure and the specified temperature
Standard state of gas	Zero pressure[c] and the specified temperature
Standard state of carbon	Graphite
Standard temperature	25°C (298.15 K)
Sign of ΔH	+ If heat is absorbed

[a]The International Union of Pure and Applied Chemistry now recommends a standard pressure of 0.1 MPa (1 bar) in place of the previously accepted standard of 101.325 kPa (1 atm). The difference in thermodynamic quantities is not significant for condensed phases, and differences in ΔH values are not significant even for gases, but the user of thermodynamic tables will have to note carefully the standard state chosen for any compilation of data. See: "The NBS Tables of Chemical Thermodynamic Properties," *J. Phys. Chem. Ref. Data* **11**, 1982, Supplement No. 2; IUPAC Division of Physical Chemistry, Commission on Symbols, Terminology and Units, "Manual of Symbols and Terminology for Physico-Chemical Quantities and Units," M. L. McGlashan, M. A. Paul, and D. H. Whiffen, Eds., *Pure and Applied Chem.* **51**, 1(1979), and Appendix IV, *Pure and Applied Chem.* **54**, 1239(1982).

[b]Thus, for some problems rhombic sulfur may be a convenient standard state, whereas for others monoclinic sulfur may be a convenient standard state.

[c]It is shown in Chapter 17 that internal consistency in the definition of standard states requires that zero pressure be the standard state for the enthalpy of a gas. Unfortunately, most reference sources use the convention of 0.1 MPa or 101.325 kPa pressure. In most cases, the difference in enthalpy between these two pressures is very small.

standard states at the specified temperature. Thus, if the standard enthalpy of formation, $\Delta Hf°$, of CO_2 at 25°C is given as -393.509 kJ mol^{-1}, the following equation is implied:

C	+	O_2	=	CO_2	(5-2)
(graphite at 298.15 K, 0.1 MPa)		(gas at 298.15 K, zero pressure)		(gas at 298.15 K, zero pressure)	

$$\Delta Hf° = -393.509 \text{ kJ mol}^{-1}$$

By this definition the enthalpy of formation of an element in its standard state is zero. In other words, elements in their standard states are taken as reference states in the tabulation of enthalpies of reaction, just as sea level is the reference point in measuring geographic heights.

A few standard enthalpies of formation have been assembled in Table 5-2. Data for other substances can be obtained from the following critical compilations:

International Critical Tables, McGraw-Hill, New York, 1933.

Landolt-Börnstein, *Physikalisch-chemische Tabellen*, 5th ed., Springer, Berlin, 1936; 6th ed., 1961, 1963, 1967, 1972.

W. M. Latimer, *Oxidation Potentials*, 2d ed., Prentice-Hall, Englewood Cliffs, N. J., 1952.

TRC Thermodynamic Tables–Non-hydrocarbons, Thermodynamics Research Center: The Texas A&M University System, College Station, Texas (Loose-leaf Data Sheets) [1].

D. R. Stull and G. C. Sinke, "Thermodynamic Properties of the Elements," Am. Chem. Soc., Washington, D.C., 1956.

D. R. Stull and H. Prophet, "JANAF Thermochemical Tables," 2d ed., National Standard Reference Data Series, Nat. Bur. Stand. (U. S.), **37**, 1971, *J. Phys. Chem. Ref. Data* **3**, 311(1974), **4**, 1(1975), **7**, 793(1978), **11**, 695(1982).

5-2 ADDITIVITY OF ENTHALPIES OF REACTION

Since the enthalpy is a thermodynamic property, the value of ΔH depends only on the nature and state of the initial reactants and final products, and not on the reactions that have been used to carry out the transformation. If we must deal with a reaction whose ΔH is not available, it is sufficient to find a series of reactions for which ΔH's are available and whose sum is the reaction in question.

Table 5-2 Standard Enthalpies[a] of Formation at 298.15 K

Substance	$\Delta Hf°$/kJ mol^{-1}	Substance	$\Delta Hf°$/kJ mol^{-1}
		$MgSiO_3$(s); enstatite	−1547.8[b]
H(g)	217.965	$CaSiO_3$(s); wollastonite	−1635.2[b]
O(g)	249.170	$CuSO_4 \cdot 5H_2O$(s); chalcanthite	−2279.7[b]
Cl(g)	121.679	SiO_2(s); α quartz	−910.74[b]
Br(g)	111.884	SiO_2(s); α cristobalite	−908.35[b]
F(g)	78.99	SiO_2(s); α tridymite	−907.49[b]
I(g)	106.838	Methane(g)	−74.475[c]
N(g)	472.704	Ethane(g)	−83.85[c]
C(g)	716.682	Propane(g)	−104.68[c]
Br_2(g)	30.907	*n*-Butane(g)	−125.65[c]
I_2(g)	62.438	Ethylene(g)	52.283[c]
H_2O(g)	−241.818	Propylene(g)	20.414[c]
H_2O(l)	−285.830	1-Butene(g)	−0.13[c]
HF(g)	−271.1	Acetylene(g)	226.748[c]
HD(g)	0.318	Benzene(g)	82.80[c]
C(diamond)	1.895	Cyclohexane(g)	−123.14[c]
HCl(g)	−92.307	Toluene(g)	50.00[c]
HBr(g)	−36.40	*o*-Xylene(g)	19.08[c]
HI(g)	26.48	*m*-Xylene(g)	17.32[c]
ICl(g)	17.78	*p*-Xylene(g)	18.03[c]
NO(g)	90.25	Methanol(l)	−238.95[c]
CO(g)	−110.525	Ethanol(l)	−276.98[c]
CO_2(g)	−393.509	Glycine(s)	−528.10
NH_3(g)	46.11	Acetic acid(l)	−484.5
Mg_2SiO_4(s);		Taurine(s)	−785.3
forsterite	−2170.4[b]	Urea(s)	−333.51

[a]Selected from "The NBS Tables of Chemical Thermodynamic Properties," *J. Phys. Chem. Ref. Data* **11**, Supplement No. 2, 1982.
[b]"Thermodynamic Properties of Minerals and Related Substances," *Geological Survey Bulletin*, 1452, 1978.
[c]TRC Thermodynamic Tables-Non-hydrocarbons, Thermodynamics Research Center: The Texas A&M University System, College Station, Texas, 1971 (Loose-leaf Data Sheets) [1]. This work uses 101.325 kPa as the standard pressure.

Calculation of Enthalpy of Formation from Enthalpy of Reaction

As an example of calculating the enthalpy of formation from the enthalpy of reaction, we will calculate the enthalpy of formation of $Ca(OH)_2$(solid) from data for other reactions, such as the following:

$$CaO(s) + H_2O(l) = Ca(OH)_2(s), \qquad \Delta H_{18^\circ C} = -63.848 \text{ kJ mol}^{-1} \quad (5\text{-}3)$$

$$H_2(g) + \tfrac{1}{2}O_2(g) = H_2O(l), \qquad \Delta H_{18^\circ C} = -285.830 \text{ kJ mol}^{-1} \quad (5\text{-}4)$$

$$Ca(s) + \tfrac{1}{2}O_2(g) = CaO(s), \qquad \Delta H_{18^\circ C} = -635.13 \text{ kJ mol}^{-1} \quad (5\text{-}5)$$

Adding these three chemical equations leads to the desired equation; hence, adding the corresponding ΔH's gives the desired enthalpy of formation:

$$Ca(s) + O_2(g) + H_2(g) = Ca(OH)_2(s),$$

$$\Delta H_{18^\circ C} = -984.81 \text{ kJ mol}^{-1} \quad (5\text{-}6)$$

Calculation of Enthalpy of Formation from Enthalpy of Combustion

Calculation of the enthalpy of formation from the enthalpy of combustion is common because with most organic compounds the experimental data are obtained for the combustion process, yet the enthalpy of formation is the more useful quantity for further thermodynamic calculations. A typical example of such a calculation is outlined by the equations:

$$C_2H_5OH(l) + 3O_2(g) = 2CO_2(g) + 3H_2O(l),$$

$$\Delta H_{298\text{ K}} = -1367.53 \text{ kJ mol}^{-1} \quad (5\text{-}7)$$

$$3H_2O(l) = 3H_2(g) + \tfrac{3}{2}O_2(g),$$

$$\Delta H_{298\text{ K}} = 857.490 \text{ kJ mol}^{-1} \quad (5\text{-}8)$$

$$2CO_2(g) = 2C(\text{graphite}) + 2O_2(g),$$

$$\Delta H_{298\text{ K}} = 787.018 \text{ kJ mol}^{-1} \quad (5\text{-}9)$$

$$C_2H_5OH(l) = 3H_2(g) + \tfrac{1}{2}O_2(g) + 2C(\text{graphite}),$$

$$\Delta H = 276.98 \text{ kJ mol}^{-1} \quad (5\text{-}10)$$

Reversing Equation 5-10, we obtain the enthalpy of formation of ethyl alcohol:

$$3H_2(g) + \tfrac{1}{2}O_2(g) + 2C(\text{graphite}) = C_2H_5OH(l),$$

$$\Delta Hf^\circ = -276.98 \text{ kJ mol}^{-1} \quad (5\text{-}11)$$

Calculation of Enthalpy of Transition from Enthalpy of Formation

Calculating the enthalpy of transition from the enthalpy of formation is particularly important when considering changes of reference state from one allotrope to another. Carbon is illustrated as an example:

$$C(\text{graphite}) + O_2(g) = CO_2(g), \qquad \Delta H_{298\text{ K}} = -393.509 \text{ kJ mol}^{-1} \quad (5\text{-}12)$$

$$C(\text{diamond}) + O_2(g) = CO_2(g), \qquad \Delta H_{298\text{ K}} = -395.404 \text{ kJ mol}^{-1} \quad (5\text{-}13)$$

Subtracting the first reaction from the second, we obtain

$$\text{C(diamond)} = \text{C(graphite)}, \qquad \Delta H_{298\text{ K}} = -1.895 \text{ kJ mol}^{-1} \qquad (5\text{-}14)$$

The enthalpy change for the transition from graphite to diamond is an essential item of information in calculating the conditions for the geological and industrial production of diamonds.

Calculation of Enthalpy of Conformational Transition of a Protein from Indirect Calorimetric Measurements

Many protein molecules exist in two (or more) different structures of the same molecular weight but with different spatial disposition of constituent atoms. A very intensively studied enzyme, aspartate transcarbamoylase (abbreviated ATCase), under some circumstances is in an enzymatically constrained steric arrangement, labeled the *T* form, and under other conditions in the enzymatically very active, structurally more swollen and open conformation, the *R* form.

The $\Delta H_{T \to R}$ of the transition is a thermodynamically valuable parameter for understanding the behavior of the enzyme. However, this quantity cannot be measured directly because the transition can only be achieved by adding a small-molecule substrate or an analog thereof. One such analog, N-(phosphonacetyl)-L-aspartate (acronym PALA), is very effective in promoting the $T \to R$ transition. Calorimetric measurements have been reported [A. Shrake, A. Ginsburg, and H. K. Schachman, *J. Biol. Chem.* **256**, 5005–5015 (1981)] for the mixed process of binding the PALA and the accompanying $T \to R$ transition. The observed ΔH values (per mole of enzyme) depend on the number of moles of PALA bound:

	Change in Moles PALA Bound per Mole ATCase			
	$0.0 \to 1.8$	$1.8 \to 3.4$	$3.4 \to 5.2$	$0.0 \to 6.0$
$\Delta H/(\text{kJ mol}^{-1})$	−63.2	−62.3	−59.0	−209.2

The different transitions for which calorimetric data have been obtained and the relationships of the experimental ΔH's to the enthalpy of transition, $\Delta H_{T \to R}$, and to the enthalpy of binding of PALA, ΔH_{PALA}, are shown in Figure 5-1, a thermodynamic cycle that permits calculation of ΔH for one step if all the others are known. Each one of the heats ΔH_{I} to ΔH_{V}, in principle, contains some contribution from the enthalpy of binding and some from the enthalpy of conformational transition. It has been assumed that the enthalpy of binding per mole of ligand, ΔH_{PALA}, is identical for each of the six PALA molecules taken up sequentially by the enzyme. Thus

$$\Delta H_{\text{VI}} = 6\,\Delta H_{\text{PALA}} \qquad (5\text{-}15)$$

However ΔH_{VI} is not accessible directly by calorimetric measurements. In each of the steps ΔH_{I} to ΔH_{V}, the contribution from $\Delta H_{T\to R}$ cannot be established without some knowledge of the extent of the conformational transition, which is known *not* to be proportional to the moles of bound PALA.

Extrathermodynamic (ultracentrifugation) experiments have established that 43% of the conformational transition occurs when the first 1.8 moles of ligand is added. Therefore, we can write

$$\Delta H_{\mathrm{I}} = 1.8\ \Delta H_{\mathrm{PALA}} + 0.43\ \Delta H_{T\to R} \tag{5-16}$$

Similarly, it is known that when the bound PALA is increased from 3.4 to 5.2, 23% of the conformational transition takes place. Therefore, for the additional binding of 1.8 moles of PALA, in the third step of the left-hand path of Figure 5.1, we can state that

$$\Delta H_{\mathrm{III}} = 1.8\ \Delta H_{\mathrm{PALA}} + 0.23\ \Delta H_{T\to R} \tag{5-17}$$

Using the experimentally determined values for ΔH_{I} and ΔH_{III}, -63.2 kJ mol^{-1} and -59.0 kJ mol^{-1}, respectively, shown in the table above, we can solve Equations 5-16 and 5-17 for the ΔH's of the two inaccessible steps in Figure 5-1. In this way it has been shown that

$$\Delta H_{T\to R} = -21 \text{ kJ/mol}$$

$$\Delta H_{\mathrm{PALA}} = -38 \text{ kJ/mol}$$

Alternatively, we can use the experimental calorimetric data for the diagonal path in Figure 5-1, that is, for the addition of 6 moles of ligand in one step. According to the table, ΔH_{V} is -209.2 kJ mol^{-1}. Consequently, we can write

$$\Delta H_{\mathrm{V}} = 6\ \Delta H_{\mathrm{PALA}} + \Delta H_{T\to R} \tag{5-18}$$

This equation can be combined with either Equation 5-16 or 5-17, and in each procedure alternative values can be computed for $\Delta H_{T\to R}$ and for ΔH_{PALA}.

Calculation of Enthalpy of Solid State Reaction from Measurements of Enthalpy of Solution

Many reactions of geological interest have crystalline solids as reactants and products. Even when such reactions have large values of ΔH, the enthalpy of reaction cannot be measured directly because the reaction is very slow. An important reaction of this kind is that of MgO (periclase) and SiO_2 (quartz) to form forsterite (Mg_2SiO_4).

$$2\mathrm{MgO(c)} + \mathrm{SiO_2(c)} = \mathrm{Mg_2SiO_4(c)} \quad \Delta H_{\mathrm{I}} \tag{5-19}$$

Workers at the Bureau of Mines were able to obtain ΔH_{I} by measuring the value of ΔH for the reaction of each of the solids with 20% HF solutions near 75°C. The reactions observed are:

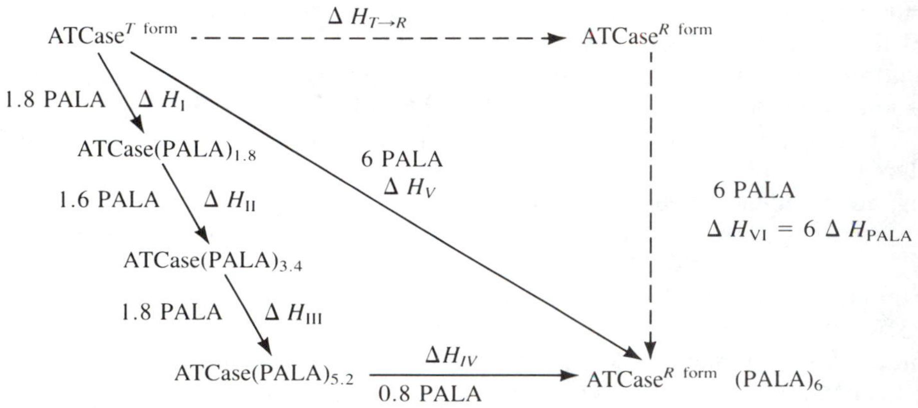

Figure 5-1. Thermodynamic cycles, illustrating states of enzyme ATCase during calorimetric measurements of heats accompanying the binding of progressively increasing quantities of the substrate analog PALA. At the outset, the ATCase is 100% in the *T* conformation; at the conclusion of the transformation, the ATCase$^{R\ \text{form}}$ (PALA)$_6$ is 100% in the *R* conformation and has six bound PALA molecules. When the extent of binding of PALA is between 0 and 6, the extent of conformational conversion is between 0 and 100%. The horizontal broken arrow at the top of the diagram indicates the process which is accompanied by the heat, $\Delta H_{T\to R}$, that we want to know. The vertical broken arrow at the right represents a pure binding step for the enzyme in the 100% *R* conformation.

$$SiO_2(c, 25°C) + 6HF\ (soln, 73.7°C) = H_2SiF_6(soln, 73.7°C) + 2H_2O(soln, 73.7°C) \quad (5\text{-}20)$$

$$\Delta H_{II} = -148.16\ kJ\ mol^{-1}\ [2]$$

$$MgO(c, 25°C) + 2HF(soln, 73.7°C) = MgF_2(soln, 73.7°C) + H_2O(soln, 73.7°C) \quad (5\text{-}21)$$

$$\Delta H_{III} = -162.13\ kJ\ mol^{-1}\ [3]$$

$$Mg_2SiO_4(c, 25°C) + 10HF(soln, 73.7°C) = 2MgF_2(soln, 73.7°C) + H_2SiF_6(soln, 73.7°C) + 4H_2O(soln, 73.7°C) \quad (5\text{-}22)$$

$$\Delta H_{IV} = -399.07\ kJ\ mol^{-1}\ [3]$$

From an inspection of the reactions, it can be seen that

$$\Delta H_I = 2(\Delta H_{III}) + \Delta H_{II} - \Delta H_{IV}$$
$$= -73.35\ kJ\ mol^{-1} \quad (5\text{-}23)$$

5-3 BOND ENERGIES

The calculation of the enthalpy of formation of a given compound depends on the determination of the enthalpy of at least one reaction of this substance.

Frequently, it is desirable to estimate the enthalpy of a chemical reaction involving a hitherto unsynthesized compound, that is, a substance for which no enthalpy data are available. For the solution of problems of this type, a system of bond energies has been established such that, if the molecular structure of the compound is known, it is possible to approximate the enthalpy of formation by adding the appropriate bond energies.

Definition of Bond Energies

We must be careful to distinguish between "bond energy" and the "bond dissociation energy" of a given bond. The latter is a definite quantity that refers to the energy required to break a given bond of some specific compound. However, bond energy is an average value of the dissociation energies of a given bond in a series of different dissociating species.

The distinction between these two terms may be more evident if described in terms of a simple example, the O—H bond. The heat of dissociation of the O—H bond depends on the nature of the molecular species from which the H atom is being separated. For example, in the water molecule

$$H_2O(g) = H(g) + OH(g), \qquad \Delta H_{298\ K} = 501.87\ kJ\ mol^{-1} \tag{5-24}$$

However, to break the O—H bond in the hydroxyl radical requires a different quantity of energy:

$$OH(g) = O(g) + H(g), \qquad \Delta H_{298\ K} = 423.38\ kJ\ mol^{-1} \tag{5-25}$$

The bond energy, ε_{O-H}, is defined as the average of these two values; that is,

$$\varepsilon_{O-H} = \frac{501.87 + 423.38}{2} = 462.63\ kJ\ mol^{-1} \tag{5-26}$$

Thus ε_{O-H} is half the value of the enthalpy change for the reaction

$$H_2O(g) = O(g) + 2H(g), \qquad \varepsilon_{O-H} = \frac{\Delta H}{2} = 462.63\ kJ\ mol^{-1} \tag{5-27}$$

In the case of diatomic elements such as H_2, the bond energy and bond dissociation energy are identical because each refers to the reaction

$$H_2(g) = 2H(g), \qquad \varepsilon_{H-H} = \Delta H_{298\ K} = 435.930\ kJ\ mol^{-1} \tag{5-28}$$

Thus, the bond energy given for any particular pair of atoms is the average value of the dissociation energy of that bond for a number of molecules in which the pair of atoms appears. The dissociation energy is given for the conversion of the compound as an ideal gas into the atomic elements, also as ideal gases.

Calculation of Bond Energies

The preceding illustration with water and the O—H bond is a relatively simple case. Since bond energies are of particular value in problems involving organic compounds, it is desirable to consider an example from this branch of chemistry, for which the fundamental data are obtained from enthalpies of combustion. We will calculate the C—H bond energy from data on the enthalpy of combustion of methane.

To find ε_{C-H}, we need to know the enthalpy of the reaction

$$CH_4(g) = C(g) + 4H(g), \qquad \varepsilon_{C-H} = \frac{\Delta H}{4} \tag{5-29}$$

The ΔH for the preceding reaction can be obtained from the summation of those for the following reactions at 298 K:

$$CH_4(g) + 2O_2(g) = CO_2(g) + 2H_2O(l), \qquad \Delta H = -890.36 \text{ kJ mol}^{-1} \tag{5-30}$$

$$CO_2(g) = C(\text{graphite}) + O_2(g), \qquad \Delta H = 393.51 \text{ kJ mol}^{-1} \tag{5-31}$$

$$2H_2O(l) = 2H_2(g) + O_2(g), \qquad \Delta H = 571.70 \text{ kJ mol}^{-1} \tag{5-32}$$

$$2H_2(g) = 4H(g), \qquad \Delta H = 871.86 \text{ kJ mol}^{-1} \tag{5-33}$$

$$C(\text{graphite}) = C(g), \qquad \Delta H = 716.682 \text{ kJ mol}^{-1} \tag{5-34}$$

$$CH_4(g) = C(g) + 4H(g), \qquad \Delta H = 1663.39 \text{ kJ mol}^{-1} \tag{5-35}$$

Thus at 298 K

$$\varepsilon_{C-H} = \frac{1663.39}{4} = 415.85 \text{ kJ mol}^{-1} \tag{5-36}$$

This value of the C—H bond energy does not correspond to the dissociation energy of the carbon–hydrogen bond in methane [4], which is 427 kJ mol^{-1}. However, the latter energy refers to the equation

$$CH_4(g) = CH_3(g) + H(g) \tag{5-37}$$

and not to one-fourth of the ΔH associated with Equation 5-35.

In the preceding calculation of ε_{C-H}, enthalpy values at 298 K were used; hence, the bond energy also refers to this temperature. For some purposes, it is the practice to calculate bond energies at 0 K, rather than 298 K. For ε_{C-H} we would obtain 410.9 kJ mol^{-1} at 0 K, a value slightly lower than that at 298 K. Generally, the differences between the bond energies at the two temperatures are small. A list of bond energies at 298 K is given in Table 5-3.

Table 5-3 Bond Energies[a] at 298 K

Bond	kJ mol^{-1}	Bond	kJ mol^{-1}	Bond	kJ mol^{-1}
H—H	435.89	Te=Te	222*	O—Cl	218
Li—Li	105*	I—I	150.88	F—Cl	253.1*
C—C	345.6	Cs—Cs	43.5*	Na—Cl	410 *
C=C	610.0	Li—H	243	Si—Cl	381
C≡C	835.1	C—H	413.0	P—Cl	326
N—N	163	N—H	390.8	S—Cl	255
N≡N	944.7	O—H	462.8	K—Cl	423*
O—O	146	F—H	565	Cu—Cl	368*
O=O	498.3	Na—H	197*	As—Cl	293
F—F	155	Si—H	318	Se—Cl	243
Na—Na	71*	P—H	322	Br—Cl	218*
Si—Si	222	S—H	347	Rb—Cl	428*
P—P	201	Cl—H	431.4	Ag—Cl	301*
S—S	226	K—H	180*	Sn—Cl	318
Cl—Cl	242.13	Cu—H	276*	Sb—Cl	310
K—K	49.4*	As—H	247	I—Cl	209*
Ge—Ge	188	Se—H	276	Cs—Cl	423*
As—As	146	Br—H	365.7	C—N	304.6
As≡As	381*	Rb—H	163*	C≡N	889.5
Se—Se	209	Ag—H	243*	C—O	357.7
Se=Se	272*	Te—H	238	C=O	745
Br—Br	192.80	I—H	298.7	C—S	272
Rb—Rb	45.2*	Cs—H	176*	C=S	536
Sn—Sn	163	Li—Cl	481*	P≡N	577*
Sb—Sb	121	C—Cl	339	S=O	498
Sb≡Sb	289*	N—Cl	192	C≡O	1046

[a]T. L. Cottrell, *The Strengths of Chemical Bonds*, 2d ed., Butterworths, London, 1958, pp. 270–289. A. G. Gaydon, *Dissociation Energies*, 3d ed., Chapman and Hall, Ltd., London, 1968. When values at 298 K have not been available, those at 0 K have been listed instead, and these are marked with an asterisk.

In some cases the value given in the table depends on that calculated previously for some other bond. For example, to obtain ε_{C-C} we combine the heat of combustion of ethane with the proper multiples of the ΔH's in Equations 5-31 through 5-34 to obtain the enthalpy change for the reaction

$$C_2H_6(g) = 2C(g) + 6H(g) \tag{5-38}$$

From our definition of bond energy

$$\varepsilon_{C-C} = \Delta H_{\text{Equation 5-38}} - 6\,\varepsilon_{C-H} \tag{5-39}$$

Thus, the value of 345.6 kJ mol^{-1} listed in Table 5-3 is based on an ε_{C-H} of 413.0 kJ mol^{-1}. Other estimates of ε_{C-H} would lead to different values for ε_{C-C}.

Estimation of the Enthalpy of Reaction from Bond Energies

The primary significance of bond energies lies in the calculation of the enthalpy of a reaction involving a compound for which no enthalpy data are available. For example, the enthalpy of formation of $Se_2Cl_2(g)$ can be calculated from bond energies by the following steps. Since the bond energy refers to the *dissociation* of Cl-Se-Se-Cl gas into gaseous atoms, the enthalpy change for the *formation* of this gaseous molecule should be given by

$$2Se(g) + 2Cl(g) = Se_2Cl_2(g),$$

$$\Delta H = -[\varepsilon_{Se-Se} + 2\varepsilon_{Se-Cl}] = -695 \text{ kJ mol}^{-1} \quad (5\text{-}40)$$

However, to estimate the heat of formation it is necessary to add two reactions to Equation 5-40, since by definition the heat of formation refers to the elements in their standard states. Therefore we introduce the following enthalpy changes to convert the elements from their standard states to the gaseous atoms at 298 K:

$$Cl_2(g) = 2Cl(g), \qquad \Delta H = 243.36 \text{ kJ mol}^{-1} \quad (5\text{-}41)$$

$$2Se(\text{hexagonal}) = 2Se(g), \qquad \Delta H = 2 \times 227.07 \text{ kJ mol}^{-1} \quad (5\text{-}42)$$

The addition of Equations 5-40 through 5-42 leads to the expression

$$2Se(\text{hexagonal}) + Cl_2(g) = Se_2Cl_2(g), \qquad \Delta H = +3 \text{ kJ mol}^{-1} \quad (5\text{-}43)$$

If we wish to know the enthalpy of formation of liquid Se_2Cl_2, we can estimate the enthalpy of condensation (perhaps from Trouton's rule, or by comparison with related sulfur compounds) and add it to the value of ΔH obtained in Equation 5-43.

By these methods it is possible to obtain fairly reliable estimates of enthalpies of formation of many compounds. Since the bond energies used are average values, they cannot be expected to result in highly accurate results for enthalpies of formation. More complex procedures also have been developed that will provide greater accuracy [5].

5-4 ENTHALPY OF REACTION AS A FUNCTION OF TEMPERATURE

In the preceding sections we discussed methods of obtaining enthalpies of reaction at a fixed temperature (generally 298.15 K). In particular, we pointed out that it is possible to tabulate enthalpies of formation and bond energies and to use these to calculate enthalpies of reaction. Such tables of enthalpies of formation are available for only a few standard temperatures. Frequently, however, it is necessary to know the enthalpy of a reaction at a temperature different from those available in a reference table. Therefore, we consider now the procedures that can be used to calculate the enthalpy of reaction (at constant pressure) at one temperature from data at another temperature.

Analytic Method

Since we are interested in the variation of enthalpy with temperature, we recall from Equation 4-19 that

$$\left(\frac{\partial H}{\partial T}\right)_P = C_P$$

Such an equation can be integrated at constant pressure:

$$\int dH = \int C_p dT$$

and

$$H = \int C_p dT + H_0 \tag{5-44}$$

In Equation 5-44, H_0 is an integration constant. If we are considering a chemical transformation, represented in general terms by

$$A + B + \cdots = M + N + \cdots \tag{5-45}$$

we can write a series of equations of the form

$$\left.\begin{aligned} H_A &= \int C_{pA} dT + H_{0A} \\ H_B &= \int C_{pB} dT + H_{0B} \\ H_M &= \int C_{pM} dT + H_{0M} \\ H_N &= \int C_{pN} dT + H_{0N} \end{aligned}\right\} \tag{5-46}$$

For the chemical reaction 5-45 the enthalpy change, ΔH, is given by

$$\begin{aligned} \Delta H &= H_M + H_N + \cdots - H_A - H_B - \cdots \\ &= (H_{0M} + H_{0N} + \cdots - H_{0A} - H_{0B} - \cdots) + \int C_{pM} dT \\ &\quad + \int C_{pN} dT + \cdots - \int C_{pA} dT - \int C_{pB} dT - \cdots \end{aligned} \tag{5-47}$$

If we define the quantities inside the parentheses as ΔH_0, and if we group the integrals together, we obtain

$$\Delta H = \Delta H_0 + \int (C_{pM} + C_{pN} + \cdots - C_{pA} - C_{pB} - \cdots)dT \quad (5\text{-}48)$$

or

$$\Delta H = \Delta H_0 + \int \Delta C_p dT \quad (5\text{-}49)$$

in which ΔC_p represents the expression in the parentheses, that is, the integrand of Equation 5-48. Thus, to obtain ΔH as a function of the temperature, it is necessary to know the dependence of the heat capacities of the reactants and products on the temperature, as well as one value of ΔH so that ΔH_0 can be evaluated.

As an example, let us consider the enthalpy of formation of CO_2(g):

$$\text{C(graphite)} + O_2(g) = CO_2(g), \qquad \Delta H_{298.15\ K} = 393{,}509\ \text{J mol}^{-1}$$

The heat capacities of the substances involved can be expressed by the equations

$$C_{p(C)} = -5.293 + 58.609 \times 10^{-3}T - 432.25 \times 10^{-7}T^2 \quad (5\text{-}50)$$

$$C_{p(O_2)} = 25.723 + 12.979 \times 10^{-3}T - 38.62 \times 10^{-7}T^2 \quad (5\text{-}51)$$

$$C_{p(CO_2)} = 21.556 + 63.697 \times 10^{-3}T - 405.05 \times 10^{-7}T^2 + 9.678 \times 10^{-9}T^3 \quad (5\text{-}52)$$

Hence the difference in heat capacities of products and reactants is given by the equation

$$\Delta C_p = 1.126 - 7.891 \times 10^{-3}T + 65.82 \times 10^{-7}T^2 + 9.678 \times 10^{-9}T^3 \quad (5\text{-}53)$$

Consequently

$$\Delta H = \Delta H_0 + \int (1.126 - 7.891 \times 10^{-3}T + 65.82 \times 10^{-7}T^2 + 9.678 \times 10^{-9}T^3)dT \quad (5\text{-}54)$$

$$= \Delta H_0 + 1.126T - 3.946 \times 10^{-3}T^2 + 21.94 \times 10^{-7}T^3 + 2.420 \times 10^{-9}T^4 \quad (5\text{-}55)$$

Since ΔH is known at 298.15 K, it is possible to substitute the value into the preceding equation and to calculate ΔH_0:

$$\Delta H_0 = -393{,}571\ \text{J mol}^{-1} \quad (5\text{-}56)$$

Now we can write a completely explicit equation for the enthalpy of formation of CO_2 as a function of the temperature:

$$\Delta H = -393{,}571 + 1.126T - 3.946 \times 10^{-3}T^2 + 21.94 \times 10^{-7}T^3 + 2.420 \times 10^{-9}T^4 \quad (5\text{-}57)$$

However, this expression is valid only in the temperature range for which Equations 5-50 through 5-52 represent the heat capacities of the reactants and products (see Table 4-1). These equations are empirical equations that are fitted to experimental data in a limited temperature range. In particular, ΔH_0 *is not* the value of ΔH at 0 K.

Arithmetic Method

A second procedure, which is fundamentally no different from the analytic method, involves adding suitable equations to get the desired equation and makes use explicitly of the property that ΔH is a state function. For example, if we consider the freezing of water, the enthalpy of the reaction is known at 0°C (T_1), but may be required at −10°C (T_2). We can obtain the desired ΔH by adding the following equations (assuming that C_p is constant over the temperature range):

$$H_2O(l, 0°C) = H_2O(s, 0°C), \qquad \Delta H = -6008 \text{ J mol}^{-1} \quad (5\text{-}58)$$

$$H_2O(s, 0°C) = H_2O(s, -10°C), \qquad \Delta H = \int_{0°C}^{-10°C} C_p(s)\, dT$$
$$= C_p(s)(T_2 - T_1)$$
$$= -364 \text{ J mol}^{-1} \quad (5\text{-}59)$$

$$H_2O(l, -10°C) = H_2O(l, 0°C), \qquad \Delta H = \int_{-10°C}^{0°C} C_p(l)\, dT$$
$$= C_p(l)(T_1 - T_2)$$
$$= 753 \text{ J mol}^{-1} \quad (5\text{-}60)$$

$$H_2O(l, -10°C) = H_2O(s, -10°C), \qquad \Delta H = -6008 - 364 + 753$$
$$= -5619 \text{ J mol}^{-1} \quad (5\text{-}61)$$

Graphical or Numerical Methods [6]

If analytic equations for the heat capacities of reactants and products are unavailable, we still can carry out the integration required by Equation 5-49 by graphical or numerical methods. In essence we replace Equation 5-49 by the expression

$$\Delta H \cong \Delta H_0 + \Sigma(\Delta \ddot{C}_p)(\Delta T) \quad (5\text{-}62)$$

EXERCISES

1. Find the enthalpy of formation of ethyl alcohol in the International Critical Tables and the National Bureau of Standards tables. Compare the respective values.
2. According to K. Schwabe and W. Wagner [*Chem. Ber.* **91**, 686 (1958)], the enthalpies of combustion in a constant-volume calorimeter for fumaric and maleic acids are -1337.21 kJ mol^{-1} and -1360.43 kJ mol^{-1}, respectively, at approximately 25°C.
   a. Calculate the enthalpies of formation of these isomers.
   b. What is the difference in enthalpy between these isomers?
3. Standard enthalpies of formation of some sulfur compounds [W. N. Hubbard, D. R. Douslin, J. P. McCullough, D. W. Scott, S. S. Todd, J. F. Messerly, I. A. Hossenlopp, A. George, and G. Waddington, *J. Am. Chem. Soc.* **80**, 3547 (1958)] are listed below, together with that for S(g) from tables of the National Bureau of Standards:

Substance	$\Delta Hf^\circ_{298.15\ K}/(\text{kJ mol}^{-1})$
$C_2H_5—S—C_2H_5(g)$	-147.24
$C_2H_5—S—S—C_2H_5(g)$	-201.92
S(g)	278.805

   From these data alone compute $\varepsilon_{S—S}$.
4. For NF_3(g) at 25°C, ΔHf° is -124.3 kJ mol^{-1} [G. T. Armstrong, S. Marantz, and C. F. Coyle, *J. Am. Chem. Soc.* **81**, 3798 (1959)]. Using Table 5-3 for any necessary bond energies, calculate $\varepsilon_{N—F}$.
5. From mass spectrometric experiments [S. N. Foner and R. L. Hudson, *J. Chem. Phys.* **28**, 719 (1958)] it is possible to compute a value of 109 kJ mol^{-1} for $\Delta H^\circ_{298\ K}$ for the reaction
$$N_2H_4(g) = N_2H_2(g) + H_2(g)$$
   Knowing in addition that $\Delta Hf^\circ_{298\ K}$ of hydrazine gas, N_2H_4, is 95.0 kJ mol^{-1}, and assuming that the structure of N_2H_2 is HN=NH, calculate $\varepsilon_{N=N}$.
6. Estimate ΔHf° of the gaseous N—H radical by using appropriate values from the table of bond energies.
7. Calculate the bond energy of the I—Cl bond at 25°C from the data listed in Table 5-2. Compare it with the bond energy given in Table 5-3.
8. Taking the enthalpy of combustion of ethane as -1559.8 kJ mol^{-1}, calculate the C—C bond energy.
9. Find data for the standard enthalpies of formation of Cl(g), S(g), S_8(g), and S_2Cl_2(g) from appropriate sources.
   a. Calculate the energy of the S—S bond. Assume that S_8 consists of eight such linkages.
   b. Calculate the energy of the S—Cl bond.
   c. Estimate the enthalpy of formation of SCl_2(g).

10. Derive an equation for the dependence of ΔH on temperature for the reaction

$$CO(g) + \tfrac{1}{2}O_2(g) = CO_2(g)$$

Appropriate data can be found in the tables of Chapters 4 and 5.

11. Enthalpies of formation of solid alloy phases can be calculated from enthalpies of solution of these phases in a suitable liquid metal. The pure metals and alloys listed below, all originally at 31°C, have each been dropped into liquid tin at 250°C and the heat of this process measured. The results [P. D. Anderson, *J. Am. Chem. Soc.* **80**, 3171 (1958)] computed for 1 mole of material are:

Phase	$\Delta H/(\mathrm{J\ mol^{-1}})$
Ag	20418
Cd	18995
$\zeta(Ag_{0.5}Cd_{0.5})$	27447
$\gamma(Ag_{0.412}Cd_{0.588})$	27949

a. Calculate $\Delta Hf°$ of the ζ phase at 31°C.

b. Calculate $\Delta Hf°$ of the γ phase at 31°C.

12. Calculate the enthalpy change at 25°C for the following reaction using standard heats of formation:

$$Mg_2SiO_4(\text{forsterite}) + SiO_2(\text{quartz}) = 2\ MgSiO_3(\text{enstatite})$$

13. The "proton affinity," $\mathscr{P}$, of a substance such as NH_3 is defined as the change in energy for the reaction

$$NH_4^+(g) = NH_3(g) + H^+(g)$$

$\mathscr{P}_{NH_3}$ at 0 K can be computed from other thermal data through consideration of the Born–Haber cycle (all substances except NH_4Cl being gases):

$$
\begin{array}{c}
\overset{U}{NH_4Cl(s) \rightarrow NH_4^+ + Cl^-} \\
\uparrow \qquad\qquad \downarrow \mathscr{P}HN_3 \\
\qquad\quad NH_3 + H^+ + Cl^- \\
\qquad\qquad\quad \downarrow {-I_H + E_{Cl}} \\
\tfrac{1}{2}N_2 + 2H_2 + \tfrac{1}{2}Cl_2 \leftarrow NH_3 + H + Cl
\end{array}
$$

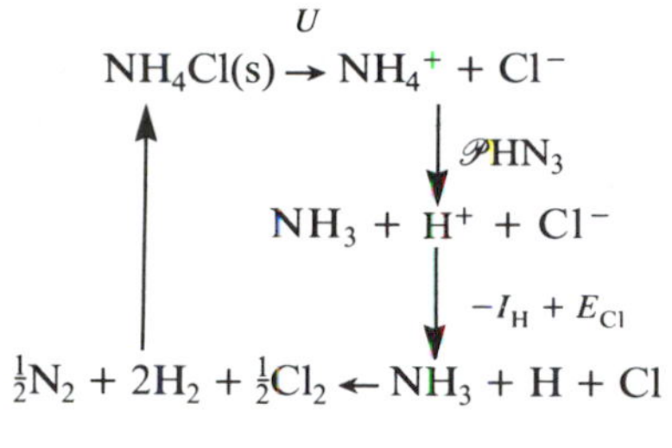

in which U represents the lattice energy per mole of crystalline NH_4Cl, I_H the ionization energy of a mole of hydrogen atoms, and E_{Cl} the electron affinity of chlorine. The values of these quantities are (in $\mathrm{kJ\ mol^{-1}}$) 640, 1305, and 387.0, respectively. Using $-314.2\ \mathrm{kJ\ mol^{-1}}$ as the enthalpy of formation of $NH_4Cl(s)$ and $-45.6\ \mathrm{kJ\ mol^{-1}}$ as the enthalpy of formation of $NH_3(g)$, and finding any other quantities you need from tables in Chapter 5, calculate $\mathscr{P}_{NH_3}$.

14. The adiabatic flame temperature is the temperature that would be attained if a compound were burned completely under adiabatic conditions so that all the heat evolved would go into heating the product gases. Calculate the adiabatic flame temperature for the burning of ethane in an air mixture containing originally twice as much air as is necessary for complete combustion to $CO_2(g)$ and $H_2O(g)$. Assume that air is composed of 20% O_2 and 80% N_2 by volume. In using heat capacity equations neglect all terms containing T^2 or higher powers of T. Assume also that the combustion occurs at constant pressure.
15. Mass spectrometry is one of the experimental methods for determining bond dissociation energies. The mass spectrometer can provide a measure of the appearance potential for a given reaction, that is, the threshold energy necessary to produce a particular set of particles. The appearance potential for the following reaction of H_2,

$$H_2 = H^+ + H + e^-$$

is 18.0 eV. The ionization energy of hydrogen is 13.6 eV. Calculate the bond energy of H_2. (1 eV = 96.44 kJ mol^{-1})
16. W. D. Good [*J. Chem. Thermodynamics,* **4**, 709(1972)] measured at 298.15 K the standard enthalpy of combustion of *n*-octane(1) as -5470.29 kJ mol^{-1} and the standard enthalpy of combustion of 2,2,3,3-tetramethyl butane(s) as -5451.46 kJ mol^{-1}. Calculate the corresponding standard enthalpies of formation.
17. G. K. Johnson and W. V. Steele [*J. Chem. Thermodynamics* **13**, 717 (1981)] determined the standard enthalpy of combustion in gaseous fluorine of $UO_2(s)$. The products of combustion are $UF_6(s)$ and $O_2(g)$. The measured value of the enthalpy of combustion is -1112.6 kJ mol^{-1}, and the value of the standard enthalpy of formation of $UF_6(s)$ is -2197.7 kJ mol^{-1}. Calculate the standard enthalpy of formation of $UO_2(s)$.
18. R. N. McDonald, A. K. Chowdhury, and W. D. McGhee [*J. Am. Chem. Soc.* **106**, 4112(1984)] used a flowing afterglow apparatus to measure the proton and electron affinities of hypovalent ion radicals in the gas phase.
    a. They obtained a value of 364 kcal mol^{-1} for the proton affinity of $(CH_3)_2CH^-$. Calculate $\Delta Hf°$ for this anion, using the value of 367.2 kcal mol^{-1} for $\Delta Hf°$ of $H^+(g)$ and -326.9 kcal mol^{-1} for $\Delta Hf°$ of $(CF_3)_2CH_2(g)$.
    b. With the results of part a and their determination of the $\Delta H°$ of dissociation of $(CF_3)_2CH^-$ to $(CF_3)_2C^{\overline{\cdot}} + H\cdot$, equal to 100 kcal mol^{-1}, calculate $\Delta Hf°$ for $(CF_3)_2C^{\overline{\cdot}}$.

REFERENCES

1. These tables are revised and expanded continually.
2. G. L. Humphrey and E. G. King, *J. Am. Chem. Soc.* **74**, 2041 (1952).
3. D. R. Torgerson and Th. G. Sahama, *J. Am. Chem. Soc.* **70**, 2156 (1948).

4. G. B. Kistiakowsky and E. R. Van Artsdalen, *J. Chem. Phys.* **12**, 469 (1944).

5. T. L. Cottrell, *The Strengths of Chemical Bonds*, 2d ed., Butterworths, London, 1958; C. T. Mortimer, *Reaction Heats and Bond Strengths*, Addison-Wesley, Reading, Mass., 1963; G. J. Janz, *Thermodynamic Properties of Organic Compounds*, Chap. 7, rev. ed., Academic Press, New York, 1967; S. W. Benson, *J. Chem. Ed.* **42**, 502 (1965); J. D. Cox and G. Pilcher, *Thermochemistry of Organic and Organometallic Compounds*, Chap. 7, Academic Press, London, 1970.

6. See Section 2-3.

CHAPTER 6

Application of the First Law to Gases

As a prelude to the development of the second law of thermodynamics, it is of interest to consider the information obtainable on the behavior of gases by the application of the first law of thermodynamics and the associated definitions that have been developed so far. In addition, the relationships developed for gases that are based on the first law will be useful in developing the second law of thermodynamics and in applying the second law to specific systems. Since the behavior of many gases at low pressure can be approximated by the simple equation of state for the ideal gas, and since the ideal equation of state describes accurately the behavior of real gases at the limit of zero pressure, we will begin our discussion with a consideration of ideal gases.

6-1 IDEAL GASES

Definition

An ideal gas is one (1) that obeys the equation of state

$$PV = nRT \tag{6-1}$$

in which n is the number of moles and R is a universal constant; and (2) for which the internal energy, E, is a function of the temperature only, that is,

$$\left(\frac{\partial E}{\partial V}\right)_T = \left(\frac{\partial E}{\partial P}\right)_T = 0 \tag{6-2}$$

Equation 6-1 can be derived from Boyle's law and Charles's law by using the total differential. Boyle's law can be expressed by the relationship

$$V = \frac{nk_T}{P} \text{ (}T\text{ constant, }n\text{ constant)} \tag{6-3}$$

or

$$\left(\frac{\partial V}{\partial P}\right)_T = -\frac{nk_T}{P^2} \tag{6-4}$$

in which k_T is a constant at a fixed temperature. Similarly, Charles's law can be expressed by either of the two expressions:

$$V = nk_P T \text{ (}P\text{ constant, }n\text{ constant)} \tag{6-5}$$

or

$$\left(\frac{\partial V}{\partial T}\right)_P = nk_P \tag{6-6}$$

If we consider the total differential of the volume, $V = f(T, P)$, we obtain

$$dV = \left(\frac{\partial V}{\partial T}\right)_P dT + \left(\frac{\partial V}{\partial P}\right)_T dP \tag{6-7}$$

Equations 6-4 and 6-6 can be used to substitute for the partial derivatives in Equation 6-7 to obtain

$$dV = nk_P\,dT - \frac{nk_T}{P^2}\,dP \tag{6-8}$$

The constants k_P and k_T can be replaced by introducing Equations 6-3 and 6-5:

$$dV = \frac{V}{T}\,dT - \frac{V}{P}\,dP \tag{6-9}$$

Rearranging Equation 6-9 we obtain

$$\frac{dV}{V} = \frac{dT}{T} - \frac{dP}{P} \tag{6-10}$$

which can be integrated to give

$$PV = k'T \tag{6-11}$$

in which k' is a constant of integration. If we identify k' with nR we have Equation 6-1, the ideal gas law.

It follows from Equation 6-2 that if an ideal gas undergoes any isothermal transformation, its energy remains fixed.

Enthalpy a Function of Temperature Only

We can show that the enthalpy, as well as the internal energy, is constant in any isothermal change of an ideal gas as follows:

$$H = E + PV \tag{4-1}$$

and

$$\left(\frac{\partial H}{\partial V}\right)_T = \left(\frac{\partial E}{\partial V}\right)_T + \left[\frac{\partial(PV)}{\partial V}\right]_T \tag{6-12}$$

But from Equations 6-1 and 6-2 it is evident that each term on the right side of Equation 6-12 is zero. Consequently

$$\left(\frac{\partial H}{\partial V}\right)_T = 0 \tag{6-13}$$

By an analogous procedure it also can be shown that

$$\left(\frac{\partial H}{\partial P}\right)_T = 0 \tag{6-14}$$

Relationship between C_P and C_V

In Chapter 4 the following expression (Equation 4-25) was shown to be a general relationship between the heat capacity at constant pressure and that at constant volume:

$$C_P = C_V + \left[P + \left(\frac{\partial E}{\partial V}\right)_T\right]\left(\frac{\partial V}{\partial T}\right)_P$$

For 1 mole of an ideal gas

$$\left(\frac{\partial E}{\partial V}\right)_T = 0 \tag{6-2}$$

and

$$\left(\frac{\partial V}{\partial T}\right)_P = \frac{R}{P} \tag{6-15}$$

The substitution of Equations 6-2 and 6-15 into Equation 4-25 leads to the familiar expression

$$C_p = C_v + R \tag{6-16}$$

Calculation of Thermodynamic Changes in Expansion Processes

Isothermal. As pointed out in Equation 3-3, work done is measured by the product of the external force and its displacement. Thus for a finite gaseous expansion (see Equation 3-5)

$$W = -\int_{V_1}^{V_2} P'dV \tag{6-17}$$

in which P' is the external pressure.

Any finite expansion that occurs in a finite time is irreversible. A reversible expansion can be approximated as closely as desired, and the values of the thermodynamic changes can be calculated for the limiting case of a reversible process. In the limiting case the process must be carried out infinitely slowly so that the pressure, P, is always a well-defined quantity. A reversible process is a succession of states, each of which is an equilibrium state or a quasi-static process.

As the actual work done approaches the reversible work, the pressure, P, of the gas and the external pressure, P', differ infinitesimally and the direction of change can be reversed by an infinitesimal change in the external pressure. Under these conditions P' is essentially equal to P, thus Equation 6-17 can be rewritten as

$$W = -\int_{V_1}^{V_2} PdV \text{ (reversible)} \tag{6-18}$$

Because the reversible process is a succession of equilibrium states, P is given by the equation of state, which is Equation 6-1 for an ideal gas. Substituting from Equation 6-1 into Equation 6-18 we obtain

$$W = -\int_{V_1}^{V_2} \frac{nRT}{V} dV = -nRT \ln \frac{V_2}{V_1} \tag{6-19}$$

for the case in which the gas is in thermal equilibrium with the surroundings, which are maintained at a constant temperature, T. The reversible expansion can be visualized as in Figure 6-1 and Figure 6-2. The curve in Figure 6-2 represents the succession of equilibrium states in the expansion, and the area under the curve between the dashed lines is the negative of the work done in an expansion from V_1 to V_2.

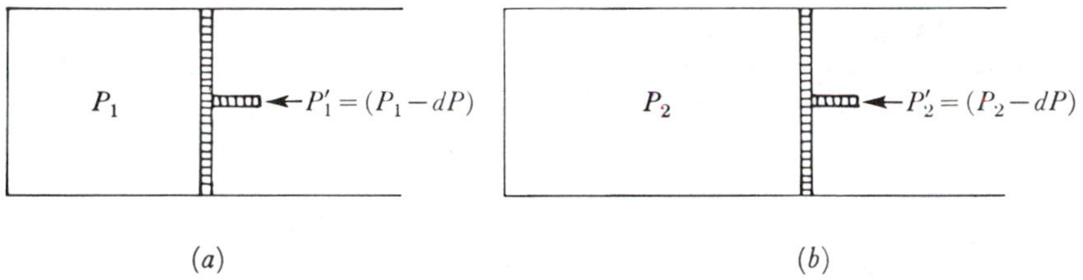

Figure 6-1. Schematic representation of an isothermal reversible expansion. The external pressure is maintained only infinitesimally below the internal pressure.

Since the process is isothermal, and since E depends only on the temperature,

$$\Delta E = 0 \tag{6-20}$$

With this information and the use of the first law of thermodynamics, we can calculate the heat absorbed from the surroundings in the process:

$$Q = \Delta E - W = nRT \ln \frac{V_2}{V_1} \tag{6-21}$$

Finally, ΔH also is equal to zero, since

$$\Delta H = \Delta E + \Delta(PV) = 0 + \Delta(nRT) = 0 \tag{6-22}$$

If the gas is allowed to expand against zero external pressure (a free expansion; Figure 6-3), then, from Equation 6-17, W equals zero. Although the temperature of the gas may change during the free expansion (indeed, the

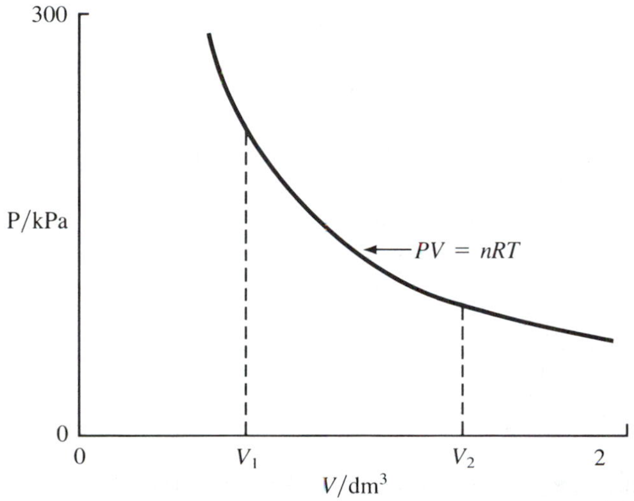

Figure 6-2. The path taken in a reversible isothermal expansion of an ideal gas, and the area representing the work done.

temperature is not a well-defined quantity during an irreversible change), the temperature of the gas will return to that of the surroundings with which it is in thermal contact when the system has reached a new equilibrium. Thus the process can be described as isothermal and, for the gas,

$$W = 0 \tag{6-23}$$

$$\Delta E = 0 \tag{6-20}$$

$$Q = \Delta E - W = 0 \tag{6-24}$$

and

$$\Delta H = \Delta E + \Delta(PV) = 0 \tag{6-25}$$

In an intermediate irreversible expansion in which the external pressure is not zero but is less than the pressure of the gas by a finite amount, some work would be obtained. However, since P' is always less than P, the work done always will be less in magnitude than that for the reversible expansion. That is,

$$0 < |W| < \left| -nRT \ln \frac{V_2}{V_1} \right| \tag{6-26}$$

since W is negative in an expansion by our convention.

For an ideal gas when isothermal conditions are maintained:

$$\Delta E = 0 \tag{6-20}$$

and Q is given by the first law as

$$Q = \Delta E - W = -W \tag{6-27}$$

Then,

$$0 < Q < nRT \ln \frac{V_2}{V_1} \tag{6-28}$$

The enthalpy change is calculated from the relationship

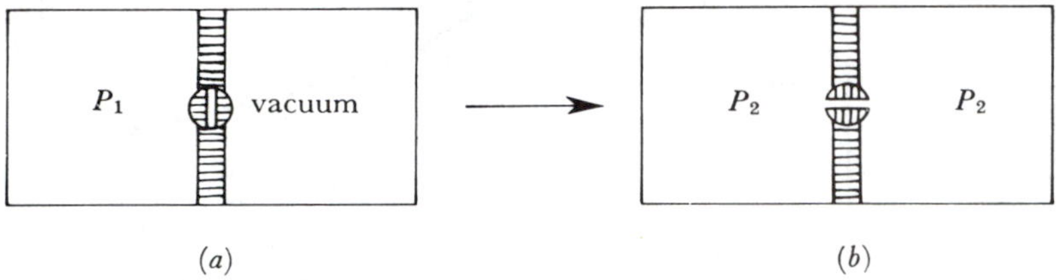

Figure 6-3. Schematic representation of a free expansion. A small valve separating the two chambers in (a) is opened so that the gas can rush in from left to right. The initial volume of the gas is V_1, and the final volume V_2.

Table 6-1 Thermodynamic Quantities for Isothermal Changes in an Ideal Gas

Expansion	Compression	Cycle
	Reversible	
$W = -nRT \ln \frac{V_2}{V_1}$	$W = -nRT \ln \frac{V_1}{V_2}$	$W = 0$
$\Delta E = 0$	$\Delta E = 0$	$\Delta E = 0$
$Q = nRT \ln \frac{V_2}{V_1}$	$Q = nRT \ln \frac{V_1}{V_2}$	$Q = 0$
$\Delta H = 0$	$\Delta H = 0$	$\Delta H = 0$
	Free	
$W = 0$	$W = -P'(V_1 - V_2) > -nRT \ln \frac{V_1}{V_2}$	$W = -P'(V_1 - V_2) > 0$
$\Delta E = 0$	$\Delta E = 0$	$\Delta E = 0$
$Q = 0$	$Q = P'(V_1 - V_2) < nRT \ln \frac{V_1}{V_2}$	$Q = P'(V_1 - V_2) < 0$
$\Delta H = 0$	$\Delta H = 0$	$\Delta H = 0$
	Intermediate	
$0 < \lvert W \rvert < \left\lvert -nRT \ln \frac{V_2}{V_1} \right\rvert$	$W = -P'(V_1 - V_2) > -nRT \ln \frac{V_1}{V_2}$	$W > 0$
$\Delta E = 0$	$\Delta E = 0$	$\Delta E = 0$
$0 < Q < nRT \ln \frac{V_2}{V_1}$	$Q = P'(V_1 - V_2) < nRT \ln \frac{V_1}{V_2}$	$Q < 0$
$\Delta H = 0$	$\Delta H = 0$	$\Delta H = 0$

$$\Delta H = \Delta E + \Delta(PV) = 0 \tag{6-29}$$

The thermodynamic changes for reversible, free, and intermediate expansions are compared in the first column of Table 6-1. This table emphasizes the difference between an exact differential and an inexact differential. Thus E and H, whose differentials are exact, undergo the same change in each of the three different paths used for the transformation. They are thermodynamic properties. However, the work and heat quantities depend on the particular path chosen, even though the initial and final values of the temperature, pressure, and volume, respectively, are the same in all these cases. Thus, heat and work are not thermodynamic properties; rather, they are energies in transfer between system and surroundings.

It is permissible to speak of the energy (or enthalpy) of a system, even though we can measure only differences in these functions, because their differences are characteristic of the initial and final states and are independent of the path. In contrast, heat and work are not properties of the system alone, but also of the path followed when the system goes from one state to another. Since the work or heat obtained in going from State A to State B may be different from that required to return it from State B to State A, it is misleading to speak of the work or heat contained in the system.

If we consider the reverse of the changes described in the first column of Table 6-1, we can examine the net results of a complete cycle.

For a reversible isothermal compression from V_2 to V_1 the work done is

$$W = -\int_{V_2}^{V_1} P dV \tag{6-30}$$

$$= -\int_{V_2}^{V_1} \frac{nRT}{V} dV$$

$$= -nRT \ln \frac{V_1}{V_2} \tag{6-31}$$

As we have seen, $\Delta E = 0$, and

$$Q = -W = nRT \ln \frac{V_1}{V_2} \tag{6-32}$$

Similarly,

$$\Delta H = \Delta E + \Delta(PV) = 0 \tag{6-33}$$

for both expansion and compression.

There is no compression equivalent to a free expansion. We shall consider that the free expansion is reversed by an irreversible compression at a constant external pressure, P', that is greater than the final pressure of the gas, as shown in Figure 6-4. Thus

$$W = -\int_{V_2}^{V_1} P' dV$$

$$= -P'(V_1 - V_2) > -nRT \ln \frac{V_1}{V_2} \tag{6-34}$$

and the work done in the irreversible compression is a greater positive quantity than the work done in the reversible compression. Again $\Delta E = 0$, so

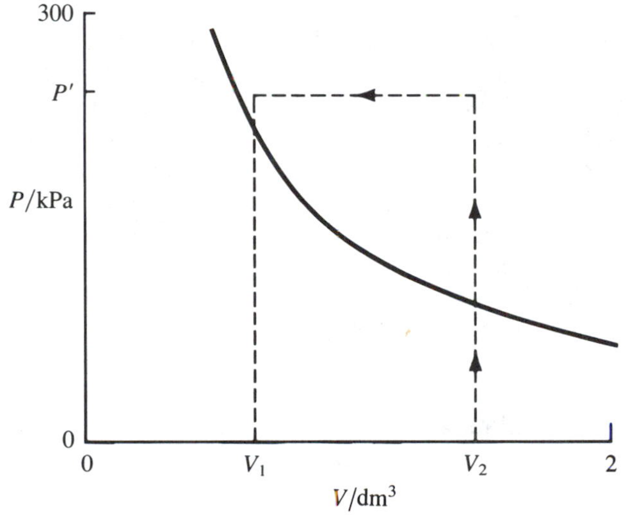

Figure 6-4. The irreversible compression at pressure P' used to return a gas to its initial state after a free expansion or an intermediate expansion. The area bounded by dashed lines represents the work done.

$$Q = -W = P'(V_1 - V_2) < nRT \ln \frac{V_1}{V_2} \tag{6-35}$$

and the heat lost to the surroundings, a negative quantity, is greater in magnitude than in the reversible compression. As in the reversible compression,

$$\Delta H = \Delta E + \Delta(PV) = 0 \tag{6-36}$$

Since there is no meaningful distinction in a compression process between the free and the intermediate expansion, the reverse of the intermediate irreversible expansion can be chosen to be the same as the reverse of the free expansion. The work done in the intermediate expansion is less in magnitude than the reversible work of expansion. Even if the compression were carried out reversibly, the work done on the gas would be numerically equal (but opposite in sign) to that done by the gas in the reversible expansion. Hence, the work done in the compression is greater in magnitude than the work obtained in the expansion, so the net work in the cycle is positive; that is, work is done by the surroundings on the system. Similarly, the heat liberated on compression is greater in magnitude than the heat absorbed during expansion; thus, the net heat exchanged is negative, and heat is transferred from the system to the surroundings.

Table 6-1 summarizes the thermodynamic changes in the ideal gas for expansions, compressions, and the complete cycles, all at constant temperature.

For both the reversible and the irreversible changes in the ideal gas ΔE and ΔH are zero for the complete cycle, since the system has been restored to its initial state and E and H are state functions. In contrast, Q and W for the cycle depend on the way in which the change is carried out. Q and W are zero for the reversible cycle, thereby indicating that for a reversible cycle the *surroundings*, as well as the system, return to their initial state. For the irreversible cycles, Q and W are not zero. A net amount of work has been done by the surroundings on the system, and a net amount of heat has been transferred to the surroundings from the system. It is this change in the surroundings that is characteristic of irreversible cycles and of all real processes, and that will be important in our consideration of the second law of thermodynamics.

Adiabatic. By definition, an adiabatic expansion is one that is not accompanied by a transfer of heat. Therefore

$$DQ = 0 \tag{6-37}$$

It follows from the definition of E that

$$dE = DW = dW(\text{adiabatic}) \tag{6-38}$$

This equality then can be used to specify more explicitly the work done, because if ΔE is known, W is obtained immediately. For an ideal gas, ΔE in the adiabatic expansion can be obtained by the following procedure.

Considering the energy as a function of temperature and volume, $E = f(T, V)$, we can write an equation for the total differential (Equation 4-23):

$$dE = \left(\frac{\partial E}{\partial T}\right)_V dT + \left(\frac{\partial E}{\partial V}\right)_T dV$$

Since we are dealing with an ideal gas,

$$\left(\frac{\partial E}{\partial V}\right)_T = 0 \tag{6-2}$$

hence Equation 4-23 can be reduced to

$$dE = \left(\frac{\partial E}{\partial T}\right)_V dT = C_v\, dT = dW \tag{6-39}$$

Then the work done, W, and the energy change, ΔE, can be obtained by integrating Equation 6-39:

$$W = \int_{T_1}^{T_2} C_v\, dT = \Delta E \tag{6-40}$$

Since dE is a perfect differential and $(\partial E/\partial V)_T$ is equal to zero, the cross derivative property of a perfect differential leads to the conclusion that $(\partial C_v/\partial V)_T$ is equal to zero, so C_v is a function of T only. If C_v is independent of temperature for an ideal gas, then

$$W = \Delta E = C_v(T_2 - T_1) \tag{6-41}$$

Using Equation 4-2 we also can calculate ΔH for the adiabatic expansion:

$$\Delta H = \Delta E + \Delta(PV)$$

For an ideal gas

$$\Delta(PV) = \Delta(nRT) \tag{6-42}$$

Substituting for (PV) from Equation 6-42 into Equation 4-2 we obtain

$$\begin{aligned} \Delta H &= \Delta E + \Delta(nRT) \\ &= C_v(T_2 - T_1) + (nRT_2 - nRT_1) \end{aligned}$$

or

$$\begin{aligned} \Delta H &= (C_v + nR)(T_2 - T_1) \\ &= (nc_v + nR)(T_2 - T_1) \\ &= n(c_v + R)(T_2 - T_1) = nc_P(T_2 - T_1) = C_P(T_2 - T_1) \end{aligned} \tag{6-43}$$

So far we have not specified whether the adiabatic expansion under consideration is reversible. The Equations 6-37, 6-41, and 6-43, for the calculation of the thermodynamic changes in this process, apply for the reversible expansion, free expansion, or intermediate expansion, so long as we are dealing with an ideal gas. However, the numerical values of W, ΔE, and ΔH will not be the same for each of the three types of adiabatic expansion because T_2, the final temperature of the gas, will differ in each case, even though the initial temperature may be identical in all cases.

If we consider the free expansion first, it is apparent from Equation 6-41 that, since no work is done, there is no change in temperature; that is, $T_2 = T_1$. Thus ΔE and ΔH also must be zero for this process. A comparison with the results for a free expansion in Table 6-1 shows that an adiabatic free expansion and an isothermal free expansion are two different names for the same process.

For the reversible adiabatic expansion, a definite expression can be derived to relate the initial and final temperatures to the respective volumes or pressures. Again we start with Equation 6-38. Recognizing the restriction of reversibility we obtain

$$dE = dW = -PdV \tag{6-44}$$

Since we are dealing with an ideal gas, Equation 6-39 is valid. Substitution from Equation 6-39 into 6-44 leads to

$$C_v dT = -P dV \tag{6-45}$$

For an ideal gas, from Equation 6-1, $P = nRT/V$ and C_v equals nc_v. Thus Equation 6-45 becomes

$$nc_v\, dT = -\frac{nRT}{V}\, dV$$

Separating the variables, we obtain

$$\frac{c_v}{T}\, dT = -\frac{R}{V}\, dV \tag{6-46}$$

which can be integrated within definite limits when c_v is constant to give

$$c_v \ln \frac{T_2}{T_1} = -R \ln \frac{V_2}{V_1} \tag{6-47}$$

This equation can be converted to

$$\left(\frac{T_2}{T_1}\right)^{c_v} = \left(\frac{V_2}{V_1}\right)^{-R} = \left(\frac{V_1}{V_2}\right)^{R}$$

or

$$\frac{T_2}{T_1} = \left(\frac{V_1}{V_2}\right)^{R/c_v} \tag{6-48}$$

Hence

$$T_2 V_2^{R/c_v} = T_1 V_1^{R/c_v} \tag{6-49}$$

Equation 6-49 states that the particular temperature–volume function shown is constant during a reversible adiabatic expansion. Hence we can write

$$TV^{R/c_v} = \text{constant} \tag{6-50}$$

or

$$VT^{c_v/R} = \text{constant}' \tag{6-51}$$

Any one of the equations 6-48 through 6-51 can be used to calculate a final temperature from the initial temperature and the observed volumes.

If we measure pressures instead of temperatures, it is possible to use the following equation instead:

$$PV^{c_p/c_v} = \text{constant}'' \tag{6-52}$$

Equation 6-52 can be derived from Equation 6-51 by substituting from the equation of state for the ideal gas [1].

We can see from Equation 6-48 that the final temperature, T_2, in the reversible adiabatic expansion must be less than T_1, since V_1 is less than V_2 and both R and c_v are positive numbers. Thus the adiabatic reversible expansion is accompanied by a temperature drop, which can be calculated from the measured volumes or pressures at the beginning and the end of the process. Knowing T_2 and T_1, we then can evaluate W, ΔE, and ΔH by substitution into Equations 6-41 and 6-43.

For an irreversible adiabatic expansion in which some work is done, the work done is less in magnitude than that in the reversible process because the external pressure is less than the pressure of the gas by a finite amount. Thus, if the final volume V_2 is the same as that in the reversible process, T_2 will not be as low in the actual expansion since, according to Equation 6-41, the temperature drop, ΔT, depends directly on the work done by the expanding gas. Similarly, from Equations 6-41 and 6-43, ΔE and ΔH, respectively, also must be numerically smaller in the intermediate expansion than in the reversible one.

In the adiabatic expansion, from a common set of initial conditions to the same final volume, the values of ΔE and ΔH, as well as the values of the work done, appear to depend on the path (see summary in Table 6-2). At first glance, such behavior may appear to contradict the assumption of exactness of dE and dH. However, careful consideration shows that the basis of the difference lies in the different endpoint for each of the three paths. Despite the fact that the final volume can be made the same in each case, the final temperature depends on whether the expansion is free, reversible, or intermediate (Table 6-2).

Table 6-2 Thermodynamic Changes in Adiabatic Expansions of an Ideal Gas

Reversible	Free	Intermediate
V_1	V_1	V_1
V_2	V_2	V_2
$W = C_v(T_2 - T_1) < 0$	$W' = C_v(T_2' - T_1) = 0$	$W'' = C_v(T_2'' - T_1) < 0$
$\therefore T_2 < T_1$	$\therefore T_2' = T_1$	$\therefore T_2'' < T_1$
$\Delta E = C_v(T_2 - T_1) < 0$	$\Delta E' = 0$	$\Delta E'' = C_v(T_2'' - T_1) < 0$
$\Delta H = C_p(T_2 - T_1) < 0$	$\Delta H' = 0$	$\Delta H'' = C_p(T_2'' - T_1) < 0$

$$T_2 < T_2'' < T_2' = T_1$$
$$W < W'' < W' = 0$$
$$\Delta E < \Delta E'' < \Delta E' = 0$$
$$|\Delta E| > |\Delta E''| > |\Delta E'| = 0$$
$$\Delta H < \Delta H'' < \Delta H' = 0$$

6-2 REAL GASES

Equations of State

Now we discuss substances for which the ideal gas laws do not give an adequate description. Numerous representations have been used to describe real gases. For our purposes it will be sufficient to consider only four of the more common ones.

Van der Waals Equation. The van der Waals equation was one of the first introduced to describe deviations from ideality. The argument behind the equation is discussed adequately in elementary textbooks. Generally, it is stated in the form

$$\left(P + \frac{a}{v^2}\right)(v - b) = RT \qquad (6\text{-}53)$$

in which v is the volume per mole and a and b are constants (Table 6-3). Alternative methods of expression will be considered later.

Table 6-3 Van der Waals Constants for Some Gases[a]

Substance	$a/(\text{kPa dm}^6\text{ mol}^{-2})$	$b/(\text{dm}^3\text{ mol}^{-1})$
Acetylene	445	0.0514
Ammonia	423	0.0371
Argon	137	0.0322
Carbon dioxide	364	0.0427
Carbon monoxide	151	0.0399
Chlorine	658	0.0562
Ethyl ether	1760	0.1344
Helium	3.5	0.0237
Hydrogen	24.7	0.0266
Hydrogen chloride	372	0.0408
Methane	228	0.0428
Nitric oxide	136	0.0279
Nitrogen	141	0.0391
Nitrogen dioxide	535	0.0442
Oxygen	138	0.0318
Sulfur dioxide	680	0.0564
Water	553	0.0305

[a]Values for other substances may be calculated from data in Landolt-Börnstein, *Physikalisch-chemische Tabellen*, 5th ed., Vol. 1, Springer, Berlin, 1923, pp. 253–263.

Virial Function. A useful form of expression of deviations from the ideal gas law is the virial equation,

$$Pv = A + BP + CP^2 + \cdots \qquad (6\text{-}54)$$

in which A, B, and C, are constants at a given temperature and are known as *virial coefficients* (Table 6-4). The term A is equal to RT since at very low pressures all gases approach ideal gas behavior. A critical discussion of the virial equation can be found in the references [2].

Berthelot Equation. This equation is too unwieldy to be used generally as an equation of state. However, it is convenient in calculations of deviations from ideality near pressures of 1 atm; hence, it has been used extensively in the determination of entropies from the third law of thermodynamics. This aspect of the equation will receive further attention in subsequent discussions.

The Berthelot equation can be expressed as

$$Pv = RT\left[1 + \frac{9}{128}\frac{P}{P_c}\frac{T_c}{T}\left(1 - 6\frac{T_c^2}{T^2}\right)\right] \qquad (6\text{-}55)$$

in which P_c and T_c are the critical pressure and critical temperature, respectively.

Compressibility Factor. The behavior of real gases can be represented with fair precision by a single chart of the compressibility factor, Z, which is defined as

Table 6-4 Virial Coefficients for Some Gases[a]

Substance	t/°C	A/(kPa dm^3 mol^{-1})	B/(10^{-2} dm^3 mol^{-1})	C/(10^{-5} dm^3 (Pa)$^{-1}$ mol^{-1})
Hydrogen[b]	0	2271.1	1.374	0.8806
	100	3102.5	1.567	0.1418
Nitrogen[c]	0	2271.1	−1.027	6.3009
	100	3102.5	0.656	3.3393
Carbon monoxide[d]	0	2271.1	−1.419	6.9570
	100	3102.5	0.449	4.2515

[a]Taken from J. H. Dymond and E. B. Smith, *The Virial Coefficients of Gases, A Critical Compilation*, The Clarendon Press, Oxford, 1969.

[b]A. Michels, W. de Graaf, T. Wassenaar, J. M. H. Levett, and P. Louwerse, *Physica's Grav.* **25**, 25 (1959); Dymond and Smith, *Virial Coefficients of Gases*, p. 164.

[c]A. Michels, H. Wouters, and T. de Boer, *Physica's Grav.* **1**, 587 (1934) and **3**, 585 (1936); Dymond and Smith, *Virial Coefficients of Gases*, p. 190.

[d]A. Michels, J. M. Lupton, T. Wassenaar, and W. de Graaf, *Physica's Grav.* **18**, 121 (1952); Dymond and Smith, *Virial Coefficients of Gases*, p. 34.

$$Z = \frac{Pv}{RT} \tag{6-56}$$

If Z is plotted as a function of the reduced pressure, $P_r = P/P_c$, then all gases approximately fit a single curve for a given value of the reduced temperature, $T_r = T/T_c$. At another reduced temperature, a new curve is obtained for Z versus P_r, but it too fits all gases. Therefore it becomes possible to condense into a single chart of compressibility factors a quantitative graphical representation of the behavior of real gases for a wide range of pressure and temperature [3, 4].

Joule–Thomson Effect

One method of measuring deviations from ideal behavior quantitatively is by determining the change in temperature in the Joule–Thomson porous-plug experiment (Figure 6-5). The enclosed gas, initially of volume V_1, flows very slowly from the left chamber through a porous plug into the right chamber. The pressure on the left side is maintained constant at P_1, while that on the right side also is constant, but at a lower value, P_2. The apparatus is jacketed with an insulator so that no heat is exchanged with the surroundings. Generally, it is observed that the final temperature, T_2, differs from the initial temperature, T_1.

Isenthalpic Nature. Since the Joule–Thomson experiment is carried out adiabatically, we can write

$$Q = 0$$

However, it does not follow that ΔH also is zero, because the process involves a change in pressure. Nevertheless, it can be shown that the process is an isenthalpic one; that is, ΔH is zero.

The work done by the gas is that accomplished in the right chamber,

$$W_2 = \int_0^{V_2} - P_2 dV = -P_2 V_2 \tag{6-57}$$

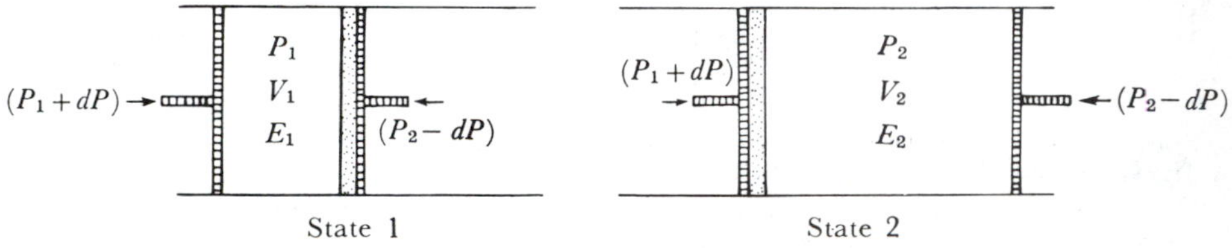

Figure 6-5. Schematic representation of a Joule–Thomson porous-plug experiment. The entire apparatus is kept in a jacket and is well insulated from the surrounding environment.

plus that done in the left chamber,

$$W_1 = \int_{V_1}^{0} -P_1 dV = P_1 V_1 \tag{6-58}$$

Hence the net work is

$$W = W_1 + W_2 = -P_2V_2 + P_1V_1 \tag{6-59}$$

Similarly, the net gain in energy, ΔE, is

$$\Delta E = E_2 - E_1 \tag{6-60}$$

Since $Q = 0$, it follows from the first law of thermodynamics that

$$E_2 - E_1 = W = P_1V_1 - P_2V_2 \tag{6-61}$$

Therefore

$$E_2 + P_2V_2 = E_1 + P_1V_1$$

or

$$H_2 = H_1$$

and

$$\Delta H = 0 \tag{6-62}$$

Thus we have proved that the Joule–Thomson experiment is isenthalpic as well as adiabatic.

Joule–Thomson Coefficient. Knowing that a process is isenthalpic, we can formulate the Joule–Thomson effect quantitatively.

Since it is the change in temperature that is observed as the gas flows from a higher to a lower pressure, the data are summarized in terms of a quantity, $\mu_{J.T.}$, which is defined by

$$\mu_{J.T.} = \left(\frac{\partial T}{\partial P}\right)_H \tag{6-63}$$

The Joule–Thomson coefficient, $\mu_{J.T.}$, is positive when a cooling of the gas (a temperature drop) is observed; since dP is always negative, $\mu_{J.T.}$ will be positive when dT is negative. Conversely, $\mu_{J.T.}$ is a negative quantity when the gas warms on expansion because dT then is a positive quantity.

Values of the Joule–Thomson coefficient for argon and nitrogen at several pressures and temperatures are listed in Table 6-5.

It frequently is necessary to express the Joule–Thomson coefficient in terms of other partial derivatives. Considering the enthalpy as a function of temperature and pressure, $H(T, P)$, we can write the total differential

Table 6-5 Joule–Thomson Coefficients[a] for Argon and Nitrogen

t/°C		$\mu_{J.T.}$/(K(MPa)$^{-1}$) $P = 101$ kPa	$P = 10.1$ MPa	$P = 20.2$ MPa
300	Ar	0.635	0.439	0.272
	N_2	0.138	−0.074	−0.169
0	Ar	4.251	2.971	1.858
	N_2	2.621	1.657	0.879
−150	Ar	17.88	−0.273	−0.632
	N_2	12.493	0.199	−0.280

[a]*The Smithsonian Physical Tables*, The Smithsonian Institution, Washington, D.C., 1954, p. 279. Based on data of Roebuck and Osterberg, *Phys. Rev.* **46**, 785 (1934) and **48**, 450 (1935).

$$dH = \left(\frac{\partial H}{\partial P}\right)_T dP + \left(\frac{\partial H}{\partial T}\right)_P dT \tag{6-64}$$

Placing a restriction of constant enthalpy on Equation 6-64, we obtain

$$0 = \left(\frac{\partial H}{\partial P}\right)_T + \left(\frac{\partial H}{\partial T}\right)_P \left(\frac{\partial T}{\partial P}\right)_H$$

which can be rearranged to give

$$\left(\frac{\partial T}{\partial P}\right)_H = -\frac{(\partial H/\partial P)_T}{(\partial H/\partial T)_P} \tag{6-65}$$

or

$$\mu_{J.T.} = -\frac{1}{C_p}\left(\frac{\partial H}{\partial P}\right)_T \tag{6-66}$$

Another relationship of interest can be obtained by substituting the fundamental definition of H into Equation 6-66:

$$\mu_{J.T.} = -\frac{1}{C_p}\left(\frac{\partial E}{\partial P}\right)_T - \frac{1}{C_p}\left[\frac{\partial(PV)}{\partial P}\right]_T \tag{6-67}$$

From either of these last two expressions it is evident that $\mu_{J.T.} = 0$ for an ideal gas, since each partial derivative is zero for such a substance.

Joule–Thomson Inversion Temperature. The Joule–Thomson coefficient is a function of temperature and pressure. Figure 6-6 shows the locus of points on a temperature–pressure diagram for which $\mu_{J.T.}$ is zero. Those points are at the Joule–Thomson inversion temperature, T_i. It is only inside the envelope of this curve that $\mu_{J.T.}$ is positive, that is, that the gas cools on expansion. This property is crucial to the problem of liquefaction of gases. Indeed, hydrogen and helium were considered "permanent" gases for many years, until it was discovered that their upper inversion temperatures are below room temperature; thus, they have to be precooled before being cooled by expansion.

For conditions under which the van der Waals equation is valid, the Joule–Thomson inversion temperature can be calculated from the expression

$$\frac{2a}{RT_i} - \frac{3abP}{R^2T_i^2} - b = 0 \tag{6-68}$$

(The derivation of Equation 6-68 is dependent on the second law of thermodynamics and will be done in Section 13-4.) Using Figure 6-6, we can see that Equation 6-68 (a quadratic equation in T_i) should have two distinct real roots for T_i at low pressures, two identical real roots at P_{max}, and two imaginary roots above P_{max}. At low pressure and high temperature, which are conditions corresponding to the upper inversion temperature, the second term in Equation 6-68 can be neglected and the result is

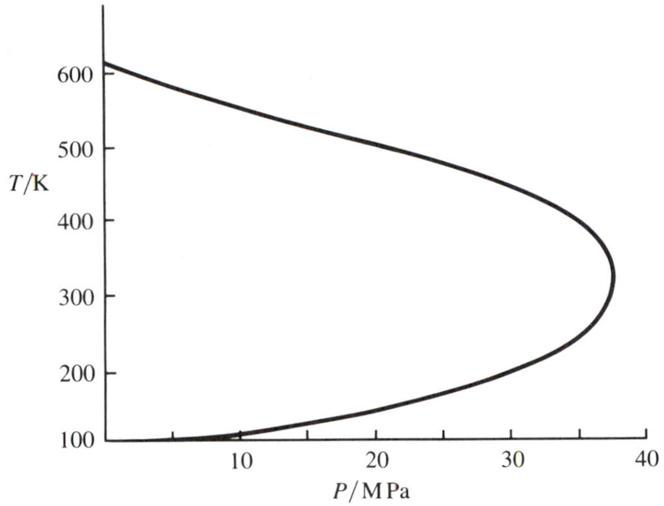

Figure 6-6. Locus of Joule–Thomson inversion temperatures for nitrogen. From J. R. Roebuck and H. Osterberg, *Phys. Rev.* **48**, 450 (1935).

$$T_i = \frac{2a}{Rb} \tag{6-69}$$

Hence, if the van der Waals constants are known, T_i can be calculated. For all gases except hydrogen and helium this inversion temperature is above common room temperatures.

Calculation of Thermodynamic Quantities in Reversible Expansions

Isothermal. The procedure used to calculate the work and energy quantities in an isothermal reversible expansion of a real gas is similar to that used for the ideal gas. Into the expression for the work done,

$$W = -\int_{V_1}^{V_2} PdV \tag{6-18}$$

we can substitute for P (or for dV) from the equation of state of the gas and then carry out the required integration. For example, for 1 mole of gas that obeys the van der Waals equation

$$W = -\int_{V_1}^{V^2} \left(\frac{RT}{V - b} - \frac{a}{V^2}\right) dV = -RT \ln\left(\frac{V_2 - b}{V_1 - b}\right) - \frac{a}{V_2} + \frac{a}{V_1} \tag{6-70}$$

The change in energy in an isothermal expansion cannot be expressed in a simple form without introducing the second law of thermodynamics. Nevertheless, we will anticipate this second basic postulate and use one of the deductions obtainable from it:

$$\left(\frac{\partial E}{\partial V}\right)_T = T\left(\frac{\partial P}{\partial T}\right)_V - P \tag{6-71}$$

For a gas that obeys the van der Waals equation, Equation 6-71 reduces to

$$\left(\frac{\partial E}{\partial V}\right)_T = \frac{a}{V^2} = \left(\frac{\partial E}{\partial V}\right)_T \tag{6-72}$$

ΔE can be obtained by integrating this equation:

$$\Delta E = \int_{V_1}^{V_2} \frac{a}{V^2}\, dV = -\frac{a}{V_2} + \frac{a}{V_1} \tag{6-73}$$

From the first law of thermodynamics we now can calculate the heat absorbed in the isothermal reversible expansion:

$$Q = \Delta E - W = RT \ln \frac{V_2 - b}{V_1 - b} \tag{6-74}$$

ΔH can be obtained from Equation 4-2:

$$\Delta H = \Delta E + \Delta(PV)$$

The PV product can be calculated from the van der Waals equation in the form

$$P = \frac{RT}{V - b} - \frac{a}{V^2} \tag{6-75}$$

by multiplying each term by V:

$$PV = RT \frac{V}{V - b} - \frac{a}{V} \tag{6-76}$$

With a few algebraic manipulations we can show that

$$(PV)_2 - (PV)_1 = \Delta(PV) = bRT \left[\frac{1}{V_2 - b} - \frac{1}{V_1 - b} \right] - \frac{a}{V_2} + \frac{a}{V_1} \tag{6-77}$$

Adding Equations 6-73 and 6-77 we obtain

$$\Delta H = bRT \left[\frac{1}{V_2 - b} - \frac{1}{V_1 - b} \right] - \frac{2a}{V_2} + \frac{2a}{V_1} \tag{6-78}$$

Adiabatic. In an adiabatic change of state, Q equals zero. However, a calculation of the work and energy quantities depends on the solution of Equation 4-23 for the change in energy. Furthermore, specifying the equation of state for the gas does not give automatically an expression for the dependence of C_v on temperature. When adequate equations, empirical or theoretical, for the variation of E and C_v with T and V are available, they can be used in Equation 4-23. If the resulting expression is integrable analytically, the energy and work quantities can be calculated as for an ideal gas. If the expression is not integrable analytically, numerical or graphical procedures can be used.

EXERCISES

When derivations or proofs of equations are called for, start from fundamental definitions and principles.

1. Derive an explicit equation for the reversible work of an isothermal expansion for each of the following cases:

   a. P is given by the equation of state of an ideal gas.

   b. P is obtained from the van der Waals equation of state.

   c. dv is obtained from the equation of state, $Pv = RT + BP + CP^2$.

   d. dv is obtained from the Berthelot equation.

2. Derive an explicit expression for the work done in the irreversible expansion of a gas from volume V_1 to volume V_2 against a constant external pressure P' that is less than the pressure of the gas throughout the expansion.

3. Rozen [*J. Phys. Chem. (USSR)* **19**, 469 (1945), and *Chem. Abstr.* **40**, 1712 (1946)] characterizes gases by "deviation coefficients" such as $T(\partial P/\partial T)_V/P$, $P(\partial v/\partial T)_P/R$, and $P^2(\partial v/\partial P)_T/RT$. Calculate the values of these coefficients for (a) an ideal gas, (b) a gas that obeys the van der Waals equation, and (c) a gas that obeys the Dieterici equation of state,

$$P = \frac{RT}{v - b} e^{-a/RTv}$$

4. Derive expressions for W, ΔE, Q, and ΔH in an isothermal reversible expansion of 1 mole of a gas that obeys the equation of state $Pv = RT + BP$. Use Equation 6-71 to calculate ΔE.

5. A gas obeys the equation of state $Pv = RT + BP$ and has a heat capacity c_v that is independent of the temperature.

   a. Derive an expression relating T and V in an adiabatic reversible expansion.

   b. Derive an equation for ΔH in an adiabatic reversible expansion.

   c. Derive an equation for ΔH in an adiabatic free expansion.

6. a. Given the equation

$$C_p = C_v + \left[V - \left(\frac{\partial H}{\partial P} \right)_T \right] \left(\frac{\partial P}{\partial T} \right)_V$$

   derive the relationship

$$C_v = C_p \left[1 - \mu_{J.T.} \left(\frac{\partial P}{\partial T} \right)_V \right] - V \left(\frac{\partial P}{\partial T} \right)_V$$

   b. To what expression can this equation be reduced at the inversion temperature?

7. For a certain ideal gas $c_v = \frac{3}{2}R$. Calculate the work done in an adiabatic reversible expansion of 1 mole of this gas by integrating Equation 6-40.

8. An ideal gas absorbs 9410 J of heat when it is expanded isothermally (at 25°C) and reversibly from 1.5 dm³ to 10 dm³. How many moles of the gas are present?

9. Derive the following relationship for an ideal gas:

$$\left(\frac{\partial E}{\partial V}\right)_P = \frac{c_v P}{R}$$

10. Keeping in mind that dE is an exact differential, prove that for an ideal gas

$$\left(\frac{\partial C_v}{\partial V}\right)_T = 0$$

$$\left(\frac{\partial C_p}{\partial P}\right)_T = 0$$

11. With the aid of a mathematical expansion it is possible to relate the constants of the virial equation to molecular properties, in particular to the van der Waals constants a and b.

a. Rearranging the van der Waals equation to the form

$$Pv = \frac{RTv}{v - b} - \frac{a}{v}$$

and applying Maclaurin's theorem from calculus, convert the van der Waals equation into the form

$$Pv = A + BP + CP^2$$

Hint: Consider P the independent variable and T fixed; then make use of the relationship

$$\frac{d(Pv)}{dP} = \frac{d(Pv)}{dv}\frac{dv}{dP}$$

b. Show that

$$B = b - \frac{a}{RT}$$

$$C = \frac{2RTab - a^2}{RT^3}$$

c. The Boyle temperature is defined as that at which

$$\lim_{P \to 0}\left(\frac{\partial(Pv)}{\partial P}\right)_T = 0$$

Using only the first two terms in the virial form of the van der Waals equation show that

$$T_{\text{Boyle}} = \frac{a}{Rb}$$

12. Using T and V as coordinates, draw a graph showing the three adiabatic expansions of Table 6-2.

13. According to the theory of acoustics, the velocity of propagation of sound, w, through a gas is given by the equation

$$w^2 = \frac{\partial P}{\partial \rho}$$

in which ρ is the density of the gas.

a. If the propagation of sound is assumed to occur adiabatically, and if the transmitting gas acts as if it were ideal and as if it were undergoing reversible compressions and rarefactions, show that

$$w^2 = \frac{C_p}{C_v} \frac{RT}{M}$$

(Hint: Take 1 mole of gas and start by finding a relationship between $d\rho$ and dV.)

b. Calculate w for sound in air assuming C_p/C_v is 7/5.

c. If the propagation of sound is assumed to occur isothermally, show that

$$w^2 = \frac{RT}{M}$$

14. Derive an equation for the coefficient of thermal expansion, β,

$$\beta = \frac{1}{V}\left(\frac{\partial V}{\partial T}\right)_P$$

for a gas that obeys the van der Waals equation.

(*Hint:* Equation 6-53 could be solved explicitly for v and then differentiated. However, an equation explicit in v would be cubic and unwieldy; Equation 6-75 is much less laborious to differentiate.)

15. For a rubber band the internal energy, E, equals $f(T, L)$; L is the length of the band. For a particular type of rubber

$$\left(\frac{\partial E}{\partial L}\right)_T = -\frac{1}{L^2}$$

and

$$\left(\frac{\partial E}{\partial T}\right)_L = C_L = \text{constant}$$

A stretched rubber band of length L_2 is allowed to snap back to length L_1 under adiabatic conditions and without doing any work. Will its temperature change? Show clearly the reason for your conclusion.

REFERENCES

1. Equation 6-52 can be derived from Boyle's law and the definitions of C_p and C_v without assuming the validity of the first law of thermodynamics. See D. Shanks, *Am. J. Phys.* **24**, 352 (1956).
2. E. A. Mason and T. H. Spurling, *The Virial Equation of State*, Pergamon Press, Oxford, 1969; J. D. Cox and I. J. Lawrenson, "The *P, V, T* Behavior of Single Gases," in *Chemical Thermodynamics, A Specialist Periodical Report*, Vol. 1, M. L. McGlashan, ed., The Chemical Society, London, 1973, pp. 162–203.
3. O. A. Hougen, K. M. Watson, and R. A. Ragatz, *Chemical Process Principles, Part II*, Wiley, New York, 1959, Chap. XII; L. C. Nelson and E. F. Obert, *Chem. Eng.* (July 1954), pp. 203–208.
4. The compressibility factor can be represented to high precision if it is expressed as a function of another parameter in addition to the reduced temperature and reduced pressure. See K. S. Pitzer, D. Z. Lippman, R. F. Curl, Jr., C. M. Huggins, and D. E. Petersen, *J. Am. Chem. Soc.* **77**, 3433 (1955).

CHAPTER 7

The Second Law of Thermodynamics

7-1 THE NEED FOR A SECOND LAW

For a scientist, the primary interest in thermodynamics is in predicting the *spontaneous* direction of natural processes, chemical or physical, where, by "spontaneous," we mean those changes that occur irreversibly in the absence of restraining forces—for example, the free expansion of a gas or the vaporization of a liquid above its boiling point. The first law of thermodynamics, which is useful in keeping account of heat and energy balances, makes no distinction between reversible and irreversible processes and makes no statement about the natural direction of a chemical or physical transformation.

As we noticed in Table 6-1, $\Delta E = 0$ both for the free expansion and for the reversible expansion of an ideal gas. We used an ideal gas as a convenient example because we could calculate easily the heat and work exchanged. Actually, for any gas, ΔE has the same value for a free and a reversible expansion between the corresponding initial and final states. Furthermore, ΔE for a compression is equal in magnitude and opposite in sign to ΔE for an expansion; there is no indication from the first law of which process is the spontaneous one.

A clue to the direction that needs to be followed to reach a criterion of spontaneity can be obtained by noticing in Table 6-1 that Q and W are equal to zero for the *reversible cycle* but are *not zero* for the *irreversible cycle*. In other words, it is changes in the *surroundings* as well as changes in the system that must be considered in distinguishing a reversible from an irreversible transformation. Evidently, then, we need to find a concept or quantity that incorporates some treatment of the surroundings to serve as a criterion of spontaneity. Ideally, we should find a state function, instead of path-dependent quantities such as heat or work. Such a function would provide the foundation for the formulation of the general principle (or law) that we are seeking.

7-2 THE NATURE OF THE SECOND LAW

Natural Tendencies Toward Equilibrium

The natural tendency of systems to proceed toward a state of equilibrium is exhibited in many familiar forms. When a hot object is placed in contact with a cold object, they reach a common temperature. We describe the change by saying that heat has flowed from the hot object to the cold object. However, we never observe that two objects in contact and at the same temperature spontaneously attain a state in which one has a high temperature and the other a low temperature. Similarly, if a vessel containing a gas is connected to an evacuated vessel, the gas will effuse into the evacuated space until the pressures in the two vessels are equal. Once this equilibrium has been reached, it never is observed that a pressure difference between the two vessels is produced spontaneously. Solutes diffuse from a more concentrated solution to a more dilute solution; concentration gradients never develop spontaneously. Magnets spontaneously become demagnetized; their magnetism never increases spontaneously. When a concentrated protein solution such as egg white is placed in a vessel of boiling water, the egg white coagulates, but we never observe that coagulated egg white at the temperature of boiling water returns spontaneously to a liquid state.

It is desirable to find some common measure (preferably a quantitative measure) of the tendency to change and of the direction in which change can occur. In 1850, Clausius introduced the entropy function, S (from the Greek word τροπή, meaning transformation), as a measure of the "transformational content," or capacity for change. In this chapter we will develop the properties of this function and its relationship to the direction and extent of natural processes as expressed in the second law of thermodynamics.

Statement of the Second Law

There are numerous equivalent statements of the second law. We will use the following statement proposed by Clausius:

> *It is impossible to construct a machine that is able to convey heat by a cyclical process from one reservoir (at a lower temperature) to another at a higher temperature unless work is done on the machine by some outside agency.*

The second law, like the first law, is a postulate that has not been derived from any prior principles. It is accepted because deductions from the postulate correspond to experience. Except in submicroscopic phenomena, to which classical thermodynamics does not apply, no exceptions to the second law have been found.

The statement that we have chosen to use as the fundamental expression of the second law of thermodynamics is in a form that resembles some other fundamental principles of physical science. It is expressed as a "principle of impotence" [1], that is, an assertion of the impossibility of carrying out a particular process. In physics, such principles of impotence occur frequently. For example, the impossibility of sending a signal with a speed greater than that of light in a vacuum provides the basis for the theory of relativity. Also, wave mechanics may be considered to be a consequence of the impossibility of measuring simultaneously the position and velocity of an elementary particle. Similarly, we can state the first law of thermodynamics in terms of humanity's impotence to construct a machine capable of producing energy, a so-called "perpetual-motion machine of the first kind." The second law describes an additional incapacity, "the impossibility of constructing a perpetual-motion machine of the second kind." Such a machine operates without contradicting the first law, but still provides an inexhaustible supply of work.

Mathematical Counterpart of the Verbal Statement

In the form in which it has been expressed thus far, the second law is not a statement that can be applied conveniently to chemical problems. We wish to use the second law of thermodynamics to establish a criterion by which we can determine whether a chemical reaction or a phase change will proceed spontaneously. Such a criterion would be available if we could obtain a function that had the following two characteristics.

1. It should be a thermodynamic property; that is, its value should depend only on the state of the system and not on the particular path by which the state has been reached.
2. It should change in a characteristic manner (for example, always increase) when a reaction proceeds spontaneously.

The entropy function of Clausius satisfies these requirements. Although the subject of heat engines at first may seem unpromising in searching for a general principle to predict which processes are spontaneous, they can also be viewed as general devices for the interconversion of heat and work. Their properties will provide a useful basis for developing the mathematical properties of the entropy function and a mathematical statement of the second law. Therefore, we shall consider the properties of the ideal heat engine described by Carnot.

7-3 THE CARNOT CYCLE

The Carnot engine is a device by which a working substance can exchange mechanical work with its surroundings and can exchange heat with two heat reservoirs, one at a high temperature, t_2, and one at a low temperature, t_1, as shown schematically in Figure 7-1. (Since we wish to derive the properties of

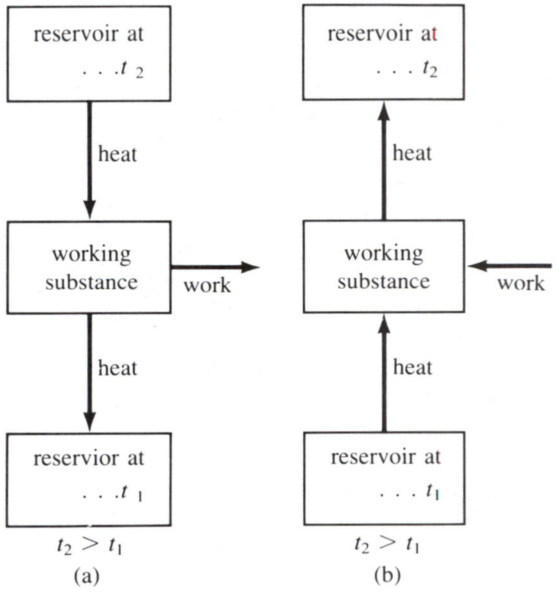

Figure 7-1. Scheme of a Carnot engine as (*a*) a heat engine, and (*b*) a refrigerator or heat pump.

the entropy function without reference to the properties of the ideal gas, we shall use any convenient empirical scale, rather than the ideal gas temperature scale, as a preliminary measure of temperature.)

The Forward Cycle

The Carnot cycle is a series of four steps that the working substance undergoes in the operation of the engine. At the completion of the four steps, the working substance has been returned to its initial state. In the forward direction, in which the engine transfers a net amount of heat to the working substance and does a net amount of work on the surroundings, the four steps are:

I. A reversible isothermal expansion in thermal contact with the high-temperature reservoir at t_2.

II. A reversible adiabatic expansion in which the temperature of the working substance decreases to t_1.

III. A reversible isothermal compression in thermal contact with the low-temperature reservoir at t_1.

IV. A reversible adiabatic compression in which the temperature of the working substance increases to t_2 and the substance is returned to its initial state.

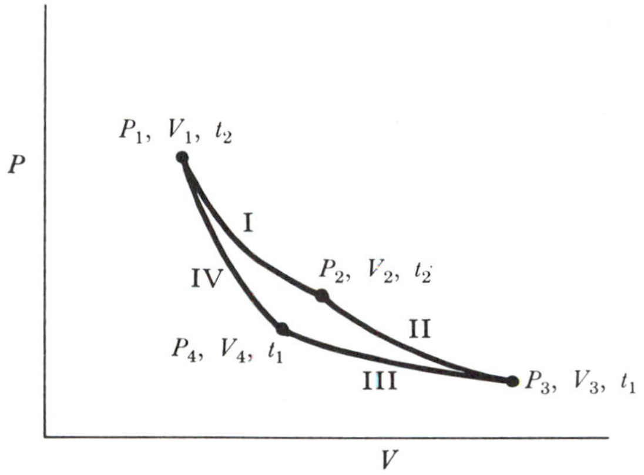

Figure 7-2. Carnot cycle; pressure–volume diagram.

Such a cycle is represented as a pressure–volume diagram in Figure 7-2. The representation of a temperature–volume diagram in Figure 7-3 emphasizes the isothermal nature of Steps I and III.

Let Q_2 (a positive quantity) represent the heat exchanged with the high-temperature reservoir in Step I, let Q_1 (a negative quantity) represent the heat exchanged with the low-temperature reservoir in Step III, and let W be the net work done (by the system). W is represented in Figure 7-2 by the negative of the area within the cycle, that is,

$$W = -\oint P dV \tag{7-1}$$

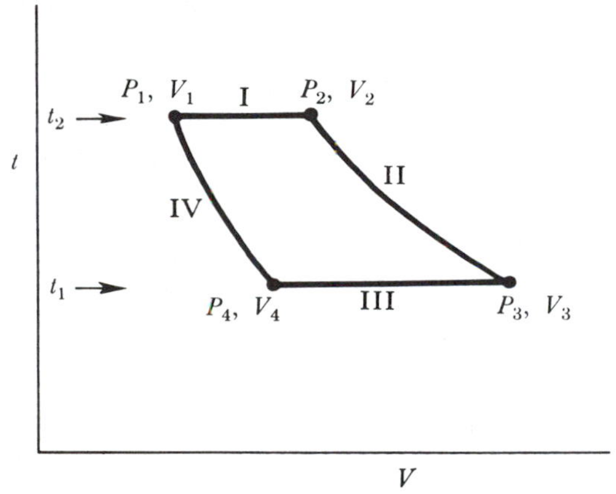

Figure 7-3. Carnot cycle; temperature–volume diagram.

Since the working substance returns to its initial state at the end of the cycle

$$\Delta E = 0 \tag{7-2}$$

Then from the first law

$$-W = Q_1 + Q_2 \tag{7-3}$$

The proportion, ε, of the heat absorbed at the high temperature that is converted to work represents a useful characteristic of a heat engine:

$$\varepsilon = -\frac{W}{Q_2} \tag{7-4}$$

and the negative sign ensures that ε is positive, since W and Q_2 are always opposite in sign. Substituting for W from Equation 7-3 into 7-4 we obtain

$$\varepsilon = \frac{Q_1 + Q_2}{Q_2} \tag{7-5}$$

$$= 1 + \frac{Q_1}{Q_2} \tag{7-6}$$

The quantity ε usually is called the *efficiency* of the engine.

The Reverse Cycle

Since the Carnot engine is a reversible device, it can be operated in the opposite direction, each step traversing the same path and characterized by the same heat and work quantities, but opposite in sign [see Figure 7-1*b*]. The reverse engine absorbs heat from the low-temperature reservoir, has work done on it by the surroundings, and liberates heat to the high-temperature reservoir. Thus it acts as a heat pump or a refrigerator. Its operation does not contradict the Clausius statement of the second law, because the surroundings do work to transport heat from the low-temperature reservoir to the high-temperature reservoir.

If we designate the heat and work quantities for the reverse cycle by primed symbols, the reverse and forward relationships are

$$\begin{aligned} W' &= -W \\ Q_2' &= -Q_2 \\ Q_1' &= -Q_1 \end{aligned} \tag{7-7}$$

and

$$\varepsilon = \frac{-W}{Q_2} = \frac{-W'}{Q_2'} \tag{7-8) [2}$$

$$= \frac{Q_1' + Q_2'}{Q_2'} = 1 + \frac{Q_1'}{Q_2'} \tag{7-9}$$

Thus ε is the same for both the forward and reverse cycles.

Alternative Statement of the Second Law

In addition to the statement we have been using, there are several alternative ways to express the second law. One that will be particularly useful is the Kelvin–Planck statement:

It is impossible to construct a machine that, operating in a cycle, will take heat from a reservoir at constant temperature and convert it into work without accompanying changes in the reservoir or its surroundings.

If such a machine could be constructed it would be a "perpetual-motion machine of the second kind."

To prove this statement we shall rely on the technique of assuming that it is false—that heat can be converted into work in a cyclic isothermal process without other changes occurring in the reservoir or the surroundings. Suppose that we carry out a cyclic process in which heat from the reservoir at a constant temperature t_2 is converted completely into work. Then $\Delta E = 0$ since the system is returned to its initial state. The work obtained can be used to operate a reversible Carnot engine in the opposite direction. This cycle, a refrigeration cycle, transfers a quantity of heat from a reservoir at the lower temperature, t_1, to the reservoir at the higher temperature, t_2, of the original isothermal cycle. The amount of heat first removed from the high-temperature reservoir is equivalent to the work obtained from the isothermal cycle. The heat Q now added to the high-temperature reservoir is the sum of the heat $(-Q_1)$ removed from the low-temperature reservoir and the heat equivalent of the work done by the isothermal cycle. The net work done in the combined cycles is zero. The final result of the combined cycles is the transfer of heat from the lower-temperature reservoir at t_1 to the higher-temperature reservoir at t_2 in a cyclic process with no other changes in the system or surroundings. This result contradicts the Clausius statement of the second law. Therefore, the assumption that heat can be converted into work in a cyclic isothermal process is false. Thus, the Kelvin–Planck statement is as valid a form of the second law as the Clausius statement.

The Kelvin–Planck statement is particularly pertinent to the application of the second law to biological systems and to the current concern about the effects of power production on the environment. Although biological systems are not heat engines, they operate isothermally, and they use chemical energy directly to produce mechanical and osmotic work and to carry out non-spontaneous synthesis of biological macromolecules such as proteins, nucleic acids, and polysaccharides. According to the Kelvin–Planck statement, isothermal conversions cannot be carried out without other changes occurring in the system or the surroundings.

In a biological system, in addition to the heat given off to the surroundings, a small number of complex food molecules is converted to a much larger number of simpler waste products, a process that involves an increase in the entropy of the surroundings, as we will see later. Since this production of entropy occurs at every level of the food chain, one ecological implication of the second law is clear: Entropy production can be minimized by consuming food from the lower levels of the food chain instead of producing and consuming animals at higher levels of the food chain.

The present concern about the thermal effects of power production are related directly to the Kelvin–Planck statement. The production of mechanical work from the heat of combustion of coal or from the heat generated by a nuclear reactor cannot be carried out without adding waste heat to the environment, which acts as the low-temperature reservoir. In the idealized examples that we have been discussing, the reservoirs are large enough to exchange the required amounts of heat without a change in temperature. In the real world, local temperature changes occur as a result of power production, and the environmental consequences may be important. As demands for power consumption increase, the problem of "thermal pollution" also will increase.

Carnot's Theorem

Carnot stated that the efficiency of a reversible Carnot engine depends only on the temperatures of the heat reservoirs and is independent of the nature of the working substance. This theorem can be proved by showing that the assumption of a reversible engine with any but the known efficiency of a reversible Carnot engine leads to a contradiction of the Clausius statement of the second law.

Let us illustrate the principle of this proof with a specific numerical example. If the ideal Carnot engine A has an efficiency $\varepsilon_A = 0.5$, dependent only on the temperatures of the reservoirs, let us assume that engine B has a higher efficiency, $\varepsilon_B = 0.6$. If 100 joules of heat are absorbed from the high-temperature reservoir by engine B operating in the forward cycle, then the engine will do 60 J of work on the surroundings and 40 J of heat will be transferred to the low-temperature reservoir, as listed in the first column of Table 7-1.

If the work done by engine B is used to operate engine A in reverse, as illustrated in Figure 7-4, then W for engine A is 60 J, Q_2 is equal to -120 J and Q_1 is equal to 60 J, as listed in the second column in Table 7-1. Thus, the net result of the coupling of the two cycles is that 20 J of heat has been removed from the low-temperature reservoir and added to the high-temperature reservoir, with no work done and no other changes occurring, since both

Table 7-1 Coupling of Two Carnot Cycles with Different Efficiencies

	Engine B (Forward cycle)	Engine A (Reverse cycle)	Net Change for Working Substances	Net Change for Reservoirs
Q_2	100 J	−120 J	−20 J	20 J
W	−60 J	60 J	0	
Q_1	−40 J	60 J	20 J	−20 J

working substances have been returned to their initial states. This result contradicts the Clausius statement of the second law. Consequently, engine B cannot have a higher efficiency than that of the ideal Carnot engine, A.

Now we can demonstrate the proof in a general way. If the efficiency of a Carnot engine A, is ε_A, we can assume a second engine, B, with efficiency $\varepsilon_B > \varepsilon_A$. When B is operated in the forward direction, thus exchanging heat Q_{2B} (a positive quantity) with the high-temperature reservoir, doing work W_B (a negative quantity) on the surroundings, and exchanging heat Q_{1B} (a negative quantity) with the low-temperature reservoir, we obtain

$$\varepsilon_B = \frac{-W_B}{Q_{2B}} \tag{7-10}$$

The Carnot engine, A, operated in a reverse cycle between the same heat reservoirs, exchanges heat Q'_{1A} with the low-temperature reservoir, is made to use all the work available from engine B so that work $W'_A = -W_B$, and exchanges heat Q'_{2A} with the high-temperature reservoir. The two engines are coupled through the exchange of work.

Since ε_B is assumed greater than ε_A

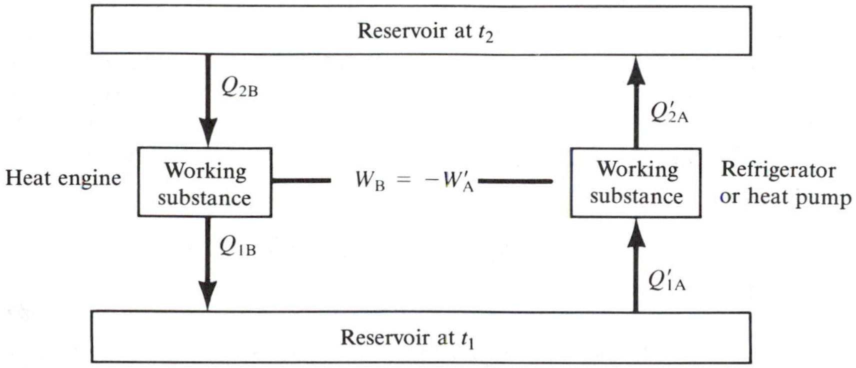

Figure 7-4. Scheme of two coupled Carnot engines—one acting as a heat engine, the other as a refrigerator or heat pump.

$$\frac{-W_B}{Q_{2B}} > \frac{-W'_A}{Q'_{2A}} \tag{7-11}$$

Because we will compare the magnitudes of the quantities rather than their value, we shall use their absolute values. Since W_B and Q_{2B} have opposite signs, as do W'_A and Q'_{2A}, the ratios in Equation 7-11 are equal to the ratios of the absolute values. Thus

$$\frac{|W_B|}{|Q_{2B}|} > \frac{|W'_A|}{|Q'_{2A}|} \tag{7-12}$$

Since $W_B = -W'_A$

$$|W_B| = |W'_A| \tag{7-13}$$

Therefore

$$\frac{1}{|Q_{2B}|} > \frac{1}{|Q'_{2A}|}$$

or

$$|Q_{2B}| < |Q'_{2A}| \tag{7-14}$$

Thus the amount of heat returned to the high-temperature reservoir in the reverse cycle is greater than the amount removed from it in the forward cycle.

By a similar argument it can be shown that the amount of heat added to the low-temperature reservoir in the forward cycle is less than the amount removed from it in the reverse cycle. The working substances of both engines have returned to their initial states, so $\Delta E = 0$ for the entire process. The net work done is zero since the work produced by the forward engine was used entirely to operate the reverse engine. Thus heat was transported to the high-temperature reservoir from the low-temperature reservoir in a cyclic process without any work being done, which contradicts the Clausius statement of the second law.

By a similar argument it can be shown that the assumption of an efficiency less than ε_A also leads to a contradiction of the second law. Thus any reversible Carnot engine operating between the same pair of reservoirs has the same efficiency, and that efficiency must be a function only of the *temperatures* of the *reservoirs*.

7-4 THE THERMODYNAMIC TEMPERATURE SCALE [3]

Equation 7-15, which includes Equation 7-6, is a mathematical statement of Carnot's theorem:

$$\varepsilon = 1 + \frac{Q_1}{Q_2} = f(t_1, t_2) \tag{7-15}$$

Solving for Q_1/Q_2 we obtain

$$\frac{Q_1}{Q_2} = -\,[1 - f(t_1, t_2)] \tag{7-16}$$

$$= -g(t_1, t_2) \tag{7-17}$$

Since Q_1 and Q_2 have opposite signs, their ratio is opposite in sign to the ratio of their absolute values. Thus

$$\frac{|Q_1|}{|Q_2|} = g(t_1, t_2) \tag{7-18}$$

Now consider a group of three heat reservoirs at temperatures $t_1 < t_2 < t_3$ and a reversible Carnot engine that operates successively between any pair of reservoirs. According to Equation 7-18

$$\frac{|Q_1|}{|Q_2|} = g(t_1, t_2) \tag{7-19}$$

$$\frac{|Q_1|}{|Q_3|} = g(t_1, t_3) \tag{7-20}$$

and

$$\frac{|Q_2|}{|Q_3|} = g(t_2, t_3) \tag{7-21}$$

If Equation 7-20 is divided by Equation 7-21 the result is

$$\frac{|Q_1|}{|Q_2|} = \frac{g(t_1, t_3)}{g(t_2, t_3)} \tag{7-22}$$

Equating the right sides of Equation 7-19 and Equation 7-22 we obtain

$$g(t_1, t_2) = \frac{g(t_1, t_3)}{g(t_2, t_3)} \tag{7-23}$$

Since the quantity on the left side of Equation 7-23 depends only on t_1 and t_2, the quantity on the right side also must be a function only of t_1 and t_2. Therefore, t_3 must appear in the numerator and denominator of the right side in such a way as to cancel, thereby giving

$$g(t_1, t_2) = \frac{h(t_1)}{h(t_2)} \tag{7-24}$$

Substituting from Equation 7-24 into Equation 7-19 we have

$$\frac{|Q_1|}{|Q_2|} = \frac{h(t_1)}{h(t_2)} \tag{7-25}$$

Since the magnitude of the heat exchanged in an isothermal step of a Carnot cycle is proportional to a function of an empirical temperature scale, the magnitude of the heat exchanged can be used as a thermometric property. An important advantage of this approach is that the measurement is independent of the properties of any particular material, since the efficiency of a Carnot cycle is independent of the working substance in the engine. Thus we define a thermodynamic temperature scale (symbol T) such that

$$\frac{|Q_2|}{|Q_1|} = \frac{T_2}{T_1} \tag{7-26}$$

with the units of the scale defined by setting $T = 273.16$ K (kelvin) at the triple point of water.

Removing the absolute value signs from Equation 7-26 we obtain

$$\frac{T_2}{T_1} = -\frac{Q_2}{Q_1} \tag{7-27}$$

Substitution from Equation 7-27 into Equation 7-6 yields

$$\varepsilon = 1 - \frac{T_1}{T_2} \tag{7-28}$$

$$= \frac{T_2 - T_1}{T_2} \tag{7-29}$$

Thus we have obtained the specific functional relationship between the efficiency of a reversible Carnot engine and the thermodynamic temperatures of the heat reservoirs.

The relationship between the thermodynamic temperature scale and the ideal gas temperature scale can be derived by calculating the thermodynamic quantities for a Carnot cycle with an ideal gas as the working substance. For this purpose we shall use θ to represent the ideal gas temperature.

In Step I of Figure 7-1 the work done is

$$W_1 = -\int_{V_1}^{V_2} P\,dV = -\int_{V_1}^{V_2} \frac{nR\theta_2}{V}\,dV$$

$$= -nR\theta_2 \ln \frac{V_2}{V_1} \tag{7-30}$$

Since Step I is isothermal $\Delta E = 0$ and

$$Q_2 = -W_{\text{I}} = nR\theta_2 \ln \frac{V_2}{V_1} \tag{7-31}$$

Similarly, for Step III, which is also isothermal, $\Delta E = 0$ and

$$Q_1 = -W_{\text{III}} = nR\theta_1 \ln \frac{V_4}{V_3} \tag{7-32}$$

Since Steps II and IV are adiabatic, Equation 6-46 applies, and

$$C_v \frac{d\theta}{\theta} = -nR \frac{dV}{V} \tag{7-33}$$

For an ideal gas C_v is a function only of θ, and both sides of Equation 7-33 can be integrated. Thus for Step II

$$\int_{\theta_2}^{\theta_1} C_v \frac{d\theta}{\theta} = -nR \int_{V_2}^{V_3} \frac{dV}{V}$$

$$= -nR \ln \frac{V_3}{V_2} \tag{7-34}$$

and for Step IV

$$\int_{\theta_1}^{\theta_2} C_v \frac{d\theta}{\theta} = -nR \int_{V_4}^{V_1} \frac{dV}{V}$$

$$= -nR \ln \frac{V_1}{V_4} \tag{7-35}$$

Since the integrals on the left sides of Equations 7-34 and 7-35 are equal and opposite in sign, the same must be the case for the integrals on the right sides of the two equations. Thus

$$nR \ln \frac{V_3}{V_2} = -nR \ln \frac{V_1}{V_4}$$

or

$$\ln \frac{V_3}{V_2} = \ln \frac{V_4}{V_1} \tag{7-36}$$

Therefore

$$\frac{V_3}{V_2} = \frac{V_4}{V_1} \tag{7-37}$$

According to Equation 6-40, W_{II} and W_{IV} are equal in magnitude and opposite in sign since the limits of integration are reversed in calculating these two quantities. Because these quantities are equal and opposite, the work done in the adiabatic steps does not contribute to the net work, which is

$$\begin{aligned} W &= W_I + W_{III} \\ &= -nR\theta_2 \ln \frac{V_2}{V_1} - nR\theta_1 \ln \frac{V_4}{V_3} \end{aligned} \tag{7-38}$$

The efficiency ε then is

$$\begin{aligned} \varepsilon &= \frac{-W}{Q_2} \\ &= \frac{nR\theta_2 \ln \dfrac{V_2}{V_1} + nR\theta_1 \ln \dfrac{V_4}{V_3}}{nR\theta_2 \ln \dfrac{V_2}{V_1}} \\ &= \frac{\theta_2 \ln \dfrac{V_2}{V_1} + \theta_1 \ln \dfrac{V_4}{V_3}}{\theta_2 \ln \dfrac{V_2}{V_1}} \end{aligned} \tag{7-39}$$

From Equation 7-37

$$\frac{V_3}{V_2} = \frac{V_4}{V_1}$$

so

$$\frac{V_2}{V_1} = \frac{V_3}{V_4}$$

and

$$\ln \frac{V_4}{V_3} = -\ln \frac{V_2}{V_1} \tag{7-40}$$

Substituting from Equation 7-40 into Equation 7-39 we obtain

$$\begin{aligned} \varepsilon &= \frac{\theta_2 \ln \dfrac{V_2}{V_1} - \theta_1 \ln \dfrac{V_2}{V_1}}{\theta_2 \ln \dfrac{V_2}{V_1}} \\ &= \frac{\theta_2 - \theta_1}{\theta_2} \end{aligned} \tag{7-41}$$

$$= 1 - \frac{\theta_1}{\theta_2} \tag{7-42}$$

Since the efficiency of a Carnot engine is independent of the working substance, the efficiency given in Equation 7-42 for an ideal gas must be equal to that given in Equation 7-28 for any reversible Carnot engine operating between the same heat reservoirs. Thus

$$1 - \frac{\theta_1}{\theta_2} = 1 - \frac{T_1}{T_2}$$

or

$$\frac{\theta_1}{\theta_2} = \frac{T_1}{T_2} \tag{7-43}$$

and the two temperature scales are proportional to one another. With the choice of the same reference point (the triple point of water)

$$T_{\text{tr}} = \theta_{\text{tr}} = 273.16 \text{ K} \tag{7-44}$$

the two scales become identical. Therefore we use T to represent both the ideal gas temperature scale and the thermodynamic temperature scale.

Although the two scales are identical numerically, their conceptual bases are quite different. The ideal gas scale is based on the properties of gases in the limit of zero pressure, whereas the thermodynamic scale is based on the properties of heat engines in the limit of reversible operation. That we can relate them so satisfactorily is an illustration of the usefulness of the concepts so far defined.

7-5 THE DEFINITION OF S, THE ENTROPY OF A SYSTEM

According to Equation 7-27

$$\frac{Q_2}{Q_1} = -\frac{T_2}{T_1}$$

Rearranging the equation, we obtain

$$\frac{Q_2}{T_2} = -\frac{Q_1}{T_1} \tag{7-45}$$

or

$$\frac{Q_2}{T_2} + \frac{Q_1}{T_1} = 0 \tag{7-46}$$

Since the isothermal steps in the Carnot cycle are the only steps in which heat is exchanged, we also can write Equation 7-46 as

$$\sum_{\text{cycle}} \frac{Q}{T} = 0 \tag{7-47}$$

That we again have a quantity whose sum over a closed cycle is zero suggests that Q/T is a thermodynamic property, even though we know that Q is not a thermodynamic property. Acting on this suggestion, we define the entropy function by the equation

$$dS = \frac{DQ_{\text{rev}}}{T} \tag{7-48}$$

As was the case with energy, the definition of entropy permits only a calculation of differences, not an absolute value. Integration of Equation 7-48 provides an expression for the finite difference in entropy between two states:

$$\Delta S = S_2 - S_1 = \int_1^2 dS = \int_1^2 \frac{DQ_{\text{rev}}}{T} \tag{7-49}$$

We will demonstrate in three steps that ΔS in Equation 7-49 is independent of the path by showing that:

1. dS is an exact differential for any substance carried through a Carnot cycle.
2. dS is an exact differential for any substance carried through any reversible cycle.
3. Entropy is a function only of the state of the system.

7-6 THE PROOF THAT S IS A THERMODYNAMIC PROPERTY

Any Substance in a Carnot Cycle

If we integrate Equation 7-48 for the steps of a reversible Carnot cycle, the results are

$$\Delta S_{\text{I}} = \int \frac{DQ_{\text{rev}}}{T_2}$$

$$= \frac{1}{T_2} \int DQ_{\text{rev}} = \frac{Q_2}{T_2} \tag{7-50}$$

and since Step II is adiabatic

$$\Delta S_{\text{II}} = \int \frac{DQ_{\text{rev}}}{T} = 0 \tag{7-51}$$

Similarly,

$$\Delta S_{\text{III}} = \frac{Q_1}{T_1} \tag{7-52}$$

and

$$\Delta S_{\text{IV}} = 0 \tag{7-53}$$

Thus for the complete cycle

$$\Delta S_{\text{cycle}} = \oint dS = \frac{Q_2}{T_2} + \frac{Q_1}{T_1} \tag{7-54}$$

From Equation 7-46 we see that the right side of Equation 7-54 is equal to zero. Therefore

$$\Delta S_{\text{cycle}} = \oint dS = 0 \tag{7-55}$$

and S is a thermodynamic property for any substance carried through a reversible Carnot cycle.

Any Substance in Any Reversible Cycle

For S to be a generally useful function we must remove any specifications as to the nature of the reversible cycle through which the substance is being carried. Let us represent a general reversible cycle by the example illustrated in Figure 7-5(*a*). This cycle also can be approximated in Carnot cycles, as illustrated in Figure 7-5(*b*), if we proceed along the heavy-lined path. It can be shown by the following procedure that

$$\oint dS = 0$$

To obtain this integral we must evaluate

$$\left(\frac{DQ}{T}\right)_{G\to B} + \left(\frac{DQ}{T}\right)_{B\to A} + \cdots + \left(\frac{DQ}{T}\right)_{M\to D} + \left(\frac{DQ}{T}\right)_{D\to N} + \cdots = \oint dS \tag{7-56}$$

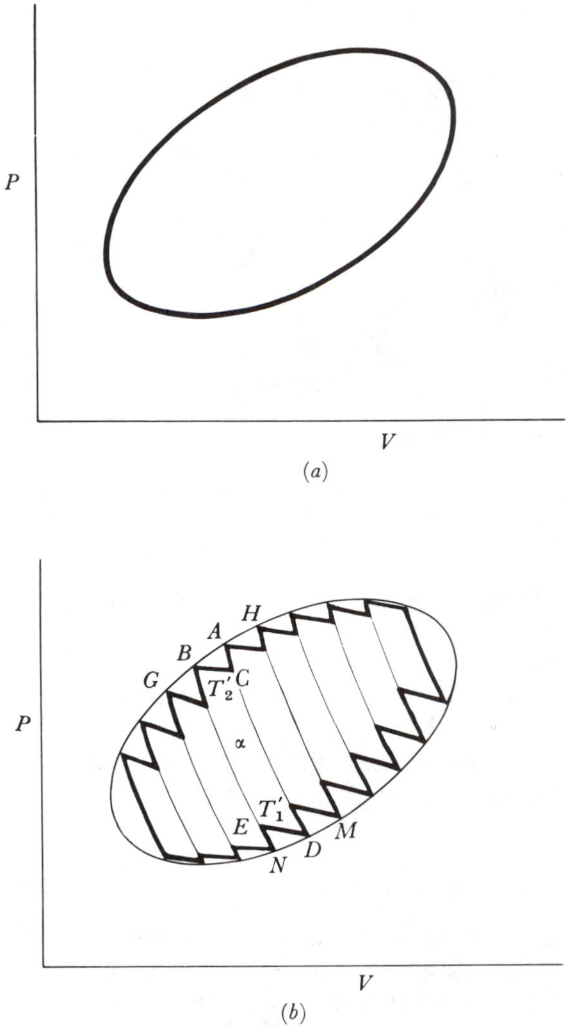

Figure 7-5. (*a*) A reversible cycle; (*b*) a reversible cycle approximated by Carnot cycles.

We can do this by examining the small Carnot cycles in more detail. For example, for the cycle labeled α we can state definitely, since the adiabatic steps contribute nothing to DQ/T, that

$$\left(\frac{DQ_2'}{T_2'}\right)_{B\to C} + \left(\frac{DQ_1'}{T_1'}\right)_{D\to E} = \left(\sum \frac{DQ}{T}\right)_{\text{cycle }\alpha} = 0 \tag{7-57}$$

in which the primes are used to emphasize that the quantities refer to the approximate Carnot cycle, not to the actual path of Figure 7-5(*a*). In the *P–V* diagram of Figure 7-5(*b*) we also note that for the small area *BACB*

$$\oint_{BACB} PdV = \text{area } BACB = -\oint_{BACB} DW = \oint_{BACB} DQ \qquad (7\text{-}58)$$

The last equality follows from the fact that, for a cycle, $\Delta E = 0$. Hence

$$\text{area } BACB = \oint DQ = (DQ)_{B\to A} + 0 + (DQ)_{C\to B} \qquad (7\text{-}59)$$

Noticing that

$$(DQ)_{C\to B} = -(DQ)_{B\to C} \qquad (7\text{-}60)$$

We conclude that

$$(DQ)_{B\to A} = (DQ_2')_{B\to C} + \text{area } BACB \qquad (7\text{-}61)$$

A better approximation to the actual cycle of Figure 7-5(*a*) would be a larger number of Carnot cycles in Figure 7-5(*b*). In each such approximation Equation 7-61 would be valid, but as the number of cycles used for the approximation is increased, the area *BACB* becomes smaller and smaller. In the limit of an infinite number of cycles, for which the approximation to the actual cycle becomes perfect,

$$\text{area } BACB \to 0 \qquad (7\text{-}62)$$

and

$$(DQ)_{B\to A} = (DQ_2')_{B\to C} \qquad (7\text{-}63)$$

For every pair of sections, *BA* and *DN*, of the actual path we have corresponding (isothermal) pairs, *BC* and *DE*, which are parts of an approximate Carnot cycle. Hence we can write

$$\oint_{\text{actual}} DS = \left(\frac{DQ}{T}\right)_{B\to A} + \cdots + \left(\frac{DQ}{T}\right)_{D\to N} + \cdots = \left(\frac{DQ_2'}{T_2'}\right)_{B\to C} + \cdots + \left(\frac{DQ_1'}{T_1'}\right)_{D\to E} + \cdots = 0 \qquad (7\text{-}64)$$

We conclude that dS is an exact differential for any reversible cycle.

Entropy S Depends Only on the State of the System

Figure 7-6 is a representation of two possible reversible paths for reaching State b from State a. We have just shown that over a reversible closed path the entropy change is zero, so

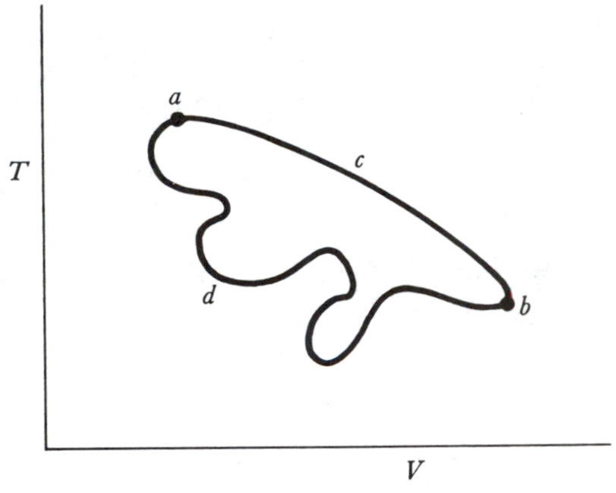

Figure 7-6. Two reversible paths from one state to another.

$$\int_{\substack{\text{Path}\\acb}} dS + \int_{\substack{\text{Path}\\bda}} dS = 0 \tag{7-65}$$

Hence

$$S_b - S_a = \int_{\substack{\text{Path}\\acb}} dS = -\int_{\substack{\text{Path}\\bda}} dS = \int_{\substack{\text{Path}\\adb}} dS \tag{7-66}$$

Notice that the order of limits is reversed in the right equality. Equation 7-66 emphasizes that the entropy change is the same for all arbitrary reversible paths from a to b. Thus the entropy change, ΔS, is a function only of State a and State b of the system.

7-7 ENTROPY CHANGES IN REVERSIBLE PROCESSES

Having established that dS is an exact differential, let us consider the value of the entropy change for several noncyclic reversible changes.

General Statement

According to the definition of heat, whenever a system absorbs a quantity of heat, DQ, the surroundings lose an equal quantity of heat. Thus

$$DQ_{\text{sys}} = -DQ_{\text{surr}} \tag{7-67}$$

It follows that

$$\frac{DQ_{\text{sys}}}{T} + \frac{DQ_{\text{surr}}}{T} = 0 \tag{7-68}$$

Consequently, for a reversible process

$$\int dS = 0 \tag{7-69}$$

(1) for the system plus surroundings undergoing a noncyclic process and (2) for the system undergoing a cyclic process. Several specific examples follow.

Isothermal Reversible Changes

For isothermal reversible changes the entropy change for the system is given by

$$\Delta S_{\text{sys}} = \int dS = \int \frac{DQ}{T} = \frac{1}{T}\int DQ = \frac{Q}{T} \tag{7-70}$$

For the specific case of the expansion of an ideal gas, since $\Delta E = 0$,

$$Q = -W = nRT \ln \frac{V_2}{V_1}$$

in which V_2 is the final volume and V_1 the initial volume. Hence

$$\Delta S_{\text{sys}} = \frac{Q}{T} = nR \ln \frac{V_2}{V_1} \tag{7-71}$$

If Q is the heat absorbed by the system, then $-Q$ must be the heat absorbed by the surroundings. Therefore

$$\Delta S_{\text{surr}} = -\frac{Q}{T} \tag{7-72}$$

Hence for the system plus surroundings

$$\Delta S_{\text{total}} = \Delta S_{\text{sys}} + \Delta S_{\text{surr}} = 0 \tag{7-73}$$

Adiabatic Reversible Changes

In any adiabatic reversible change DQ_{rev} equals zero. Thus

$$\Delta S_{\text{sys}} = \Delta S_{\text{surr}} = \Delta S_{\text{total}} = 0 \tag{7-74}$$

Reversible Phase Transitions

A change from one phase to another—for example, from ice to water—can be carried out reversibly and at a constant temperature. Under these conditions

Equation 7-70 is applicable. Generally, equilibrium phase transitions also are carried out at a fixed pressure. Since no work is expended in these transitions except against the atmosphere, Q is given by the enthalpy of transition, and

$$\Delta S_{\text{subs}} = \frac{\Delta H_{\text{trans}}}{T} \tag{7-75}$$

In these isothermal reversible phase transitions, for every infinitesimal quantity of heat absorbed by the substance, an equal quantity of heat is released by the surroundings. Consequently

$$\Delta S_{\text{surr}} = -\Delta S_{\text{subs}} \tag{7-76}$$

and again the entropy change for system plus surroundings is zero.

As a specific example of a calculation of ΔS for a phase transition, we consider the data for the fusion of ice at 0°C:

$$H_2O(s, 0°C) = H_2O(l, 0°C), \qquad \Delta H = 6008 \text{ J mol}^{-1}$$

$$\Delta S_{\text{water}} = \frac{6008}{273.15} = 21.996 \text{ J mol}^{-1} \text{ K}^{-1}$$

$$\Delta S_{\text{surr}} = -21.996 \text{ J mol}^{-1} \text{ K}^{-1}$$

$$\Delta S_{\text{total}} = 0$$

Earlier conventions expressed entropy changes as cal mol^{-1} K^{-1} or entropy units (eu) and sometimes as gibbs mol^{-1}.

Isobaric Reversible Temperature Change

The reversible expansion of a gas (a reversible flow of work) requires that the pressure of the gas differ only infinitesimally from the pressure of the surroundings. Similarly, a reversible flow of heat requires that the temperature of the system differ only infinitesimally from the temperature of the surroundings. If the temperature of the system is to change by a finite amount, then the temperature of the surroundings must change infinitely slowly. Thus, the reversible flow of heat, like the reversible expansion of a gas, is a limiting case that can be approached as closely as desired, but can never be reached.

If an isobaric temperature change is carried out reversibly, the heat exchanged in the process can be substituted into the expression for the entropy change, and the equations at constant pressure when no work is done other than PV work are

$$\Delta S_{\text{sys}} = \int_{T_1}^{T_2} \frac{DQ}{T} = \int_{T_1}^{T_2} \frac{dH}{T} = \int_{T_1}^{T_2} \frac{C_p dT}{T} = \int_{T_1}^{T_2} C_p \, d \ln T \tag{7-77}$$

If C_p is constant

$$\Delta S = C_p \ln \frac{T_2}{T_1} \tag{7-78}$$

in which T_2 is the final temperature and T_1 is the initial temperature.

If the system is heated reversibly, the change in the surroundings is equal and opposite in sign to that for the system, and

$$\Delta S_{\text{sys}} + \Delta S_{\text{surr}} = \Delta S_{\text{total}} = 0$$

Isochoric Reversible Temperature Change

The entropy changes for temperature changes at constant volume are analogous to those at constant pressure except that C_v replaces C_p. Thus, since $PdV = 0$,

$$\Delta S_{\text{sys}} = \int_{T_1}^{T_2} \frac{DQ}{T} = \int_{T_1}^{T_2} \frac{dE}{T} = \int_{T_1}^{T_2} \frac{C_v dT}{T} = \int_{T_1}^{T_2} C_v d \ln T \tag{7-79}$$

Again, the entropy change for the system plus surroundings is zero. Strictly speaking, Equations 7-77, 7-78, and 7-79 are applicable only when no phase changes or chemical reactions occur.

7-8 ENTROPY CHANGES IN IRREVERSIBLE PROCESSES

The definition of entropy requires that information about a reversible path be available to calculate an entropy change. To obtain the change of entropy in an irreversible process, it is necessary to discover a reversible path between the same initial and final states. Since S is a property of a system, ΔS is the same for the irreversible as for the reversible process.

Irreversible Isothermal Expansion of an Ideal Gas

It has been shown (Equation 7-71) that in the reversible isothermal expansion of an ideal gas

$$\Delta S_{\text{sys}} = nR \ln \frac{V_2}{V_1}$$

Since S is a thermodynamic property, ΔS_{sys} is the same in an irreversible isothermal process from the same initial volume, V_1, to the same final volume,

V_2. However, the change in entropy of the surroundings differs in the two types of processes. First let us consider an extreme case, a free expansion into a vacuum with no work being done. Since the process is isothermal, ΔE for the perfect gas must be zero; consequently, the heat absorbed by the gas, Q, also is zero:

$$Q = \Delta E - W = 0$$

Thus the surroundings have given up no heat. Consequently,

$$\Delta S_{\text{surr}} = 0 \tag{7-80}$$

and

$$\Delta S_{\text{total}} = nR \ln \frac{V_2}{V_1} + 0 > 0 \tag{7-81}$$

In other words, for the system plus surroundings this irreversible expansion has been accompanied by an increase in entropy.

We may contrast this result for ΔS_{total} with that for ΔE_{total} for an ideal gas, mentioned in Section 7-1. In the irreversible expansion of an ideal gas, $\Delta E_{\text{sys}} = 0$; the surroundings undergo no change of state (Q and W are both equal to zero), and hence $\Delta E_{\text{total}} = 0$. If we consider the reversible expansion of the ideal gas, ΔE_{sys} is also equal to zero, and ΔE_{surr} is equal to zero because $Q = -W$, so again $\Delta E_{\text{total}} = 0$. Clearly, in contrast to ΔS, ΔE does not discriminate between a reversible and an irreversible transformation.

In any intermediate isothermal expansion, the work done by the gas is not zero, but is less in magnitude than that obtained by reversible means (see Table 6-1). Since ΔE is zero and since

$$|W_{\text{irrev,sys}}| < nRT \ln \frac{V_2}{V_1} \tag{7-82}$$

it follows that

$$Q_{\text{irrev,sys}} < nRT \ln \frac{V_2}{V_1} \tag{7-83}$$

Nevertheless, the entropy change for the gas still is given by Equation 7-71 since it is equal to that for the reversible process between the same endpoints. If we divide both sides of Equation 7-83 by T and apply Equation 7-71 we obtain

$$\Delta S_{\text{sys}} = nR \ln \frac{V_2}{V_1} > \frac{Q_{\text{irrev}}}{T} \tag{7-84}$$

We can combine Equation 7-70 and Equation 7-84 into the compact form

$$\Delta S \geqslant \frac{Q}{T} \tag{7-85}$$

in which the equality applies to the reversible isothermal change and the inequality applies to the irreversible isothermal change. In the corresponding differential form the result is not limited to isothermal changes:

$$dS \geqslant \frac{DQ}{T} \tag{7-86}$$

Equation 7-86 is a condensed mathematical statement of the second law; the inequality applies to any real process, necessarily irreversible, and the equality applies to the limiting case of the reversible process.

For the heat exchange with the surroundings to occur reversibly (so that we can calculate the entropy change in the surroundings), we can imagine the gas to be in a vessel immersed in a large two-phase system (for example, solid–liquid) at equilibrium at the desired temperature. The heat lost by the surroundings must be numerically equal, but opposite in sign, to that gained by the gas:

$$Q_{\text{surr}} = -Q_{\text{irrev,sys}} \tag{7-87}$$

However, for the two-phase mixture at constant pressure and temperature, the change in entropy depends only on the quantity of heat evolved:

$$\Delta S_{\text{surr}} = \frac{Q_{\text{surr}}}{T} \tag{7-88}$$

since the change in state of the two-phase mixture during this process is a reversible one. Thus

$$\begin{aligned} \Delta S_{\text{total}} &= \Delta S_{\text{sys}} + \Delta S_{\text{surr}} = nR \ln \frac{V_2}{V_1} + \frac{Q_{\text{surr}}}{T} \\ &= nR \ln \frac{V_2}{V_1} - \frac{Q_{\text{irrev,sys}}}{T} \end{aligned} \tag{7-89}$$

From Equation 7-84

$$nR \ln \frac{V_2}{V_1} > \frac{Q_{\text{irrev,sys}}}{T}$$

and

$$\Delta S_{\text{total}} > 0 \tag{7-90}$$

Irreversible Adiabatic Expansion of an Ideal Gas

Points a and b in Figure 7-7 represent the initial and final states of an irreversible adiabatic expansion of an ideal gas. The path between is not represented because the temperature has no well-defined value in such a change; different parts of the system may have different temperatures. The

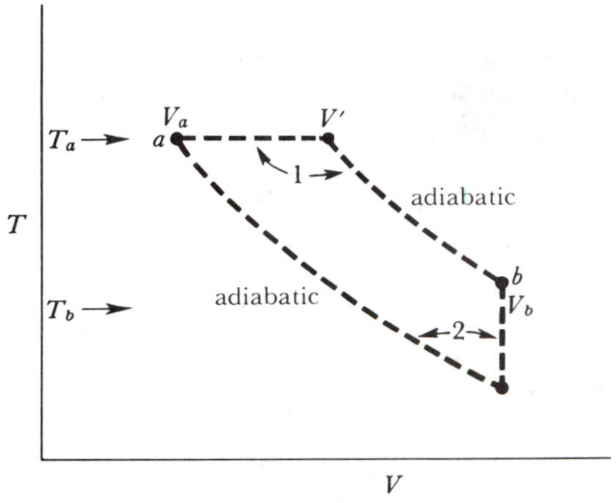

Figure 7-7. Irreversible change from State *a* to State *b*. The dashed lines represent two possible reversible paths through the same endpoints.

inhomogeneities in the system that develop during the irreversible change do not disappear until a new equilibrium is reached at *b*.

To determine the entropy change in this irreversible adiabatic process, it is necessary to find a reversible path from *a* to *b*. An infinite number of reversible paths are possible, and two are illustrated by the dashed lines in Figure 7-7.

The first consists of two steps: (1) an isothermal reversible expansion at the temperature T_a until the volume V' is reached, and (2) an adiabatic reversible expansion from V' to V_b. The entropy change for the gas is given by the sum of the entropy changes for the two steps:

$$\Delta S_{\text{gas}} = nR \ln \frac{V'}{V_a} + 0 \tag{7-91}$$

Since $V' > V_a$, the entropy change for the gas is clearly positive for the reversible path and, therefore, also for the irreversible change.

The second path consists of (1) an adiabatic reversible expansion to V_b and a temperature T' less than T_b (see Table 6-2), and (2) an isochoric temperature rise from T' to T_b. The entropy change for the gas is again given by the sum of the changes for the two steps:

$$\Delta S_{\text{gas}} = 0 + \int_{T'}^{T_b} C_v \frac{dT}{T} \tag{7-92}$$

Since T_b is greater than T', ΔS_{gas} is positive.

A reversible adiabatic expansion of an ideal gas has a zero entropy change, and an irreversible adiabatic expansion of the same gas from the same initial state to the same final *volume* has a positive entropy change. This statement may seem to be inconsistent with the statement that S is a thermodynamic property. The resolution of the discrepancy is that the two changes do not constitute the same change of state; the final temperature of the reversible adiabatic expansion is lower than the final temperature of the irreversible adiabatic expansion (as in path 2 in Figure 7-7).

Irreversible Flow of Heat from a Higher to a Lower Temperature

Imagine the flow of heat, by means of a conductor, from a very large reservoir at a higher temperature, T_2, to a very large reservoir at a lower temperature, T_1. By the use of large reservoirs we may consider the heat sources to be at constant temperature despite the gain or loss of a small quantity of heat Q.

To calculate the change in entropy in this irreversible flow, it is necessary to consider a corresponding reversible process. One process would be to allow an ideal gas to absorb reversibly the quantity of heat Q at the temperature T_2. The gas then can be expanded adiabatically and reversibly (therefore with no change in entropy) until it reaches the temperature T_1. At T_1 the gas is compressed reversibly and evolves the quantity of heat Q. During this reversible process, the reservoir at T_2 loses heat and undergoes the entropy change

$$\Delta S_{\text{hot reservoir}} = -\frac{Q}{T_2} \tag{7-93}$$

Since the same change in state occurs in the irreversible process, ΔS for the hot reservoir still is given by Equation 7-93. During the reversible process, the reservoir T_1 absorbs heat and undergoes the entropy change

$$\Delta S_{\text{cold reservoir}} = \frac{Q}{T_1} \tag{7-94}$$

Since the same change in state occurs in the irreversible process, ΔS for the cold reservoir still is given by Equation 7-94. In the irreversible process, the two reservoirs are the only substances that undergo any changes. Since $T_2 > T_1$, the entropy change for the system as a whole is positive:

$$\Delta S_{\text{sys}} = \Delta S_{\text{hot reservoir}} + \Delta S_{\text{cold reservoir}} = -\frac{Q}{T_2} + \frac{Q}{T_1} > 0 \tag{7-95}$$

Irreversible Phase Transition

A convenient illustration of an irreversible phase transition is the crystallization of water at $-10°C$ and constant pressure:

$$H_2O(\text{l}, -10°\text{C}) = H_2O(\text{s}, -10°\text{C}) \tag{7-96}$$

To calculate the entropy changes it is necessary to consider a series of reversible steps leading from liquid water at −10°C to solid ice at −10°C. One such series might be the following: (1) Heat supercooled water at −10°C very slowly (reversibly) to 0°C, (2) convert the water at 0°C very slowly (reversibly) to ice at 0°C, and (3) cool the ice very slowly (reversibly) from 0°C to −10°C. Since each of these steps is reversible, the entropy changes can be calculated by the methods discussed previously. Since S is a thermodynamic property, the sum of these entropy changes is equal to ΔS for the process indicated by Equation 7-96. The necessary calculations are summarized in Table 7-2, in which T_2 represents 0°C and T_1 represents −10°C.

Notice that there has been a decrease in the entropy of the water (that is, ΔS is negative) upon crystallization at −10°C despite the fact that the process is irreversible. This example emphasizes again that the sign of the entropy change *for the system plus surroundings*, and not merely for any component, is related to irreversibility. To obtain ΔS for the combination we must consider the entropy change in the surroundings, since the process described by Equation 7-96 occurs irreversibly. If we consider the water as being in a large reservoir at −10°C, then the crystallization will evolve a certain quantity of heat, Q, which will be absorbed by the reservoir without a significant rise in temperature. The change in state of the reservoir is the same as would occur if it were heated reversibly; hence, ΔS is given by

$$\Delta S_{\text{reservoir}} = \int \frac{DQ}{T} = -\frac{\Delta H}{T} = -\frac{(-5619\ \text{J})}{263.15\ \text{K}} = 21.353\ \text{J mol}^{-1}\ \text{K}^{-1} \quad (7\text{-}97)$$

in which ΔH represents the heat of crystallization of water at −10°C. Clearly, for the system plus surroundings the entropy has increased:

$$\Delta S_{\text{total}} = \Delta S_{\text{H}_2\text{O}} + \Delta S_{\text{reservoir}} = -20.545 + 21.353 = 0.808\ \text{J mol}^{-1}\ \text{K}^{-1} \quad (7\text{-}98)$$

Irreversible Chemical Reaction

As a final specific example let us examine the particular chemical reaction

$$\text{H}_2(\text{g}) + \tfrac{1}{2}\text{O}_2(\text{g}) = \text{H}_2\text{O}(\text{l}) \quad (7\text{-}99)$$

The formation of water from gaseous hydrogen and oxygen is a spontaneous reaction at room temperature, although its rate may be unobservably small in the absence of a catalyst. At 298.15 K the heat of the irreversible reaction at constant pressure is −285,830 J mol^{-1}. To calculate the entropy change, we must carry out the same transformation reversibly, which can be done electrochemically with a suitable set of electrodes. Under reversible conditions the heat of reaction for Equation 7-99 is −48,647 J mol^{-1}. Hence, for the irreversible or reversible change

$$\Delta S_{\text{chem}} = \frac{-48{,}647\ \text{J mol}^{-1}}{298.15\ \text{K}} = -163.16\ \text{J mol}^{-1}\ \text{K}^{-1} \quad (7\text{-}100)$$

Table 7-2 Entropy Change in Spontaneous Crystallization of Water

$H_2O(l, -10°C) = H_2O(l, 0°C)$

$$\Delta S_1 = \int \frac{DQ}{T} = \int \frac{C_p dT}{T} = C_p \ln \frac{T_2}{T_1} = 2.807 \text{ J mol}^{-1} \text{ K}^{-1}$$

$H_2O(l, 0°C) = H_2O(s, 0°C)$

$$\Delta S_2 = \int \frac{DQ}{T} = \frac{\Delta H}{T} = \frac{-6008 \text{ J mol}^{-1}}{273.15 \text{ K}} = -21.996 \text{ J mol}^{-1} \text{ K}^{-1}$$

$H_2O(s, 0°C) = H_2O(s, -10°C)$

$$\Delta S_3 = \int \frac{DQ}{T} = \int \frac{C_p dT}{T} = C_p \ln \frac{T_1}{T_2} = -1.356 \text{ J mol}^{-1} \text{ K}^{-1}$$

Adding:

$H_2O(l, -10°C) = H_2O(s, -10°C)$

$$\Delta S_{H_2O} = \Delta S_1 + \Delta S_2 + \Delta S_3 = -20.545 \text{ J mol}^{-1} \text{ K}^{-1}$$

The heat absorbed by the surrounding reservoir during the irreversible reaction is 285,830 J, and this heat produces the same change in state of the reservoir as the absorption of an equal amount of heat supplied reversibly. If the surrounding reservoir is large enough to keep the temperature essentially constant, its entropy change is

$$\Delta S_{\text{reservoir}} = \frac{285{,}830 \text{ J mol}^{-1}}{298.15 \text{ K}} = 958.68 \text{ J mol}^{-1} \text{ K}^{-1} \tag{7-101}$$

In the spontaneous formation of water, the system plus surroundings, chemicals plus environment, increases in entropy:

$$\Delta S_{\text{total}} = -163.16 + 958.68 = 795.52 \text{ J mol}^{-1} \text{ K}^{-1} \tag{7-102}$$

General Statement

In the preceding examples irreversible processes are accompanied by an increase in total entropy. It remains to be shown that such an increase occurs generally for isolated systems. By an isolated system we mean a region large enough to include all the changes under consideration, so that no matter or heat or work (thus, no energy) is exchanged between this region and the environment. In other words, an isolated system is at constant E and V. Thus, the isolated region is adiabatically insulated during the course of any spontaneous processes that occur within its boundaries.

Consider an irreversible process in which an isolated system goes from State a to State b. Since the process is irreversible, Figure 7-8 indicates only the initial and final states and not the path. To calculate the entropy change in going from a to b let us complete a cycle by going from b to a by the series of reversible steps indicated by the dashed lines in Figure 7-8. The adiabatic path bc is followed to some temperature T_c, which may be higher or lower than T_a. The only requirement in fixing T_c is that it be a temperature at which an isothermal reversible process can be carried out from State c to State d. State d is chosen in such a way that a reversible adiabatic change will return the system to its initial state, a. By means of these three reversible steps the system is returned from State b to State a. Since the first and third steps in this reversible process are adiabatic, the entropy change for those steps is zero. Consequently, the entropy of State c is the same as that of State b, namely, S_b. Similarly, the entropy of State d is S_a. However, an entropy change does occur along the path cd. Since this is an isothermal reversible process

$$S_d - S_c = S_a - S_b = \frac{Q}{T_c} \tag{7-103}$$

Since in the complete cycle (irreversible adiabatic process from a to b followed by the three reversible steps)

$$\Delta E_{\text{cycle}} = 0$$

it follows that

$$Q_{\text{cycle}} = -W_{\text{cycle}} \tag{7-104}$$

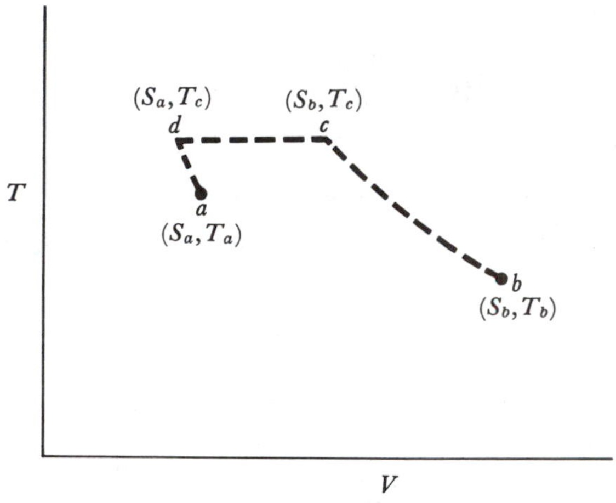

Figure 7-8. Schematic diagram of general irreversible change. The dashed line represents one possible reversible path between the endpoints.

Furthermore, in the four steps of the cycle (Figure 7-8) three are adiabatic (one irreversible, two reversible). Hence Q_{cycle} is identical with Q of the isothermal step, that is, Q of Equation 7-103. If $Q > 0$, then $W < 0$; that is, work would have been *done* by the system. In other words, if Q were positive, we would have carried out a cyclical process in which heat at a constant temperature T_c had been converted completely into work. According to the Kelvin–Planck statement of the second law, such a process cannot be carried out. Hence Q cannot be a positive number. Since Q must be either negative or zero, it follows from Equation 7-103 that

$$S_a - S_b \leqslant 0$$

therefore

$$\Delta S = S_b - S_a \geqslant 0 \tag{7-105}$$

Thus the entropy change for an irreversible process occurring in an isolated system is greater than or equal to zero, with the equal sign applying to the limiting case of a reversible process.

We have thus been able to show that for a closed section of space including all the changes under observation,

$$\Delta S \geqslant 0$$

the equality sign applying to isolated systems at equilibrium, and the inequality to all isolated systems capable of undergoing spontaneous changes, hence, systems in which changes occur irreversibly. Reversible changes, as has been emphasized, are only idealizations of certain actual changes. In many cases actual transformations can be carried out under conditions exceedingly close to the limiting ideal case; nevertheless, strictly speaking, all observable changes are really irreversible; that is, for all observable changes, ΔS is positive for the system plus surroundings.

7-9 GENERAL EQUATIONS FOR THE ENTROPY OF GASES

Entropy of the Ideal Gas

We can obtain an explicit equation for the entropy of an ideal gas from the mathematical statements of the two laws of thermodynamics. It is convenient to derive this equation for reversible changes in the gas. However, the final result will be perfectly general because entropy is a state function.

In a system in which only reversible expansion work is possible, the first law can be stated as

$$dE = DQ - PdV \tag{7-106}$$

For a reversible transformation (Equation 7-48)

$$dS = \frac{DQ}{T}$$

Substituting for DQ from Equation 7-106 into Equation 7-48 we obtain

$$dS = \frac{dE}{T} + \frac{PdV}{T} \tag{7-107}$$

For one mole of an ideal gas

$$\frac{P}{T} = \frac{R}{v}$$

and, from Equation 6-39,

$$d\varepsilon = c_v dT$$

thus

$$ds = \frac{c_v dT}{T} + \frac{R dv}{v} \tag{7-108}$$

If c_v is constant, this expression can be integrated to give

$$s = c_v \ln T + R \ln v + s_0 \tag{7-109}$$

in which s_0 is an integration constant characteristic of the gas. This integration constant cannot be evaluated by classical thermodynamic methods. However, it can be evaluated with the aid of kinetic molecular theory and statistical methods. For a monatomic gas, s_0 was formulated explicitly originally by Tetrode [4] and by Sackur [5].

Entropy of a Real Gas

In deriving an equation for the entropy of a real gas we can start with Equation 7-107, which is general and not restricted to ideal gases. A suitable substitution for dE in Equation 7-107 can be obtained from the total differential of E as a function of V and T (Equation 4-23):

$$dE = \left(\frac{\partial E}{\partial T}\right)_V dT + \left(\frac{\partial E}{\partial V}\right)_T dV$$

Thus

$$dS = \frac{1}{T}\left(\frac{\partial E}{\partial T}\right)_V dT + \frac{1}{T}\left(\frac{\partial E}{\partial V}\right)_T dV + \frac{P}{T} dV \tag{7-110}$$

The entropy, S, also can be considered to be a function of V and T; thus, a second equation for the total differential, dS, is

$$dS = \left(\frac{\partial S}{\partial T}\right)_V dT + \left(\frac{\partial S}{\partial V}\right)_T dV \qquad (7\text{-}111)$$

A comparison of the coefficients of the dT terms in Equations 7-110 and 7-111 leads to the following equality:

$$\left(\frac{\partial S}{\partial T}\right)_V = \frac{1}{T}\left(\frac{\partial E}{\partial T}\right)_V = \frac{1}{T} C_v \qquad (7\text{-}112)$$

It can be shown also, by a procedure to be outlined in Chapter 8, that the following relationship is true:

$$\left(\frac{\partial S}{\partial V}\right)_T = \left(\frac{\partial P}{\partial T}\right)_V \qquad (7\text{-}113)$$

Substituting from Equations 7-112 and 7-113 into Equation 7-111 we obtain

$$dS = \frac{C_v}{T} dT + \left(\frac{\partial P}{\partial T}\right)_V dV \qquad (7\text{-}114)$$

Equation 7-114 can be integrated to give

$$S = \int C_v d \ln T + \int \left(\frac{\partial P}{\partial T}\right)_V dV + \text{constant} \qquad (7\text{-}115)$$

To integrate this expression, it is necessary to know the equation of state of the gas and the dependence of C_v on temperature.

If a gas obeys the van der Waals equation of state it can be shown that

$$s = \int c_v d \ln T + R \ln (v - b) + \text{constant} \qquad (7\text{-}116)$$

7-10 TEMPERATURE–ENTROPY DIAGRAM

In making diagrams of various reversible cycles it is a common practice to plot pressure as a function of volume because the area under the curve, $\int PdV$, gives the negative of the work done in any step. Instead, we have used temperature and volume as coordinates because a diagram on this basis emphasizes the constancy of temperature in an isothermal process. However, it has the disadvantage that the area is not related to the work. Gibbs [6] pointed out that a diagram with temperature and entropy as coordinates is particularly useful since it illustrates graphically not only the work involved in a reversible cycle but also the heats. In addition, this type of diagram emphasizes the isentropic nature of an adiabatic reversible process, as well as the constancy of temperature in isothermal stages. A typical diagram for a simple Carnot cycle is illustrated in Figure 7-9. The four stages in a forward cycle are labeled by roman numerals. In Step I the temperature is constant, heat Q_2 is absorbed by the working substance, and the entropy increases from S_1 to S_2. Since this stage is reversible and isothermal

$$\Delta S_{\mathrm{I}} = S_2 - S_1 = \frac{Q_2}{T_2} \tag{7-50}$$

and

$$Q_2 = T_2(S_2 - S_1) = \text{area under line I} \tag{7-117}$$

In Step II there is a drop in temperature in the adiabatic reversible expansion, but no change in entropy. The isentropic nature of II is emphasized by the vertical line. Step III is an isothermal reversible compression, with a heat numerically equal to Q_1 being evolved. Since this step is reversible and isothermal

$$\Delta S_{\mathrm{III}} = S_1 - S_2 = -(S_2 - S_1) = \frac{Q_1}{T_1} \tag{7-52}$$

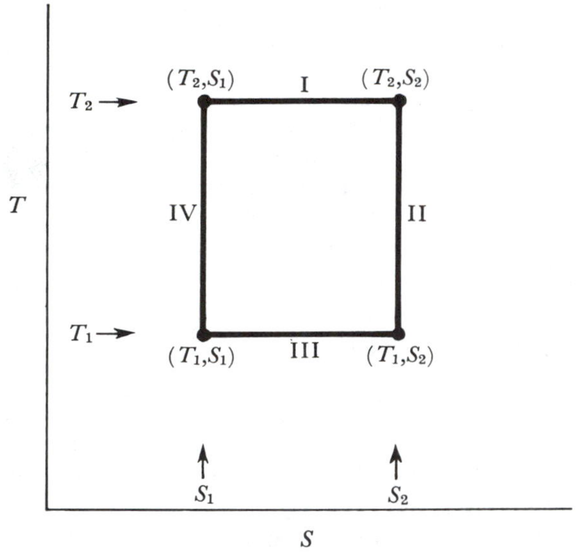

Figure 7-9. Gibbs temperature–entropy diagram for a Carnot cycle.

and

$$Q_1 = -T_1(S_2 - S_1) = \text{negative of area under line III} \qquad (7\text{-}118)$$

In the fourth step, which is adiabatic and reversible, there is no entropy change, but the temperature rises to the initial value, T_2. Since the process is cyclic

$$\Delta E = 0$$

and

$$Q_2 + Q_1 = -W$$

Therefore

$$\begin{aligned} -W &= T_2(S_2 - S_1) - T_1(S_2 - S_1) \\ &= (T_2 - T_1)(S_2 - S_1) = \text{area enclosed by cycle} \qquad (7\text{-}119) \end{aligned}$$

Thus the work and heats involved in the cycle are illustrated clearly by a T–S diagram, and the nature of the isothermal and isentropic steps is emphasized.

7-11 ENTROPY AS AN INDEX OF EXHAUSTION

To obtain a better grasp of the essential character of the entropy concept, let us examine in more detail a few transformations that occur spontaneously despite the fact that $\Delta E = 0$ in each case (Figure 7-10).

If two blocks of metal, one at a high temperature T_3 and the other at a low temperature T_1 [Figure 7-10(*a*)] are separated by a perfect heat insulator, and the system as a whole is surrounded by a thermal blanket that permits no transfer of heat in or out, then there can be no change in internal energy with time. If the insulator between the blocks is removed, the hotter block will drop in temperature and the cooler one will rise in temperature until the uniform temperature T_2 is reached. This is a spontaneous transformation. ΔE is zero. But there has been a loss of *capacity to do work*. In the initial state, we could insert a thermocouple in the block at T_3 and another in that at T_1 and obtain electrical work. Insertion of thermocouples in the same positions in the double block at T_2 cannot generate any work. At the conclusion of the spontaneous transformation without the thermocouples, the internal energy is still the same as that at the outset, but it is no longer in a condition where it has the capacity to do work [7].

Similarly, if two ideal gases at the same temperature, one at a high pressure P_3 and the other at a low pressure P_1, are separated by a partition, the high-pressure gas will move spontaneously into the low-pressure chamber if the separating barrier is removed [Figure 7-10(*b*)]. Yet $\Delta E = 0$. Again there has been a loss of *capacity to do work*. In the initial state, we could arrange a piston

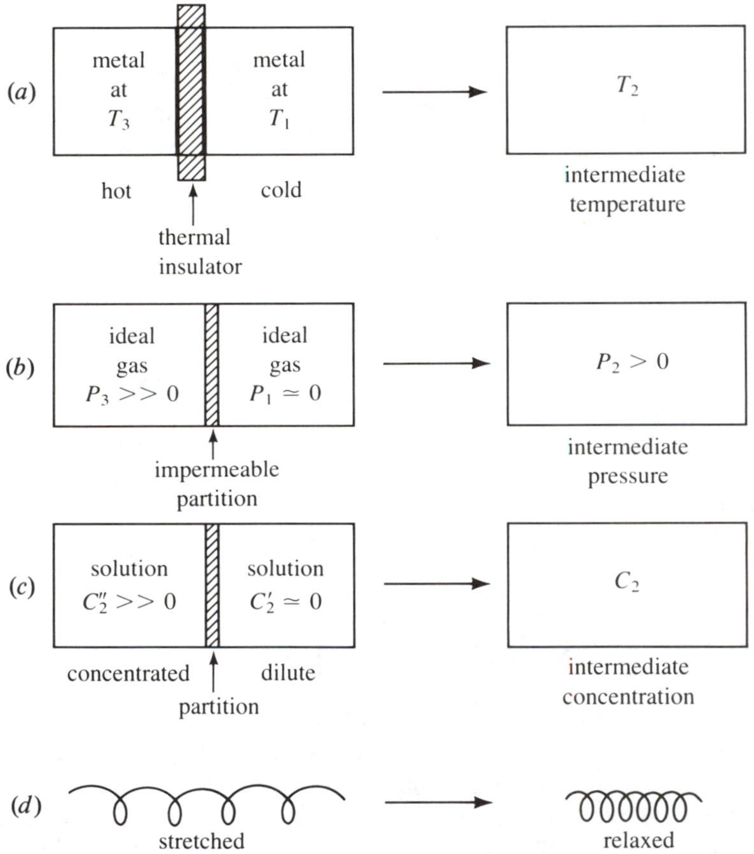

Figure 7-10. Some processes that occur spontaneously (with $\Delta E = 0$) in which there is a loss in capacity to do work.

with a rod extending to the outside of the containers, have the gas at P_3 on one side of the piston and the gas at P_1 on the other, and the piston could perform work (for example, by lifting a weight against gravity). That capacity to do work is no longer present at the conclusion of the spontaneous transformation (with no piston present); the character of the energy has changed [7].

A similar analysis can be made for two solutions with a solute (for example, an electrolyte such as $CuSO_4$) at two different concentrations being allowed to mix spontaneously upon removal of a separating partition [Figure 7-10(*c*)]. A copper metal electrode inserted into solutions C_2'' and C_2' will generate electrical energy, but such a pair of electrodes in the same physical position after spontaneous mixing has been completed to give solution C_2, can produce no work. In a very different type of system, a stretched rubber band [Figure 7-10(*d*)], release of the constraint will spontaneously lead to the relaxed rubber band. In each system $\Delta E = 0$; but again the capacity to do work has been diminished.

This loss in capacity to do work is a property of each of the systems illustrated in Figure 7-10, whether or not it has actually done any work during the transformation.

Thus, we ought to view entropy as an index of condition or character of a system (perhaps somewhat analogous to a cost-of-living index or to pH as an index of acidity). It is an index of the state of differentiation [7] of the energy, an index of the capacity to do work, an index of the tendency toward spontaneous change. The more a system exhausts its capacity for spontaneous change, the larger the entropy index. Hence, we should preferably say that *entropy is an index of exhaustion*; the more a system has lost its capacity for spontaneous change—the more this capacity has been exhausted—the greater is the entropy.

Thus, the second law of thermodynamics provides us with a measure of this exhaustion, the entropy change, ΔS, to be used as the fundamental criterion of spontaneity. For a closed region of space (for which, therefore, $\Delta E = 0$) including all changes under observation,

$$\Delta S \geqslant 0 \tag{7-120}$$

the equality sign applying to systems at equilibrium, the inequality to all systems capable of undergoing spontaneous changes.

Spontaneous transformations occur all around us all the time. Hence ΔS, for a section of space encompassing each such transformation and its affected surroundings, is a positive number. This realization led Clausius to his famous aphorism:

> Die Energie der Welt ist konstant; die Entropie der Welt
> strebt einem Maximum zu [8]

To a beginning student, this form of statement is frequently the source of more perplexity than enlightenment. The constancy of energy causes no difficulty of course. Since energy is conserved, it fits into the category of concepts to which we attribute permanence. In thought, we usually picture energy as a kind of material fluid, and hence, even if it flows from one place to another, its conservation may be visualized readily. However, when we carry over an analogous mental picture to the concept of entropy, we immediately are faced with the bewildering realization that entropy is "being created out of nothing" whenever there is an increase in entropy in an isolated system undergoing a spontaneous transformation.

The heart of the difficulty in "understanding" the concept of increase in entropy is a verbal one. It is difficult to dissociate the unconscious verbal implications of a word that we have used all of our lives in other contexts without critical analysis. In speaking of "increase in entropy," we are using language appropriate for the description of material bodies. Automatically, therefore, we associate with entropy other characteristics of material bodies that are at variance with the nature of entropy and hence that are a source of confusion.

Ultimately, we must realize that entropy is essentially a mathematical function. It is a concise function of the variables of experience, such as temperature, pressure, and composition. Natural processes tend to occur only in certain directions; that is, the variables pressure, temperature, and composition change only in certain—but very complicated—ways, which are most concisely described by the change in a single function, the entropy function ($\Delta S > 0$).

Some of the historical reluctance to assimilate the entropy concept into general scientific thinking, and much of the introductory student's bewilderment, might have been avoided if Clausius had defined entropy (as would have been perfectly legitimate to do) as

$$dS' = -\frac{DQ_{\text{rev}}}{T} \tag{7-121}$$

with a negative sign instead of the positive one of Equation 7-48. With this definition all of the thermodynamic consequences that have been derived from the entropy function would be just as valid except that some relations would change in sign. Thus, in place of Equation 7-120 we would find that for an isolated system,

$$\Delta S' \leqslant 0 \tag{7-122}$$

the equality sign applying to reversible changes in isolated systems, the inequality to irreversible changes in isolated systems. Now, however, we would recognize that for all isolated sections of space undergoing actual changes, ΔS is a negative number; that is, the entropy decreases. Paraphrasing Clausius, we would say, "Die Entropie der Welt strebt einem *Minimum* zu." This statement would accord more obviously with our experience that observable spontaneous changes go in the direction that *decreases* the capacity for further spontaneous change and that *decreases* the capacity to do work, and that the universe (or at least the solar system) changes in time toward a state in which (ultimately) no further spontaneous change will be possible. We need merely reiterate a few examples: Solutes always diffuse from a more concentrated solution to a dilute one; clocks tend to run down; magnets become self-demagnetized; heat always flows from a warm body to a colder one; gases always effuse into a vacuum; aqueous solutions of NaCl and $AgNO_3$ if mixed always form AgCl. Although some of these individual changes can be reversed by an outside agency, this outside agent must itself undergo a transformation that decreases its capacity for further spontaneous change. It is impossible to restore every system back to its original condition. On earth, our ultimate sources of energy for work are the sun, or nuclear power; in either case, these ultimate nuclear reactions proceed unidirectionally and toward the loss of capacity for further spontaneous change.

In some respects, especially pedagogical ones, it might have been better to change the sign of the original definition of the index so that it would measure residual capacity rather than loss of capacity. However, with the development of molecular-statistical energetics and the identification of entropy (in terms of kinetic-molecular theory) with the degree of disorder of a system, the original sign chosen by Clausius turns out to be the more convenient one. The universal tendency of all changes to reduce everything to a state of equilibrium may be correlated with the rearrangements of molecules from orderly to disorderly configurations. And since there are more disorderly arrangements than orderly ones, it is appropriate that the entropy index increase with the approach of all things to a state of equilibrium.

EXERCISES

1. One mole of an ideal monatomic gas ($c_v = \frac{3}{2}R$) at 101 kPa (1 atm) and 273.1 K is to be transformed to 55.5 kPa (0.5 atm) and 546.2 K. Consider the following four reversible paths, each consisting of two parts, A and B:
   1. Isothermal expansion and isobaric temperature rise
   2. Isothermal expansion and isochoric temperature rise
   3. Adiabatic expansion and isobaric temperature rise
   4. Adiabatic expansion and isochoric temperature rise

   a. Determine *P*, *V*, and *T* of the gas after the initial step of each of the four paths. Represent the paths on a *T*–*V* diagram. To facilitate plotting, some of the necessary data for the adiabatic expansions are given in Table 7-3. Supply the additional data required for the completion of the adiabatic curve.

Table 7-3

V/dm^3	P/kPa	T/K
22.41	101.33	273.1
	50.66	
44.82	31.92	172.0
67.23	16.24	131.3
89.65		

   b. For each portion of each path and for each complete path, calculate the following: *W*, the work done; *Q*, the heat absorbed by the gas; ΔE of the gas; ΔH of the gas; ΔS of the gas. Tabulate your results.

   c. Note which functions in (b) have values that are independent of the path used in the transformation.

   Note: *R* in joules should be used in all parts of the calculation. Check units carefully.

2. An ideal gas is carried through a Carnot cycle. Draw diagrams of this cycle using each of the following sets of coordinates:

   a. P, V
   b. T, P
   c. T, S
   d. E, S
   e. S, V
   f. T, H

3. a. By a procedure analogous to that used to obtain Equation 7-112 show that

$$\left(\frac{\partial S}{\partial V}\right)_T = \frac{P + (\partial E/\partial V)_T}{T} \tag{7-123}$$

   b. Starting with Equation 7-123 demonstrate the validity of Equation 7-113. Rearrange Equation 7-123 to

$$P = T\left(\frac{\partial S}{\partial V}\right)_T - \left(\frac{\partial E}{\partial V}\right)_T \tag{7-124}$$

   Differentiate with respect to temperature at constant volume to obtain

$$\left(\frac{\partial P}{\partial T}\right)_V = \left(\frac{\partial S}{\partial V}\right)_T + T\frac{\partial^2 S}{\partial V \partial T} - \frac{\partial^2 E}{\partial V \partial T} \tag{7-125}$$

   Show also that appropriate differentiation of Equation 7-112 leads to the relationship

$$\frac{\partial^2 S}{\partial V \partial T} = \frac{1}{T}\frac{\partial^2 E}{\partial V \partial T} \tag{7-126}$$

   and proceed to obtain

$$\left(\frac{\partial S}{\partial V}\right)_T = \left(\frac{\partial P}{\partial T}\right)_V \tag{7-113}$$

   c. Combining the results of parts (a) and (b), show that

$$\left(\frac{\partial E}{\partial V}\right)_T = T\left(\frac{\partial P}{\partial T}\right)_V - P \tag{7-127}$$

   d. Prove that $(\partial E/\partial V)_T = 0$ for any gas obeying a general gas law of the form

$$Pf(V) = RT$$

   in which $f(V)$ is any continuous function of volume.

e. Derive the expression

$$\left(\frac{\partial E}{\partial V}\right)_P = C_v\left(\frac{\partial T}{\partial V}\right)_P + T\left(\frac{\partial P}{\partial T}\right)_V - P \tag{7-128}$$

4. A gas obeys the equation of state

$$Pv = RT + BP$$

in which B is a constant at all temperatures.

a. Show that the internal energy, E, is a function of the temperature only.

b. Compute $(\partial E/\partial V)_P$. Compare with the value obtained for this same partial derivative for an ideal gas.

c. Derive an equation for the entropy of this gas that is analogous to Equation 7-109 for the ideal gas.

5. Show that the efficiency of a Carnot cycle in which any step is carried out irreversibly cannot be greater than that of a reversible Carnot cycle.

6. The heat quantities, Q_2 and Q_1, absorbed by an engine during a completed Carnot cycle (in which Q_2 refers to a higher temperature and Q_1 to a lower temperature) can be plotted against each other using the coordinates shown in the following figure. Any conceivable Carnot cycle for an engine then can be characterized by a point (Q_1, Q_2) on this plane. The figure then can be divided into eight octants. In which octant or octants would useful Carnot cycles fall?

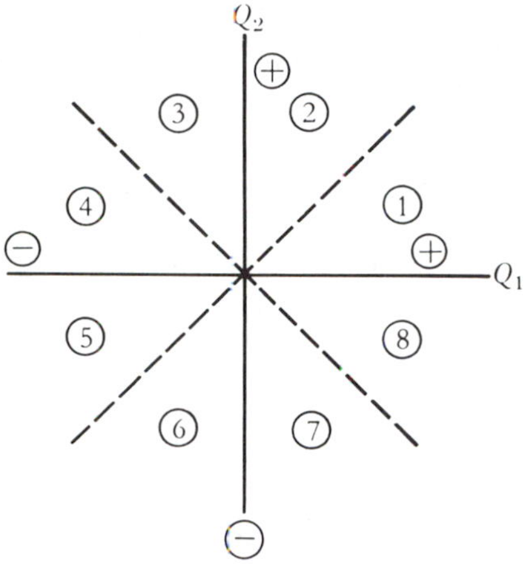

7. Gibbs [9] has suggested that the equation

$$E = V^{-R/C_v} \exp(S/C_v) \tag{7-129}$$

(C_v and R constants) be regarded as the fundamental thermodynamic equation of an ideal gas. With the aid of the two laws of thermodynamics, show that Equations 6-1 and 6-2 are contained implicitly in Equation 7-129.

8. A (reversible) Joule cycle consists of the following four steps: isobaric increase in volume, adiabatic expansion, isobaric decrease in volume, and adiabatic compression. Helium gas, with the equation of state

$$PV = RT + BP \tag{7-130}$$

(in which $B = 15\ \text{cm}^3\ \text{mol}^{-1}$), is carried through a Joule cycle. Draw diagrams of this cycle using each of the following sets of coordinates:

   a. P, V

   b. E, V

   c. T, S

9. A (reversible) Sargent cycle consists of the following four steps: isochoric increase in pressure, adiabatic expansion, isobaric decrease in volume, and adiabatic compression. A gas obeying Equation 7-130 is carried through a Sargent cycle. Draw diagrams of this cycle using each of the following sets of coordinates:

   a. P, V
   b. T, V
   c. E, V
   d. S, V
   e. S, T
   f. H, T

10. A reversible cycle also can be completed in three steps, such as: isothermal expansion (at T_2) from V_1 to V_2, cooling (at constant V_2) from T_2 to T_1, and adiabatic compression back to the initial state.

   a. Draw a diagram of this cycle using T and V as coordinates.

   b. A nonideal gas obeying Equation 7-130 is carried through this cycle. Compute ΔS for each step and show that $\oint dS = 0$ for the nonideal gas in this cycle. Assume that C_v for this gas is a constant. Some of its other characteristics in an adiabatic process have been worked out as Exercise 4, Chapter 6.

11. In Figure 7-11 a series of adiabatic reversible paths are drawn ($1 - 2$, $1' - 2'$, and so on), each one starting at the temperature T_2 and ending at the temperature T_1. The points $1, 1'$, and so on, are labeled in an order such that, for a process proceeding to the right along the isothermal T_2 (for example, $1' - 1$) heat is absorbed by the system.

    An essential step in the Caratheodory formulation of the second law of thermodynamics is a proof of the following statement: Two adiabatics (such as a and b in Figure 7-11) cannot intersect. Prove that a and b cannot intersect. (Suggestion: Assume a and b do intersect at the temperature T_1 and show that this assumption permits you to violate the Kelvin–Planck statement of the second law.)

12. For an isolated (adiabatic) system, $\Delta S > 0$ for any natural (spontaneous) process from State a to State b, as was proved in Section 7-8. An alternative and probably simpler proof of this proposition can be obtained if we use a temperature–entropy diagram (Figure 7-12) instead of Figure 7-8. In Figure 7-12, a reversible adiabatic process is represented as a vertical line since $\Delta S = 0$ for this process. In terms of Figure 7-12 we can state our proposition

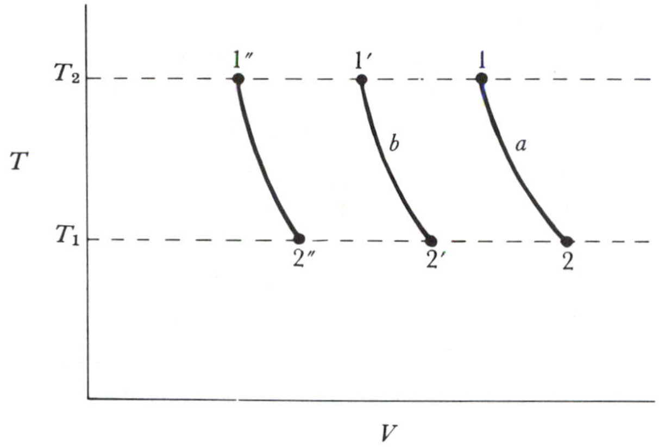

Figure 7-11.

as follows: For an isolated system a spontaneous process from a to b must lie to the right of the reversible one, since $\Delta S = S_b - S_a > 0$.

Prove that b cannot be to the left of b'; that is, that ΔS cannot be negative for the isolated spontaneous process. (Suggestion: Assume that b is to the left of b' and then complete a suitable cycle back to a that allows you to violate the Kelvin–Planck statement of the second law.)

13. A spring is placed in a large thermostat at 27°C and stretched isothermally and reversibly from its equilibrium length L_0 to $10L_0$. During this reversible stretching 1.00 J of heat is absorbed by the spring. The stretched spring, still in the constant-temperature thermostat, then is released without any restrain-

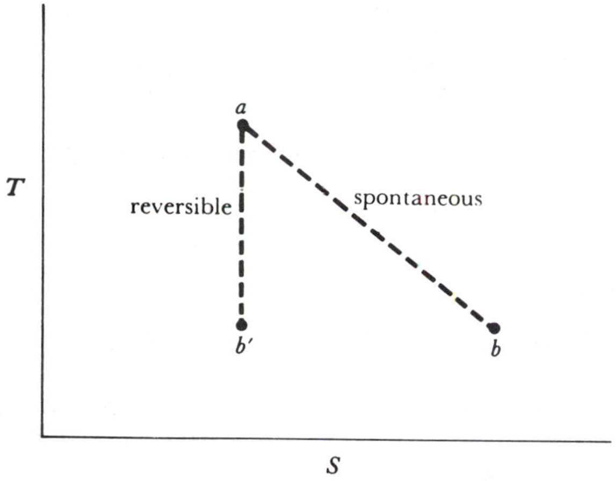

Figure 7-12.

ing back-tension and allowed to jump back to its initial length L_0. During this spontaneous process the spring evolves 2.50 J of heat.

a. What is the entropy change for the stretching of the spring?

b. What is the entropy change for the collapse of the spring?

c. What is the entropy change for the universe (spring plus surrounding thermostat) for the total process, stretching plus return collapse to initial L_0?

d. How much work was done on the spring in the stretching process?

14. It has been suggested that biological systems may constitute exceptions to the second law of thermodynamics, since they carry out irreversible processes that result in a decrease in the entropy of the biological system. Comment on this suggestion.

15. Hydrogen atoms at 25°C and 101 kPa pressure can spontaneously form H_2 gas:

$$2H = H_2$$

However, $\Delta S = -90.4 \text{ J mol}^{-1} \text{ K}^{-1}$ for this system. Since the change occurs despite the fact that ΔS is negative, this reaction apparently violates the second law of thermodynamics. How do you explain this anomaly in terms of the second law?

16. In carrying out a (reversible) Carnot cycle, we can place a two-phase system instead of an ideal gas into a cylinder. A suitable two-phase system is

$$H_2O(l) = H_2O(g)$$

In the first, isothermal, step with this two-phase system, one mole of liquid water is vaporized at 400 K with an absorption of 39.3 kJ. In the second, adiabatic, step the system is further expanded, with an accompanying drop in temperature to 300 K. At 300 K an isothermal compression is carried out, which is followed by an adiabatic compression to return the system to its starting point.

Assuming that this two-phase system obeys the first and second laws of thermodynamics, and given that the heat of vaporization of water at 300 K is 43.5 kJ mol^{-1}, how many grams of liquid water must condense out of the vapor in the isothermal compression step? Show your reasoning in your answer.

REFERENCES

1. E. Whittaker, *From Euclid to Eddington*, Dover, New York, 1958, pp. 58–60.
2. It is sometimes convenient to define a coefficient of performance for a refrigerator as $\beta = Q_1'/W'$, which is equal to $(1/\varepsilon) - 1$. Similarly, a coefficient of performance for a heat pump can be defined as $\gamma = -Q_2'/W'$, which is equal to $1/\varepsilon$.
3. K. Denbigh, *The Principles of Chemical Equilibrium*, 3d ed., Cambridge University Press, Cambridge, 1971, pp. 29–32.
4. H. Tetrode, *Ann. Physik.* [4] *38*, 434; *39*, 225 (1912).

5. O. Sackur, *Ann. Physik.* [4] *40*, 67 (1913).

6. *The Collected Works of J. Willard Gibbs*, Yale University Press, New Haven, 1928, p. 9; reprinted 1957.

7. The energy is "differentiated" in the separated bodies, "dedifferentiated" after thermal equilibration. The entropy is different in the differentiated state of this system than it is in the dedifferentiated state; the entropy is an index of extent of dedifferentiation. Corresponding statements can be phrased for two gases initially at different pressures, two solutions with different concentrations of Cu^{++} [Figure 7-10(*b,c*)]. In a molecular visualization, a system also has a larger entropy when it is "dedifferentiated." For example, let us place three layers of black balls at the bottom of a cubic box and carefully place three layers of white balls on top of the black layers. If the box is then buffeted around, in time the balls will achieve one of many possible random arrangements of black balls and white balls. At the outset, the system was very differentiated; at the end of the transformation, it reached a highly dedifferentiated state, one that Boltzmann associated with a larger entropy.

8. That is:

   The energy of the universe is constant; the entropy tends toward a maximum.

   Today, we would hesitate to comment on the entropy or energy of the universe, and would refer only to those surroundings that are observed to interact with the system, since we have no operation that can measure the energy or entropy of the universe as a whole. Cosmological theorists have suggested that the entropy constraint may change when the universe reaches 10^{20} years of age [see S. Frantschi, *Science,* **217**, 593 (1982)]. At present, the universe is thought to be 10^{10} years old.

9. *Collected Works of J. Willard Gibbs*, p. 13.

CHAPTER 8

Equilibrium and Spontaneity for Systems at Constant Temperature: The Free Energy Functions

8-1 REVERSIBILITY, SPONTANEITY, AND EQUILIBRIUM

The second law of thermodynamics was stated in Equation 7-105 as

$$\Delta S \geqslant 0$$

for an isolated system, in which the equality refers to a system undergoing a reversible change and the inequality refers to a system undergoing an irreversible change.

An irreversible change is always spontaneous in an isolated system because there is no external force that can interact with the system. Only at equilibrium can a change in an isolated system be conceived to occur reversibly. At equilibrium any infinitesimal fluctuations away from equilibrium are opposed by the natural tendency to return to equilibrium. Therefore the criterion of reversibility is a criterion of equilibrium, and the criterion of irreversibility is a criterion of spontaneity.

When a system is in thermal contact with a reservoir, allowing the exchange of heat, and when the system can exchange work with its surroundings, the system and the surroundings with which it interacts can together be considered an isolated system. Then Equation 7-105 can be expressed as

$$\Delta S_{\text{total}} = \Delta S_{\text{sys}} + \Delta S_{\text{surr}} \geqslant 0 \tag{8-1}$$

or, for an infinitesimal change,

$$dS_{\text{total}} = dS_{\text{sys}} + dS_{\text{surr}} \geqslant 0 \tag{8-2}$$

As we indicated in the discussion of an irreversible isothermal expansion of an ideal gas (Section 7-8), the heat transfer between the system and the reservoir can be considered reversible even if the change in the system is irreversible.

For example, the reservoir can be a two-phase system (for example, solid-liquid) in equilibrium, sufficiently large that its temperature does not change as a result of an exchange of heat with the system. Then, in view of the first law,

$$\frac{DQ_{\text{surr}}}{T} = -\frac{DQ_{\text{sys}}}{T} \tag{8-3}$$

Consequently,

$$dS_{\text{total}} = dS_{\text{sys}} + dS_{\text{surr}} = dS_{\text{sys}} - \frac{DQ_{\text{sys}}}{T} \geqslant 0 \tag{8-4}$$

or

$$dS_{\text{sys}} \geqslant \frac{DQ_{\text{sys}}}{T} \tag{8-5}$$

Equation 8-5 is of the same form as Equation 7-86, but now derived in a general way and not dependent on the properties of an ideal gas.

It will be convenient to use criteria of spontaneity and equilibrium, such as Equation 8-5, that omit explicit references to changes in the surroundings. Nevertheless, we should remember that such changes are included implicitly. Equation 8-5 will be the starting point from which to obtain criteria to decide whether a given system is at equilibrium or can be expected to change spontaneously, given sufficient time or an appropriate catalyst.

Systems at Constant Temperature and Volume

When the value of DQ obtained from the first law is substituted into Equation 8-5, the result is

$$dS \geqslant \frac{dE - DW}{T} \tag{8-6}$$

If the only restraint on the system is the pressure of the environment, then the only work is mechanical work against the external pressure P'. Therefore DW is equal to $-P'dV$ and Equation 8-6 becomes

$$dS \geqslant \frac{dE + P'dV}{T} \tag{8-7}$$

or

$$dE + P'dV - TdS \leqslant 0 \tag{8-8}$$

Since the volume is constant, $P'dV$ equals zero and can be omitted. Because the temperature is constant, $(-SdT)$ can be added to the left side of Equation 8-8 without changing its value. Thus

$$dE - TdS - SdT \leqslant 0$$

or

$$dE - (TdS + SdT) \leqslant 0 \tag{8-9}$$

The terms in parentheses in Equation 8-9 are equal to the differential of the function *TS* and Equation 8-9 can be written

$$dE - d(TS) \leqslant 0 \tag{8-10}$$

or

$$d(E - TS) \leqslant 0 \tag{8-11}$$

If the temperature and volume are constant, and if the only constraint on the system is the pressure of the environment, Equation 8-5 and Equations 8-6 through 8-11 provide criteria of equilibrium and spontaneity. The equality in Equation 8-6 applies to a reversible change and, since there is no exchange of work with the environment, the reversible change must be in a system at equilibrium. Similarly, the inequality in Equation 8-6 applies to an irreversible change and, in the absence of any constraint other than the pressure of the environment, this change must be spontaneous.

Since E, T, and S are state functions, the quantity $(E - TS)$ also must be a state function. This quantity is sufficiently important that it is given the name *Helmholtz free energy*, or *Helmholtz function*, defined as

$$A = E - TS \tag{8-12}$$

Thus we can state concisely that in a system at constant temperature and volume

$$dA < 0 \tag{8-13}$$

for a spontaneous change and

$$dA = 0 \tag{8-14}$$

for an infinitesimal change at equilibrium.

Systems at Constant Temperature and Pressure

The most familiar transformations occur under conditions of constant temperature and pressure, so it will be particularly useful to have a criterion of spontaneity and equilibrium that applies to these conditions.

We can start with Equation 8-8, in which the only restriction is that no work is done except mechanical work against an external pressure, P'. If the change is carried out at constant pressure, the pressure P of the system must equal P', so Equation 8-8 can be written

$$dE + PdV - TdS \leqslant 0 \tag{8-15}$$

If the pressure and temperature are constant, dP and dT are zero, so we can add $-SdT$ and VdP to the left side of Equation 8-15 without changing its value:

$$dE + PdV + VdP - TdS - SdT \leqslant 0$$

or

$$d(E + PV - TS) \leqslant 0 \tag{8-16}$$

The function in parentheses in Equation 8-16 is a state function and is called the *Gibbs free energy*, or *Gibbs function*, symbolized by G [1]. Relationships for G are

$$G = E + PV - TS \tag{8-17}$$

$$= H - TS \tag{8-18}$$

and

$$dG \leqslant 0 \text{ (const } T, P) \tag{8-19}$$

If the temperature and pressure are constant, and if the only constraint on the system is the pressure of the environment, Equation 8-19 provides criteria of equilibrium and spontaneity. The equality in Equation 8-19 applies to a reversible change; since the only restraint is the *constant* pressure of the environment, equal to the pressure of the system, the reversible change must be taking place in a system at equilibrium. Similarly, the inequality in Equation 8-19 refers to an irreversible change; in the absence of any constraint other than the constant pressure of the environment, equal to the pressure of the system, this change must be spontaneous.

As we mentioned previously, changes in the environment are included implicitly in Equations 8-13, 8-14, and 8-19, even though they are not mentioned explicitly. For example, from Equation 8-19 for a system undergoing change at constant temperature we can write

$$dH_{sys} - TdS_{sys} \leqslant 0 \tag{8-20}$$

Since the pressure is constant and only mechanical work is done (Equation 4-5),

$$dH_{sys} = DQ_{sys}$$

can be used to give

$$DQ_{sys} - TdS_{sys} \leqslant 0 \tag{8-21}$$

From Equation 7-67,

$$DQ_{sys} = -DQ_{surr}$$

and since the entropy change in the surroundings can be assumed to be equal to DQ_{surr}/T we obtain

$$-TdS_{surr} - TdS_{sys} \leqslant 0$$

or

$$TdS_{surr} + TdS_{sys} \geqslant 0 \tag{8-22}$$

Equation 8-22 is a restatement of the criterion for an isolated system. The choice of dG as a criterion included an implicit consideration of changes in the environment, although only functions of the state of the system were used.

Planck [2] used a function Y, which he defined as

$$Y = S - \frac{E + PV}{T}$$

$$= S - \frac{H}{T} \tag{8-23}$$

as a criterion of equilibrium and spontaneity at constant temperature and pressure. From Equations 4-5 and 7-67, we can see that

$$Y = S_{sys} + S_{surr} \tag{8-24}$$

and from Equation 8-18 we can see that

$$Y = -\frac{G}{T} \tag{8-25}$$

It can be seen from Equation 8-23 that S is the dominant term at high temperatures, whereas H/T is the dominant term at low temperatures.

Heat of Reaction as an Approximate Criterion of Spontaneity

For many years it was thought, purely on an empirical basis, that if the enthalpy change for a given reaction were negative, that is, if heat were evolved at constant pressure, the transformation could occur spontaneously. This rule was verified for many reactions. Nevertheless, there are numerous exceptions, such as the polymorphic transformation of α quartz to β quartz at 848 K and atmospheric pressure:

$$SiO_2(\alpha \text{ quartz}) = SiO_2(\beta \text{ quartz}),$$

$$\Delta H = 0.63 \text{ kJ mol}^{-1}$$

which is spontaneous even though ΔH is positive.

Since, at a fixed temperature, Equation 8-18 yields the differential expression

$$dG = dH - TdS \tag{8-26}$$

or, for a macroscopic change,

$$\Delta G = \Delta H - T\Delta S \tag{8-27}$$

ΔH and ΔG will be nearly equal if $T\Delta S$ is small compared with ΔH. Usually, $T\Delta S$ is of the order of magnitude of a few thousand joules. If ΔH is sufficiently large, perhaps above 40 kJ, the sign of ΔH will be the same as that of ΔG. For such relatively large values of ΔH, the heat of reaction may be a reliable criterion of spontaneity, since if ΔH were negative ΔG probably would be negative also. However, ΔH is not the fundamental criterion, and judgments based on its sign frequently are misleading, particularly when the magnitudes involved are small.

8-2 PROPERTIES OF THE FREE ENERGY FUNCTIONS

Free Energy, a Thermodynamic Property

Since G and A are defined by explicit equations in which the variables are functions of the state of the system, both of these functions are thermodynamic properties and their differentials are exact. Thus we can write

$$\oint dG = 0$$

$$\oint dA = 0$$

Relationship between G and A

From the definitions for G and A (Equations 8-12 and 8-17) we can write

$$G = H - TS = E + PV - TS = (E - TS) + PV$$

The relationship between G and A is then

$$G = A + PV \tag{8-28}$$

Free Energy Changes for Isothermal Conditions

Transformations at constant temperature are of frequent interest. For finite changes at a fixed temperature T

$$\begin{aligned}\Delta G = G_2 - G_1 &= (H_2 - TS_2) - (H_1 - TS_1)\\ &= H_2 - H_1 - (TS_2 - TS_1)\\ &= H_2 - H_1 - T(S_2 - S_1)\\ &= \Delta H - T\Delta S\end{aligned}$$

For an infinitesimal change for which $dT = 0$

$$dG = dH - TdS - SdT$$
$$= dH - TdS \quad (8\text{-}29)$$

The equations for ΔA and dA can be derived similarly to obtain

$$\Delta A = \Delta E - T\Delta S \quad (8\text{-}30)$$

and

$$dA = dE - TdS \quad (8\text{-}31)$$

Equations for Total Differentials

Since the procedure is the same for both free energy functions, we shall consider in detail only the derivation for the Gibbs function, G. After Equation 8-15 we obtained the differential of the function, later to be defined as G, as

$$dG = dE + PdV + VdP - TdS - SdT \quad (8\text{-}32)$$

Substituting from the first law expression for dE we have

$$dG = DQ + DW + PdV + VdP - TdS - SdT \quad (8\text{-}33)$$

If the change is carried out reversibly and the only work done is PdV work, then from Equation 7-48,

$$DQ = TdS$$

and from Equations 6-38 and 6-44,

$$DW = -PdV$$

Substituting from Equations 7-48 and 6-44 into Equation 8-33 and canceling terms we obtain

$$dG = VdP - SdT \quad (8\text{-}34)$$

By an analogous procedure, it can be shown that the total differential of the Helmholtz function is given by the expression

$$dA = -PdV - SdT \quad (8\text{-}35)$$

Although the condition of reversibility was used, for convenience, in deriving Equations 8-34 and 8-35, the result applies also to irreversible changes, since G and A are state functions. The limitation to PdV work, however, applies to the final equations. We shall consider later circumstances in which other than PdV work is done.

Pressure and Temperature Coefficients of the Free Energy

Since dG is a criterion for equilibrium and spontaneity at constant T and P, it is useful to express the total differential of G as a function of T and P. That is,

$$dG = \left(\frac{\partial G}{\partial T}\right)_P dT + \left(\frac{\partial G}{\partial P}\right)_T dP \qquad (8\text{-}36)$$

If we compare the coefficients of dP and dT in Equations 8-36 and 8-34, it is clear that

$$\left(\frac{\partial G}{\partial P}\right)_T = V \qquad (8\text{-}37)$$

and

$$\left(\frac{\partial G}{\partial T}\right)_P = -S \qquad (8\text{-}38)$$

In addition to determining the variation of G with pressure for a substance, it is useful to determine the change of ΔG for a transformation with pressure. To do so let us represent a transformation by the equation

$$\mathrm{A + B = C + D} \qquad (8\text{-}39)$$

and use Equation 8-37 to write

$$\left.\begin{aligned} \left(\frac{\partial G_{\mathrm{A}}}{\partial P}\right)_T &= V_{\mathrm{A}} \\ \left(\frac{\partial G_{\mathrm{B}}}{\partial P}\right)_T &= V_{\mathrm{B}} \\ \left(\frac{\partial G_{\mathrm{C}}}{\partial P}\right)_T &= V_{\mathrm{C}} \\ \left(\frac{\partial G_{\mathrm{D}}}{\partial P}\right)_T &= V_{\mathrm{D}} \end{aligned}\right\} \qquad (8\text{-}40)$$

Subtracting the sum of the pressure coefficients for the reactants from that for the products we obtain the desired relationship:

$$\left(\frac{\partial G_{\mathrm{C}}}{\partial P}\right)_T + \left(\frac{\partial G_{\mathrm{D}}}{\partial P}\right)_T - \left(\frac{\partial G_{\mathrm{A}}}{\partial P}\right)_T - \left(\frac{\partial G_{\mathrm{B}}}{\partial P}\right)_T = V_{\mathrm{C}} + V_{\mathrm{D}} - V_{\mathrm{A}} - V_{\mathrm{B}} \qquad (8\text{-}41)$$

or

$$\left(\frac{\partial \Delta G}{\partial P}\right)_T = \Delta V \tag{8-42}$$

Similarly it can be shown that

$$\left(\frac{\partial \Delta G}{\partial T}\right)_P = -\Delta S \tag{8-43}$$

If the same operations are carried out using the Helmholtz function, the results are

$$\left(\frac{\partial A}{\partial V}\right)_T = -P \tag{8-44}$$

$$\left(\frac{\partial \Delta A}{\partial V}\right)_T = -\Delta P \tag{8-45}$$

$$\left(\frac{\partial A}{\partial T}\right)_V = -S \tag{8-46}$$

$$\left(\frac{\partial \Delta A}{\partial T}\right)_V = -\Delta S \tag{8-47}$$

Equations Derived from the Reciprocity Relationship

Since the Gibbs free energy, $G(T, P)$, is a thermodynamic property, the reciprocity relationship (Equation 2-24) applies. Thus we may write

$$\frac{\partial}{\partial T}\frac{\partial G}{\partial P} = \frac{\partial}{\partial P}\frac{\partial G}{\partial T} \tag{8-48}$$

According to Equations 8-37 and 8-38

$$\left(\frac{\partial G}{\partial P}\right)_T = V$$

and

$$\left(\frac{\partial G}{\partial T}\right)_P = -S$$

Hence

$$\frac{\partial}{\partial T}\frac{\partial G}{\partial P} = \left(\frac{\partial V}{\partial T}\right)_P = \frac{\partial}{\partial P}\frac{\partial G}{\partial T} = -\left(\frac{\partial S}{\partial P}\right)_T$$

or

$$\left(\frac{\partial S}{\partial P}\right)_T = -\left(\frac{\partial V}{\partial T}\right)_P \tag{8-49}$$

By a similar set of operations on the A function we can show that

$$\left(\frac{\partial S}{\partial V}\right)_T = \left(\frac{\partial P}{\partial T}\right)_V \tag{8-50}$$

8-3 FREE ENERGY AND THE EQUILIBRIUM CONSTANT

In the preceding sections we established the properties of the free energy functions as criteria for equilibrium and spontaneity of transformations. Thus, from the sign of ΔG it is possible to predict whether a given chemical transformation can proceed spontaneously. Further considerations show that ΔG can given even more information. In particular, from the value of the free energy change under certain standard conditions, it is possible to calculate the equilibrium constant for a given reaction.

Definitions

Like E and H, G can only be determined relative to some reference state or standard state. The standard states that have been agreed upon are given in Table 8-1. The most stable form is the form of lowest free energy.

Another important concept is that of the standard free energy of formation, $\Delta Gf°$, of a substance (Table 8-2). By this we shall mean the change in free energy that accompanies the formation of one mole of a substance in its standard state from its elements in their standard states, all of the substances being at the specified temperature. For example, the standard free energy of formation of CO_2 given in Table 8-2 refers to the reaction

Table 8-1 Standard States for Free Energies of Reaction

Standard state of solid	Pure solid in most stable form at 1 bar pressure (100 kPa) and the specified temperature
Standard state of liquid	Pure liquid in most stable form at 1 bar pressure (100 kPa) and the specified temperature
Standard state of gas	Pure gas at unit fugacity;[a] for ideal gas, fugacity is unity when pressure is 1 bar (100 kPa); at specified temperature

[a]The term *fugacity* has yet to be defined. Nevertheless, it is used in this table because reference is made to it in future problems. For now, the standard state of a gas may be considered to be that of an ideal gas—1 bar pressure.

$$\text{C(graphite, } P^\circ) + \text{O}_2(\text{g}, P^\circ) = \text{CO}_2(\text{g}, P^\circ),$$

$$\Delta G = \Delta Gf^\circ = -394.359 \text{ kJ mol}^{-1} \quad (8\text{-}51)$$

It is a consequence of this definition that ΔGf° for any element is zero.

The standard free energy change, ΔG°, for any reaction is also an important quantity. ΔG° is the change in free energy that accompanies the conversion of reactants in their standard states to products in their standard states. Since the free energy is a thermodynamic property and does not depend on the path used to carry out a transformation, simple addition can be used to obtain

$$\Delta G^\circ = \Sigma\Delta Gf^\circ \text{ (products)} - \Sigma\Delta Gf^\circ \text{ (reactants)} \quad (8\text{-}52)$$

Relationship between ΔG° and the Equilibrium Constant for Gaseous Reactions

From the definition of ΔG°, we can derive the expression relating the standard free energy of a reaction to its equilibrium constant. For now we shall treat the systems we discuss as if each reactant and product were a pure phase and as if gases were ideal, because we are not yet able to consider the free energy functions for mixtures or real gases.

For an ideal gas we can derive an equation for the change in free energy in an isothermal expansion by recognizing that Equation 8-34 reduces at a fixed temperature to

$$dG = VdP \quad (8\text{-}53)$$

or, for an ideal gas,

$$dG = \frac{nRT}{P}\,dP \quad (8\text{-}54)$$

For a macroscopic change in pressure ΔG can be obtained by integrating Equation 8-54:

Table 8-2 Standard Gibbs Free Energies of Formation[a] at 298.15 K

Substance	$\Delta_G f°/(\text{kJ mol}^{-1})$	Substance	$\Delta_G f°/(\text{kJ mol}^{-1})$
H(g)	203.247	Propylene(g)	62.718
O(g)	231.731	1-Butene(g)	71.50
Cl(g)	105.680	Acetylene(g)	209.200
Br(g)	82.396	Benzene(g)	129.658
Br_2(g)	3.110	Toluene(g)	122.290
I(g)	70.250	*o*-Xylene(g)	121.826
I_2(g)	19.327	*m*-Xylene(g)	118.846
H_2O(g)	−228.512	*p*-Xylene(g)	121.135
H_2O(l)	−237.129	Methanol(l)	−166.73
HF(g)	−273.2	Ethanol(l)	−174.26
HCl(g)	−95.299	Glycine(s)	−368.44
HBr(g)	−53.45	Acetic acid(l)	−389.9
HI(g)	1.70	Taurine(s)	−561.7
ICl(g)	−5.46	Urea(s)	−197.33
NO(g)	86.56	α-D-glucose(s)	−910.27
CO(g)	−137.168	Succinic acid(s)	−747.60
CO_2(g)	−394.359	SiO_2(s); α quartz	−856.29
NH_3(g)	−16.45	SiO_2(s); α cristobalite	−854.51
Methane(g)	−50.794	SiO_2(s); α tridymite	−853.81
Ethane(g)	−32.886	$CaSO_4$(s); anhydrite	−1321.70
Propane(g)	−23.489	$CaSO_4 \cdot 2H_2O$(s); gypsum	−1797.20
n-Butane(g)	−17.15	Fe_2SiO_4(s); fayalite	−1379.38
Ethylene(g)	68.124	Mg_2SiO_4(s); forsterite	−2051.33

[a]Selected from "The NBS Tables of Chemical Thermodynamic Properties," *J. Phys. Chem. Reference Data,* **11**, Supplement No. 2, 1982; TRC Thermodynamic Tables–Hydrocarbons, Thermodynamics Research Center: The Texas A & M University System, College Station, Texas, 1971 (Loose-leaf sheets, continuously updated); "Thermodynamic Properties of Minerals and Related Substances," *Geological Survey Bulletin* **1452**, 1978.

$$\Delta G = G_2 - G_1 = \int_1^2 dG = \int_{P_1}^{P_2} \frac{nRT}{P}\, dP \qquad (8\text{-}55)$$

or

$$\Delta G = nRT \ln \frac{P_2}{P_1} \qquad (8\text{-}56)$$

Let us represent a chemical transformation by the equation (in which lowercase letters are the stoichiometric coefficients):

$$a\mathrm{A}(\mathrm{g}, P_\mathrm{A}) + b\mathrm{B}(\mathrm{g}, P_\mathrm{B}) = c\mathrm{C}(\mathrm{g}, P_\mathrm{C}) + d\mathrm{D}(\mathrm{g}, P_\mathrm{D}) \tag{8-57}$$

Each substance is assumed to be a pure ideal gas at a given partial pressure. This reaction is accompanied by a free energy change, ΔG. We can calculate ΔG° for Reaction 8-57 by adding to the ΔG for the reaction the values of ΔG that accompany the change for each reactant and product from the given partial pressure to P°, equal to 100 kPa. Thus we add the following:

$$a\mathrm{A}(P_\mathrm{A}) + b\mathrm{B}(P_\mathrm{B}) = c\mathrm{C}(P_\mathrm{C}) + d\mathrm{D}(P_\mathrm{D}), \qquad \Delta G$$

$$a\mathrm{A}(P'_\mathrm{A} = P^\circ) = a\mathrm{A}(P_\mathrm{A}), \qquad \Delta G_\mathrm{A} = aRT \ln\left(\frac{P_\mathrm{A}}{P^\circ}\right) \tag{8-58}$$

$$b\mathrm{B}(P'_\mathrm{B} = P^\circ) = b\mathrm{B}(P_\mathrm{B}), \qquad \Delta G_\mathrm{B} = bRT \ln\left(\frac{P_\mathrm{B}}{P^\circ}\right) \tag{8-59}$$

$$c\mathrm{C}(P_\mathrm{C}) = c\mathrm{C}(P'_\mathrm{C} = P^\circ), \qquad \Delta G_\mathrm{C} = cRT \ln\left(\frac{P^\circ}{P_\mathrm{C}}\right) \tag{8-60}$$

$$d\mathrm{D}(P_\mathrm{D}) = d\mathrm{D}(P'_\mathrm{D} = P^\circ), \qquad \Delta G_\mathrm{D} = dRT \ln\left(\frac{P^\circ}{P_\mathrm{D}}\right) \tag{8-61}$$

$$a\mathrm{A}(P'_\mathrm{A} = P^\circ) + b\mathrm{B}(P'_\mathrm{B} = P^\circ) = c\mathrm{C}(P'_\mathrm{C} = P^\circ) + d\mathrm{D}(P'_\mathrm{D} = P^\circ),$$

$$\begin{aligned}\Delta G^\circ &= \Delta G + \Delta G_\mathrm{A} + \Delta G_\mathrm{B} + \Delta G_\mathrm{C} + \Delta G_\mathrm{D} \\ &= \Delta G + aRT \ln\left(\frac{P_\mathrm{A}}{P^\circ}\right) + bRT \ln\left(\frac{P_\mathrm{B}}{P^\circ}\right) + cRT \ln\left(\frac{P^\circ}{P_\mathrm{C}}\right) + dRT \ln\left(\frac{P^\circ}{P_\mathrm{D}}\right)\end{aligned} \tag{8-62}$$

Equation 8-62 can be simplified by combining logarithmic terms to yield

$$\Delta G^\circ = \Delta G + RT \ln \frac{(P_\mathrm{A}/P^\circ)^a\,(P_\mathrm{B}/P^\circ)^b}{(P_\mathrm{C}/P^\circ)^c\,(P_\mathrm{D}/P^\circ)^d}$$

Solving for ΔG and inverting the ratio, we obtain

$$\Delta G = \Delta G^\circ + RT \ln \frac{(P_\mathrm{C}/P^\circ)^c\,(P_\mathrm{D}/P^\circ)^d}{(P_\mathrm{A}/P^\circ)^a\,(P_\mathrm{B}/P^\circ)^b} \tag{8-63}$$

The transformation is isobaric (that is, P_A, P_B, P_C, and P_D are fixed) and is isothermal. Then, if the pressures in Equation 8-63 represent equilibrium pressures,

$$\Delta G = 0 \tag{8-64}$$

and

$$\Delta G^\circ = -RT \ln \left[\frac{(P_C/P^\circ)^c \, (P_D/P^\circ)^d}{(P_A/P^\circ)^a \, (P_B/P^\circ)^b} \right]_{\text{equil}} \tag{8-65}$$

Since ΔG° is the change in free energy under standard conditions, it must have a fixed value at a given temperature. Because R and T are constants as well, the ratio of pressures in the brackets is also a constant. This constant is the thermodynamic equilibrium constant and is symbolized by K. That is,

$$K = \left[\frac{(P_C/P^\circ)^c \, (P_D/P^\circ)^d}{(P_A/P^\circ)^a \, (P_B/P^\circ)^b} \right]_{\text{equil}} \tag{8-66}$$

Equation 8-65 becomes

$$\Delta G^\circ = -RT \ln K \tag{8-67}$$

The numerical value of K can be evaluated if ΔG° for the reaction is known, or vice versa. The choice depends on which quantity is easier to obtain for a given reaction. Although K refers to equilibrium pressures, it is calculated from data for ΔG° that refer to the reaction occurring with reactants and products in their standard states.

If we use the symbol $\mathcal{Q}$ for the quotient of pressures on the right side of Equation 8-63, then we can write

$$\begin{aligned} \Delta G &= \Delta G^\circ + RT \ln \mathcal{Q} \\ &= -RT \ln K + RT \ln \mathcal{Q} \\ &= -RT \ln (\mathcal{Q}/K) \end{aligned}$$

Thus, if the initial quotient of pressures is greater than K, ΔG is positive and the reaction will be spontaneous to the left, whereas if the initial quotient of pressures is less than K, ΔG is negative and the reaction will be spontaneous to the right.

The form of the equilibrium constant in Equation 8-66 is different from that used conventionally. It has the advantages that: (1) It is explicit that K is a dimensionless quantity; (2) it is explicit that the numerical value of K depends on the choice of standard state, but not on the units used to describe the standard state pressure; the equilibrium constant has the same value whether P° is expressed as 750.062 Torr, 0.98692 atm, 0.1 MPa, or 1 bar.

Dependence of K on Temperature

From the value of ΔG° at a single temperature it is possible to calculate the equilibrium constant, K. It is also desirable to be able to calculate K as a function of the temperature, so that it is not necessary to have extensive tables of ΔG° values at frequent temperature intervals. The derivation of the necessary functional relationship requires a direct relationship between ΔG and ΔH.

Direct Relationship between ΔG and ΔH. Starting from the definition of G (Equation 8-18) we have

$$G = H - TS \tag{8-68}$$

By dividing through by T we can obtain

$$\frac{G}{T} = \frac{H}{T} - S \tag{8-69}$$

Equation 8-69 can be differentiated with respect to temperature at constant pressure to give

$$\left[\frac{\partial(G/T)}{\partial T}\right]_P = \left[\frac{\partial(H/T)}{\partial T}\right]_P - \left(\frac{\partial S}{\partial T}\right)_P = H\left[\frac{\partial(1/T)}{\partial T}\right]_P + \frac{1}{T}\left(\frac{\partial H}{\partial T}\right)_P - \left(\frac{\partial S}{\partial T}\right)_P \tag{8-70}$$

The first two terms on the right in Equation 8-70 can be evaluated to give the following expression:

$$H\left[\frac{\partial(1/T)}{\partial T}\right]_P = -\frac{H}{T^2} \tag{8-71}$$

and

$$\frac{1}{T}\left(\frac{\partial H}{\partial T}\right)_P = \frac{C_P}{T} \tag{8-72}$$

From the definition of entropy (Equation 7-48),

$$dS = \frac{DQ_{\text{rev}}}{T}$$

and from the definition of heat capacity (Equation 4-17),

$$DQ_p = C_p dT$$

we obtain

$$dS_p = \frac{C_p dT}{T} \tag{8-73}$$

and

$$\left(\frac{\partial S}{\partial T}\right)_P = \frac{C_p}{T} \tag{8-74}$$

Substitution of Equations 8-71, 8-72, and 8-74 into Equation 8-70 leads to

$$\left[\frac{\partial(G/T)}{\partial T}\right]_P = -\frac{H}{T^2} \tag{8-75}$$

By a procedure analogous to that used to derive Equation 8-42 we can use Equation 8-75 to obtain

$$\left[\frac{\partial(\Delta G/T)}{\partial T}\right]_P = -\frac{\Delta H}{T^2} \tag{8-76}$$

When Equation 8-76 is applied to the temperature dependence of ΔG°, where ΔG° applies to an isothermal transformation, the ΔH° that is used is the enthalpy change at zero pressure for gases and at infinite dilution for substances in solution (see Chapter 17).

ΔG As a Function of Temperature. We have seen in Chapter 5 that a general expression for ΔH as a function of temperature can be written in the form (Equation 5-49)

$$\Delta H = \Delta H_0 + \int \Delta C_p \, dT$$

If the heat capacities of the substances involved in the transformation can be expressed in the form of a simple power series (Equation 4-35)

$$C_p = a + bT + cT^2 + \cdots$$

in which *a, b*, and *c* are constants, then Equation 5-49 becomes

$$\Delta H = \Delta H_0 + \Delta a T + \frac{\Delta b}{2} T^2 + \frac{\Delta c}{3} T^3 + \cdots \tag{8-77}$$

in which the Δ's refer to the sums of the coefficients for the products minus the sums of the coefficients for the reactants. Equation 8-77 can be inserted into Equation 8-76, which then can be integrated at constant pressure. If terms higher than T^3 are neglected, the result is

$$\int d\left(\frac{\Delta G}{T}\right) = -\int \frac{\Delta H}{T^2} \, dT = -\int \left(\frac{\Delta a}{T} + \frac{\Delta b}{2} + \frac{\Delta c}{3} T + \frac{\Delta H_0}{T^2}\right) dT \tag{8-78}$$

If the constant of integration is I, the result of the integration can be written

$$\frac{\Delta G}{T} = I - \Delta a \ln T - \frac{\Delta b}{2} T - \frac{\Delta c}{6} T^2 + \frac{\Delta H_0}{T} \tag{8-79}$$

Solving for ΔG we have

$$\Delta G = \Delta H_0 - \Delta a T \ln T + IT - \frac{\Delta b}{2} T^2 - \frac{\Delta c}{6} T^3 \tag{8-80}$$

The constant ΔH_0 can be evaluated as described in Chapter 5 if one value of the heat of reaction is known. Similarly, the constant I can be determined if ΔH_0 and one value of ΔG are known.

The use of Equation 8-80 can be illustrated by an example. Consider the reaction

$$\text{C(graphite)} + O_2(g) = CO_2(g)$$

From the heat capacity equations and $\Delta Hf°$ we showed in Chapter 5 (Equation 5-57) that

$$\Delta H = -393{,}571 + 1.126T - 3.946 \times 10^{-3}T^2 + 21.94 \times 10^{-7}T^3 + 2.420 \times 10^{-9}T^4$$

Substituting into Equation 8-78 and integrating we obtain the following expression for $\Delta G/T$:

$$\frac{\Delta G}{T} = I - 1.126 \ln T + 1.973 \times 10^{-3}T - 3.657 \times 10^{-7}T^2 - \frac{393{,}571}{T} - 0.3025 \times 10^{-9}T^3 \tag{8-81}$$

At 298.15 K the standard free energy of formation of CO_2, $\Delta G°$, is -394.359 J mol^{-1} (Table 8-2). Using this value for ΔG in Equation 8-81 we can evaluate the integration constant, I:

$$I = 3.225 \text{ J mol}^{-1} \text{ K}^{-1}$$

Thus, the explicit equation for the standard free energy change for the formation of CO_2 is

$$\Delta G° = 3.225T - 1.126T \ln T + 1.973 \times 10^{-3}T^2 - 3.657 \times 10^{-7}T^3 - 393{,}571 - 0.3025 \times 10^{-9}T^4 \tag{8-82}$$

K *As a Function of Temperature.* The equilibrium constant can be related to the temperature through either of two thermodynamic functions.

The differential relationship. Rearranging Equation 8-67 we obtain

$$\frac{\Delta G°}{T} = -R \ln K \tag{8-83}$$

Differentiation of Equation 8-83 and substitution into Equation 8-76 yields

$$\left[\frac{\partial(\Delta G^\circ/T)}{\partial T}\right]_P = -R\left(\frac{\partial \ln K}{\partial T}\right)_P = -\frac{\Delta H^\circ}{T^2} \tag{8-84}$$

From this equation we can obtain

$$\frac{d \ln K}{dT} = \frac{\Delta H^\circ}{RT^2} \tag{8-85}$$

Since ΔG° does not depend on pressure, K does not depend on pressure. Hence we need not use partial derivative notation in Equation 8-85.

The integral relationship. If the heat capacities fit the power series specified in Equation 4-35, ΔG can be expressed by Equation 8-80. Using Equation 8-67 we obtain

$$\ln K = -\frac{\Delta H_0}{RT} + \frac{\Delta a}{R}\ln T - \frac{I}{R} + \frac{\Delta b}{2R}T + \frac{\Delta c}{6R}T^2 \tag{8-86}$$

In the general case, when a function for the heat capacities is unavailable, it is necessary to integrate Equation 8-85 graphically or numerically:

$$\int_{\ln K_1}^{\ln K_2} d\ln K = \int_{T_1}^{T_2} \frac{\Delta H^\circ}{RT^2}\, dT \tag{8-87}$$

Pressure and Temperature Dependence of ΔG

Particularly in geological problems we are interested in reactions that take place in a wide range of pressures and temperatures. Therefore, we are interested in the pressure and temperature dependence of ΔG. According to Equations 8-42 and 8-43:

$$\left(\frac{\partial \Delta G}{\partial P}\right)_T = \Delta V \qquad \text{and} \qquad \left(\frac{\partial \Delta G}{\partial T}\right)_P = -\Delta S$$

Thus the total differential for ΔG can be written

$$d(\Delta G) = \left(\frac{\partial \Delta G}{\partial T}\right)_P dT + \left(\frac{\partial \Delta G}{\partial P}\right)_T dP = -(\Delta S)dT + (\Delta V)dP \tag{8-88}$$

If Equation 8-88 is integrated from a reference temperature and pressure of 298 K and $P = P^\circ$ to any temperature T' and pressure P', we obtain

$$\int_{\substack{P=P^\circ\\T=298}}^{P',T'} d(\Delta G) = \int_{\substack{P=P^\circ\\T=298}}^{\substack{P=P^\circ\\T'}} (-\Delta S)dT + \int_{\substack{P=P^\circ\\T'}}^{P',T'} (\Delta V)dP \tag{8-89}$$

From Equation 8-89 we can see that the first integral on the right side is integrated with respect to temperature at constant pressure P° and that the second integral on the right side is integrated with respect to pressure at constant temperature T'. Thus the temperature and pressure dependence can be dealt with independently.

Temperature Dependence. A simple expression for the temperature dependence is found if we remember that (Equation 8-27)

$$\Delta G(P = P^\circ, T') = \Delta H(P = P^\circ, T') - T'\Delta S(P = P^\circ, T')$$

and that from Equation 4-19

$$\left(\frac{\partial \Delta H}{\partial T}\right)_P = \Delta C_P \tag{8-90}$$

and from Equation 8-74

$$\left(\frac{\partial \Delta S}{\partial T}\right)_P = \frac{\Delta C_P}{T} \tag{8-91}$$

In view of Equation 8-90 and Equation 8-91 we can write Equation 8-27 as

$$\Delta G(P = P^\circ, T') = \Delta H(P = P^\circ, 298\text{ K}) + \int_{T=298}^{T'} (\Delta C_p)dT$$

$$-T'\Delta S(P = P^\circ, 298\text{ K}) - T'\int_{T=298}^{T'} \frac{\Delta C_p}{T}\,dT \tag{8-92}$$

in which the integrations with respect to temperature are carried out at $P = P^\circ$.

Pressure Dependence. Since (Equation 8-42)

$$\left(\frac{\partial \Delta G}{\partial P}\right)_T = \Delta V_T$$

in which the subscript T indicates that ΔV is a function of T, we can write

$$\Delta G(P', T') = \Delta G(P = P°, T') + \int_{P=P°}^{P'} (\Delta V)_{T'}\, dP \qquad (8\text{-}93)$$

in which integration with respect to P is carried out at temperature T' and $(\Delta V)_{T'}$ must be known at every temperature T' at which the integration is carried out.

General Expression. If we add Equation 8-92 and Equation 8-93 we obtain

$$\Delta G(P', T') = \Delta H(P = P°, 298\text{ K}) - T'\Delta S(P = P°, 298\text{ K})$$
$$+ \int_{T=298}^{T'} (\Delta C_p)dT - T'\int_{T=298}^{T'} \frac{\Delta C_p}{T}\, dT + \int_{P=P°}^{P'} (\Delta V)_{T'} dP \qquad (8\text{-}94)$$

It can be seen from Equation 8-94 that to calculate ΔG at any temperature and pressure we need to know values of ΔH and ΔS at standard conditions (P = 100 kPa, T = 298 K), the value of ΔC_P as a function of temperature at the standard pressure, and the value of ΔV_T as a function of pressure at each temperature T'. Thus the temperature dependence of ΔC_P and the temperature and pressure dependence of ΔV_T are needed. If such data are available in the form of empirical equations, the required integrations can be carried out analytically. If the data are available in tabular form, graphical or numerical integration can be used. If the data are not available, an approximate result can be obtained by assuming ΔC_P and ΔV_T are constant over the range of interest. The approximate result if both are assumed constant is

$$\Delta G(P', T') = \Delta H(P = P°, 298\text{ K}) - T'\Delta S(P = P°, 298\text{ K})$$
$$+ \Delta C_p(P = P°)[T' - 298] - T'\Delta C_p(P = P°)[\ln T' - \ln (298)]$$
$$+ (\Delta V)(P' - P°) \qquad (8\text{-}95)$$

8-4 USEFUL WORK AND FREE ENERGY

Thus far we have seen that the free energy functions provide a criterion for spontaneity and equilibrium in isothermal changes of state and also provide a basis for calculating the equilibrium yields of chemical reactions. If we extend our analysis to systems in which other constraints are placed on the system, and therefore work other than mechanical work can be done, we shall find that the free energy functions also provide a means for calculating the maximum magnitude of work obtainable from an isothermal change.

Isothermal Changes

We can begin with Equation 8-6 as a combined statement of the first and second laws, rearranged as

$$dE - DW - TdS \leqslant 0 \tag{8-96}$$

Since we are concerned with isothermal changes, $-SdT$ can be added to the left side of Equation 8-96 without changing its value:

$$dE - TdS - SdT - DW \leqslant 0$$

or

$$d(E - TS) - DW \leqslant 0$$

or

$$dA - DW \leqslant 0$$

or

$$dA \leqslant dW \tag{8-97}$$

Since constraints other than the constant pressure of the environment are now considered, the one-to-one relationships between reversibility and equilibrium on the one hand and irreversibility and spontaneity on the other hand are no longer valid. A spontaneous change of state, or the reverse change, a nonspontaneous change, can be carried out reversibly by the appropriate adjustment of a constraint, such as an electrical voltage. As before, the inequality applies to an irreversible process and the equality to a reversible process. If the change of state is spontaneous, dA is negative, work can be done on the surroundings, and DW is negative. The value for dA is the same for a change of state whether it is carried out reversibly or irreversibly. The reversible work, DW_{rev}, then is equal to dA, whereas the irreversible work, DW_{irrev}, is algebraically greater than dA, but smaller in magnitude. For a macroscopic change we can write

$$\left.\begin{aligned} \Delta A &= W_{rev} \\ \Delta A &< W_{irrev} \end{aligned}\right\} \tag{8-98}$$

The change in the Helmholtz function thus provides a *limiting* value for the magnitude of the *total* work (including work against the pressure of the atmosphere) obtainable in any isothermal process. That is,

$$|W_{rev}| > |W_{irrev}| \tag{8-99}$$

and the magnitude of the reversible work is a maximum. If the change of state is not spontaneous, dA is positive, work must be done on the system to produce the change, and DW is positive. Then W_{rev} is the minimum work required to carry out the change of state.

An interesting alternative demonstration of Equation 8-99 can be carried out on the basis of isothermal cycles and the Kelvin–Planck statement of the second law. Consider two possible methods of going from State a to State b, a spontaneous change of state, (Figure 8-1) in an isothermal fashion: (1) a reversible process and (2) an irreversible process. For each path the first law of thermodynamics is valid:

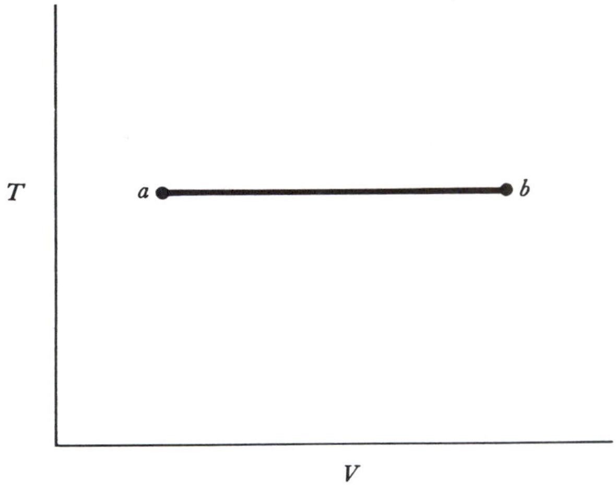

Figure 8-1. An isothermal process.

$$\Delta E_{\text{rev}} = Q_{\text{rev}} + W_{\text{rev}}$$

$$\Delta E_{\text{irrev}} = Q_{\text{irrev}} + W_{\text{irrev}}$$

Since E is a state function, and since both processes have the same starting and endpoints,

$$\Delta E_{\text{rev}} = \Delta E_{\text{irrev}}$$

and

$$Q_{\text{rev}} + W_{\text{rev}} = Q_{\text{irrev}} + W_{\text{irrev}}$$

or

$$Q_{\text{rev}} - Q_{\text{irrev}} = -(W_{\text{rev}} - W_{\text{irrev}}) \tag{8-100}$$

Let us *assume* that the spontaneous irreversible process gives work of a greater magnitude than the spontaneous reversible one. In that case

$$|W_{\text{irrev}}| > |W_{\text{rev}}|$$

and, from Equation 8-100, since W is negative for both alternatives,

$$Q_{\text{irrev}} > Q_{\text{rev}}$$

Let us use the irreversible process (*which goes in only one direction*) to carry the system from State a to State b, and the reversible process to return the system to its initial state. We can construct a table for the various steps (Table 8-3). As we can see from Table 8-3 the net result is that a positive amount of heat has been absorbed and work has been done on the surroundings in an isothermal cycle.

Table 8-3 Isothermal Cycle

	Irreversible Process (Forward)	Reversible Process (Backward)	Net for Both Processes
Heat absorbed	Q_{irrev}	$-Q_{rev}$	$Q_{irrev} - Q_{rev} > 0$
Work done	W_{irrev}	$-W_{rev}$	$W_{irrev} - W_{rev} < 0$

However, such a consequence is in contradiction to the Kelvin–Planck statement of the second law of thermodynamics, which denies the possibility of converting heat from a reservoir at constant temperature into work without some accompanying changes in the reservoir or its surroundings. In the postulated cyclical process, no such accompanying changes have occurred. Hence, the original assumption is incorrect and the irreversible work cannot be greater in magnitude than the reversible work:

$$|W_{rev}| \geqslant |W_{irrev}| \tag{8-101}$$

Thus the reversible work is a limiting maximum value for the magnitude of work obtainable in an isothermal change, with the equality applying to the limit when the process becomes reversible.

Changes at Constant Temperature and Pressure

Equation 8-96 can be rewritten to include explicit reference to DW_{net}, the net useful (non-PdV) work, by substituting $-P'dV + DW_{net}$ for DW. That is,

$$dE + P'dV - DW_{net} - TdS \leqslant 0 \tag{8-102}$$

For a constant-pressure process PdV can be substituted for $P'dV$, and VdP can be added without changing the value of the expression. Since the temperature is constant, $-SdT$ also can be added. With these additions and substitutions Equation 8-102 becomes

$$dE + PdV + VdP - TdS - SdT \leqslant DW_{net}$$

or (see Equation 8-16)

$$dG \leqslant DW_{net} \tag{8-103}$$

For a spontaneous change at constant T and P, dG is negative, work can be obtained, and DW_{net} is negative. The value of dG is the same for a given change of state whether it proceeds irreversibly in the absence of additional constraints, or whether it follows a reversible path or proceeds irreversibly when subjected to additional constraints (for example, electrical). If the process is reversible, the equality in Equation 8-103 applies. For the irreversible change the inequality applies. Thus $DW_{net,rev}$ is equal to dG, whereas $DW_{net,irrev}$ is greater algebraically than dG, but smaller in magnitude. For a macroscopic change we can write

$$\left.\begin{aligned} \Delta G &= W_{\text{net,rev}} \\ \Delta G &< W_{\text{net,irrev}} \end{aligned}\right\} \tag{8-104}$$

If the change of state is spontaneous, $|\Delta G|$ is equal to the maximum magnitude of non-PdV work that can be obtained. If the change of state is nonspontaneous, ΔG is equal to the minimum non-PdV work that must be done to carry out the change.

Relationship between ΔH_P and Q_P When Useful Work Is Done

We repeatedly have used the relationship (Equation 4-6)

$$\Delta H_P = Q_P$$

but always with the stipulation that pressure on the system is constant and that there is no work other than expansion work. Most chemical reactions are carried out under these conditions; hence, the heat of a reaction has been valuable as a measure of the enthalpy change. If nonatmospheric work also is being obtained, Equation 4-6 is no longer valid. The value of Q_P under these conditions can be obtained as follows. From the first law

$$\begin{aligned} dE &= DQ + DW \\ &= DQ - P'dV + DW_{\text{net}} \end{aligned}$$

For a constant-pressure process, PdV can be substituted for $P'dV$, so that

$$dE + PdV = DQ + DW_{\text{net}}$$

But at constant pressure, from Equation 4-3,

$$dH = dE + PdV$$

and we have, explicitly indicating the constancy of pressure,

$$\left.\begin{aligned} dH_P &= DQ_P + DW_{\text{net}} \\ &\text{or} \\ \Delta H_P &= Q_P + W_{\text{net}} \end{aligned}\right\} \tag{8-105}$$

Only when W_{net} is equal to zero is Equation 4-6 applicable.

Application to Electrical Work

Electrical work is among the most common kinds of nonmechanical work obtained from chemical transformations. The ordinary storage battery and the electric eel are examples of systems in which electrical work is produced from chemical transformations. In both cases the change in the Gibbs function gives the limiting value of the magnitude of electrical work; the actual value is always less in magnitude than the decrease in the Gibbs free energy.

An example in which the relationships can be explored in some detail is the formation at constant T and P of one mole of aqueous HCl from the gaseous elements H_2 and Cl_2:

$$\tfrac{1}{2}H_2(g) + \tfrac{1}{2}Cl_2(g) = HCl(aq) \tag{8-106}$$

If the gaseous mixture is exposed to a photochemical stimulus in the absence of any other constraints, the reaction proceeds spontaneously and irreversibly. Thus

$$dG < 0$$

and

$$DW_{net} = 0$$

The reaction also can be carried out reversibly if additional constraints are placed on the system, as in the cell illustrated by Figure 8-2. The H_2 and Cl_2 electrodes are connected to a potentiometer. If the electromotive force of the cell is opposed by the emf of the potentiometer, maintained at an infinitesimally lower value than that of the H_2–Cl_2 cell, then the conversion to HCl can be carried out reversibly, although it would take an infinitely long time to obtain one mole. In either the reversible or the explosively spontaneous path for carrying out the transformation, the change in free energy is the same since the states of the initial and of the final substances are the same in both methods. However, the amount of useful (electrical) work is different, and for the reversible path

$$DW_{net} \neq 0$$

When electrical work is obtained from the reaction under reversible conditions, that is, against a counterpotential only infinitesimally smaller than that of the cell, then

$$\begin{aligned} W_{elec} = W_{net,rev} &= (\text{potential difference}) \times (\text{charge transferred}) \\ &= (\mathscr{E})(-n\mathscr{F}) \end{aligned} \tag{8-107}$$

in which $\mathscr{E}$ is the counterpotential, which is equal to the cell potential under reversible conditions, $\mathscr{F}$ is the charge of one mole of protons, and n is the stoichiometric coefficient of the electron in each half-reaction of the cell reaction. It follows from Equation 8-104 and Equation 8-107 that

$$\Delta G = -n\mathscr{F}\mathscr{E} \tag{8-108}$$

The potentiometer also can be kept at a finitely lower potential than the cell. In this case the reaction would proceed spontaneously, but the work

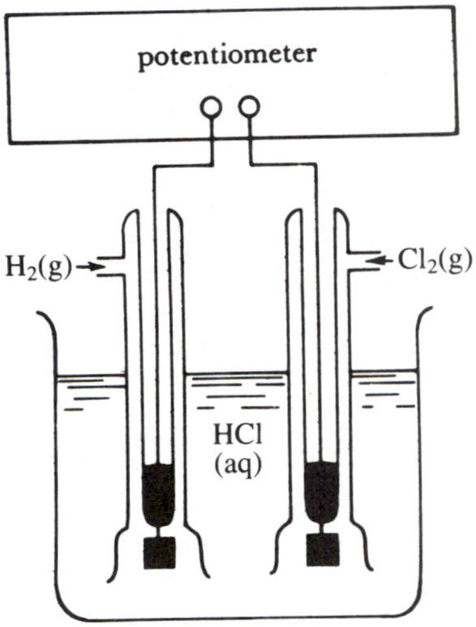

Figure 8-2. Formation of aqueous HCl in a reversible manner.

obtained would be less in magnitude than the value given in Equation 8-107. If the potentiometer is kept at a potential greater than that of the cell, the reverse of Equation 8-106 will occur, either reversibly or irreversibly, with ΔG positive and W_{net} positive. Then $W_{\text{net,rev}}$ is the minimum work required to reverse the process in Equation 8-106.

Gibbs–Helmholtz Equation

In Section 8-3 we developed equations for the temperature dependence of the free energy change of a reaction. An alternative approach that frequently is applied to the temperature dependence of the potential of an electrochemical cell will now be derived.

For an isothermal reaction (Equation 8-27)

$$\Delta G = \Delta H - T\Delta S$$

From Equation 8-43

$$\Delta S = -\left(\frac{\partial \Delta G}{\partial T}\right)_P$$

It follows that

$$\Delta G = \Delta H + T\left(\frac{\partial \Delta G}{\partial T}\right)_P \tag{8-109}$$

Equation 8-109 frequently is called the *Gibbs–Helmholtz equation*. From it, the temperature coefficient of the free energy change, $(\partial \Delta G_{P,T}/\partial T)_P$, can be obtained if ΔG and ΔH are known. By differentiating Equation 8-108 we obtain

$$\left(\frac{\partial \Delta G}{\partial T}\right)_P = -n\mathcal{F}\left(\frac{\partial \mathcal{E}}{\partial T}\right)_P \tag{8-110}$$

since n and $\mathcal{F}$ are temperature-independent quantities. Substitution of Equations 8-108 and 8-110 into Equation 8-109 gives

$$-n\mathcal{F}\mathcal{E} = \Delta H - n\mathcal{F}T\left(\frac{\partial \mathcal{E}}{\partial T}\right)_P \tag{8-111}$$

or, on rearrangement, an alternative form of the Gibbs–Helmholtz equation,

$$\Delta H = n\mathcal{F}\left[T\left(\frac{\partial \mathcal{E}}{\partial T}\right)_P - \mathcal{E}\right] \tag{8-112}$$

Free Energy and Useful Work in Biological Systems

Since biological systems operate at constant temperature and pressure, the free energy change of a reaction occurring in a biological system is a measure of the maximum magnitude of the net useful work that can be obtained from the reaction.

Biosynthetic Work. One of the primary functions of reactions that occur with a decrease in free energy (exergonic reactions) is to provide free energy for synthetic reactions that occur with an increase in free energy (endergonic reactions). The statement about maximum work can be paraphrased for this case: An exergonic reaction can make an endergonic reaction feasible if the increase in free energy of the endergonic reaction is smaller than the decrease in free energy of the exergonic reaction. In fact, such a coupling of exergonic and endergonic reactions can occur only if the two reactions have a common intermediate.

The molecule adenosine triphosphate (ATP) frequently acts as a coupling agent between exergonic and endergonic reactions in biological systems. For example, one exergonic reaction step that occurs in the overall oxidation of glucose in the cell is the oxidation of 3-phosphoglyceraldehyde to 3-phosphoglycerate by pyruvate, for which $\Delta G° = -29{,}300$ J mol^{-1}.

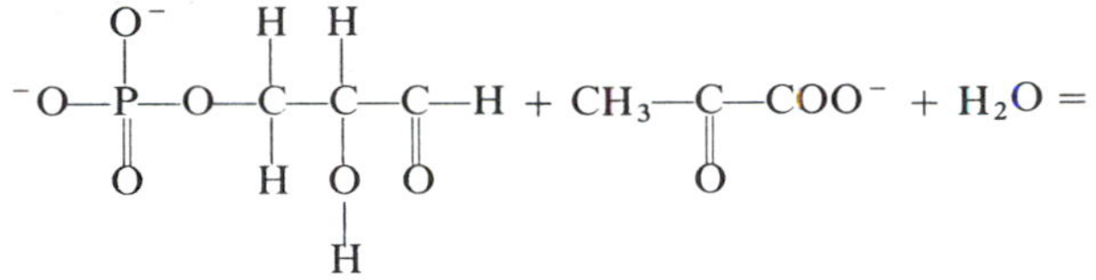

3-phosphoglyceraldehyde pyruvate

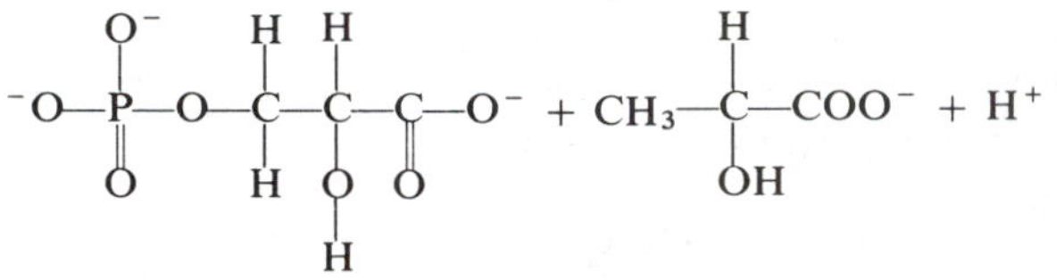

$$\text{}^-\text{O}-\overset{\text{O}^-}{\underset{\text{O}}{\overset{|}{\underset{\|}{\text{P}}}}}-\text{O}-\overset{\text{H}}{\underset{\text{H}}{\overset{|}{\underset{|}{\text{C}}}}}-\overset{\text{H}}{\underset{\text{O}}{\overset{|}{\underset{|}{\text{C}}}}}-\overset{}{\underset{\text{O}}{\overset{|}{\underset{\|}{\text{C}}}}}-\text{O}^- + \text{CH}_3-\overset{\text{H}}{\underset{\text{OH}}{\overset{|}{\underset{|}{\text{C}}}}}-\text{COO}^- + \text{H}^+ \qquad (8\text{-}113)$$

3-phosphoglycerate   lactate

The free energy decrease in this reaction can be utilized in an endergonic reaction if the two can be coupled.

As it occurs in the cell, the reaction (Equation 8-113) involves a mole of inorganic phosphate:

3-phosphoglyceraldehyde + HPO_4^{2-} + pyruvate =

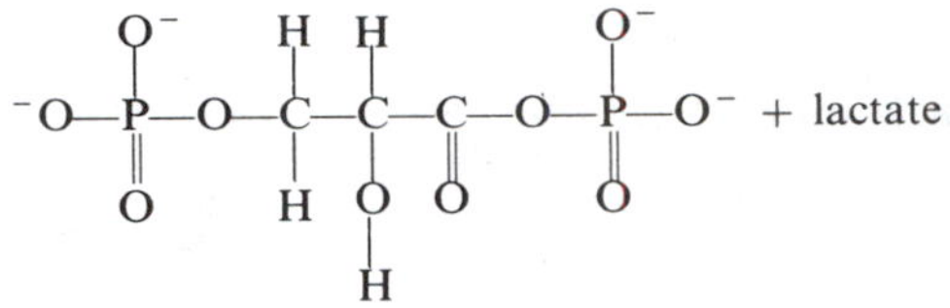

$$\text{}^-\text{O}-\overset{\text{O}^-}{\underset{\text{O}}{\overset{|}{\underset{\|}{\text{P}}}}}-\text{O}-\overset{\text{H}}{\underset{\text{H}}{\overset{|}{\underset{|}{\text{C}}}}}-\overset{\text{H}}{\underset{\text{O}}{\overset{|}{\underset{|}{\text{C}}}}}-\overset{}{\underset{\text{O}}{\overset{|}{\underset{\|}{\text{C}}}}}-\text{O}-\overset{\text{O}^-}{\underset{\text{O}}{\overset{|}{\underset{\|}{\text{P}}}}}-\text{O}^- + \text{lactate} \qquad (8\text{-}114)$$

1,3-diphosphoglycerate

The 1,3-diphosphoglycerate then reacts with adenosine diphosphate to form adenosine triphosphate and 3-phosphoglycerate:

1,3-diphosphoglycerate + adenosine$-\text{O}-\overset{\text{O}^-}{\underset{\text{O}}{\overset{|}{\underset{\|}{\text{P}}}}}-\text{O}-\overset{\text{O}^-}{\underset{\text{O}}{\overset{|}{\underset{\|}{\text{P}}}}}-\text{O}^- =$

ADP

3-phosphoglycerate + adenosine$-\text{O}-\overset{\text{O}^-}{\underset{\text{O}}{\overset{|}{\underset{\|}{\text{P}}}}}-\text{O}-\overset{\text{O}^-}{\underset{\text{O}}{\overset{|}{\underset{\|}{\text{P}}}}}-\text{O}-\overset{\text{O}^-}{\underset{\text{O}}{\overset{|}{\underset{\|}{\text{P}}}}}-\text{O}^- \qquad (8\text{-}115)$

ATP

The net result of Equations 8-114 and 8-115 is the same as the sum of Equation 8-113, for which $\Delta_G° = -29,300$ J mol^{-1}, and the formation of ATP from ADP,

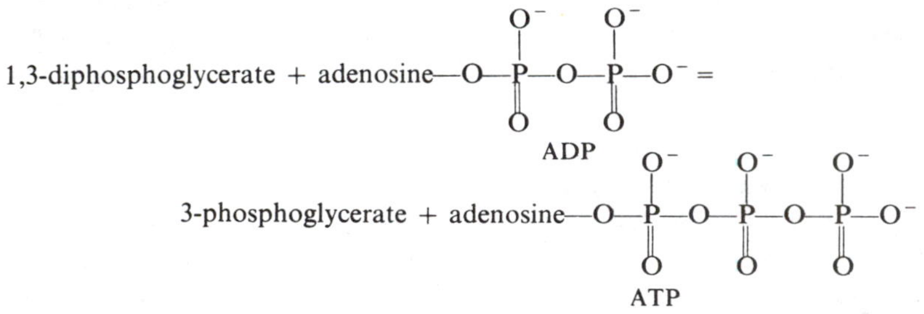

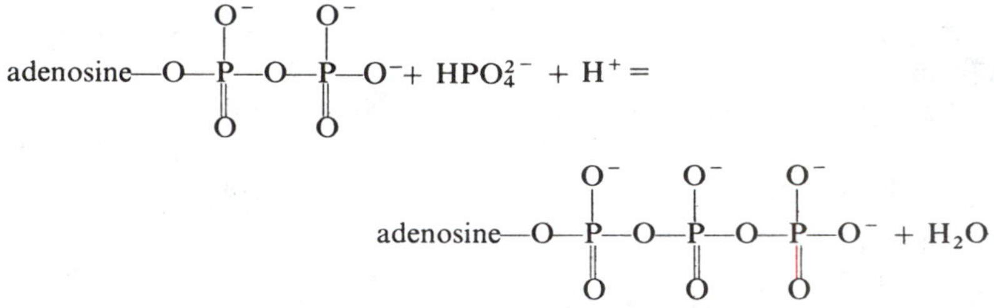

adenosine$-\text{O}-\overset{\text{O}^-}{\underset{\text{O}}{\overset{|}{\underset{\|}{\text{P}}}}}-\text{O}-\overset{\text{O}^-}{\underset{\text{O}}{\overset{|}{\underset{\|}{\text{P}}}}}-\text{O}^- + HPO_4^{2-} + \text{H}^+ =$

adenosine$-\text{O}-\overset{\text{O}^-}{\underset{\text{O}}{\overset{|}{\underset{\|}{\text{P}}}}}-\text{O}-\overset{\text{O}^-}{\underset{\text{O}}{\overset{|}{\underset{\|}{\text{P}}}}}-\text{O}-\overset{\text{O}^-}{\underset{\text{O}}{\overset{|}{\underset{\|}{\text{P}}}}}-\text{O}^- + \text{H}_2\text{O} \qquad (8\text{-}116)$

for which $\Delta_G° = +29,300$ J mol^{-1}.

The ATP, acting as an intermediate, then can be used to carry out a synthetic reaction that is endergonic, such as the synthesis of sucrose:

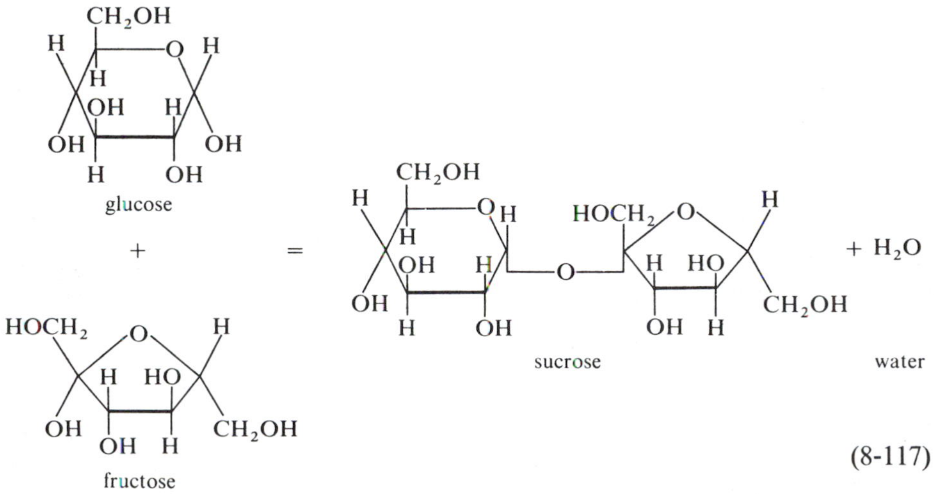

(8-117)

for which $\Delta G^\circ = 23{,}000 \text{ J mol}^{-1}$. The molecular coupling is carried out by ATP reacting first with glucose to form glucose 1-phosphate:

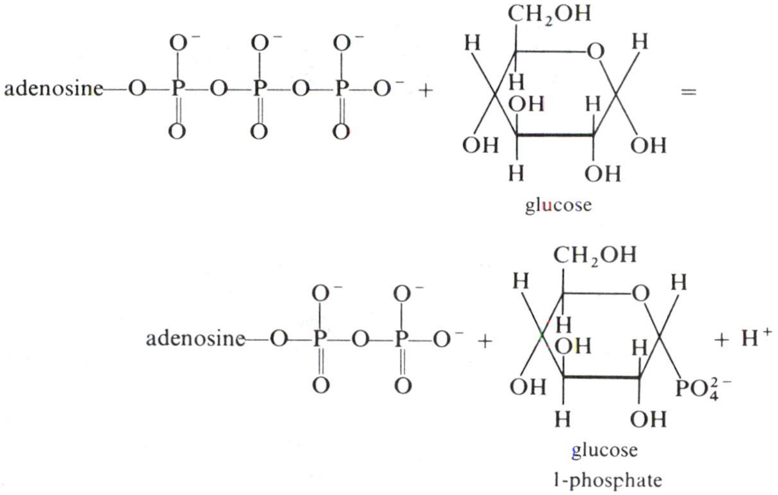

(8-118)

Following the reaction in Equation 8-118, the glucose 1-phosphate reacts with fructose to form sucrose:

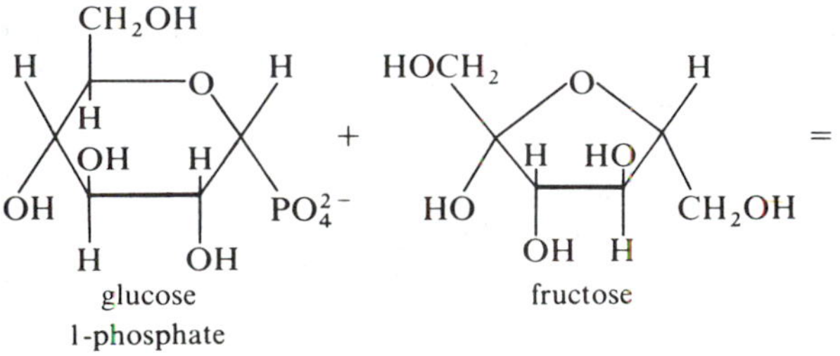

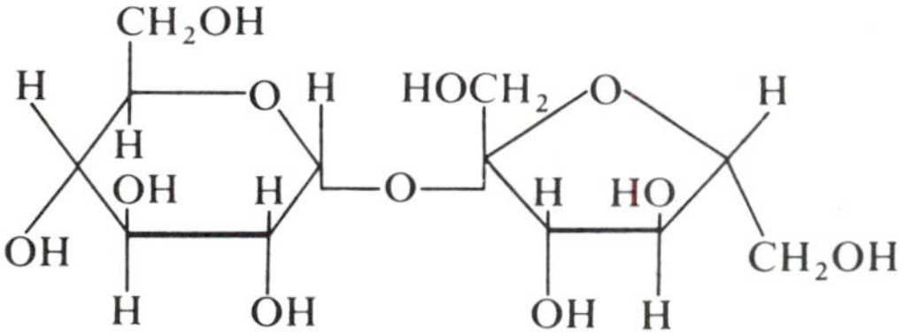

sucrose

$$+ \cdot HPO_4^{2-} \tag{8-119}$$

The sum of Equations 8-118 and 8-119 is the same as the sum of Equation 8-117 and the reverse of Equation 8-116, so the net free energy change is

$$\Delta G° = 23{,}000 \text{ J mol}^{-1} - 29{,}300 \text{ J mol}^{-1}$$
$$= -6{,}300 \text{ J mol}^{-1}$$

Thus the sequence of the two reactions is spontaneous when reactants and products are in their standard states.

As an alternative to considering that an endergonic reaction may be made to occur by coupling it to a sufficiently exergonic reaction, one might consider that a reaction with an equilibrium constant less than 1 can be made to occur by coupling it to a reaction with an equilibrium constant sufficiently greater than 1. To say that the sum of the Gibbs free energies of the coupled reactions must be negative is equivalent to saying that the product of the equilibrium constants of the coupled reactions must be greater than 1.

Mechanical Work. All cells exhibit motile and contractile properties. The remarkable thing about these activities of cells is that they are based on the direct coupling of chemical to mechanical action, in contrast to the heat engines that we have developed to do our work for us. The mechanisms by which this coupling of chemical to mechanical processes takes place is not well understood, but the hydrolysis of adenosine triphosphate is known to be an important part of the mechanism. Although thermodynamic studies cannot provide information about the molecular mechanisms involved, any mechanism that is proposed must be consistent with thermodynamic data [3].

Osmotic Work. It is characteristic of living cells that they are able to maintain nonequilibrium values of the concentrations of certain solutes, particularly ions such as Na^+ and K^+. It is this nonequilibrium distribution of ions that probably is responsible for the electrical potentials developed by living organisms. Again, although thermodynamic data do not lead to deductions about molecular mechanisms, they provide limiting values with which any mechanism must be consistent. We shall be able to discuss the thermodynamic aspects of osmotic work in detail when we have developed the methods required to deal with solutions.

EXERCISES

1. Prove the validity of Equations 8-35 and 8-50.
2. Derive the following expressions:

a.
$$\left[\frac{\partial(\Delta G/T)}{\partial(1/T)}\right]_P = \Delta H \tag{8-120}$$

b.
$$dG = V\left(\frac{\partial P}{\partial V}\right)_T dV + \left[V\left(\frac{\partial P}{\partial T}\right)_V - S\right] dT \tag{8-121}$$

3. a. Rearrange the definition of A to read

$$E = A + TS$$

Then prove that

$$\left(\frac{\partial E}{\partial V}\right)_T = -P + T\left(\frac{\partial P}{\partial T}\right)_V$$

which also has been proved as Equation 7-127.

b. With similar operations on G show that

$$\left(\frac{\partial H}{\partial P}\right)_T = V - T\left(\frac{\partial V}{\partial T}\right)_P$$

This relationship will be useful later as Equation 11-58.

4. If a rubber band is stretched, the reversible work is given by

$$DW = \tau dL$$

in which τ is the tension on the band and L its length.

a. If the stretching is carried out at constant pressure show that

$$dG = \tau dL - SdT$$

b. Show further that

$$\left(\frac{\partial G}{\partial L}\right)_T = \tau$$

c. Prove that

$$\left(\frac{\partial \tau}{\partial T}\right)_L = -\left(\frac{\partial S}{\partial L}\right)_T$$

d. Assuming that the volume of the rubber band does not change during stretching, derive the following equation from fundamental thermodynamic principles:

$$\left(\frac{\partial E}{\partial L}\right)_T = \tau + T\left(\frac{\partial S}{\partial L}\right)_T = \tau - T\left(\frac{\partial \tau}{\partial T}\right)_L$$

e. For an ideal gas it can be shown that

$$\frac{1}{P}\left(\frac{\partial P}{\partial T}\right)_V = \frac{1}{T}$$

Show that the corresponding equation for an "ideal" rubber band is

$$\frac{1}{\tau}\left(\frac{\partial \tau}{\partial T}\right)_L = \frac{1}{T}$$

5. One mole of an ideal gas at 273.15 K is allowed to expand isothermally from 10.0 MPa to 1 MPa.

   a. Calculate (and arrange in tabular form) the values of W, Q, ΔE, ΔH, ΔS, ΔG, and ΔA of the gas if the expansion is reversible.

   b. Calculate (and arrange in tabular form adjacent to the preceding table) the values of W, Q, ΔE, ΔH, ΔG, and ΔA of the entire isolated system (gas plus its environment) if the expansion is reversible.

   c. Calculate (and arrange in tabular form adjacent to the preceding tables) the values of the same thermodynamic quantities in (a) for the gas if it is allowed to expand freely so that no work whatever is done by it.

   d. Calculate (and arrange in tabular form adjacent to the preceding tables) the values of the same thermodynamic quantities in (b) for the entire isolated system if the expansion is free.

6. A mole of steam is condensed reversibly to liquid water at 100°C and 101.325 kPa (constant) pressure. The heat of vaporization of water is 2256.8 J g^{-1}. Assuming that steam behaves as an ideal gas, calculate W, Q, ΔE, ΔH, ΔS, ΔG, and ΔA for the condensation process.

7. Using thermal data available in this and preceding chapters derive an expression for $\Delta G°$ as a function of temperature for the reaction

$$CO(g) + \tfrac{1}{2}O_2(g) = CO_2(g)$$

8. If the heat capacities of reactants and products are expressed by equations of the form

$$c_p = a + bT - \frac{c'}{T^2}$$

in which a, b, and c' are constants, what will be the form of the equation for ΔG as a function of temperature?

9. Consider a reaction such as 8-57, in which A is a pure solid at 1 bar pressure, and the other substances are gases, as indicated. Derive an expression corresponding to Equation 8-63.

10. In theories of electrolytes it is customary to regard the free energy of the solution as composed of two parts: G_u, the free energy the particles would have if uncharged, and G_e, the additional free energy resulting from charging the particles to form ions. G_e can be given by the equation

$$G_e = -\frac{2\pi^{1/2}N_1^{3/2}\varepsilon^3 V(\nu_+ Z_+^2 + \nu_- Z_-^2)^{3/2}}{3D^{3/2}(kT)^{1/2}} \tag{8-122}$$

in which N_1 = number of molecules per unit volume of solution
ε = charge on electron
V = volume of solution
ν_+ = number of positive ions per molecule
Z_+ = number of charges on each positive ion
ν_- = number of negative ions per molecule
Z_- = number of charges on each negative ion
D = dielectric constant of solution
k = Boltzmann constant
T = absolute temperature

a. Assume that V and D do not change with temperature. Show that

$$H_e = \tfrac{3}{2}G_e$$

and that

$$S_e = \frac{G_e}{2T}$$

b. It is obvious from Equation 8-122 that G_e is a negative quantity. Hence S_e must be negative. What does this mean about the degree of order in a solution of ions as compared to that in an equivalent solution of uncharged particles? How would you interpret this difference in terms of the molecular structure of the solution?

11. A spring obeys Hooke's law, $\tau = -Kx$, in which τ is the tension and x the displacement from the equilibrium position. For a particular spring at 25°C, $K = 2.0 \times 10^{-6}$ N m^{-1} and $dK/dT = -1.0 \times 10^{-8}$ N m^{-1} K^{-1}.

a. The spring is placed in a thermostat at 25°C and stretched in a reversible manner from $x = 0$ m to $x = 0.010$ m. How much heat is given to or absorbed from the thermostat by the spring?

b. The spring then is allowed to snap back to its original position without doing any work. How much heat would it deliver into the thermostat?

12. Consider as an example the equilibrium

$$H_2O(l) = H_2O(g)$$

at some fixed temperature. Let n represent the number of moles of $H_2O(g)$ and let G and V represent the total free energy and the total volume of all the substances involved. Equilibrium exists if

$$\left(\frac{\partial G}{\partial n}\right)_P = 0$$

a. Show that

$$\left(\frac{\partial G}{\partial n}\right)_V = \left(\frac{\partial G}{\partial n}\right)_P + \left(\frac{\partial G}{\partial P}\right)_n \left(\frac{\partial P}{\partial n}\right)_V$$

b. Prove therefrom that

$$\left(\frac{\partial A}{\partial n}\right)_{V,T} = \left(\frac{\partial G}{\partial n}\right)_{P,T}$$

13. For stretching a film of water at constant pressure and temperature until its area is increased by 1 m^2, the free energy change, ΔG, is given by the equation

$$\Delta G = 75.64 \times 10^{-3} - 1.4 \times 10^{-4} t \text{ (in joules)}$$

in which t is the temperature in °C at which the stretching is carried out. When the film is stretched, the total volume of the water is not changed measurably.

a. How much work must be done to increase the area of the film reversibly by 1 m^2 at 10°C?

b. How much heat will be absorbed in the process in (a)?

c. Calculate ΔE, ΔH, ΔS, and ΔA for (a).

d. After the film has been stretched 1 m^2 reversibly, it is allowed to contract spontaneously and irreversibly to its original area. No work is regained in this process. What is ΔG for this step?

e. Calculate Q, ΔE, ΔH, ΔS, and ΔA for the process in (d).

14. In the historical development of thermodynamics, functions other than those of Gibbs and Helmholtz also have been worked with. Thus

$$J = -\frac{E}{T} + S$$

is called the Massieu function and

$$Y = -\frac{H}{T} + S \tag{8-23}$$

is known as the Planck function. Show that

$$dJ = \frac{E}{T^2}\, dT + \frac{P}{T}\, dV$$

and

$$dY = \frac{H}{T^2}\, dT - \frac{V}{T}\, dP$$

15. An electrochemical cell is placed in a thermostated bath at 25°C and 101.325 kPa, under which conditions it produces an emf of 0.100 volt. For this cell $\partial \mathscr{E}/\partial T$ is 1×10^{-4} volt K^{-1}.

    a. If electrical work is done on this cell and it is charged reversibly until 1 faraday of charge has been passed through it, how much heat is given to or absorbed from the thermostat?

    b. If the charged cell is short-circuited, so that no electrical work is obtained from it, and allowed to return to its initial state in (a), how much heat is given to or absorbed from the thermostat?

16. In the synthesis of sucrose 23,000 J of the 29,300 J available from the hydrolysis of ATP are used for synthetic work. If we call 23,000/29,300 the efficiency of this pair of reactions carried out at 37°C, and if we consider 37°C equivalent to the temperature of the high-temperature reservoir of a heat engine, what would the temperature of the low-temperature reservoir have to be to attain a comparable efficiency for a reversible Carnot engine?

17. S. A. Hawley [*Biochemistry* **10**, 2436 (1971)] has measured the free energy change for the reaction between native and denatured chymotrypsinogen as a function of temperature and pressure. The reaction can be described as

$$N(\text{native}) = D(\text{denatured})$$

The following values were found for the given thermodynamic functions at 0°C and 101.325 kPa:

$$\begin{aligned} \Delta G^\circ &= 10{,}600 \text{ J mol}^{-1} \\ \Delta S^\circ &= -950 \text{ J mol}^{-1} \text{ K}^{-1} \\ \Delta V^\circ &= -14.3 \times 10^{-6} \text{ m}^3 \text{ mol}^{-1} \\ (\partial \Delta V^\circ/\partial T)_P &= 1.32 \times 10^{-6} \text{ m}^3 \text{ mol}^{-1} \text{ K}^{-1} \\ (\partial \Delta V^\circ/\partial P)_T &= -.296 \times 10^{-12} \text{ m}^6 \text{ J}^{-1} \text{ mol}^{-1} \\ \Delta C_p &= 16{,}000 \text{ J mol}^{-1} \text{ K}^{-1} \end{aligned}$$

Calculate the value of ΔG at 35°C and 300 MPa assuming that ΔC_p, $(\partial \Delta V/\partial T)_P$, and $(\partial \Delta V/\partial P)_T$ are constants.

18. The oxygen-binding protein hemerythrin exists as an octamer in equilibrium with its monomers:

$$8\text{Hr} = \text{Hr}_8$$

At pH 7.0 and 25°C, Langerman and Klotz [*Biochemistry* **8**, 4746 (1969)] found $\Delta G^\circ = -24.3$ kJ mol^{-1} (monomer). Langerman and Sturtevant [*Biochemistry* **10**, 2809 (1971)] found $\Delta H = 4 \pm 2$ kJ mol^{-1} (monomer). Calculate ΔS for the formation of one mole of the octamer.

19. The deamination of aspartic acid,

$$^{-}\text{OOC—CH}_2\text{—CH(NH}_3^+\text{)—COO}^- = {}^{-}\text{OOC—CH═CH—COO}^- + \text{NH}_4^+ \qquad (8\text{-}123)$$

is a reversible reaction catalyzed by the enzyme aspartase. For D,L-aspartic acid the equilibrium constant in a range of temperature [J. L. Bada and S. L. Miller, *Biochemistry* **7**, 3403 (1968)] can be expressed by the equation

$$\log K_{\text{D,L}} = 8.188 - \frac{2315.5}{T} - 0.01025T \tag{8-124}$$

a. What is the value of ΔG° at 25°C?

b. Derive an equation for ΔH° as a function of the temperature.

c. Calculate the value of ΔH° at 25°C.

d. What is the value of ΔS° at 25°C?

e. Using the thermodynamic relationship $\Delta C_p^\circ = (\partial\, \Delta H^\circ/\partial T)_P$ calculate ΔC_p° at 25°C for the chemical transformation in Equation 8-123.

20. Show that the addition of ΔG for two reactions to obtain ΔG for the sum reaction is equivalent to the multiplication of the corresponding equilibrium constants.

21. T. F. Young, C. R. Singleterry, and I. M. Klotz [*J. Phys. Chem.* **82**, 671 (1978)] found that the ionization constant of the bisulfate ion could be described by the equation

$$\log_{10} K = 61.378 - \frac{1857.1}{T} - 23.093 \log_{10} T$$

between 5°C and 55°C. Calculate values of ΔG°, ΔH°, ΔS°, and ΔC_p° for the ionization at 50°C.

REFERENCES

1. The letter F also has been commonly used for the Gibbs function, particularly in the United States. Some tabulations of chemical thermodynamic data use F for the function of Equation 8-18.

2. M. Planck, *Treatise on Thermodynamics*, Dover Publications, New York; L. E. Strong and H. F. Halliwell, *J. Chem. Educ.*, **47**, 347 (1970).

3. R. S. Edelstein and E. Eisenberg, *Ann. Rev. of Biochemistry,* **49**, 921–956 (1980).

CHAPTER 9

Application of the Gibbs Free Energy Function to Some Phase Changes

Now that we have developed convenient criteria for equilibrium and for spontaneity we are able to apply the laws of thermodynamics to problems of interest. In this chapter we will deal with changes of phase in one-component systems, which are transformations of concern to the chemist and of particular concern to the geologist.

9-1 TWO PHASES AT EQUILIBRIUM AT GIVEN PRESSURE AND TEMPERATURE

The equations that describe equilibrium conditions between two phases of the same substance are derivable from the two laws of thermodynamics with the aid of the free energy functions that we defined in the preceding chapter. Let us represent the equilibrium in a closed system at any given temperature and pressure by the equation

$$\text{Phase A} = \text{Phase B} \tag{9-1}$$

Since the system is at equilibrium at constant temperature and pressure, any infinitesimal transfer of matter between Phase A and Phase B occurs with a free energy change of zero. That is,

$$dG = G_A dn_A + G_B dn_B = 0 \tag{9-2}$$

in which G_A and G_B are the molar free energies of A and B and dn_A and dn_B are the infinitesimal changes in the number of moles of A and B. Since the system is closed, that is, $dn_B = -dn_A$,

$$(G_A - G_B)\, dn_A = 0 \tag{9-3}$$

Since Equation 9-3 holds for any infinitesimal transfer dn_A whatsoever, the quantity in parentheses must equal zero and

$$G_A = G_B \tag{9-4}$$

If the temperature and pressure are changed by amounts dT and dP such that the system reaches a new state of equilibrium, then the molar free energies of A and B change by amounts dG_A and dG_B such that

$$G_A + dG_A = G_B + dG_B$$

or

$$dG_A = dG_B \tag{9-5}$$

Clapeyron Equation

If we apply Equation 8-34 for the total differential to the Phases A and B, the result is

$$dG_A = v_A dP - s_A dT \tag{9-6}$$

and

$$dG_B = v_B dP - s_B dT \tag{9-7}$$

in which v_A and v_B are the molar volumes of A and B and s_A and s_B are the molar entropies. Substituting from Equations 9-6 and 9-7 into Equation 9-5 we obtain

$$v_B dP - s_B dT = v_A dP - s_A dT$$

which can be rearranged to give

$$(v_B - v_A)dP = (s_B - s_A)dT$$

Consequently

$$\frac{dP}{dT} = \frac{s_B - s_A}{v_B - v_A} = \frac{\Delta s}{\Delta v} \tag{9-8}$$

From Equation 9-8 we conclude that T and P cannot be varied independently and still have a system at equilibrium. Once a value of T or P is chosen, the value of the other is fixed by Equation 9-8 and its integrated form.

We are interested in the value of the derivative dP/dT at a specified temperature and pressure such as is indicated by point a in Figure 9-1. For an isothermal, reversible (that is, equilibrium) condition at constant pressure

$$s_B - s_A = \Delta s = \int \frac{DQ_P}{T} = \frac{1}{T}\int DQ_P = \frac{\Delta H}{T}$$

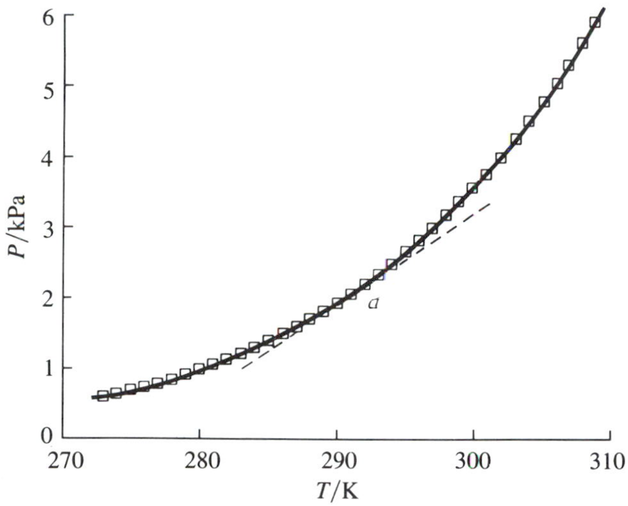

Figure 9-1. Equilibrium vapor-pressure curve for water. The broken line gives the slope at a specified pressure and temperature.

Therefore Equation 9-8 can be converted to

$$\frac{dP}{dT} = \frac{\Delta H}{T\,\Delta v} \tag{9-9}$$

which generally is known as the *Clapeyron equation*.

So far we have made no special assumptions as to the nature of the Phases A and B in deriving Equation 9-9. Evidently the Clapeyron equation is applicable to equilibrium between any two phases of one component at the same temperature and pressure, and describes the functional relationship between the equilibrium pressure and the equilibrium temperature.

Clausius–Clapeyron Equation

The Clapeyron equation can be reduced to a particularly convenient form when the equilibrium between A and B is that of a gas and a condensed phase (liquid or solid). In this situation

$$V_B - V_A = V_{gas} - V_{cond}$$

Generally, the molar volume of a gas, v_{gas}, is much larger than the molar volume of the condensed phase, v_{cond}; that is,

$$V_{gas} \gg V_{cond}$$

For example, the molar volume of liquid H_2O near the boiling point is about 18 cm^3, whereas that for water vapor is near 30,000 cm^3.

If v_{cond} is neglected with respect to v_{gas} in Equation 9-9 (with the condensed phase as Phase A), the result is

$$\frac{dP}{dT} = \frac{\Delta H}{T v_{gas}} \tag{9-10}$$

Furthermore, if we assume that the gas behaves ideally, then

$$\frac{dP}{dT} = \frac{\Delta H}{T(RT/P)} = \frac{P(\Delta H)}{RT^2}$$

or

$$\frac{1}{P}\frac{dP}{dT} = \frac{d \ln P}{dT} = \frac{\Delta H}{RT^2} \tag{9-11}$$

For many substances in a moderate temperature range, the heat of vaporization is substantially constant. Equation 9-11 then can be integrated as follows:

$$d \ln P = \frac{\Delta H}{R}\frac{dT}{T^2} = -\frac{\Delta H}{R} d\left(\frac{1}{T}\right)$$

and

$$\ln \frac{P_2}{P_1} = -\frac{\Delta H}{R}\left(\frac{1}{T_2} - \frac{1}{T_1}\right) \tag{9-12}$$

or, written as the indefinite integral,

$$\ln P = -\frac{\Delta H}{RT} + \text{constant} \tag{9-13}$$

or,

$$P = (\text{constant}) \exp(-\Delta H/RT) \tag{9-14}$$

Any one of Equations 9-11, 9-12, 9-13, or 9-14 is known as the *Clausius–Clapeyron equation* and can be used either to obtain ΔH from known values of the vapor pressure as a function of temperature, or to predict vapor pressures of a liquid (or a solid) when the heat of vaporization (or sublimation) and one vapor pressure are known. The same equations also represent the variation in the boiling point of a liquid with pressure.

If we do not wish to be limited to the assumption of gas ideality, we can use Equation 6-56 to substitute for v_{gas} in Equation 9-10, so that

$$\frac{dP}{dT} = \frac{\Delta H}{T(ZRT/P)}$$

or

$$\frac{1}{P}\frac{dP}{dT} = \frac{d \ln P}{dT} = \frac{\Delta H}{ZR}\frac{1}{T^2}$$

Rearranging, we have

$$d \ln P = \frac{\Delta H}{ZR}\frac{dT}{T^2} = -\frac{\Delta H}{ZR}\, d\,\frac{1}{T}$$

or

$$\frac{d \ln P}{d(1/T)} = -\frac{\Delta H}{ZR} \tag{9-15}$$

Thus, the slope of a plot of ln P against $1/T$ is $-\Delta H/ZR$, and numerical differentiation (Section 2-3) of experimental vapor-pressure data will provide values of $\Delta H/Z$ as a function of temperature and pressure. If Z is known, ΔH can be calculated.

9-2 THE EFFECT OF AN INERT GAS ON VAPOR PRESSURE

Liquid–vapor equilibria commonly are observed when the system is exposed to the atmosphere (as in Figure 9-2) rather than only to the vapor itself. Therefore, it is of interest to derive the equations that are applicable to such a situation. We will assume that air is essentially insoluble in the liquid phase and that atmospheric pressure is represented by P. The saturation vapor pressure of the liquid in the absence of any foreign gas such as air can be shown to differ from that of the partial pressure of the vapor, p, in the presence of air.

Since the liquid and vapor are in equilibrium at a given temperature and total pressure, P, we can write

$$G_{\text{gas}} = G_{\text{liquid}} \tag{9-16}$$

Let us assume that the vapor behaves as an ideal gas, even in the presence of the foreign gas. From Equation 8-54 we see that for one mole of an ideal gas at constant T

$$dG_{\text{gas}} = \frac{RT}{P}\, dP$$

Integration of this equation from any pressure, p, to the standard state for an ideal gas, a pressure of 0.1 MPa, $P°$,

$$\int_{G}^{G°} dG_{\text{gas}} = \int_{p}^{P°} RT\, d \ln P$$

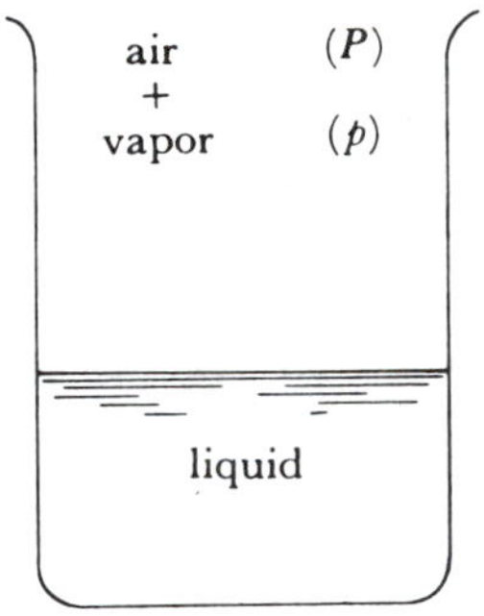

Figure 9-2. Liquid–vapor equilibrium in the presence of an inert gas.

leads to

$$G_{\text{gas}} = G^{\circ}_{\text{gas}} + RT \ln p/P^{\circ} \tag{9-17}$$

in which G°_{gas} is the standard free energy of the vapor at P° and the given temperature.

If Equation 9-17 is substituted into Equation 9-16, we can rearrange the result to

$$RT \ln \frac{p}{P^{\circ}} = G_{\text{liquid}} - G^{\circ}_{\text{gas}}$$

Thus

$$R \ln \frac{p}{P^{\circ}} = \frac{G_{\text{liquid}}}{T} - \frac{G^{\circ}_{\text{gas}}}{T} \tag{9-18}$$

Variable Total Pressure at Constant Temperature

At constant temperature Equation 9-18 can be differentiated with respect to the total pressure to give

$$R\left(\frac{\partial \ln p}{\partial P}\right)_T - R\left(\frac{\partial \ln P^{\circ}}{\partial P}\right)_T = \frac{1}{T}\left(\frac{\partial G_{\text{liquid}}}{\partial P}\right)_T - \frac{1}{T}\left(\frac{\partial G^{\circ}_{\text{gas}}}{\partial P}\right)_T \tag{9-19}$$

But G°_{gas} is independent of pressure since it is the standard molar Gibbs function at a definite fixed pressure—P°. Thus the second term on the right side of Equation 9-19 is equal to zero. Similarly, the second term on the left is also equal to zero. From Equation 8-37, $(\partial G/\partial P)_T$ is equal to v. Thus Equation 9-19 reduces to

$$\left(\frac{\partial \ln p}{\partial P}\right)_T = \frac{V_{\text{liquid}}}{RT} \tag{9-20}$$

in which V_{liquid} is the molar volume of the liquid phase.

The change of vapor pressure with change in total pressure of an inert atmosphere is small. For example, for water, where v_{liquid} is 18 cm^3, the right side of Equation 9-20 reduces to less than 0.001 per bar, at room temperature (for R = 8.31 J mol^{-1} K^{-1} and T = 298 K).

Variable Temperature at Constant Total Pressure

At constant total pressure Equation 9-18 can be differentiated with respect to T to give the temperature dependence of the vapor pressure of a liquid in equilibrium with its vapor in the presence of air at a fixed atmospheric pressure:

$$R\left(\frac{\partial \ln p}{\partial T}\right)_P = \left[\frac{\partial(G_{\text{liquid}}/T)}{\partial T}\right]_P - \left[\frac{\partial(G^\circ_{\text{gas}}/T)}{\partial T}\right]_P \qquad (9\text{-}21)$$

From Equation 8-75 we can substitute for each term on the right side of Equation 9-21 to obtain

$$R\left(\frac{\partial \ln p}{\partial T}\right)_P = -\frac{H_{\text{liquid}}}{T^2} + \frac{H^\circ_{\text{gas}}}{T^2} \qquad (9\text{-}22)$$

However, the enthalpy, H°_{gas}, is independent of pressure since the standard state is at a fixed pressure at any given temperature. Consequently, the enthalpy terms of Equation 9-22 may be combined into the heat of vaporization, $\Delta H = H_{\text{gas}} - H_{\text{liquid}}$, to give

$$\left(\frac{\partial \ln p}{\partial T}\right)_P = \frac{\Delta H}{RT^2} \qquad (9\text{-}23)$$

which is a result comparable to Equation 9-11.

9-3 TEMPERATURE DEPENDENCE OF ENTHALPY OF PHASE TRANSITION

We are accustomed to think of the temperature coefficient of ΔH as given from Equation 4-19 by the expression

$$\left(\frac{\partial\, \Delta H}{\partial T}\right)_P = \Delta C_p \qquad (9\text{-}24)$$

However, in any phase transition the equilibrium pressure does not remain constant as the temperature is varied. Hence to obtain $(d\ \Delta H/dT)_{\text{equil}}$, the temperature coefficient of the latent heat, we must find the dependence of ΔH on pressure as well as temperature. Thereafter, since pressure is also a function of the temperature, we can obtain $(d\Delta H/dT)_{\text{equil}}$.

Let us start with an equation for the total differential:

$$d(\Delta H) = \left(\frac{\partial \Delta H}{\partial T}\right)_P dT + \left(\frac{\partial \Delta H}{\partial P}\right)_T dP$$

$$= \Delta C_p\, dT + \left(\frac{\partial H_B}{\partial P} - \frac{\partial H_A}{\partial P}\right)_T dP \tag{9-25}$$

in which H_B and H_A are the molar enthalpies of Phase B and Phase A. It can be shown from properties of the Gibbs function (Exercise 3, chapter 8) that

$$\left(\frac{\partial H}{\partial P}\right)_T = V - T\left(\frac{\partial V}{\partial T}\right)_P$$

With this relationship Equation 9-25 can be converted to

$$d(\Delta H) = \Delta C_p\, dT + \left[\Delta V - T\left(\frac{\partial \Delta V}{\partial T}\right)_P\right] dP$$

Since dT and dP are not independent if equilibrium between phases is maintained, we can use the Clapeyron equation (9-9) to substitute for dP and obtain

$$d(\Delta H) = \Delta C_p\, dT + \left[\Delta V - T\left(\frac{\partial \Delta V}{\partial T}\right)_P\right] \frac{\Delta H}{T\,\Delta V} dT$$

From this it follows that

$$\frac{d(\Delta H)}{dT} = \Delta C_p + \frac{\Delta H}{T} - \Delta H\left(\frac{\partial \ln \Delta V}{\partial T}\right)_P \tag{9-26}$$

So far in the derivation we have made no assumption as to the nature of Phases A or B; thus, Equation 9-26 is applicable to all types of phase transitions. When both A and B are condensed phases, the third term on the right side of Equation 9-26 is small compared with the others and the equation reduces to

$$\frac{d(\Delta H)}{dT} = \Delta C_p + \frac{\Delta H}{T} \tag{9-27}$$

If the phase transition is a vaporization or sublimation, an alternative approximation applies:

$$\Delta V \cong V_{\text{gas}} \cong \frac{RT}{P}$$

From this approximation it follows that the third term of Equation 9-26 cancels the second and

$$\frac{d(\Delta H)}{dT} \cong \Delta C_p \qquad (9\text{-}28)$$

Although Equations 9-24 and 9-28 are formally alike, they refer to different types of processes. The former is strictly true for a process occurring at a constant pressure throughout a temperature range. Vaporization or sublimation does not fulfill this restriction, but nevertheless, Equation 9-28 is approximately correct because the molar volume of the condensed phase is small compared to that of the gas.

9-4 CALCULATION OF ΔG FOR SPONTANEOUS PHASE CHANGE

Thus far we have restricted our attention to phase changes in which equilibrium is maintained. It also is useful, however, to find procedures for calculating the change in free energy in transformations that are known to be spontaneous—for example, the freezing of supercooled water at $-10°C$:

$$H_2O(l, -10°C) = H_2O(s, -10°C)$$

At 0°C and 101.325 kPa pressure the process is at equilibrium. Hence

$$\Delta G_{0°C} = 0$$

At $-10°C$ the supercooled water can freeze spontaneously. Therefore we can say that

$$\Delta G_{-10°C} < 0$$

Now we wish to evaluate ΔG numerically.

Arithmetic Method

The simplest procedure to calculate the free energy change at $-10°C$ uses the relationship (Equation 8-27) for one mole,

$$\Delta G = \Delta H - T\,\Delta S$$

for any isothermal process. ΔH and ΔS at $-10°C$ (T_2) are calculated from the known values at 0°C (T_1) and from the temperature coefficients of these thermodynamic quantities. Since the procedure can be represented by the sum

Table 9-1 Change in Gibbs Function for Freezing of Supercooled Water

$H_2O(l, 0°C) = H_2O(s, 0°C)$	$\Delta H = -6008 \text{ J mol}^{-1}$
	$\Delta S = \dfrac{-6008}{273.15} = -21.995 \text{ J mol}^{-1} \text{ K}^{-1}$
$H_2O(s, 0°C) = H_2O(s, -10°C)$	$\Delta H = \int_{T_1}^{T_2} C_p \, dT = 36.4(-10)$
	$= 364 \text{ J mol}^{-1}$
	$\Delta S = \int_{T_1}^{T_2} \dfrac{C_p}{T} \, dT = C_p \ln \dfrac{T_2}{T_1}$
	$= -1.358 \text{ J mol}^{-1} \text{ K}^{-1}$
$H_2O(l, -10°C) = H_2O(l, 0°C)$	$\Delta H = \int_{T_2}^{T_1} C_p' \, dT = 75(10)$
	$= 750 \text{ J mol}^{-1}$
	$\Delta S = \int_{T_2}^{T_1} \dfrac{C_p'}{T} \, dT = 2.797 \text{ J mol}^{-1} \text{ K}^{-1}$
$H_2O(l, -10°C) = H_2O(s, -10°C)$	$\Delta H = -5622 \text{ J mol}^{-1}$
	$\Delta S = -20.556 \text{ J mol}^{-1} \text{ K}^{-1}$

of a series of equations, the method may be called an arithmetic one. The series of equations is given in Table 9-1. From the values calculated for ΔH and ΔS

$$\Delta G = -5622 + (263.15)(20.556) = -213 \text{ J mol}^{-1}$$

Analytic Method

The proposed problem also could be solved by integrating Equation 8-76 for one mole,

$$\left[\frac{\partial(\Delta G/T)}{\partial T}\right]_P = -\frac{\Delta H}{T^2}$$

As in the arithmetic method we can assume that the heat capacities of ice and water are substantially constant throughout the small temperature range under consideration. Thus, from Equation 9-24,

$$\left(\frac{\partial\ \Delta H}{\partial T}\right)_P = \Delta C_p$$

and upon integration we obtain

$$\Delta H = \Delta H_0 + \int (C_{p(\text{ice})} - C_{p(\text{water})})\ dT$$
$$= \Delta H_0 - 38.9T \qquad (9\text{-}29)$$

Since at 0°C ΔH is -6008 J mol^{-1} we can determine ΔH_0:

$$\Delta H_0 = -6008 + 38.9(273.15) = 4617.54$$

(More significant figures are retained in these numbers than can be justified by the precision of the data upon which they are based. However, such a procedure is necessary in calculations involving small differences between large numbers.) Thus

$$\Delta H = 4617.54 - 38.9T$$

and

$$\frac{\Delta G}{T} = -\int \frac{(4617.54 - 38.9T)}{T^2}\ dT$$

$$= \frac{4617.54}{T} + 38.9 \ln T + I \qquad (9\text{-}30)$$

in which I is a constant. Rearrangement of Equation 9-30 leads to

$$\Delta G = 4617.54 + 38.9T \ln T + IT$$

Since ΔG is known to be zero at 0°C the constant I can be evaluated:

$$I = \frac{-4617.54 - 38.9(273.15) \ln (273.15)}{273.15}$$
$$= -235.135$$

Thus we have an explicit equation for ΔG as a function of temperature:

$$\Delta G = 4617.54 + 38.9T \ln T - 235.135T \qquad (9\text{-}31)$$

At -10°C this equation leads to

$$\Delta G_{-10°\text{C}} = -213 \text{ J mol}^{-1}$$

This is the same result as was obtained by the arithmetic method.

EXERCISES

1. Examine each of the following seven transformations:
   a. $H_2O(s, -10°C, P°) = H_2O(l, -10°C, P°)$ (*Note:* No specification is made that this process is carried out isothermally, isobarically, or reversibly.)
   b. Same as part (a), but restricted to a reversible change
   c. Same as part (a), but restricted to isothermal and isobaric conditions
   d. The two-step, isobaric, reversible transformation

$$H_2O(l, 25°C, 1\text{ atm}) = H_2O(l, 100°C, 1\text{ atm})$$

$$H_2O(l, 100°C, 1\text{ atm}) = H_2O(g, 100°C, 1\text{ atm})$$

   e. Ideal gas (25°C, 10 MPa) = ideal gas (25°C, 100 kPa), reversible
   f. Ideal gas (25°C, 10 MPa) = ideal gas (25°C, 1 MPa), no work done
   g. Adiabatic reversible expansion of an ideal gas from 10 MPa to 1 MPa

   Consider each of the following equations:

   1. $\displaystyle\int \frac{DQ}{T} = \Delta S$
   2. $Q = \Delta H$
   3. $\dfrac{\Delta H}{T} = \Delta S$
   4. ΔG = actual net work
   5. ΔG = maximum net work

   For each transformation, list the equations of the group (1) to (5) that are valid. If your decision depends on the existence of conditions not specified, state what these conditions are.

2. Calculate $\Delta G°$ for each of the following transformations:
   a. $H_2O(l, 100°C) = H_2O(g, 100°C)$.
   b. $H_2O(l, 25°C) = H_2O(g, 25°C)$. The vapor pressure of H_2O at 25°C is 3.17 kPa.

3. The vapor pressure of pure bromine at 25°C is 28.4 kPa. The vapor pressure of bromine in a dilute aqueous solution at 25°C obeys the equation $p = 147\ m_2$, in which m_2 is molality, and p is expressed in kPa.
   a. Calculate ΔG for the transformation

$$Br_2(l, 25°C) = Br_2(m_2 = 0.01, \text{aq. soln.}, 25°C)$$

   b. What would be the molality of bromine in a saturated solution in water at 25°C?

4. An equation for ΔG for the freezing of supercooled water can be obtained also by integrating the equation

$$\left(\frac{\partial\ \Delta G}{\partial T}\right)_P = -\Delta S$$

An expression for ΔS as a function of temperature for substitution into the preceding equation can be derived from

$$\left(\frac{\partial\ \Delta S}{\partial T}\right)_P = \frac{\Delta C_p}{T}$$

a. Show that

$$\Delta G = I' - \Delta C_p(T \ln T) + (\Delta C_p - \Delta S_0)T$$

in which I' and ΔS_0 are constants.

b. Evaluate the constants from data for the freezing process at 0°C.

c. Calculate ΔG at -10°C, and compare the result with the values calculated by the methods described in the text.

5. The transition

$$\text{sulfur(rhombic)} = \text{sulfur(monoclinic)}$$

is at equilibrium at 101.325 kPa at 95.5°C. The entropies (in J mol^{-1} K^{-1}) of the allotropic forms are the following functions of temperature:

$$S_{rh} = -61.13 + 14.98 \ln T + 26.11 \times 10^{-3}T$$
$$S_{mono} = -60.88 + 14.90 \ln T + 29.12 \times 10^{-3}T$$

Compute the free energy change for this allotropic transition at 25°C.

6. By convention, the free energy of formation, $\Delta G f^\circ$, of graphite is assigned the value of zero. On this basis, $\Delta G f^\circ_{298}$ of diamond is 2900 J mol^{-1}. Entropies and densities also are listed in Table 9-2. Assuming that the entropies and densities are approximately constant, determine the possibilities for the manufacture of diamonds from graphite by the use of high temperatures and pressures. [See F. P. Bundy, H. T. Hall, H. M. Strong, and R. H. Wentorf, Jr., *Nature* **176**, 51 (1955).]

Table 9-2

	Graphite	Diamond
$\Delta G f^\circ_{298}$/(J mol^{-1})	0	2900
S°_{298}/(J mol^{-1} K^{-1})	5.740	2.377
Density/(g cm^{-3})	2.22	3.51

7. The melting points of carbon tetrachloride at various pressures are given in Table 9-3 together with ΔV of fusion. Calculate ΔH and ΔS of fusion at (a) 0.1 MPa and (b) 600 MPa.

Table 9-3

P/MPa	t/°C	Δv/cm^3 mol^{-1}
0.1	−22.6	3.97
101	15.3	3.06
203	48.9	2.51
507	130.8	1.52
709	176.2	1.08

8. For liquid thiacyclobutane the vapor pressure, in millimeters of mercury, can be expressed by the equation [D. W. Scott, H. L. Finke, W. N. Hubbard, J. P. McCullough, C. Katz, M. E. Gross, J. F. Messerly, R. E. Pennington, and G. Waddington, *J. Am. Chem. Soc.* **75**, 2795 (1953)]

$$\log_{10} P = 7.01667 - \frac{1321.331}{t + 224.513}$$

in which t is in °C. Calculate ΔH of vaporization at 298 K.

9. The vapor pressure of liquid helium can be expressed by the equation

$$P = AT^{5/2}e^{-[(a/T)+bT^{5.5}]}$$

in which A, a, and b are constants. Derive an equation for ΔH of vaporization as a function of temperature.

10. Compute $\Delta G°$ for the transformation

$$H_2O(l, -5°C) = H_2O(s, -5°C)$$

given that the vapor pressure of supercooled liquid water at −5°C is 421.7 Pa and that of ice is 401.6 Pa.

11. What would be the form of the integrated Clausius–Clapeyron equation if the heat capacity of the vapor were given by the equation

$$c_p = a + bT$$

and that of the condensed phase by

$$c'_p = a' + b'T$$

in which the a's and b's are constants?

12. Rhombic sulfur is the stable form at room temperature and monoclinic the metastable form. The transition temperature is 95.5°C. The melting point of monoclinic sulfur is 120°C. How would you evaluate a report that 77°C is the melting point of rhombic sulfur? Answer in terms of a diagram of Gibbs free energy versus temperature for this system.

13. In the free-volume theory of liquids, the molar Helmholtz free energy, A, is defined by the equation

$$A = A^{\ddagger}(T) - RT \ln v_f - \Lambda$$

in which $A^{\ddagger}(T)$ is the volume independent term of the free energy, v_f is the free volume in the liquid, and Λ is the contribution to the potential energy from intermolecular forces. Assuming that

$$v_f = v - b$$

and

$$\Lambda = \frac{a}{v}$$

in which v is the geometric volume and a and b are the van der Waals constants, prove that this liquid would obey the van der Waals equation of state.

14. The following data [D. Ambrose and C. H. Sprake, *J. Chem. Thermodynamics* **4**, 603 (1972)] represent the vapor pressure of mercury as a function of temperature. Plot ln P as a function of $1/T$ to a scale consistent with the precision of the data. If the resultant plot is linear, calculate $\Delta H/z$ from the slope obtained by a least squares fit to the line. If the plot is curved, use a numerical differentiation procedure to obtain the value of $\Delta H/z$ as a function of T, and calculate ΔC_p.

T/K	P/kPa
400.371	0.139
417.129	0.293
426.240	0.424
432.318	0.538
439.330	0.706
441.757	0.774
447.720	0.964
451.420	1.101
454.160	1.213
456.359	1.309
462.673	1.627
469.222	2.024
474.605	2.414
479.080	2.784
485.190	3.369
491.896	4.128
497.570	4.882

CHAPTER 10

Application of the Gibbs Free Energy Function to Chemical Changes

Now let us consider the application of the free energy criterion to chemical transformations. Since most chemical reactions are carried out at constant pressure and temperature, with no restraints other than the pressure of the atmosphere, it is the Gibbs free energy, G, with which we shall be most interested. For application to chemical transformations tables of free energy data generally are assembled in terms of $\Delta G°$ so that the equilibrium constant of a reaction can be calculated.

The standard free energy change for a reaction can be obtained by several procedures. It will be convenient to discuss all of them briefly so that their advantages and limitations can be compared.

10-1 ADDITION OF KNOWN $\Delta G°$'s FOR SUITABLE CHEMICAL EQUATIONS LEADING TO THE DESIRED EQUATION

Since the free energy is a thermodynamic property, values of ΔG do not depend on the intermediate chemical reactions that have been used to transform a set of reactants, under specified conditions, to a series of products. Thus one can add known free energies to obtain values for reactions for which direct data are not available.

Let us consider the determination of ΔG°_{298} for the reaction

$$CO_2(g) + H_2(g) = H_2O(l) + CO(g) \tag{10-1}$$

(We will continue to use the ideal gas standard state of 1 bar for the present. See Chapter 13 for a treatment adequate for real gases.) We will assume that the standard free energies of formation of CO(g), CH_4(g), and H_2O(l) are known and also that standard free energies are available for the condensation of water vapor at 25°C and for the reaction

$$CO_2(g) + 4H_2(g) = CH_4(g) + 2H_2O(g) \tag{10-2}$$

The solution of the problem then can be obtained by the summation process shown in Table 10-1.

Once the standard free energy change is known, it is possible to calculate the equilibrium constant for Reaction 10-1:

$$\begin{aligned} \Delta G^\circ &= -RT \ln K \\ 20{,}062 \text{ J mol}^{-1} &= -(8.314 \text{ J mol}^{-1} \text{ K}^{-1})(298.15 \text{ K})(298.15 \text{ K}) \ln K \\ K &= 3.06 \times 10^{-4} \end{aligned} \tag{10-3}$$

Table 10-1 Summation of ΔG°'s

$CO_2(g) + 4H_2(g) = CH_4(g) + 2H_2O(g)$	$\Delta G^\circ_{298} = -113{,}505 \text{ J mol}^{-1}$
$CH_4(g) = C(\text{graphite}) + 2H_2(g)$	$\Delta G^\circ_{298} = 50{,}720 \text{ J mol}^{-1}$
$C(\text{graphite}) + \frac{1}{2}O_2(g) = CO(g)$	$\Delta G^\circ_{298} = -137{,}168 \text{ J mol}^{-1}$
$2H_2O(g) = 2H_2O(l)$	$\Delta G^\circ_{298} = -17{,}114 \text{ J mol}^{-1}$
$H_2O(l) = H_2(g) + \frac{1}{2}O_2(g)$	$\Delta G^\circ_{298} = 237{,}129 \text{ J mol}^{-1}$
$CO_2(g) + H_2(g) = H_2O(l) + CO(g)$	$\Delta G^\circ = 20{,}062 \text{ J mol}^{-1}$

10-2 DETERMINATION OF ΔG° FROM EQUILIBRIUM MEASUREMENTS

Frequently, the standard free energies required to calculate ΔG° for a specified reaction are not available. Then it is necessary to resort to more direct relationships between ΔG° and experimental measurements.

One of these direct methods depends on the determination of the equilibrium constant of a given reaction. As an example we shall consider the dissociation of isopropyl alcohol to form acetone and hydrogen:

$$(CH_3)_2CHOH(g) = (CH_3)_2CO(g) + H_2(g) \tag{10-4}$$

With a suitable catalyst, equilibrium pressures can be measured for this dissociation. At 452.2 K and a total pressure, P, of 95.9 kPa the degree of dissociation, α, at equilibrium has been found [1] to be 0.564.

If we start with 1 mole of isopropyl alcohol, α moles each of acetone and hydrogen are formed. The quantity of alcohol remaining at equilibrium must be $1 - \alpha$. The total number of moles of all three gases is

$$\text{total moles} = (1 - \alpha) + \alpha + \alpha = 1 + \alpha$$

Hence the mole fraction, X, of each substance is

$$X_{(CH_3)_2CHOH} = \frac{1 - \alpha}{1 + \alpha}$$

$$X_{(CH_3)_2CO} = \frac{\alpha}{1+\alpha}$$

$$X_{H_2} = \frac{\alpha}{1+\alpha}$$

Since the equilibrium constant, K, is a function of the equilibrium partial pressures, it is given by

$$K = \frac{P_{(CH_3)_2CO}P_{H_2}}{P_{(CH_3)_2CHOH}P^\circ} = \frac{[\alpha/(1+\alpha)]P[\alpha/(1+\alpha)]P}{[(1-\alpha)/(1+\alpha)]P(P^\circ)} = \frac{\dfrac{\alpha^2}{1-\alpha^2}P}{P^\circ} \qquad (10\text{-}5)$$

and since $\alpha = 0.564$ at 95.9 kPa

$$K = 0.450$$

The standard free energy change then can be calculated from Equation 10-3:

$$\Delta G^\circ_{452.2\text{ K}} = -RT \ln (0.450) = 3000 \text{ J mol}^{-1}$$

The value of ΔG° just calculated applies to a system in which all reactants and products are at standard pressure and which is sufficiently large that one mole of reaction does not alter the pressures appreciably. Alternatively, the expression $RT \ln K$ can be equated to $(\partial G/\partial n)^\circ_{T,P}$ for a finite system, or the initial rate of change of the total free energy of the system per mole of reaction when all reactants and products are at standard pressure [2].

The positive value of ΔG° does not imply that the reaction under consideration may not proceed spontaneously under any conditions. ΔG° refers to the reaction

$$(CH_3)_2CHOH(g, P^\circ) = (CH_3)_2CO(g, P^\circ) + H_2(g, P^\circ) \qquad (10\text{-}6)$$

in which each substance is in its standard state, that is, at a partial pressure of 0.1 MPa. The positive value of ΔG° allows us to state categorically that Reaction 10-6 will not proceed spontaneously under these conditions. However, as in the experiment described, if we were to start with isopropyl alcohol at a partial pressure of 0.1 MPa and no acetone or hydrogen, the alcohol decomposes spontaneously at 452.2 K, and as the value of the equilibrium constant and the experimental data on which it is based indicate, more than 50% dissociation can occur in the presence of a suitable catalyst. Yields can be made even greater if one of the products is removed continuously.

We also might calculate ΔG for one set of conditions with the substances not all in their standard states; for example,

$$(CH_3)_2CHOH(g, P^\circ) = (CH_3)_2CO(g, P = 10.13 \text{ kPa}) + H_2(g, P = 10.13 \text{ kPa}) \qquad (10\text{-}7)$$

For this computation we refer to Equation 8-63, which relates ΔG under any pressure conditions to ΔG° and the corresponding P's and which, rearranged, is written

$$\Delta G = \Delta G^\circ + RT \ln \frac{(P_C)^c (P_D)^d}{(P_A)^a (P_B)^b} P^{\circ (a+b-c-d)} \tag{10-8}$$

Applied to Equation 10-7 this relationship gives

$$\Delta G = 3000 \text{ J mol}^{-1} + RT \ln \frac{(10.13 \text{ kPa})^2}{(100 \text{ kPa})(100 \text{ kPa})}$$

$$= -14{,}200 \text{ J mol}^{-1}$$

Just as in the calculation of ΔG°, the value of ΔG just calculated applies to a system large enough that the pressure given for reactant and product do not change when one mole of reaction occurs. Alternatively, $\Delta G^\circ + RT \ln \mathcal{Q}$ can be equated to $(\partial G/\partial n)_{T,P}$ for a finite system, or to the instantaneous rate of change of the total free energy of the system per mole of reactants as the system composition passes through the designated pressures of reactants and products.

Thus if we are considering a given reaction in connection with the preparation of some substance, it is important not to be misled by positive values of ΔG°, because ΔG° refers to the reaction under standard conditions. It is quite possible that appreciable yields can be obtained even though a reaction will not go to *completion*. Such a case is illustrated by the example of isopropyl alcohol, just cited. Only if ΔG° has very large positive values, perhaps greater than 40 kJ, can we be assured, without calculations of the equilibrium constant, that no significant degree of transformation can be obtained. If we start with pure reactant, the initial value of $(\partial G/\partial n)_{T,P}$ is always negative, since the pressures of products in the numerator of the logarithmic term in Equation 10-8 are equal to zero. As the reaction proceeds, the value of $(\partial G/\partial n)_{T,P}$ becomes less negative and equals zero at equilibrium. How far the reaction proceeds before reaching equilibrium depends on the sign and magnitude of ΔG° [3].

The value of ΔG° is obtained by considering a reaction like that of Equation 10-6, in which reactants and products are in their standard states. Nevertheless, since $\Delta G = 0$ at equilibrium, Equation 10-3 is used to calculate a value of the equilibrium constant, K, from ΔG°. In general, equilibrium states such as

$$(CH_3)_2CHOH(g, P_{equil}) = (CH_3)_2CO(g, P_{equil}) + H_2(g, P_{equil}) \tag{10-9}$$

are different from standard states. The equilibrium pressures can be calculated from K and thus from ΔG°. There is no unique value for the equilibrium pressures; rather they depend on the initial pressures of the reactants.

10-3 DETERMINATION FROM MEASUREMENTS OF CELL POTENTIALS

The method of determination from measurements of cell potentials depends on the ability of the system to undergo a transformation *reversibly* in an electrical cell. (See Figure 8-2.) In this case the system will be opposed by an opposing potential just sufficient to balance the potential obtained in the electrical cell. The potential observed under such circumstances is related to the free energy change for the reaction by Equation 8-108:

$$\Delta G = -n \mathscr{F} \mathscr{E}$$

From Equation 8-104, $\Delta G = W_{net,rev}$, so that $n \mathscr{F} \mathscr{E}$ represents the net, reversible work per mole of reaction. From the appropriate extrapolation of measured values of $\mathscr{E}$ (see Chapters 19 and 20) a value of the standard potential, $\mathscr{E}^\circ$, can be obtained. The value of $\mathscr{E}^\circ$ then can be used to calculate ΔG°:

$$\Delta G^\circ = -n \mathscr{F} \mathscr{E}^\circ$$

An example of a reaction to which this method is applicable is

$$\text{(cytochrome } c) - Fe^{II} + \text{(cytochrome } f) - Fe^{III} = \text{(cytochrome } c) - Fe^{III} + \text{(cytochrome } f) - Fe^{II} \quad (10\text{-}10)$$

Cytochrome *c* and cytochrome *f* both are involved in the electron transfer chain from glucose metabolites to molecular oxygen in aerobic organisms. From measurements of the emf of a cell in which the reaction in Equation 10-10 occurs, it is possible to calculate that $\mathscr{E}^\circ$ is 0.11 V [4]. Hence

$$\Delta G^\circ = -(1)(96{,}487)(0.11) \text{ volt coulomb mol}^{-1}$$

or

$$\Delta G^\circ = -10{,}600 \text{ J mol}^{-1}$$

10-4 CALCULATION FROM THERMAL DATA AND THE THIRD LAW OF THERMODYNAMICS

The methods described so far depend directly or indirectly on the reversible character of at least one reaction for every substance of interest. For some time it was a challenge to theoretical chemists to devise some method of calculating free energies from thermal data alone (that is, from enthalpies and heat capacities) so that the need for experiments under equilibrium conditions might be avoided. One of the equations for ΔG, Equation 8-27,

$$\Delta G = \Delta H - T\Delta S$$

is applicable to any isothermal reaction. Clearly, if it were possible to obtain ΔS from thermal data alone, it would be possible to calculate ΔG. The calculation of ΔS from thermal data alone cannot be made without the introduction of a new assumption beyond the first two laws of thermodynamics. We will discuss the nature of this assumption and the consequences of it in the next chapter.

10-5 CALCULATION FROM SPECTROSCOPIC DATA AND STATISTICAL MECHANICS

Many free energy changes, particularly for gaseous reactions, can be calculated from theoretical analyses of vibrational and rotational energies of molecules. The parameters used in these calculations are obtained from spectroscopic data. However, this method depends on assumptions beyond those of classical chemical thermodynamics and will not be discussed in this textbook. Nevertheless, the results of these calculations can be used even prior to an understanding of the methods by which they have been obtained.

EXERCISES

1. According to D. P. Stevenson and J. H. Morgan [*J. Am. Chem. Soc.* **70**, 2773 (1948)] the equilibrium constant, K, for the isomerization reaction

$$\text{cyclohexane(l)} = \text{methylcyclopentane(l)}$$

can be expressed by the equation

$$\ln K = 4.814 - \frac{2059}{T}$$

a. Derive an equation for $\Delta G°$ as a function of T.

b. Calculate $\Delta H°$ and $\Delta S°$ at 25°C.

c. Calculate $\Delta H°$ and $\Delta S°$ at 0°C.

d. Calculate $\Delta C_p°$.

2. According to D. M. Golden, K. W. Egger, and S. W. Benson [*J. Am. Chem. Soc.* **86**, 5416 (1969)] $\Delta H°$ and $\Delta S°$ for the reaction

$$\textit{cis}\text{-2-butene} = \textit{trans}\text{-2-butene}$$

have the constant values of -1.2 kcal mol^{-1} and -1.2 cal mol^{-1} K^{-1} in the range of temperatures from 400 K to 500 K. Calculate an equation for ln K as a function of T from their results.

3. According to J. L. Hales and E. F. G. Herington [*Trans. Faraday Soc.* **53**, 616 (1957)] the equilibrium constant K for the hydrogenation of pyridine to piperidine,

$$C_5H_5N(g) + 3H_2(g) = C_5H_{11}N(g)$$

in the temperature range of 140°C to 260°C can be expressed by the equation

$$\ln K_p = -46.699 + \frac{24{,}320}{T}$$

Calculate $\Delta H°$, $\Delta S°$, and $\Delta C_p°$ at 200°C. (These data are for $P° = 101.325$ kPa.)

4. Energy changes for the conversion of the chair to the boat conformation of the cyclohexane ring can be estimated from a study of the equilibrium

between *cis*- and *trans*-1,3-di-*t*-butylcyclohexane. Some analytical results of N. L. Allinger and L. A. Freiberg [*J. Am. Chem. Soc.* **82**, 2393 (1960)] are listed below:

T/K	492.6	522.0	555.0	580.0	613.0
% trans	2.69	3.61	5.09	6.42	8.23

a. Compute the equilibrium constant for the reaction *cis* = *trans* at each temperature.

b. Draw a graph of $\ln K$ versus T and of $\ln K$ versus $1/T$.

c. Calculate ΔH° and ΔS° for the conformational change.

5. The standard potentials for a galvanic cell in which the reaction

$$\tfrac{1}{2}H_2(g) + AgCl(s) = Ag(s) + HCl(aq)$$

is being carried on are given in Table 2-5.

a. By numerical differentiation construct the curve for $(\partial \mathscr{E}^\circ/\partial T)_P$ as a function of temperature.

b. Calculate ΔG°, ΔH°, and ΔS° at 10, 25, and 50°C, respectively. Tabulate the values you obtain.

c. The empirical equation for $\mathscr{E}^\circ$ as a function of temperature is

$$\mathscr{E}^\circ = 0.22239 - 645.52 \times 10^{-6}(t - 25) - 3.284 \times 10^{-6}(t - 25)^2 + 9.948 \times 10^{-9}(t - 25)^3$$

in which t is °C. Using this equation, compute ΔG°, ΔH°, and ΔS° at 50°C. Compare these values with the values obtained by the numerical method.

6. Consider the problem of calculating a value for ΔG as a function of the extent of reaction. For example, consider a case that can be represented as

$$A = B$$

in which the initial number of moles of A is n and the initial number of moles of B is zero. Let n_B represent the number of moles of B at any extent of reaction. If X represents the mole fraction of B formed at any time in the reaction, then $X = n_B/n$ and $X_A = 1 - X_B = 1 - n_B/n$. The corresponding pressures of A and B are

$$P_A = (1 - X)P \quad \text{and} \quad P_B = XP$$

in which P is the total pressure. At any extent of conversion n_B, it follows from the discussion following Equation 10-8 that

$$\left(\frac{\partial G}{\partial n_B}\right)_{T,P} = \left(\frac{1}{n}\right)\left(\frac{\partial G}{\partial X}\right)_{T,P} = \Delta G^\circ + RT \ln\left[\frac{X}{(1 - X)}\right]$$

or

$$dG = n\left\{\Delta G^\circ + RT \ln\left[\frac{X}{(1 - X)}\right]\right\} dX$$

a. Show that ΔG for raising the mole fraction of B from 0 to X' is given by

$$\int_0^{X'} dG = nX'\Delta G^\circ + nRT\left[X' \ln \frac{X'}{1 - X'} + \ln(1 - X')\right]$$

b. Prove that to reach the equilibrium state, X_{equil},

$$\Delta G = -nRT \ln (K + 1)$$

in which K is the equilibrium constant for the reaction.

7. J. Carey and O. C. Uhlenbeck, [*Biochemistry,* **22**, 2610 (1983)] found the values of -19 kcal/mol for ΔH and -30 cal/(mol K) for ΔS for the reaction of phage R17 coat protein with its 21-nucleotide RNA binding site at 2°C. In a similar study of the equilibrium between the *lac* repressor and its operator, P. de Haseth, T. M. Lohman, and M. T. Record, Jr. [*Biochemistry,* **16**, 4783 (1977)] found that $\Delta H = 8.5$ kcal/mol and $\Delta S = 81$ cal/(mol K) at 24°C. Calculate ΔG for each reaction, and speculate on the possible significance of the difference between entropy-driven and enthalpy-driven reactions.

8. B. E. Eliel, K. D. Hargrave, K. M. Pietrusiewicz, and M. Manoharan [*J. Am. Chem. Soc.,* **104**, 3635 (1982)] measured the values of ΔG° for the following conformational transitions:

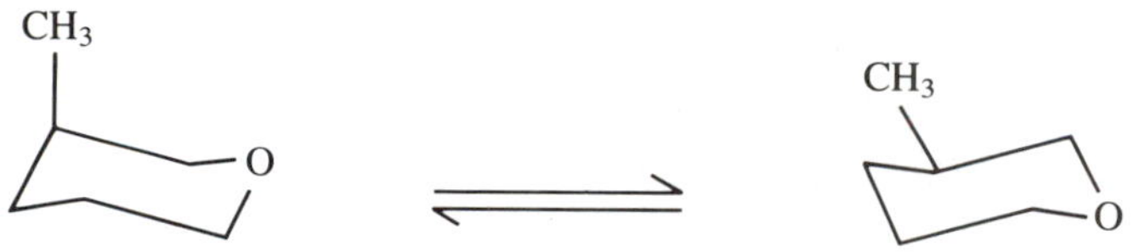

1, $-\Delta G^\circ = 1.44$ kcal/mol (173 K)

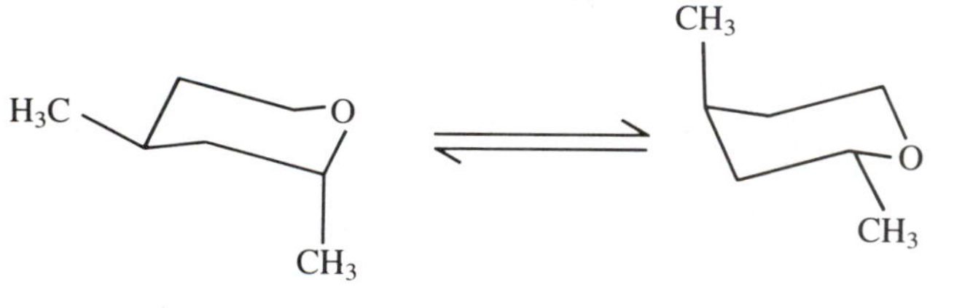

3, $-\Delta G^\circ = 0.89$ kcal/mol (173 K)

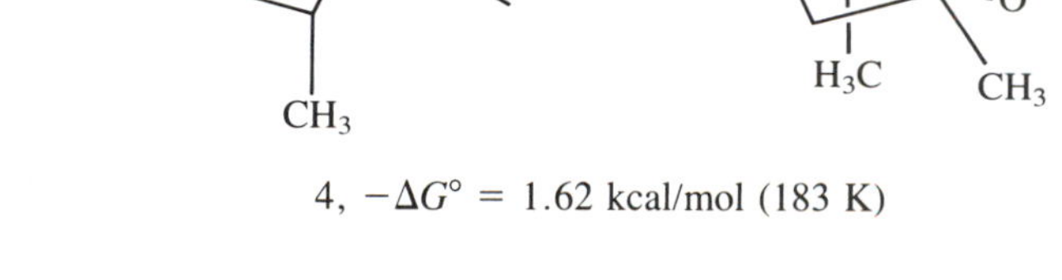

4, $-\Delta G^\circ = 1.62$ kcal/mol (183 K)

Calculate the equilibrium ratios of the conformers in each case. Reprinted with permission from B. E. Eliel, K. D. Hargrave, K. M. Pietrusiewicz, and M. Manoharan, *J. Am. Chem. Soc.* **104**, 3635 (1982). Copyright 1982 American Chemical Society.

REFERENCES

1. H. J. Kolb and R. L. Burwell, Jr., *J. Am. Chem. Soc.* **67**, 1084 (1945).
2. H. A. Bent, *J. Chem. Educ.* **50**, 323 (1973); J. N. Spencer, *J. Chem. Educ.* **51**, 577 (1974).
3. E. Hamori, *J. Chem. Educ.* **52**, 370 (1975).
4. W. M. Clark, *Oxidation-Reduction Potentials of Organic Systems*, Williams and Wilkins, Co., Baltimore, 1960.

CHAPTER 11

The Third Law of Thermodynamics

11-1 PURPOSE OF THE THIRD LAW

As we pointed out in Chapter 10, it would be an advantage to be able to calculate values of ΔG from thermal data alone, that is, from calorimetric measurements of heats of reaction and heat capacities. The limitations imposed by requiring that measurements be made for an equilibrium state or by carrying out a reaction reversibly thus could be avoided. The thermal-data method would be of particular advantage for reactions for which ΔG is very large (either positive or negative) since equilibrium measurements are most difficult for such reactions.

We saw in Chapter 5 that ΔH for a reaction at any temperature can be calculated from a value at one temperature and the values of the heat capacities of reactants and products in the temperature range of interest. Similarly, ΔS can be calculated at any temperature from the value at one temperature and the appropriate heat capacity data. However, unlike ΔH, ΔS cannot be calculated from thermal data alone (that is, without obtaining equilibrium data) on the basis of the first and second laws. For that, we require another postulate, the third law.

11-2 FORMULATION OF THE THIRD LAW

We have pointed out previously that for many reactions the contribution of the $T\Delta S$ term in Equation 8-27 is relatively small; thus, ΔG and ΔH frequently are close in value even at relatively high temperatures. In a comprehensive series of experiments on galvanic cells Richards [1] showed that as the temperature decreases, ΔG approaches ΔH more closely, in the manner indicated in Figure 11-1 or Figure 11-2. Although these results were only fragmentary evidence,

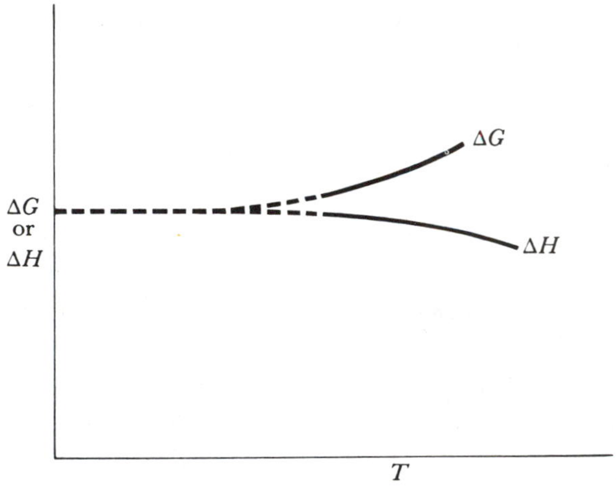

Figure 11-1. Limiting approach of ΔG and of ΔH as the temperature approaches absolute zero (selected from Richards's data).

they did furnish the clues that led Nernst to the first formulation of the third law of thermodynamics.

Nernst Heat Theorem

The trend of ΔG toward ΔH can be expressed as

$$\lim_{T \to 0} (\Delta G - \Delta H) = 0 \tag{11-1}$$

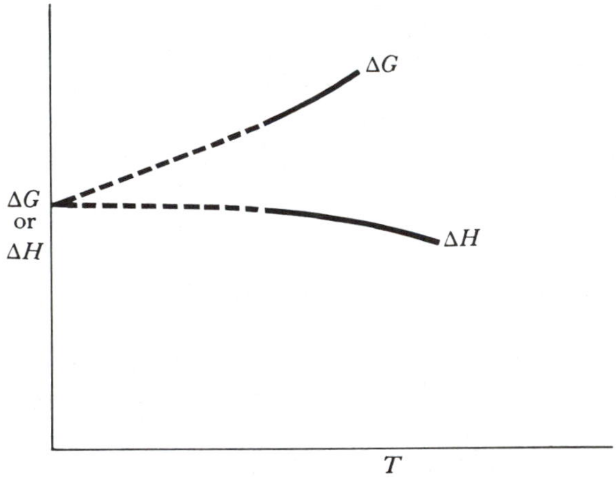

Figure 11-2. Alternative limiting approach of ΔG and of ΔH as the temperature approaches absolute zero (selected from Richard's data).

Equation 11-1 is a consequence of Equation 8-27,

$$\Delta G = \Delta H - T\Delta S$$

since

$$\lim_{T\to 0} (-T\Delta S) = 0 \tag{11-2}$$

without regard to the limit of ΔS as long as ΔS is finite. However, Nernst made the additional assumption, based on the appearance of some of Richards's curves (Figure 11-1), that the limiting value of ΔS is zero for all condensed systems:

$$\lim_{T\to 0} (-\Delta S) = \lim_{T\to 0} \left(\frac{\partial\, \Delta G}{\partial T}\right)_P = 0 \tag{11-3}$$

This assumption states that ΔG approaches ΔH as T approaches 0 K, but also that the ΔG curve (Figure 11-1) approaches a horizontal limiting tangent.

Nernst asserted his postulate although the available data were inconclusive. In fact, Richards extrapolated some of his data to give a graph such as that shown in Figure 11-2, which suggests that Equation 11-1 is valid but that Equation 11-3 is not. Numerous subsequent experiments have confirmed Nernst's postulate if it is limited to perfect crystalline systems. *Apparent* exceptions have been accounted for satisfactorily. The term *perfect* implies a single, pure substance. Other restrictions are implied by this term, but they will be discussed later.

Planck's Formulation

In Nernst's statement of the third law no comment is made on the value of the entropy of a substance at 0 K, although it follows from his hypothesis that all pure crystalline substances must have the same entropy at 0 K. Planck [2] extended Nernst's assumption by adding the postulate that *the value of the entropy of a pure solid or a pure liquid approaches zero at* 0 K:

$$\lim_{T\to 0} S = 0 \tag{11-4}$$

The assumption of any finite constant for the entropy of all pure solids and liquids at 0 K leads to Nernst's theorem (Equation 11-3) for these substances. Equation 11-4 provides a convenient value for that constant. Planck's statement asserts that $S_{0\,K}$ is zero only for *pure solids and pure liquids*, whereas Nernst assumed that his theorem was applicable to *all condensed phases*, including solutions. According to Planck, solutions at 0 K have a positive entropy equal to the entropy of mixing. (The entropy of mixing will be discussed in Chapters 13 and 15.)

Statement of Lewis and Randall

Lewis and Gibson [3] also emphasized the positive entropy of solutions at 0 K and also pointed out that supercooled liquids, such as glasses, even when composed of a single element (such as sulfur), probably retain positive entropies as the temperature approaches absolute zero. For these reasons Lewis and Randall [4] proposed the following statement of the third law of thermodynamics:

If the entropy of each element in some crystalline state be taken as zero at the absolute zero of temperature, every substance has a finite positive entropy, but at the absolute zero of temperature the entropy may become zero, and does so become in the case of perfect crystalline substances.

We will adopt this statement as the working form of the third law of thermodynamics. This statement is the most convenient formulation for making free energy calculations. Nevertheless, it should be realized that from a theoretical point of view more elegant formulations have been suggested [5].

The preceding statement of the third law has been formulated to exclude solutions and glasses from the class of substances that are assumed to have zero entropy at 0 K. Let us examine one example of each exclusion to see that this limitation is essential.

For the mixing process

$$\mathrm{AgCl(s,\ 0\ K)} + \mathrm{AgBr(s,\ 0\ K)} = \text{solid solution(s, 0 K)} \qquad (11\text{-}5)$$

the entropy change can be represented as

$$\Delta s_{0\,\mathrm{K}} = S_{\substack{\text{solid}\\ \text{soln}}} - s_{\mathrm{AgBr}} - s_{\mathrm{AgCl}}$$

and can be computed from the experimentally known ΔS_{298} for the same mixing reaction and heat capacity data from near 0 K to 298 K for each of the three solids [6].

For the formation of one mole of this solid solution, $\Delta s_{0\,\mathrm{K}}$ is 4.31 J K^{-1} mol^{-1}. Hence if s_{AgBr} and s_{AgCl} each are assigned zero at 0 K, the entropy of the solid solution at 0 K is not zero but 4.31 J K^{-1} mol^{-1}. This value is close to 4.85 J K^{-1} mol^{-1}, which is the value that would be calculated for the entropy of mixing to form an ideal solution.

Likewise, glasses do not have zero entropy at 0 K; that is, $\Delta s_{0\,\mathrm{K}}$ is not zero for a transition such as

$$\text{glycerol(crystalline, 0 K)} = \text{glycerol(glass, 0 K)} \qquad (11\text{-}6)$$

To calculate Δs for this transition, it is necessary to have heat capacity data for both solid forms of glycerol from near 0 K to the melting point and the heat of fusion of crystals. Such data [7] lead to a Δs for Equation 11-6 of 19.2 J K^{-1} mol^{-1}. Thus it follows that glassy glycerol cannot be assigned zero entropy at 0 K; rather, it possesses a residual entropy of 19.2 J K^{-1} mol^{-1}.

Many substances can exist in two or more crystalline forms at low temperatures. Of course, one form is more stable than the others. Nevertheless, if each is a *perfect crystalline substance*, its entropy at 0 K will be zero. For example, for the transition

$$\text{sulfur(monoclinic, 0 K)} = \text{sulfur(rhombic, 0 K)} \tag{11-7}$$

$\Delta s_{0\text{ K}}$ can be computed from heat capacity measurements [8] for each crystalline form from near 0 K to the transition temperature (368.6 K) and the heat of transition. The result is zero within experimental error. Hence both rhombic and monoclinic sulfur are assigned zero entropy at 0 K.

Unattainability of Absolute Zero

For purposes of computation the statement of Lewis and Randall serves very well. However, there is an alternative statement of the third law (from which the Nernst heat theorem and the Lewis and Randall statement can follow as consequences), which has more general significance. It is in the form of a principle of impotence, parallel to the statements of the first two laws:

It is impossible to attain absolute zero in a finite number of operations.

At first glance, it may seem that this is not an independent principle. The classical kinetic theory assumption that the heat capacity of a substance has a constant, nonzero value at all temperatures, together with the first and second laws of thermodynamics, leads directly to the statement of the unattainability of absolute zero. If we consider an entropy change at constant pressure, then from Equations 7-48 and 4-17,

$$dS = \frac{DQ_P}{T} = \frac{C_p}{T}\,dT \tag{11-8}$$

Since the heat capacity is assumed to have a nonzero value at all temperatures,

$$\lim_{T\to 0}\left(\frac{\partial S}{\partial T}\right)_P = \lim_{T\to 0}\frac{C_p}{T} = \infty \tag{11-9}$$

Consequently, the entropy of any substance should tend toward negative infinity as T approaches absolute zero, and for no finite series of processes could T reach absolute zero.

However, both the quantum description of matter and experimental determination indicate that heat capacities tend toward zero as $T \to 0$ K (at a rate proportional to T for metals and to T^3 for other solids). Hence unattainability of absolute zero does not follow from the first two laws of thermodynamics, and another rationale must be sought for the unattainability of absolute zero. In low-temperature cooling experiments the common experience that it becomes increasingly difficult to achieve a still lower temperature as the experimental temperature decreases is summarized in the postulate that:

No system can be reduced to 0 K.

Like the statements of the first and second laws, we accept this statement because deductions from it have not been contradicted by experiment.

The unattainability statement of the third law is no more convenient for computation than the principle of impotence form of Clausius's statement of the second law. However, it can be demonstrated that the Lewis and Randall statement follows from the unattainability principle; thus, it is not an independent theorem.

We also can show that the assumption that $\Delta S_{0\,\mathrm{K}}$ is zero leads to the unattainability statement. To show this, let us consider the problem of cooling a system below any hitherto attainable temperature. The process for doing this might be an allotropic transformation, a change in external magnetic field, or a volume change. If an allotropic change from form A at temperature T_1 to form B at temperature T_2 is examined under adiabatic conditions (since any reservoir available is not at a lower temperature than the working system), we can write

$$S_{\mathrm{A}}(T_1) = S_{\mathrm{A}}(0\ \mathrm{K}) + \int_0^{T_1} \frac{C_{\mathrm{A}}}{T}\, dT \tag{11-10}$$

and

$$S_{\mathrm{B}}(T_2) = S_{\mathrm{B}}(0\ \mathrm{K}) + \int_0^{T_2} \frac{C_{\mathrm{B}}}{T}\, dT \tag{11-11}$$

The adiabatic transformation may be either reversible or irreversible. In the former circumstance, since ΔS is zero for any adiabatic change, it follows from Equations 11-10 and 11-11 that

$$S_{\mathrm{A}}(0\ \mathrm{K}) + \int_0^{T_1} \frac{C_{\mathrm{A}}}{T}\, dT = S_{\mathrm{B}}(0\ \mathrm{K}) + \int_0^{T_2} \frac{C_{\mathrm{B}}}{T}\, dT \tag{11-12}$$

If we take as a statement of the third law

$$S_{\mathrm{A}}(0\ \mathrm{K}) = S_{\mathrm{B}}(0\ \mathrm{K}) \tag{11-13}$$

then Equation 11-12 becomes

$$\int_0^{T_1} \frac{C_A}{T}\,dT = \int_0^{T_2} \frac{C_B}{T}\,dT \tag{11-14}$$

The left side of Equation 11-14 must be positive since $T_1 > 0$ and since $C_A > 0$ for any nonzero temperature. Therefore the right side of Equation 11-14 also must be positive, and this can be true for $C_B > 0$ only if $T_2 \neq 0$, that is, if the cooling process does not attain 0 K.

If the transformation from A to B is irreversible, $\Delta S > 0$, and in place of Equation 11-12 we must write

$$S_A(0\text{ K}) + \int_0^{T_1} \frac{C_A}{T}\,dT < S_B(0\text{ K}) + \int_0^{T_2} \frac{C_B}{T}\,dT \tag{11-15}$$

Again, if we take Equation 11-13 as a statement of the third law it follows that

$$\int_0^{T_1} \frac{C_A}{T}\,dT < \int_0^{T_2} \frac{C_B}{T}\,dT \tag{11-16}$$

Since the left side of Equation 11-16 is positive it follows that the right side, and therefore T_2, is positive. Thus the assumption that ΔS (0 K) is equal to zero leads to the conclusion that absolute zero is unattainable.

11-3 THERMODYNAMIC PROPERTIES AT ABSOLUTE ZERO

From the third law of thermodynamics it is possible to derive a number of limiting relationships for the values of thermodynamic quantities at absolute zero for perfect crystalline substances.

Equivalence of *G* and *H*

It follows immediately from the Lewis and Randall statement of the third law and the definition of free energy that

$$G_{0\text{ K}} = H_{0\text{ K}} - TS_{0\text{ K}} = H_{0\text{ K}} \tag{11-17}$$

ΔC_p in an Isothermal Chemical Transformation

From Equations 11-2 and 11-3 we can see that

$$\lim_{T\to 0}\left(\frac{\partial\ \Delta G}{\partial T}\right)_P = \lim_{T\to 0}(-\Delta S) = \lim_{T\to 0}\frac{\Delta G - \Delta H}{T} = 0 \qquad (11\text{-}18)$$

The expression

$$\lim_{T\to 0}[(\Delta G - \Delta H)/T]$$

is indeterminate since both $(\Delta G - \Delta H)$ and T approach zero. To resolve an indeterminate expression, we can apply the mathematical rule of differentiating numerator and denominator, respectively, with respect to the independent variable, T. Carrying out this procedure, from Equation 11-18 we obtain

$$\lim_{T\to 0}\frac{(\partial\ \Delta G/\partial T)_P - (\partial\ \Delta H/\partial T)_P}{1} = 0$$

and

$$\lim_{T\to 0}\left(\frac{\partial\ \Delta G}{\partial T}\right)_P = \lim_{T\to 0}\left(\frac{\partial\ \Delta H}{\partial T}\right)_P = \lim_{T\to 0}\Delta C_p \qquad (11\text{-}19)$$

From Equation 11-3 and Equation 11-19 it follows that

$$\lim_{T\to 0}\Delta C_p = 0 \qquad (11\text{-}20)$$

Many investigators have shown that ΔC_p does approach zero as T approaches absolute zero. Nevertheless, these results in themselves do not constitute experimental evidence for the third law, which is a sufficient, but not a necessary, condition for Equation 11-20. If $(\partial\ \Delta G/\partial T)_P$ is a nonzero, finite number, it can be shown by a series of equations corresponding to Equations 11-18 and 11-19 that Equation 11-20 is still valid.

Limiting Values of C_p and C_v

The third law asserts that the entropy of a substance (referred to the corresponding elements) must be finite or zero at absolute zero. In view of the finite values observed for ΔS at higher temperatures it follows that the entropy of a substance must be finite at all (finite) temperatures.

According to Equation 11-8

$$dS_p = \frac{DQ_P}{T} = \frac{C_p dT}{T}$$

This differential equation can be integrated *at constant pressure* to give

$$S(T) = \int_0^T \frac{C_p dT}{T} + S(0\text{ K}) \tag{11-21}$$

Since S must be finite at all temperatures it follows that

$$\lim_{T\to 0} C_p = 0 \tag{11-22}$$

If C_p had a finite value at $T = 0$, the integral in Equation 11-21 would not converge, since T in the denominator goes to zero and S would not be finite.

By an analogous procedure we can show that

$$\lim_{T\to 0} C_v = 0 \tag{11-23}$$

Temperature Coefficients of Pressure and Volume

From Equation 11-4

$$\lim_{T\to 0} S = 0$$

It follows that in the limit of absolute zero the entropy of a perfect crystalline substance must be independent of changes in pressure or volume (or any other variable of state except T). Thus

$$\lim_{T\to 0} \left(\frac{\partial S}{\partial P}\right)_T = 0 \tag{11-24}$$

and

$$\lim_{T\to 0} \left(\frac{\partial S}{\partial V}\right)_T = 0 \tag{11-25}$$

Applying Equation 8-49,

$$\left(\frac{\partial S}{\partial P}\right)_T = -\left(\frac{\partial V}{\partial T}\right)_P$$

to Equation 11-24 we obtain

$$\lim_{T\to 0} \left(\frac{\partial V}{\partial T}\right)_P = 0 \tag{11-26}$$

Similarly, Equation 8-50,

$$\left(\frac{\partial S}{\partial V}\right)_T = \left(\frac{\partial P}{\partial T}\right)_V$$

and Equation 11-25 lead to

$$\lim_{T\to 0}\left(\frac{\partial P}{\partial T}\right)_V = 0 \tag{11-27}$$

In other words, the temperature gradients of the pressure and volume vanish as absolute zero is approached.

11-4 ENTROPIES AT 298 K

In the statement that we have adopted for the third law it is assumed (arbitrarily) that the entropy of each element in some crystalline state is zero at 0 K. Then for every perfect crystalline substance the entropy is also zero at 0 K. Consequently we can set S(0 K) in Equation 11-21 equal to zero. Thus we may write

$$S(T) = \int_0^T \frac{C_p dT}{T} \tag{11-28}$$

and we can evaluate the entropy of a perfect crystalline solid at any specified temperature by integrating Equation 11-28. The molar entropy so obtained frequently is called the "absolute" entropy and is indicated as s_T°. However, in no sense is s_T° truly an absolute entropy, because Equation 11-28 is based on the *convention* that zero entropy is assigned to each element in some state at 0 K. For example, this convention neglects any entropy associated with the nucleus, because no *changes* in nuclear entropy are expected under the conditions in which chemical reactions occur. Entropies obtained from Equation 11-28 properly are called *conventional* entropies or *standard* entropies.

To calculate the entropy of a substance at a temperature at which it is no longer a solid, it is necessary to add the entropy of transformation to a liquid or gas and the subsequent entropies of warming. The same procedure would apply to a solid that exists in different crystalline forms as the temperature is raised. The procedure can be illustrated by some sample calculations.

Typical Calculations

For Solid or Liquid. For either of these final states it is necessary to have heat capacity data for the solid down to temperatures approaching absolute zero.

The integration indicated by Equation 11-28 then is carried out in two steps. From approximately 20 K up, graphical or numerical methods can be used. However, below 20 K few data are available. Therefore it is customary to rely on the Debye equation in this region.

Use of Debye Equation at Very Low Temperatures. Generally, it is assumed that the Debye equation expresses the behavior of the heat capacity adequately below about 20 K [9]. This relationship (Equation 4-37),

$$c_p \cong c_v = 1943.8 \frac{T^3}{\theta^3} \text{ J mol}^{-1} \text{ K}^{-1}$$

contains only one constant, θ, which can be determined from a value of c_p in the region below 20 K. The integral for the entropy then becomes

$$s = \int_0^T \frac{kT^3}{T} dT = \int_0^T kT^2 \, dT = \frac{kT^3}{3} = \frac{c_p(T)}{3} \tag{11-29}$$

in which k represents $1943.8/\theta^3$.

Entropy of Methylammonium Chloride. Heat capacities for this solid in its various crystalline modifications have been determined [10] very precisely down to 12 K. Some of these data are summarized in Figure 11-3. There are three crystalline forms between 0 K and 298 K. One can calculate the entropy by integrating Equation 11-28 for each allotrope in the temperature region in which it is most stable and then adding the two entropies of transition to the integrals thus obtained. The details at a pressure of 101.3 kPa are as follows:

a. CH_3NH_3Cl(s, β form, 0 K) = CH_3NH_3Cl(s, β form, 12.04 K)

$$\Delta S_1 = \Delta s_1 = \int_{0 \text{ K}}^{12.04 \text{ K}} \frac{c_p dT}{T} = 0.280 \text{ J mol}^{-1} \text{ K}^{-1} \begin{pmatrix} \text{Debye equation} \\ \text{with } \theta = 200.5 \text{ K} \end{pmatrix}$$

b. CH_3NH_3Cl(s, β form, 12.04 K) = CH_3NH_3Cl(s, β form, 220.4 K)

$$\Delta S_2 = \Delta s_2 = \int_{12.04 \text{ K}}^{220.4 \text{ K}} \frac{c_p dT}{T} = 93.412 \text{ J mol}^{-1} \text{ K}^{-1} \text{ (numerical integration)}$$

c. CH_3NH_3Cl(s, β form, 220.4 K) = CH_3NH_3Cl(s, γ form, 220.4 K)

$$\Delta S_3 = \Delta s_3 = \frac{\Delta H}{T} = \frac{1779.0}{220.4} = 8.072 \text{ J mol}^{-1} \text{ K}^{-1}$$

d. CH_3NH_3Cl(s, γ form, 220.4 K) = CH_3NH_3Cl(s, γ form, 264.5 K)

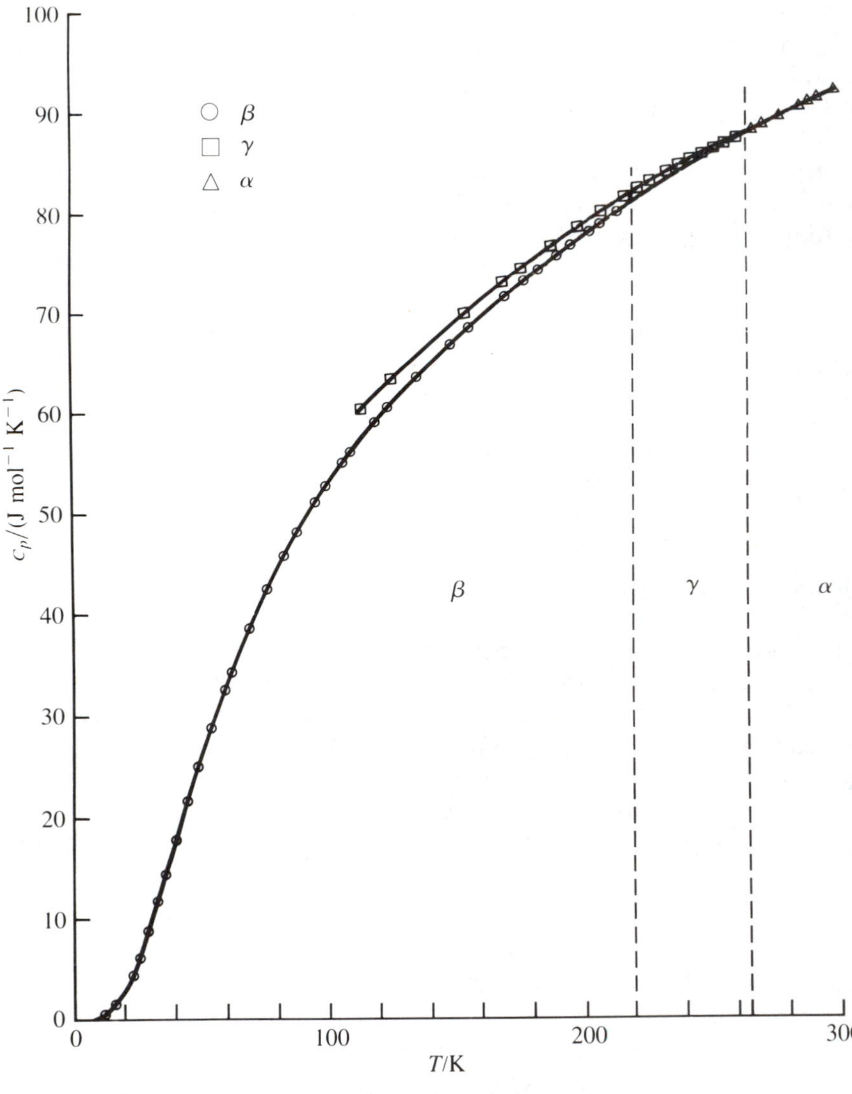

Figure 11-3. Heat capacities of the three allotropic forms, α, β, γ, of methylammonium chloride. The dashed curve represents the heat capacity of the metastable γ form, supercooled.

$$\Delta S_4 = \Delta s_4 = \int_{220.4\ \mathrm{K}}^{264.5\ \mathrm{K}} \frac{c_p'\, dT}{T} = 15.439\ \mathrm{J\ mol^{-1}\ K^{-1}}\ \text{(numerical integration)}$$

e. CH_3NH_3Cl(s, γ form, 264.5 K) = CH_3NH_3Cl(s, α form, 264.5 K)

$$\Delta S_5 = \Delta s_5 = \frac{\Delta h'}{T'} = \frac{2818.3}{264.5} = 10.655 \text{ J mol}^{-1} \text{ K}^{-1}$$

f. CH_3NH_3Cl(s, α form, 264.5 K) = CH_3NH_3Cl(s, α form, 298.15 K)

$$\Delta S_6 = \Delta s_6 = \int_{264.5 \text{ K}}^{298.15 \text{ K}} \frac{C_p'' \, dT}{T} = 10.690 \text{ J mol}^{-1} \text{ K}^{-1} \text{ (numerical integration)}$$

Addition of Steps a through f gives

g. CH_3NH_3Cl(s, β form, 0 K) = CH_3NH_3Cl(s, α form, 298.15 K)

and

$$\Delta S^\circ = \Delta s^\circ = \sum_1^6 \Delta s_i = 138.548 \text{ J mol}^{-1} \text{ K}^{-1}$$

Thus for methylammonium chloride s°_{298} is 138.548 J mol^{-1} K^{-1}.

For a Gas. The procedure for the calculation of the entropy of a gas in its standard state is substantially the same as that for a solid or liquid except for two factors. If the heat capacity data have been obtained at a pressure of 1 atm (101.325 kPa), the resultant value of s^* is appropriate for that pressure and must be corrected to the standard state pressure of 1 bar (0.1 MPa). This correction is given by the formula [11]

$$\begin{aligned} s^\circ(T) - s(T) &= R \ln (p/p^\circ) \qquad (11\text{-}30) \\ &= R \ln (101.325 \text{ kPa}/100 \text{ kPa}) \\ &= 0.1094 \text{ J mol}^{-1} \text{ K}^{-1} \end{aligned}$$

In addition to the correction for the new standard state, it will be necessary to correct for the transformation from the real gas at standard pressure to the ideal gas at standard pressure, which is defined as the standard state.

Correction for Gas Imperfection. The nature of the correction is indicated in Figure 11-4. From thermal data we obtain the entropy indicated by line *b*. However, it is the entropy indicated by line *d* that we need; that is, we want to find the entropy the substance would have if it behaved as an ideal gas at the given temperature and 0.1 MPa.

At this stage in our discussion of thermodynamic concepts we know of no direct method to get from *b* to *d*. However, since entropy is a thermodynamic property we can use any path to calculate Δs as long as the initial and final states are *b* and *d*, respectively.

If we know $s_{298.15}$ for the real gas at 0.1 MPa, then we know Δs for the reaction

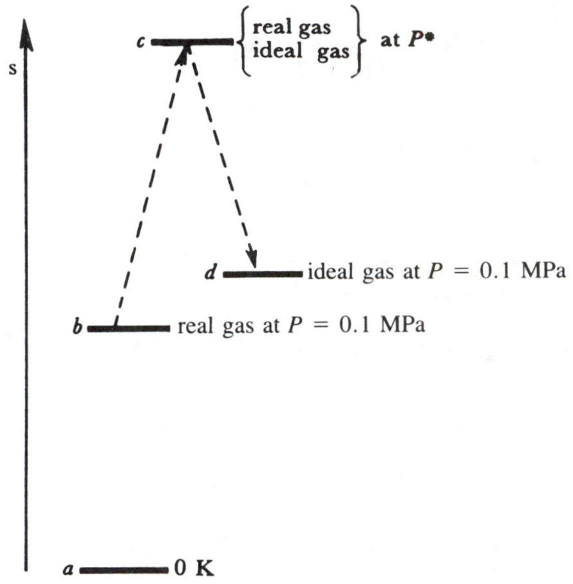

Figure 11-4. Schematic representation of various entropies in connection with correction for gas imperfections.

$$\text{A(s, 0 K)} = \text{A(g, } P = 0.1 \text{ MPa, 298.15 K)}$$

$$\Delta S = \Delta s = S_{298.15} \text{ for A(g, } P = 0.1 \text{ MPa, real, 298.15 K)} \qquad (11\text{-}31)$$

To this equation we can add the following two transformations:

$$\text{A(g, } P = 0.1 \text{ MPa, real, 298.15 K)} = \text{A(g, } P = P^*\text{, real, 298.15 K)} \qquad (11\text{-}32)$$

$$\text{A(g, } P = P^*\text{, ideal, 298.15 K)} = \text{A(g, } P = 0.1 \text{ MPa, ideal, 298.15 K)} \qquad (11\text{-}33)$$

for each of which ΔS can be calculated. If P^* approaches zero, then the real gas approaches ideality in its behavior. Hence the right side of Equation 11-32 is the same as the left side of Equation 11-33, and the addition of Equations 11-32 and 11-33 gives

$$\text{A(g, } P = 0.1 \text{ MPa, real, 298.15 K)} = \text{A(g, } P = 0.1 \text{ MPa, ideal, 298.15 K)} \qquad (11\text{-}34)$$

The sum of the entropy changes for Reactions 11-31 and 11-34 gives s°_{298} for the gas.

The entropy changes for Equations 11-32 and 11-33 can be calculated from the relationship (Equation 8-49)

$$\left(\frac{\partial S}{\partial P}\right)_T = -\left(\frac{\partial V}{\partial T}\right)_P$$

or

$$dS = -\left(\frac{\partial V}{\partial T}\right)_P dP$$

Thus for the transformation indicated in Equation 11-32

$$s(P = P^*) - s(P = 0.1\text{ MPa}) = -\int_{0.1\text{ MPa}}^{P^*} \left(\frac{\partial V_{\text{real}}}{\partial T}\right)_P dP = +\int_{P^*}^{0.1\text{ MPa}} \left(\frac{\partial v}{\partial T}\right)_P dP \tag{11-35}$$

and for that in Equation 11-33

$$s_i(P = 0.1\text{ MPa}) - s_i(P = P^*) = -\int_{P^*}^{0.1\text{ MPa}} \left(\frac{\partial V_{\text{ideal}}}{\partial T}\right)_P dP = -\int_{P^*}^{0.1\text{ MPa}} \frac{R}{P}\, dP \tag{11-36}$$

in which the subscript i refers to the ideal gas. If P^* is allowed to approach zero then

$$s_i(P = P^*) = s(P = P^*)$$

and the correction for gas imperfection becomes

$$s_i(P = 0.1\text{ MPa}) - s(P = 0.1\text{ MPa}) = \int_{P^*}^{0.1\text{ MPa}} \left(\frac{\partial v}{\partial T}\right)_P dP - \int_{P^*}^{0.1\text{ MPa}} \frac{R}{P}\, dP$$

$$= \int_{P^*}^{0.1\text{ MPa}} \left[\left(\frac{\partial v}{\partial T}\right)_P - \frac{R}{P}\right] dP \tag{11-37}$$

Equation 11-37 can be integrated if the value of $(\partial v/\partial T)_P$ is known for the gas under consideration. Generally, this coefficient is obtained from the Berthelot equation of state (Equation 6-55):

$$Pv = RT\left[1 + \frac{9}{128}\frac{P}{P_c}\frac{T_c}{T}\left(1 - 6\frac{T_c^2}{T^2}\right)\right]$$

Differentiation gives

$$\left(\frac{\partial v}{\partial T}\right)_P = \frac{R}{P} + \frac{9}{128} R \frac{1}{P_c} T_c\left(12\frac{T_c^2}{T^3}\right) = \frac{R}{P}\left[1 + \frac{27}{32}\frac{P}{P_c}\frac{T_c^3}{T^3}\right]$$

Therefore the equation for the correction becomes

$$(s_i - s)_{P=0.1\text{ MPa}} = \int_{P^*}^{0.1\text{ MPa}} \frac{27}{32}\frac{R}{P_c}\frac{T_c^3}{T^3}\,dP = \frac{27}{32}R\frac{P}{P_c}\frac{T_c^3}{T^3}\Bigg]_{P^*}^{P=0.1\text{ MPa}} \tag{11-38}$$

If P^* approaches zero

$$(s_i - s)_{P=0.1\text{ MPa}} = \frac{27}{32}\frac{R}{P_c}\frac{T_c^3}{T^3}(0.1\text{ MPa}) \tag{11-39}$$

Thus from a knowledge of the critical constants of the gas it is possible to evaluate the correction in the entropy for deviations from ideal behavior if the Berthelot equation is applicable [12].

Entropy of Gaseous Cyclopropane at its Boiling Point. Heat capacities for cyclopropane have been measured down to temperatures approaching absolute zero by Ruehrwein and Powell [13]. Their calculation of the entropy of the gas at the boiling point, 240.30 K, is summarized as follows:

a. $C_3H_6(s, 0\text{ K}) = C_3H_6(s, 15\text{ K})$

$$\Delta S_1 = \Delta s_1 = 1.017\text{ J mol}^{-1}\text{ K}^{-1}\text{ (Debye equation with } \theta = 130\text{ K)}$$

b. $C_3H_6(s, 15\text{ K}) = C_3H_6(s, 145.54\text{ K})$

$$\Delta S_2 = \Delta s_2 = 65.827\text{ J mol}^{-1}\text{ K}^{-1}\text{ (numerical integration)}$$

c. $C_3H_6(s, 145.54\text{ K}) = C_3H_6(l, 145.54\text{ K})$

$$\Delta S_3 = \Delta s_3 = \frac{\Delta H_{\text{fusion}}}{T} = 37.401\text{ J mol}^{-1}\text{ K}^{-1}$$

d. $C_3H_6(l, 145.54\text{ K}) = C_3H_6(l, 240.30\text{ K})$

$$\Delta S_4 = \Delta s_4 = 38.392\text{ J mol}^{-1}\text{ K}^{-1}\text{ (numerical integration)}$$

e. $C_3H_6(l, 240.30\text{ K}) = C_3H_6(\text{real gas}, 240.30\text{ K})$

$$\Delta S_5 = \Delta s_5 = \frac{\Delta H_{\text{vaporization}}}{T} = 83.454\text{ J mol}^{-1}\text{ K}^{-1}$$

Summing Steps a through e we obtain

f. $C_3H_6(s, 0\text{ K}) = C_3H_6(\text{real gas}, 240.30\text{ K})$

$$\Delta S = \Delta s = \sum_{1}^{5} \Delta s_i = 226.10\text{ J mol}^{-1}\text{ K}^{-1}$$

at a pressure of the real gas equal to 101.325 kPa. To correct the entropy of the real gas to a pressure of 0.1 MPa, we must add 0.11 J mol K to the value above.

The critical constants used for the correction for gas imperfection were $T_c = 375$ K and $P_c = 5.065$ MPa. Thus for the transformation

g. C_3H_6(real gas, 240.30 K) = C_3H_6(ideal gas, 240.30 K)

Equation 11-39 yields

$$\Delta S_6 = \Delta s_6 = 0.53 \text{ J mol}^{-1} \text{ K}^{-1}$$

Therefore the entropy, $s^\circ_{240.30}$, of cyclopropane in the ideal gas (that is, standard) state is 226.74 J mol^{-1} K^{-1}.

The correction for gas imperfection may seem small. However, we should keep in mind that an error of 0.4 J K^{-1} mol^{-1} affects the free energy by about 125 J mol^{-1} near room temperature since ΔS would be multiplied by T. An error of 125 J mol^{-1} would change an equilibrium constant of 1.00 to 1.05, a difference of 5%.

Apparent Exceptions to the Third Law

There are several cases in which calculations of the entropy change of a reaction from values of the entropy obtained from thermal data and the third law disagree with values calculated directly from measurements of ΔH° and determinations of ΔG° from experimental equilibrium constants. For example, for the reaction

$$H_2(g) + \tfrac{1}{2}O_2(g) = H_2O(l)$$

$$\Delta s^\circ(\text{thermal data}) = -153.6 \text{ J mol}^{-1} \text{ K}^{-1} \tag{11-40}$$

$$\Delta s^\circ(\text{equilibrium}) = \frac{\Delta H^\circ - \Delta G^\circ}{T} = -163.3 \text{ J mol}^{-1} \text{ K}^{-1} \tag{11-41}$$

Thus there is a large discrepancy between the two entropy values.

A satisfactory explanation for this discrepancy was not available until the development of statistical thermodynamics with its methods of calculating entropies from spectroscopic data, and the discovery of the existence of ortho- and parahydrogen. It then was found that the major portion of the deviation observed between Equations 11-40 and 11-41 is due to the failure to obtain a true equilibrium between these two forms of H_2 molecules (which differ in their nuclear spins) during thermal measurements at very low temperatures (Figure 11-5). If true equilibrium were established at all times, more parahydrogen would be formed as the temperature is lowered, and at 0 K all the hydrogen molecules would be in the para form and the entropy would be zero. In practice, measurements actually are made on a 3/1 mixture of ortho/para. This mixture at 0 K has a positive entropy. If the hydrogen were in contact with an appropriate catalyst for ortho–para conversion, an equilibrium mixture would be obtained, and the entropy could be calculated correctly from the integral that gives the area under the equilibrium curve in Figure 11-5.

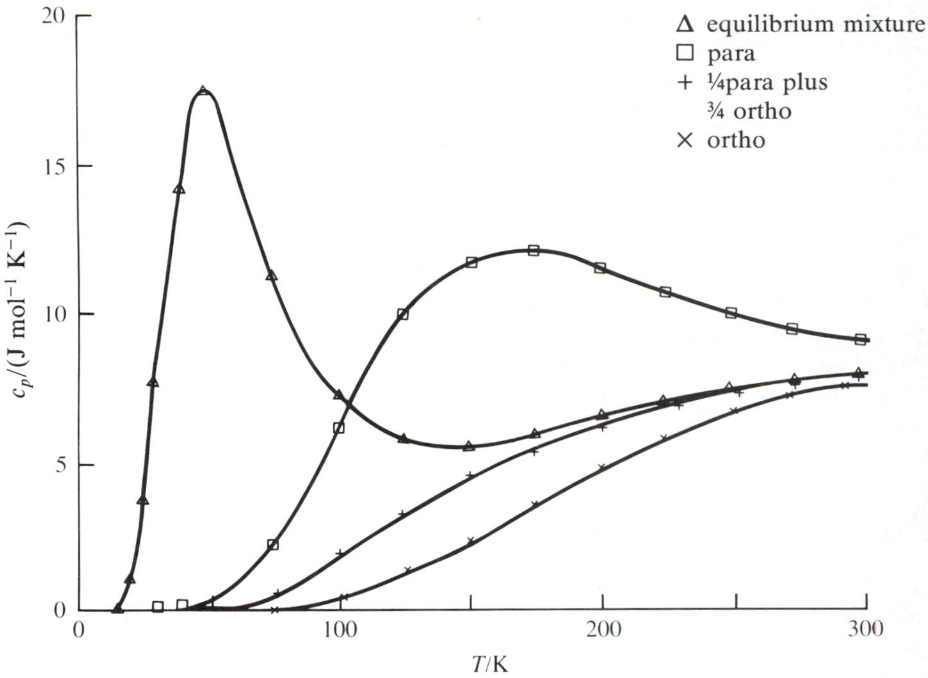

Figure 11-5. Heat capacities (excluding translation) for hydrogen gas as a function of temperature. Based on data of W. F. Giauque, *J. Am. Chem. Soc.*, **52** 4816 (1930).

When the corrections for ortho-/parahydrogen are applied to Equation 11-40 the value obtained is -163.2 J mol^{-1} K^{-1}, in agreement with the equilibrium value. Most recent critical tables list s°_{298} corrected for the effects discussed here.

With the development of statistical thermodynamics and the calculation of the entropies of many substances from spectroscopic data, several other substances in addition to hydrogen have been found to have values of molar entropies that disagree with those calculated from thermal data alone [14] (Table 11-1). The discrepancies can be accounted for on the assumption that even near absolute zero not all the molecules are in the same state and that true equilibrium has not been attained. For CO, $COCl_2$, N_2O, NO, and ClO_3F, the close similarity in the sizes of the atoms makes different orientations possible in the crystals, whereas in the case of H_2O, hydrogen bonds maintain an irregularity in the distribution of molecules in the crystal. For example, in carbon monoxide molecules may have random positions such as CO and OC in the crystal. Because of these exceptional situations it is necessary to interpret the term *perfect crystal* as excluding situations in which several orientations of the molecules are present simultaneously.

Table 11-1 Molar Entropies

Substance	Temperature/ K	$s°$ (spectroscopic)/ J mol^{-1} K^{-1}	$s°$ (calorimetric)/ J mol^{-1} K^{-1}	Deviation/ J mol^{-1} K^{-1}
CO	298.1	197.958	193.3	4.7
$COCl_2$	280.6	285.60	278.78	6.82
H_2O	298.1	188.70	185.27	3.43
N_2O	298.1	219.999	215.22	4.78
NO	121.4	183.05	179.9	3.2
ClO_3F	226.48	261.88	251.75	10.13

Ulbrich and Waldbaum [15] have pointed out that calorimetrically determined third law entropies for many geologically important minerals may be in error because site mixing among cations, magnetic spin disorder, and disorder among water molecules in the crystals is frozen in in the samples used for calorimetric measurements. They have calculated corrections based on known crystallographic data for a number of minerals.

An exceptional case of a very different type is provided by helium [16], for which the third law is valid despite the fact that He remains a liquid at 0 K. A phase diagram for helium is shown in Figure 11-6. In this case, in contrast to other substances, the solid–liquid equilibrium line at high pressures does not continue downward at low pressures until it meets the liquid–vapor pressure curve to intersect at a triple point. Rather, the solid–liquid equilibrium line takes an unusual turn toward the horizontal as the temperature drops to near 2 K. This change is due to a surprising metamorphosis in the character of liquid helium as the temperature drops below 2.2 K. Below this temperature the liquid has a heat conductivity 100 times greater than that of a metal such as copper or silver, and becomes a superfluid in its flow behavior with a viscosity less than 10^{-9} that of a liquid such as water.

This transformed liquid, labeled He^{II}, also shows unusual thermal properties. One of these, its entropy, is illustrated in Figure 11-7. An important feature of this curve is the approach of s to zero as T approaches 0. Thus liquid He^{II} possesses zero entropy at 0 K despite the fact that it is a liquid.

A confirmation of this conclusion also is provided by an examination of the solid–liquid equilibrium in the neighborhood of 0 K. As shown in Equation 9-8 a two-phase equilibrium obeys the Clapeyron equation,

$$\frac{dP}{dT} = \frac{s_{\text{liquid}} - s_{\text{solid}}}{v_{\text{liquid}} - v_{\text{solid}}}$$

The densities of liquid and solid helium are different, thus Δv of Equation 9-8 is not zero. Yet the horizontal slope of the melting line of the phase diagram shows that dP/dT is zero near 0 K. Hence it is clear that Δs of Equation 9-8

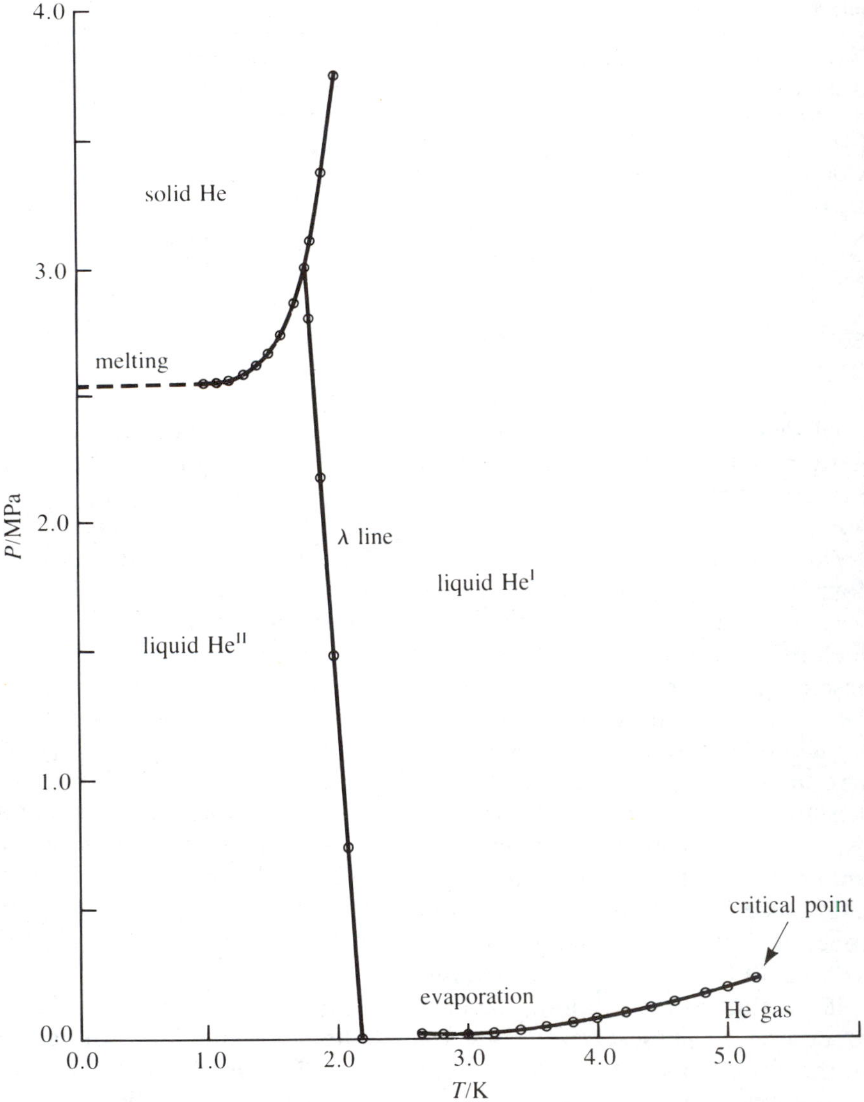

Figure 11-6. Phase diagram for ^{4}He. Data for the melting curve and the λ line from C. A. Swenson, *Phys. Rev.* **79**, 626 (1950), **89**, 541 (1953). Data for the evaporation curve from H. Van Dijk and M. Durieux, *Physica,* **24**, 920 (1958). The evaporation curve was measured down to 0.5 K, but the values of the vapor pressure were too small to be visible on the scale of this graph.

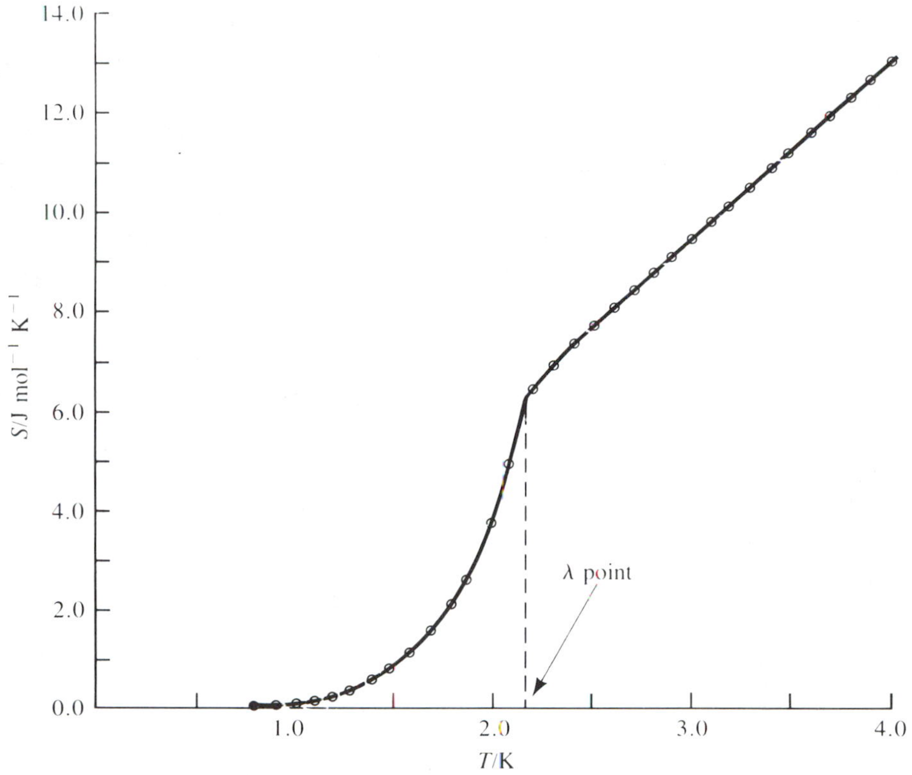

Figure 11-7. Entropy of liquid ^{4}He under its equilibrium vapor pressure. Data below 1.90 K from H. C. Kramers, J. D. Wasscher, and C. J. Gorter, *Physica,* **18**, 329 (1952). Data from 1.90 K to 4.00 K from R. W. Hill and O. V. Lounasmaa, *Phil. Mag.,* Ser. 8, **2**, 143 (1957).

must be zero at 0 K, that is, that $s_{0\ \mathrm{K}}$ is zero for liquid He as well as solid He.

The λ transition in liquid helium shown in Figures 11-6 and 11-7 is a second-order transition. Most phase transitions that follow the Clapeyron equation exhibit a nonzero value of Δs and Δv; that is, they show a discontinuity in s and v, the first derivatives of the Gibbs free energy, G. Thus they are called first-order transitions. In contrast, the λ transition shows a zero value of Δs and Δv and exhibits discontinuities in the second derivatives of G, such as the heat capacity, c_p.

Tabulation of Entropy Values

Several sources that have critical tabulations of molar entropies are available:

International Critical Tables, McGraw-Hill, New York, 1933.

Landolt-Börnstein, *Physikalisch-chemische Tabellen*, 5th ed., Springer, Berlin, 1936; 6th ed., 1961, 1963, 1967, 1972, 1980.

K. K. Kelley, *U.S. Bur. Mines Bull.* 477 (1950); K. K. Kelley and E. G. King, *U.S. Bur. Mines Bull.*, 592 (1961).

W. M. Latimer, *Oxidation Potentials*, 2d ed., Prentice-Hall, Englewood Cliffs, N.J., 1952.

"NBS Tables of Chemical Thermodynamic Properties," *J. Phys. Chem. Ref. Data,* **11**, Supplement No. 2 (1982).

R. Hultgren, P. D. Desai, D. T. Hawkins, M. Gleiser, K. K. Kelley, and D. D. Wagman, "Selected Values of Thermodynamic Properties of the Elements," American Society of Metals, Metals Park, Ohio, 1973.

R. Hultgren, P. D. Desai, D. T. Hawkins, M. Gleiser, and K. K. Kelley, "Selected Values of the Thermodynamic Properties of Binary Alloys," American Society for Metals, Metals Park, Ohio, 1973.

R. A. Robie, B. S. Hemingway, and J. R. Fisher, "Thermodynamic Properties of Minerals and Related Substances," *Geological Survey Bulletin* 1452 (1978).

TRC Thermodynamic Tables–Hydrocarbons, Thermodynamics Research Center: The Texas A & M University System, College Station, Texas (Loose-leaf Data Sheets).

TRC Thermodynamic Tables–Non-hydrocarbons, Thermodynamics Research Center: The Texas A & M University System, College Station, Texas (Loose-leaf Data Sheets).

D. R. Stull and H. Prophet, JANAF Thermochemical Tables, 2d ed., National Standard Reference Data Series, Nat. Bur. Stand. (U.S.), **37**, 1971.

M. W. Chase, J. L. Curnutt, A. T. Hu, H. Prophet, A. Syverud, and L. C. Walker, JANAF Thermochemical Tables, 1974 Supplement, *J. Phys. Chem. Ref. Data* **3**, 311 (1974).

M. W. Chase, J. L. Curnutt, H. Prophet, R. A. McDonald, and A. N. Syverud, JANAF Thermochemical Tables, 1975 supplement, *J. Phys. Chem. Ref. Data* **4**, 1 (1975).

M. W. Chase, J. L. Curnutt, R. A. McDonald, and A. N. Syverud, JANAF Thermochemical Tables, 1978 Supplement, *J. Phys. Chem. Ref. Data* **7**, 793 (1978).

M. W. Chase, J. L. Curnutt, J. R. Downey, Jr., R. A. McDonald, A. N. Syverud, and E. A. Valenzuela, JANAF Thermochemical Tables, 1982 Supplement, *J. Phys. Chem. Ref. Data* **11**, 695 (1982).

"CODATA Recommended Key Values for Thermodynamics," 1977, *J. Chem. Thermo.* **10**, 903 (1978).

A few typical values of molar entropies have been assembled in Table 11-2. Data obtained from spectroscopic studies have been included even though the methods used in their calculation have not been discussed.

11-5 CALCULATION OF STANDARD FREE ENERGIES FROM STANDARD ENTROPIES AND STANDARD ENTHALPIES

Now that we are able to calculate standard entropies let us consider their primary use, which is the calculation of standard free energies. The methods of estimating thermodynamic functions can be divided into two categories: precise and approximate. When sufficient data are available, the precise methods are used. In the absence of adequate information we still can obtain useful estimates of entropies, enthalpies, and free energies from one or more of the semiempirical approximate methods, which will be discussed in Chapter 23.

Enthalpy Calculations

As we mentioned, it is necessary to have information about the standard enthalpy change for a reaction as well as the standard entropies of the reactants and products to calculate the free energy change. At some temperature T, ΔH_T° can be obtained from ΔHf° of each of the substances involved in the transformation. Data on the standard enthalpies of formation are tabulated in either of two ways. One method is to list ΔHf° at some convenient temperature, such as 25°C, or at a series of temperatures. Table 5-2 contains values of ΔHf°. Values at temperatures not listed are calculated with the aid of heat capacity equations.

Under the stimulus of statistical thermodynamics, another method of tabulation, using $H_T^\circ - H_0^\circ$ or $(H_T^\circ - H_0^\circ)/T$, in which the subscripts refer to the absolute temperatures, has come into general use. This method of presentation, illustrated in Table 11-3 (p. 246), does not include empirical heat capacity equations in the results and allows easier comparison of data from different sources.

The following procedure is used to calculate ΔHf° at any temperature T from Table 11-3. The standard enthalpy of formation of a compound C refers to the reaction

$$\underset{\left\{\begin{array}{c}\text{element in}\\ \text{standard state}\\ \text{at temperature } T\end{array}\right\}}{\mathrm{A}} + \underset{\left\{\begin{array}{c}\text{element in}\\ \text{standard state}\\ \text{at temperature } T\end{array}\right\}}{\mathrm{B}} + \cdots = \underset{\left\{\begin{array}{c}\text{compound in}\\ \text{standard state}\\ \text{at temperature } T\end{array}\right\}}{\mathrm{C}}, \qquad \Delta Hf_T^\circ \qquad (11\text{-}42)$$

Table 11-2 Conventional Entropies at 298.15 K

Substance	$s^\circ_{298.15}$/(J mol^{-1} K^{-1})	Substance	$s^\circ_{298.15}$/(J mol^{-1} K^{-1})
	Elements[a]		
Ag(c)	42.55	Kr(g)	164.082
Al(c)	28.33	Li(c)	29.12
Ar(g)	154.843	Mg(c)	32.68
As(c)	35.1	Mn(c)	32.01
B(c)	5.86	Mo(c)	28.66
Ba(c)	62.8	N_2(g)	191.61
Be(c)	9.50[b]	Na(c)	51.21
Bi(c)	56.74	Ne(g)	146.328
Br_2(l)	152.231	Ni(c)	29.87
C(graphite)	5.740	O_2(g)	205.138
Ca(c)	41.42	P(red)	22.80
Cd(c)	51.76	Pb(c)	64.81
Cl_2(g)	222.066	Rb(c)	76.78
Co(c)	30.04	S(rhombic)	32.054[b]
Cr(c)	23.77	Sb(c)	45.69
Cu(c)	33.150	Se(black)	42.442
D_2(g)	144.960	Si(c)	18.83
F_2(g)	202.78	Sn(white)	51.18[b]
Fe(c)	27.28	Te(c)	49.71
Ge(c)	31.09	Ti(c)	30.63
H_2(g)	130.684	V(c)	28.91
He(g)	126.150	Xe(g)	169.683
Hg(l)	76.02	Zn(c)	41.63
I_2(c)	116.135	Zr(c)	38.99
K(c)	64.68[b]		
	Inorganic Compounds[a]		
BaO(c)	72.07[c]	HF(g)	173.779
$BaCl_2 \cdot 2H_2O$(c)	202.9	HI(g)	206.594
$BaSO_4$(c)	132.21[c]	ICl(g)	247.551
BF_3(g)	254.12	NO(g)	210.761
$Ca(OH)_2$(c)	83.39[c]	NaCl(c)	72.12[c]
CO(g)	197.674	SO_2(g)	248.22
CO_2(g)	213.74	SO_3(g)	256.76
CNCl(g)	236.17	SiO_2(c); α quartz	41.46[c]
CuO(c)	42.63[c]	SiO_2(c); α cristobalite	43.40[c]
H_2O(l)	69.91	SiO_2(c); α tridymite	43.93[c]
H_2O(g)	188.825	$CaSO_4$(c); anhydrite	106.69[c]

Table 11-2 Conventional Entropies at 298.15 K (*cont'd*)

Substance	$s^\circ_{298.15}$/(J mol^{-1} K^{-1})	Substance	$s^\circ_{298.15}$/(J mol^{-1} K^{-1})
	Inorganic Compounds[a] (*cont'd*)		
HBr(g)	198.695	$CaSO_4 \cdot 2H_2O$(c); gypsum	194.14[c]
HCl(g)	186.908	Fe_2OSiO_4(c); fayalite	148.32[c]
		Mg_2SiO_4(c); forsterite	95.19[c]
	Organic Compounds[d]		
Methane(g)	186.274[a]	Benzene(l)	173.45
Ethane(g)	229.12	Toluene(l)	221.03
Propane(g)	270.20	*o*-Xylene(l)	253.25
n-Butane(g)	309.91	*m*-Xylene(l)	252.17
Ethene(g)	219.45	*p*-Xylene(l)	247.36
Propene(g)	266.94	Cyclohexane(l)	204.39
1-Butene(g)	305.60	Glycine(c)	103.51[e]
Acetylene(g)	200.819	L-Leucine(c)	211.79[f]
		DL-Leucylglycine(s)	281.16[g]

[a]"NBS Tables of Chemical Thermodynamic Properties," *J. Phys. Chem. Ref. Data* **11**, Supplement No. 2 (1982).

[b]"CODATA Recommended Key Values for Thermodynamics," 1977, *J. Chem. Thermo.* **10**, 903 (1978).

[c]R. A. Robie, B. S. Hemingway, and J. R. Fisher, *Geological Survey Bulletin* 1452 (1978). The standard state for these values is also 1 bar (100.00 kPa).

[d]*TRC Thermodynamic Tables–Hydrocarbons*, Thermodynamics Research Center: The Texas A & M University System, College Station, Texas. These data are based on a standard state of 101.325 kPa.

[e]J. O. Hutchens, A. G. Cole, and J. W. Stout, *J. Am. Chem. Soc.* **82**, 4813 (1960).

[f]J. O. Hutchens, A. G. Cole, and J. W. Stout, *J. Phys. Chem.* **67**, 1128 (1963).

[g]H. Huffman, *J. Am. Chem. Soc.* **63**, 686 (1941). The heat capacity data from which the entropy is calculated acquire an extrapolation from 90 K to 0 K.

The sum of the following equations gives the required ΔHf° in terms of the functions $(H^\circ_T - H^\circ_0)/T$

$$\mathrm{A}(0\ \mathrm{K}) + \mathrm{B}(0\ \mathrm{K}) + \cdots = \mathrm{C}(0\ \mathrm{K}),\ \Delta H = \Delta Hf^\circ_0 \tag{11-43}$$

$$\mathrm{C}(0\ \mathrm{K}) = \mathrm{C}(T\ \mathrm{K}),\ \Delta H = \Delta H^\circ = (H^\circ_T - H^\circ_0)_\mathrm{C} \tag{11-44}$$

$$\mathrm{A}(T\ \mathrm{K}) = \mathrm{A}(0\ \mathrm{K}),\ \Delta H = \Delta H^\circ = (H^\circ_0 - H^\circ_T)_\mathrm{A} = -(H^\circ_T - H^\circ_0)_\mathrm{A} \tag{11-45}$$

$$\mathrm{B}(T\ \mathrm{K}) = \mathrm{B}(0\ \mathrm{K}),\ \Delta H = \Delta H^\circ = (H^\circ_0 - H^\circ_T)_\mathrm{B} = -(H^\circ_T - H^\circ_0)_\mathrm{B} \tag{11-46}$$

$$\mathrm{A}(T\ \mathrm{K}) + \mathrm{B}(T\ \mathrm{K}) + \cdots = \mathrm{C}(T\ \mathrm{K})$$

$$\Delta Hf^\circ_T = \Delta H^\circ_T = \Delta Hf^\circ_0 + (H^\circ_T - H^\circ_0)_\mathrm{compound} - \sum (H^\circ_T - H^\circ_0)_\mathrm{elements} \tag{11-47}$$

Table 11-3 Enthalpy Increment Function, $(H_T^\circ - H_0^\circ)/T$[d]

Substance	$\Delta Hf_0^\circ/(\text{kJ mol}^{-1})$	$[(H_T^\circ - H_0^\circ)/T]/(\text{J mol}^{-1}\text{ K}^{-1})$ 298.15	400	600	800	1000	1500
$H_2(g)$*[a]	0	28.401	28.564	28.794	28.958	29.146	29.836
$O_2(g)$[a]	0	29.112	29.263	29.878	30.644	31.384	32.857
C(graphite)[c]	0	3.522	4.114	7.776	10.526	12.621	16.072
CO(g)[a]	−113.801	29.083	27.991	29.355	29.811	30.359	31.681
$CO_2(g)$[a]	−393.145	31.409	33.422	37.120	40.217	42.765	47.380
$H_2O(g)$[a]	−238.915	33.213	33.380	33.999	34.865	35.882	38.664
Methane(g)[b]	−66.505	33.60	34.72	38.65	43.51	48.49	59.54
Ethane(g)[b]	−68.20	39.83	44.69	55.73	66.53	76.36	96.11
Propane(g)[b]	−81.67	49.45	58.20	76.19	92.72	107.19	135.48
Ethylene(g)[b]	60.760	35.44	38.83	46.94	54.81	61.76	75.60
Acetylene(g)[b]	227.313	33.560	37.041	42.727	47.066	50.585	57.296
Benzene(g)[b]	100.416	47.74	60.29	85.69	107.78	126.19	160.00
Toluene(g)[b]	73.220	60.42	76.02	106.94	134.01	156.69	198.74
o-Xylene(g)[b]	46.425	78.24	97.19	132.97	164.14	190.50	239.79
m-Xylene(g)[b]	45.714	74.73	93.35	129.45	161.17	187.99	237.99
p-Xylene(g)[b]	46.292	75.19	93.39	128.99	160.50	187.28	237.32

*The zero energy state at 0 K is taken as parahydrogen. H_T is for the nominal mixture of ¼ para and ¾ ortho.

[a]D. D. Wagman, *et al., Natl. Bur. Std. (U.S.)*, Technical Note 270-3, 1968; 270-4, 1969; 270-5, 1971; 270-6, 1971; 270-7, 1973; 270-8, 1981.

[b]*TRC Thermodynamic Tables–Hydrocarbons*, Thermodynamics Research Center: The Texas A & M University System, College Station, Texas.

[c]R. A. Robie, B. S. Hemingway, and J. R. Fisher, *Geological Survey Bulletin*, 1452 (1978).

[d]These values are calculated for a 101.325 kPa standard state, but are not significantly different from values at 0.1 MPa.

and

$$\Delta Hf_T^\circ/T = \Delta Hf_0^\circ/T + (H_T^\circ - H_0^\circ)_{\text{compound}}/T - \sum (H_T^\circ - H_0^\circ)_{\text{elements}}/T \qquad (11\text{-}48)$$

Each of the quantities in Equation 11-48 can be obtained from tables such as Table 11-3.

Entropy Calculations

Standard entropies for many substances are available in tables such as Table 11-2. Generally, the values listed for 25°C, but many of the original sources, such as the tables of the Thermodynamics Research Project or the Geological Survey, give values for other temperatures also. If heat capacity data are available, entropy values for one temperature can be converted to those for another temperature by the methods discussed in Section 11-4. Similar procedures can be used to obtain standard entropies from heat capacity data, if the appropriate integrations are not given in the literature.

In a reaction such as that represented by Equation 11-42 the standard entropy change, ΔS_T°, at the temperature T, is given by the expression

$$\Delta S_T^\circ = S_{T(\text{compound})}^\circ - \sum S_{T(\text{elements})}^\circ \tag{11-49}$$

Change in Standard Free Energy

When adequate enthalpy and entropy data are available, the calculation of ΔG_T° is a matter of substitution into the equation

$$\Delta G_T^\circ = \Delta H_T^\circ - T\,\Delta S_T^\circ$$

Generally, data for ΔH° and ΔS° are available at least at one temperature. The conversion of the free energy data from one temperature to another can be carried out by the methods outlined in Chapter 8.

With the development of statistical methods and the use of spectroscopic information, an alternative method of presenting free energy data as a function of temperature also has come into use. This consists of tabulating the function $(G_T^\circ - H_0^\circ)/T$, as illustrated in Table 11-4. This method of tabulation avoids the use of empirical equations, with their associated constants, and allows direct comparison of data from different sources. Although we will not discuss the methods used for calculating this new function from experimental data, we will use tables of these functions to obtain the free energy change of a reaction.

To calculate the free energy of formation of a compound C we use Equations 11-43 through 11-47, which were used to calculate ΔHf°, but with the appropriate free energy change for each. In each case we use the relationship (Equation 11-17) at 0 K:

$$G_0^\circ = H_0^\circ$$

to obtain

$$\text{A}(0\text{ K}) + \text{B}(0\text{ K}) + \cdots = \text{C}(0\text{ K}),\ \Delta G = \Delta G_0^\circ = \Delta Hf_0^\circ \tag{11-50}$$

$$\text{C}(0\text{ K}) = \text{C}(T\text{ K}),\ \Delta G = (\Delta G^\circ)_\text{C} = (G_T^\circ - G_0^\circ)_\text{C} = (G_T^\circ - H_0^\circ)_\text{C} \tag{11-51}$$

$$\begin{aligned}\text{B}(T\text{ K}) = \text{B}(0\text{ K}),\ \Delta G = (\Delta G^\circ)_\text{B} &= (G_0^\circ - G_T^\circ)_\text{B} = -(G_T^\circ - G_0^\circ)_\text{B} \\ &= -(G_T^\circ - H_0^\circ)_\text{B}\end{aligned} \tag{11-52}$$

$$\text{A}(T\text{ K}) = \text{A}(0\text{ K}),\ \Delta G = (\Delta G^\circ)_\text{A} = -(G_T^\circ - H_0^\circ)_\text{A} \tag{11-53}$$

The summation of Equations 11-50 through 11-53 leads to the expression

Table 11-4 Gibbs Free Energy Increment Function, $(G_T^\circ - H_0^\circ)/T$[d]

	$[(G_T^\circ - H_0^\circ)/T]/(\mathrm{J\ mol^{-1}\ K^{-1}})$					
Substance	298.15	400	600	800	1000	1500
$H_2(g)$[a]	−102.165	−110.550	−122.177	−130.486	−136.959	−148.904
$O_2(g)$[a]	−175.916	−184.498	−196.464	−205.167	−212.087	−225.108
C(graphite)[c]	−2.22	−4.62	−6.80	−9.34	−11.81	−17.66
CO(g)[a]	−168.473	−177.025	−188.870	−197.372	−204.083	−216.652
$CO_2(g)$[a]	−182.268	−191.778	−206.045	−217.162	−226.421	−244.705
$H_2O(g)$[a]	−155.503	−165.285	−178.925	−188.816	−196.707	−211.777
Methane(g)[b]	−152.67	−162.67	−177.44	−189.20	−199.45	−221.29
Ethane(g)[b]	−189.28	−201.67	−221.84	−239.37	−255.31	−290.20
Propane(g)[b]	−220.75	−236.48	−263.47	−287.73	−309.99	−359.15
Ethylene(g)[b]	−184.01	−195.02	−212.13	−226.73	−239.70	−267.53
Acetylene(g)[b]	−167.260	−177.615	−193.774	−206.690	−217.589	−239.455
Benzene(g)[b]	−221.45	−237.24	−266.79	−294.87	−321.24	−379.82
Toluene(g)[b]	−260.06	−280.16	−317.33	−352.29	−385.02	−457.65
o-Xylene(g)[b]	−275.38	−301.03	−347.20	−389.70	−429.14	−516.24
m-Xylene(g)[b]	−284.35	−308.68	−353.12	−394.60	−433.38	−519.51
p-Xylene(g)[b]	−278.40	−302.67	−347.18	−388.83	−427.78	−514.25

[a]D. D. Wagman, *et al., Natl. Bur. Std. (U.S.)*, Technical Note 270-3, 1968; 270-4, 1969; 270-5, 1971; 270-6, 1971; 270-7, 1973; 270-8, 1981.

[b]*TRC Thermodynamic Tables–Hydrocarbons*, Thermodynamics Research Center: The Texas A & M University System, College Station, Texas.

[c]R. A. Robie, B. S. Hemingway, and J. R. Fisher, *Geological Survey Bulletin*, 1452 (1978).

[d]The tabulated Gibbs free energy increments are for a standard state of 101.325 kPa. For corrections, see [11], p. 2–23.

$$\mathrm{A}(T\ \mathrm{K}) + \mathrm{B}(T\ \mathrm{K}) + \cdots = \mathrm{C}(T\ \mathrm{K}),$$

$$\Delta G_T^\circ = \Delta Hf_0^\circ + (G_T^\circ - H_0^\circ)_{\text{compound}} - \sum (G_T^\circ - H_0^\circ)_{\text{elements}} \tag{11-54}$$

To reduce the range of numerical values it is convenient to tabulate $(G_T^\circ - H_0^\circ)/T$ instead of $(G_T^\circ - H_0^\circ)$. Hence we have the following expression for the free energy of formation, ΔGf_T°, of any compound at some temperature T:

$$\frac{\Delta Gf_T^\circ}{T} = \left[\frac{\Delta Hf_0^\circ}{T} + \left(\frac{G_T^\circ - H_0^\circ}{T}\right)\right]_{\text{compound}} - \sum \left(\frac{G_T^\circ - H_0^\circ}{T}\right)_{\text{elements}} \tag{11-55}$$

If the free energy of formation of each substance in a reaction is known, the free energy change in any reaction involving these substances can be calculated from the equation

$$\Delta G_T^\circ = \sum \Delta Gf_{T(\text{products})}^\circ - \sum \Delta Gf_{T(\text{reactants})}^\circ \tag{11-56}$$

EXERCISES

1. Assuming that

$$\lim_{T \to 0} \left(\frac{\partial \, \Delta G}{\partial T} \right)_P = 0$$

for reactions involving perfect crystalline solids, prove that

$$\lim_{T \to 0} \left(\frac{\partial \, \Delta A}{\partial T} \right)_V = 0$$

2. Prove that $\lim_{T \to 0} \Delta C_v = 0$.
3. Assume that the limiting slope, as T approaches zero, of a graph of ΔG versus T has a finite value but not zero. Prove that ΔC_p for the reaction still would approach zero at 0 K.
4. It has been suggested [G. J. Janz, *Can. J. Res.* **25b**, 331 (1947)] that α-cyanopyridine might be prepared from cyanogen and butadiene by the reaction

$C_4H_6(g) + C_2N_2(g) =$ 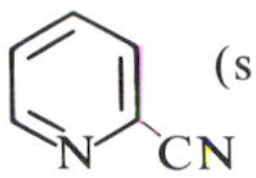 $(s) + H_2(g)$

Pertinent thermodynamic data are given in Table 11-5. From these data, would you consider it worthwhile to attempt to work out this reaction?

Table 11-5

Substance	$\Delta Hf_{298.15}^\circ/(\text{J mol}^{-1})$	$S_{298.15}^\circ/(\text{J mol}^{-1}\ \text{K}^{-1})$
Butadiene(g)	111,914	277.90
Cyanogen(g)	300,495	241.17
α-Cyanopyridine(s)	259,408	322.54
Hydrogen(g)	0	130.58

5. Methylammonium chloride exists in a number of crystalline forms, as is evident from Figure 11-3. The thermodynamic properties of the β and γ forms have been investigated by Aston and Ziemer [*J. Am. Chem. Soc.* **68**, 1405 (1946)] down to temperatures near 0 K. Some of their data are listed below. From the information given calculate the heat of transition from the β to the γ form at 220.4 K:

   $$C_p \text{ for } \beta \text{ at } 12.0 \text{ K} = 0.845 \text{ J mol}^{-1} \text{ K}^{-1}$$

   $$\int C_p \, d \ln T \text{ from } 12.0 \text{ K to } 220.4 \text{ K} = 93.412 \text{ J mol}^{-1} \text{ K}^{-1}$$

   $$C_p \text{ for } \gamma \text{ at } 19.5 \text{ K} = 5.966 \text{ J mol}^{-1} \text{ K}^{-1}$$

   $$\int C_p \, d \ln T \text{ from } 19.5 \text{ K to } 220.4 \text{ K} = 99.918 \text{ J mol}^{-1} \text{ K}^{-1}$$

   Transition temperature (that is, at $P = 1$ atm) = 220.4 K.

6. Cycloheptatriene has two different crystalline forms in the solid state. That labeled I undergoes an entropy change of 116.223 J mol^{-1} K^{-1} on being warmed from 0 K to 154 K [H. L. Finke, D. W. Scott, M. E. Gross, J. F. Messerly, and G. Waddington, *J. Am. Chem. Soc.* **78**, 5469 (1956)]. Some thermal data for form II, the more stable one at very low temperatures, as well as for I, are listed below. Is I a "perfect crystalline substance"?

   $$C_p \text{ at } 12 \text{ K for II} = 4.52 \text{ J mol}^{-1} \text{ K}^{-1}$$

   $$\int C_p \, d \ln T \text{ from } 12 \text{ K to } 154 \text{ K for II} = 99.475 \text{ J mol}^{-1} \text{ K}^{-1}$$

   Transition temperature for (II = I) = 154 K

   ΔH of transition (II = I) = 2347 J mol^{-1}

   $$\int C_p \, d \ln T \text{ from } 154 \text{ K to } 198 \text{ K for I} = 31.192 \text{ J mol}^{-1} \text{ K}^{-1}$$

   Fusion temperature for (I = liquid) = 198 K

   ΔH of fusion for (I = liquid) = 1160.6 J mol^{-1}

   $$\int C_p \, d \ln T \text{ from } 198 \text{ K to } 298 \text{ K for liquid} = 61.375 \text{ J mol}^{-1} \text{ K}^{-1}$$

   Vapor pressure of liquid at 298 K = 3.137 kPa

7. The equilibrium constant, K, for the formation of a deuterium atom from two hydrogen atoms can be defined by the equation

   $$2\text{H} = \text{D}; \qquad K = \frac{p_{\text{D}}}{(p_{\text{H}})^2} P^\circ$$

   The equation for the temperature dependence of K is

   $$\log_{10} K = 20.260 + \tfrac{3}{2}\log_{10} T + \frac{7.04 \times 10^9}{T}$$

a. Calculate K at a temperature of 10^8 K.

b. Calculate $\Delta G°$ and $\Delta H°$ at the same temperature.

c. What is the change in entropy for the conversion of atomic hydrogen into atomic deuterium at a temperature of 10^8 K?

8. The heat capacity of $Na_2SO_4 \cdot 10H_2O$ has been measured from 15 K to 300 K, and $S°_{298.15}$, which was computed from

$$\int_{0\ \mathrm{K}}^{298.15\ \mathrm{K}} C_p\, d\ln T$$

[G. Brodale and W. F. Giauque, *J. Am. Chem. Soc.* **80**, 2042 (1958)], was found to be 585.55 J mol K^{-1}. The following thermodynamic data also are known for the hydration reaction:

$$Na_2SO_4(s) + 10H_2O(g) = Na_2SO_4 \cdot 10H_2O(s)$$

$$\Delta G°_{298.15} = -91{,}190\ \mathrm{J\ mol^{-1}}$$

$$\Delta H°_{298.15} = -521{,}950\ \mathrm{J\ mol^{-1}}$$

Furthermore the entropies, $S°_{298.15}$, for anhydrous Na_2SO_4 and for water vapor are 149.49 and 188.715 J mol^{-1} K^{-1}, respectively. Is $Na_2SO_4 \cdot 10H_2O$ a perfect crystal at 0 K?

9. Some thermodynamic information for benzene and its products of hydrogenation are listed in Table 11-6 [G. J. Janz, *J. Chem. Phys.* **22**, 751 (1954)].

Table 11-6

Substance	$\Delta Hf°_{298}$/(kJ mol^{-1})	$S°_{298}$/(J mol^{-1} K^{-1})
Benzene	82.93	269.0
1,3-Cyclohexadiene	107.1	288.3
Cyclohexane	−7.20	310.5
Cyclohexane	−123.14	298.3

a. Make a graph of $\Delta G°$ for each compound, relative to benzene, versus the moles of H_2 consumed to form each compound. Show that the diene is thermodynamically unstable relative to any of the other three substances.

b. If hydrogenation of benzene were carried out with a suitable catalyst so that equilibrium was attained between benzene and the three products, what would be the relative composition of the mixture at a hydrogen pressure of 101.3 kPa?

10. Two different crystalline forms of benzothiophene have been described by H. L. Finke, M. E. Gross, J. F. Messerly, and G. Waddington [*J. Am. Chem. Soc.* **76**, 854 (1954)]. The one that is the stable form at low temperatures is labeled I, the other II. Calorimetric measurements down to 12 K have been made with each crystalline form. At the normal transition temperature, 261.6 K, the molar enthalpy of transition (I = II) is 3010 J mol^{-1}. Some additional

Table 11-7

	Crystal I	Crystal II
C_p at 12.4 K	4.469 J mol^{-1} K^{-1}	6.573 J mol^{-1} K^{-1}
$\int C_p\, d \ln T$ from 12 K to 261.6 K (numerical)	148.105 J mol^{-1} K^{-1}	152.732 J mol^{-1} K^{-1}
$\int C_p\, d \ln T$ from 261.6 K to 304.5 K (numerical)		23.142 J mol^{-1} K^{-1}
ΔH fusion (at 304.5 K)		11,827.3 J mol^{-1}

thermodynamic data obtained by these investigators are given in Table 11-7. Is Crystal II a perfect crystal at 0 K?

11. Some thermodynamic data for tin are tabulated in Table 11-8. It is possible to construct an electrochemical cell

Sn (gray), electrolyte, Sn (white)

in which the following reaction occurs during operation:

Sn (gray) = Sn (white)

a. Compute the emf of this cell at 25°C and 101.3 kPa.

b. If the cell is operated reversibly, what would be the values of W_{net}, Q, ΔE, ΔH, ΔS, and ΔG, respectively, for the conversion of one mole of gray tin to white?

c. If the cell is short-circuited so that no electrical work is obtained, what would be the values of the thermodynamic quantities listed in (b)?

Table 11-8

	Sn (gray)	Sn (white)
$\Delta H f^\circ_{298}$/(J mol^{-1})	−2090	0
S°_{298}/(J mol^{-1} K^{-1})	44.14	51.55
Density/(g cm^{-3})	5.75	7.31

12. The heats of combustion of quinone(s) and hydroquinone(s) at 1 atm and 25°C are 2745.92 and 2852.44 kJ mol^{-1}, respectively (G. Pilcher and L. E. Sutton, *J. Chem. Soc.* **1956**, 2695). Entropies have been computed from specific heat data; S°_{298} is 161.29 J mol^{-1} K^{-1} for quinone(s) and 137.11 J mol^{-1} K^{-1} for hydroquinone(s).

a. Compute $\Delta H f^\circ$, the standard heat of formation, for each substance.

b. Compute ΔH° for the reduction of quinone to hydroquinone.

c. Compute ΔS° and ΔG° for the reduction.

d. Calculate the $\mathscr{E}^\circ$ for this reaction. The value obtained from electrochemical measurements [J. B. Conant and L. F. Fieser, *J. Am. Chem. Soc.* **44**, 2480 (1922)] is 0.681 volt.

13. We have proved the unattainability statement of the third law by carrying out adiabatic transformations from A to B with substances for which we can state that

$$S_A\ (0\ K) = 0$$

Suppose State A consists of glassy glycerol, for which S_A (0 K) > 0, and State B consists of crystalline glycerol. Prove that absolute zero is still unattainable by means of a transformation from the glassy to the crystalline state.

14. The values of ΔHf°_{298} and S°_{298} for $CaCO_3$ (calcite) and $CaCO_3$ (aragonite) are given in Table 11-9 ["Thermodynamic Properties of Minerals and Related Substances," *Geological Survey Bulletin*, 1452 (1978)]. Predict the thermodynamically stable form of $CaCO_3$ at 298 K from the value of ΔG°_{298} for the transformation

$$CaCO_3\ \text{(calcite)} = CaCO_3\ \text{(aragonite)}$$

Table 11-9

	Calcite	Aragonite
$\Delta Hf^\circ_{298}/(\text{kJ mol}^{-1})$	−1207.5	−1207.9
$S^\circ_{298}/(\text{J mol}^{-1}\ \text{K}^{-1})$	+91.7	+88.0

15. The following problem illustrates the application of calculations involving the third law to a specific organic compound, *n*-heptane. The necessary data can be obtained from sources mentioned in the footnotes of Tables 11-3 and 11-4, or in Section 11-4.

a. From published tables of the enthalpy function, $(H^\circ_T - H^\circ_0)/T$ compute ΔH° for the reaction (at 298.15 K)

$$7C\ \text{(graphite)} + 8H_2\ \text{(hypothetical ideal gas)} = n\text{-}C_7H_{16}\ \text{(hypothetical ideal gas)} \tag{11-57}$$

b. The following equation can be derived for the pressure coefficient of the enthalpy (see Exercise 3, p. 189):

$$\left(\frac{\partial H}{\partial P}\right)_T = V - T\left(\frac{\partial V}{\partial T}\right)_P \tag{11-58}$$

Using the Berthelot equation of state, derive an expression to evaluate $(H_{\text{ideal}} - H_{\text{real}})$, that is, the correction in enthalpy for deviations from ideal behavior at any pressure P (analogous to Equation 11-39 for $S_{\text{ideal}} - S_{\text{real}}$).

c. Find the critical constants for *n*-heptane in the International Critical Tables and its vapor pressure at 298.15 K in the tables of the Thermodynamics Research Center, Texas A & M. Calculate the change in enthalpy for the transformation (at 298.15 K)

$$n\text{-}C_7H_{16}\text{ (hypothetical ideal gas, } P = P^\circ) = n\text{-}C_7H_{16}\text{ (real gas, } P = p_{vap}) \quad (11\text{-}59)$$

in which p_{vap} represents the pressure of the vapor in equilibrium with liquid n-heptane.

d. With the aid of data for the vaporization of n-heptane calculate $\Delta H^\circ_{298.15}$ for the reaction

$$7\text{ C (graphite)} + 8\text{ }H_2(g) = n\text{-}C_7H_{16}(l) \quad (11\text{-}60)$$

e. Calculate $S^\circ_{298.15}$ for liquid n-heptane from the heat capacity data in Table 11-10 and from those for solid n-heptane given in Table 2-7. Integrate by means of the Debye equation to obtain the entropy up to 15.14 K and carry out a numerical integration (C_p/T versus T) thereafter. Obtain ΔH° of fusion from Thermodynamics Research Center tables.

Table 11-10 Heat Capacities of Liquid n-Heptane

T/K	C_p/J mol^{-1} K^{-1}
194.60	201.12
218.73	202.88
243.25	208.24
268.40	216.35
296.51	224.60

f. Calculate ΔS° for Reaction 11-60 at 298.15 K. Use National Bureau of Standards data on graphite and hydrogen.

g. Calculate the entropy of vaporization of liquid n-heptane at 298.15 K.

h. Calculate the entropy change for the following transformation at 298.15 K:

$$n\text{-}C_7H_{16}\text{ (real gas, } P = p_{vap}) = n\text{-}C_7H_{16}\text{ (hypothetical ideal gas, } P = P^\circ) \quad (11\text{-}61)$$

i. By appropriate summation of the results of (f), (g), and (h), calculate ΔS° for Reaction 11-57 at 298.15 K. What approximation is being made? Provide the missing step and corresponding correction to ΔS°.

j. Calculate $\Delta Gf^\circ_{298.15}$ for liquid n-heptane.

k. Calculate $\Delta Gf^\circ_{298.15}$ for gaseous n-heptane in the (hypothetical ideal gas) standard state.

l. From tables of the free energy function $(G^\circ_T - H^\circ_0)/T$ calculate $\Delta Gf^\circ_{298.15}$ for gaseous n-heptane in the (hypothetical ideal gas) standard state. Compare with the value obtained in (k).

REFERENCES

1. T. W. Richards, *Z. Physik. Chem.* **42**, 129 (1902).
2. M. Planck, *Thermodynamik*, 3d ed., Veit & Co., Leipzig, 1911, p. 279; *Treatise on Thermodynamics*, 3d ed., Dover Publications, New York, p. 274.
3. G. N. Lewis and G. E. Gibson, *J. Am. Chem. Soc.* **42**, 1529 (1920).
4. G. N. Lewis and M. Randall, *Thermodynamics*, McGraw-Hill, New York, 1923, p. 448.
5. O. Stern, *Ann. Physik* **49**, 823 (1916); E. D. Eastman and R. T. Milner, *J. Chem. Phys.* **1**, 444 (1933); R. H. Fowler and E. A. Guggenheim, *Statistical Thermodynamics*, Macmillan, New York, 1939, p. 224; P. C. Cross and H. C. Eckstrom, *J. Chem. Phys.* **10**, 287 (1942); F. E. Simon, *Physica* **4**, 1089 (1937).
6. E. D. Eastman and R. T. Milner, *J. Chem. Phys.* **1**, 444 (1933).
7. F. Simon and E. Lange, *Z. Physik* **38**, 227 (1926).
8. E. D. Eastman and W. C. McGavock, *J. Am. Chem. Soc.* **59**, 145 (1937).
9. Deviations from the T^3 law and their significance have been discussed by K. Clusius and L. Schachinger, *Z. Naturforschung* **2a**, 90 (1947); by G. L. Pickard and R. E. Simon, *Proc. Phys. Soc.* **61**, 1 (1948); and by A. R. Ubbelohde, *Modern Thermodynamical Principles*, 2d ed., Clarendon Press, Oxford, 1952, p. 109.
10. J. G. Aston and C. W. Ziemer, *J. Am. Chem. Soc.* **68**, 1405 (1946).
11. "NBS Tables of Chemical Thermodynamic Properties," *J. Phys. Chem. Ref. Data,* **11**, Supplement No. 2 (1982), p. 2-23.
12. It has been pointed out [J. O. Halford, *J. Chem. Phys.* **17**, 111, 405 (1949)] that the Berthelot equation may be inadequate for the calculation of entropy corrections due to gas imperfection, especially for vapors such as water and ethyl alcohol.
13. R. A. Ruehrwein and T. M. Powell, *J. Am. Chem. Soc.* **68**, 1063 (1946).
14. W. F. Giauque and H. L. Johnston, *J. Am. Chem. Soc.* **50**, 3221 (1928); H. L. Johnston and W. F. Giauque, **51**, 3194 (1929); W. F. Giauque, **52**, 4816 (1930); J. O. Clayton and W. F. Giauque, **54**, 2610 (1932); W. F. Giauque and M. F. Ashley, *Phys. Rev.* **43**, 81 (1933); R. W. Blue and W. F. Giauque, *J. Am. Chem. Soc.* **57**, 991 (1935); W. F. Giauque and J. W. Stout, **58**, 1144 (1936); W. F. Giauque and W. M. Jones, **70**, 120 (1948); J. K. Koehler and W. F. Giauque, **80**, 2659 (1958); L. Pauling, *Phys. Rev.* **36**, 430 (1930); L. Pauling, *J. Am. Chem. Soc.* **57**, 2680 (1935).
15. H. H. Ulbrich and D. R. Waldbaum, *Geochimica et Cosmochimica Acta* **40**, 1 (1976).
16. For an interesting discussion of the properties of helium at very low temperatures see K. Mendelssohn, *Science* **127**, 218 (1958); W. F. Vinen, *Endeavour* **25**, 3 (1966).

CHAPTER 12

Thermodynamics of Systems of Variable Composition

Many of the equations we have used thus far can be applied only to closed systems of constant composition. This limitation simply means that we have been dealing with a special case. In general, to fix the state of a system the values of two intensive variables and the mole numbers of the components must be fixed. It is these latter variables that we have been able to neglect because we have discussed only systems of fixed composition. Now we will extend our discussion to the more general systems and, in succeeding chapters, apply the equations developed.

12-1 VARIABLES OF STATE FOR SYSTEMS OF CHANGING COMPOSITION

For a closed system of fixed composition the extensive thermodynamic properties such as V, E, S, A, and G are functions of any pair of convenient intensive variables, for example, T and P. Since dG is a criterion for equilibrium and spontaneity at constant temperature and pressure it is useful to consider

$$G = f(T, P)$$

and to express the total differential (Equations 8-34 and 8-36) as

$$dG = \left(\frac{\partial G}{\partial T}\right)_P dT + \left(\frac{\partial G}{\partial P}\right)_T dP$$

$$= -S(T, P)\, dT + V(T, P)\, dP$$

When the composition of a system varies, the mole numbers of the components are taken into consideration and we have

$$G = f(T, P, n_1, n_2, \ldots, n_i, \ldots) \tag{12-1}$$

so that the total differential becomes

$$dG = \left(\frac{\partial G}{\partial T}\right)_{P,n_i} dT + \left(\frac{\partial G}{\partial P}\right)_{T,n_i} dP + \sum \left(\frac{\partial G}{\partial n_i}\right)_{T,P,n_j} dn_i \tag{12-2}$$

The partial derivative $(\partial G/\partial n_i)_{T,P,n_j}$, in which $i \neq j$, represents the rate of increase in free energy per mole of component i added to the system when T, P, and the other mole numbers are held constant. (Equation 12-2 is based on the assumption that surface effects can be neglected.) The summation is over all the components of the system.

If the composition is constant, so the dn_i terms are zero, Equation 12-2 becomes Equation 8-36, and we can write

$$\left(\frac{\partial G}{\partial T}\right)_{P,n_i} = -S(T, P, n_i) \tag{12-3}$$

and

$$\left(\frac{\partial G}{\partial P}\right)_{T,n_i} = V(T, P, n_i) \tag{12-4}$$

Here we recognize explicitly that S and V also are functions of the mole numbers as well as functions of T and P. The partial derivative of G with respect to the mole number n_i at constant T and P and mole numbers $n_j \neq n_i$ is called the *partial molar Gibbs free energy*, $\bar{G}_i$, or the *chemical potential*, μ_i. We will use μ_i and rewrite Equation 12-2 as

$$dG = -S\, dT + V\, dP + \sum \mu_i\, dn_i \tag{12-5}$$

From Equation 8-28 we have

$$G = A + PV$$

and

$$dG = dA + P\, dV + V\, dP \tag{12-6}$$

If Equation 12-6 is equated with Equation 12-5

$$dA = -S\, dT - P\, dV + \sum_i \mu_i\, dn_i \tag{12-7}$$

is obtained, from which it follows that

$$\left(\frac{\partial A}{\partial n_i}\right)_{T,V,n_j} = \mu_i \tag{12-8}$$

Similarly, it can be shown that

$$\mu_i = \left(\frac{\partial H}{\partial n_i}\right)_{S,P,n_j} = \left(\frac{\partial E}{\partial n_i}\right)_{S,V,n_j} = -T\left(\frac{\partial S}{\partial n_i}\right)_{E,V,n_j} \tag{12-9}$$

However, note that

$$\left.\begin{aligned}
\left(\frac{\partial A}{\partial n_i}\right)_{T,P,n_j} &= \bar{A}_i \neq \mu_i \\
\left(\frac{\partial H}{\partial n_i}\right)_{T,P,n_j} &= \bar{H}_i \neq \mu_i \\
\left(\frac{\partial E}{\partial n_i}\right)_{T,P,n_j} &= \bar{E}_i \neq \mu_i \\
\left(\frac{\partial S}{\partial n_i}\right)_{T,P,n_j} &= \bar{S}_i \neq -\frac{\mu_i}{T}
\end{aligned}\right\} \tag{12-10}$$

since partial molar properties by definition are derivatives with respect to the mole numbers at *constant temperature and pressure*.

12-2 CRITERIA OF EQUILIBRIUM AND SPONTANEITY IN SYSTEMS OF VARIABLE COMPOSITION

The criteria for spontaneity and equilibrium developed in Chapter 8 (Equations 8-13, 8-14, and 8-19), that is

$$dA \leqslant 0 \text{ (constant } T,\ V)$$

$$dG \leqslant 0 \text{ (constant } T,\ P)$$

are valid for all closed systems in which only PdV work is done. Similarly, Equation 8-103,

$$dG \leqslant DW_{net} \text{ (constant } T, P)$$

is valid for all closed systems in which work other than pressure-volume work is done. In this expression, the equality applies to a reversible process and the inequality applies to an irreversible process, whether the change of state is spontaneous or nonspontaneous. If the change of state is spontaneous, $dG < 0$ and $DW_{net} < 0$, so that in absolute magnitude $|dG| \geqslant |DW_{net}|$. If the change of state is nonspontaneous, $dG > 0$ and $DW_{net} > 0$, so that $|dG| \leqslant |DW_{net}|$. Thus, for a spontaneous change of state, the magnitude of dG is equal to the maximum non-*PV* work that can be done by the system, whereas, for a spontaneous change of state, the magnitude of dG is equal to the minimum non-*PV* work that must be done on the system to bring about the change in state. Since G is a state function, the value of dG is the same for a given change of state, whether it is carried out reversibly or irreversibly; it is the value of dW that depends on reversibility.

If temperature and pressure are constant, Equation 12-5 becomes

$$dG = \sum_i \mu_i \, dn_i \tag{12-11}$$

which means that the criteria for spontaneity and equilibrium become (when the only constraint on the system is the constant pressure of the atmosphere and only *PdV* work is done)

$$\sum_i \mu_i \, dn_i \leqslant 0 \text{ (constant } T, P) \tag{12-12}$$

When the system is placed under additional constraints, the relationships for non-*PdV* work are

$$\sum_i \mu_i \, dn_i \leqslant DW_{net} \text{ (constant } T, P) \tag{12-13}$$

in which the equality applies to a reversible process and the inequality applies to an irreversible process. It can be shown that Equation 12-12 is a general criterion of equilibrium and spontaneity and is not limited to constant temperature and pressure. However, it is with these limitations that we will apply the criteria.

The chemical potential for chemical, biological, or geological systems is analogous to the height, or gravitational potential, for a gravitational system; the former systems change spontaneously in the direction of decreasing chemical potential just as an object in a gravitational field moves spontaneously in the direction of decreasing gravitational potential (downward).

12-3 RELATIONSHIPS AMONG PARTIAL MOLAR PROPERTIES OF A SINGLE COMPONENT

Generally, the thermodynamic relationships that we have developed for extensive thermodynamic properties also apply to partial molar properties. Thus, since (Equation 8-18)

$$G = H - TS$$

by differentiation with respect to n_i, we can obtain (at constant temperature, pressure, and other mole numbers n_j)

$$\left(\frac{\partial G}{\partial n_i}\right)_{T,P,n_j} = \left(\frac{\partial H}{\partial n_i}\right)_{T,P,n_j} - T\left(\frac{\partial S}{\partial n_i}\right)_{T,P,n_j}$$

or, from the definition of a partial molar quantity,

$$\mu_i = \bar{G}_i = \bar{H}_i - T\bar{S}_i \tag{12-14}$$

Similarly, from the relationship (Equation 4-19)

$$\left(\frac{\partial H}{\partial T}\right)_P = C_p$$

we can write

$$\left(\frac{\partial \bar{H}_i}{\partial T}\right)_P = \bar{C}_{P_i} \tag{12-15}$$

since the value of the cross derivative of a thermodynamic property is independent of the order of differentiation (Equation 2-24). That is,

$$\left[\frac{\partial^2 H}{\partial T\,\partial n_i}\right]_{P,n_j} = \left[\frac{\partial}{\partial T}\left(\frac{\partial H}{\partial n_i}\right)_{T,P,n_j}\right]_{P,n_i,n_j} = \left(\frac{\partial \bar{H}_i}{\partial T}\right)_{P,n_i,n_j}$$

and also

$$\left[\frac{\partial^2 H}{\partial n_i\,\partial T}\right]_{P,n_j} = \left[\frac{\partial}{\partial n_i}\left(\frac{\partial H}{\partial T}\right)_{P,n_i,n_j}\right]_{T,P,n_j} = \left(\frac{\partial C_p}{\partial n_i}\right)_{T,P,n_j} = \bar{C}_{P_i}$$

On the same basis we can state

$$\left(\frac{\partial \mu_i}{\partial T}\right)_P = \left(\frac{\partial \bar{G}_i}{\partial T}\right)_P = -\bar{S}_i \tag{12-16}$$

and

$$\left(\frac{\partial \mu_i}{\partial P}\right)_T = \left(\frac{\partial \bar{G}_i}{\partial P}\right)_T = \bar{V}_i \tag{12-17}$$

where the derivatives are taken at constant composition as well as constant pressure or constant temperature.

12-4 RELATIONSHIPS BETWEEN PARTIAL MOLAR QUANTITIES OF DIFFERENT COMPONENTS

Extensive thermodynamic properties at constant temperature and pressure are homogeneous functions of degree 1 of the mole numbers. From Euler's theorem (Equation 2-34) for a homogeneous function of degree n

$$x\left(\frac{\partial f}{\partial x}\right)_y + y\left(\frac{\partial f}{\partial y}\right)_x = nf(x, y)$$

For a general two-component system and any extensive thermodynamic property J we can write

$$J = f(n_1, n_2)$$

and

$$n_1\left(\frac{\partial J}{\partial n_1}\right)_{T,P,n_2} + n_2\left(\frac{\partial J}{\partial n_2}\right)_{T,P,n_1} = J \tag{12-18}$$

From the definition of partial molar quantities Equation 12-18 can be written

$$n_1\bar{J}_1 + n_2\bar{J}_2 = J \tag{12-19}$$

Like J, both $\bar{J}_1$ and $\bar{J}_2$ are functions of T and P and the system composition.

Although the function J is a homogeneous function of the mole numbers of degree 1, the partial molar quantity, $\bar{J}_i$, is a homogeneous function of degree 0; that is, the partial molar quantities are *intensive* variables. This statement can be proved by the following procedure. Differentiate both sides of Equation 2-33 with respect to x:

$$\frac{\partial f(\lambda x, \lambda y, \lambda z)}{\partial x} = \lambda^n \frac{\partial f(x, y, z)}{\partial x}$$

Divide both sides by λ:

$$\frac{\partial f(\lambda x, \lambda y, \lambda z)}{\partial(\lambda x)} = \lambda^{n-1} \frac{\partial f(x, y, z)}{\partial x}$$

and rewrite this resultant equation in the common alternative notation for derivatives:

$$f'(\lambda x, \lambda y, \lambda z) = \lambda^{n-1} f'(x, y, z)$$

The last equation defines the degree of homogeneity of the partial derivative function, f', and states that its degree of homogeneity is $n - 1$, that is, one less than the degree of homogeneity, n, of the original function f. The partial molar quantities, though still functions of n_1 and n_2, are functions only of the ratio n_1/n_2, and thus are independent of the size of the system.

Differentiation of Equation 12-19 leads to

$$dJ = n_1\, d\bar{J}_1 + \bar{J}_1\, dn_1 + n_2\, d\bar{J}_2 + \bar{J}_2\, dn_2 \tag{12-20}$$

Since at constant pressure and temperature J is a function of two variables, $f(n_1, n_2)$, the following equation is valid for the total differential:

$$dJ = \left(\frac{\partial J}{\partial n_1}\right)_{n_2} dn_1 + \left(\frac{\partial J}{\partial n_2}\right)_{n_1} dn_2$$

$$= \bar{J}_1\, dn_1 + \bar{J}_2\, dn_2 \tag{12-21}$$

Equating Equations 12-20 and 12-21 we obtain

$$n_1 d\bar{J}_1 + \bar{J}_1 dn_1 + n_2 d\bar{J}_2 + \bar{J}_2 dn_2 = \bar{J}_1 dn_1 + \bar{J}_2 dn_2$$

or

$$n_1 d\bar{J}_1 + n_2 d\bar{J}_2 = 0 \tag{12-22}$$

Equation 12-22 is one of the most useful relationships between partial molar quantities. When applied to the chemical potential it becomes

$$n_1 d\mu_1 + n_2 d\mu_2 = 0 \tag{12-23}$$

which is called the *Gibbs–Duhem equation*. This equation shows that in a two-component system only one of the chemical potentials can vary independently at constant T and P.

It follows from Equation 12-22 that

$$n_1 \frac{d\bar{J}_i}{dn_1} + n_2 \frac{d\bar{J}_2}{dn_1} = 0 \tag{12-24}$$

This equation is very useful in deriving certain relationships between the partial molar quantity for a solute and that for the solvent. An analogous equation can be written with dn_2 in place of dn_1.

Partial Molar Quantities for Pure Phase

If a system is a single, pure phase, a graph of any extensive thermodynamic property plotted against mole number at constant temperature and pressure gives a straight line passing through the origin (again neglecting surface

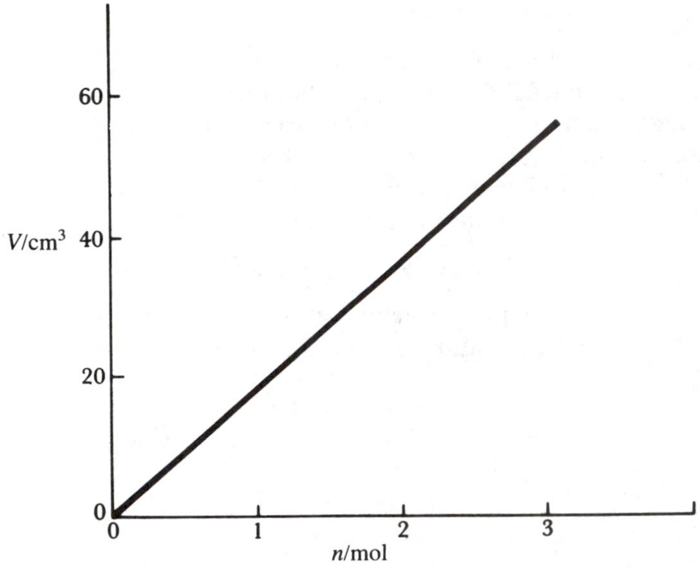

Figure 12-1. Volume of a pure phase at a specified temperature and pressure. Data are for water at 25°C and 101.3 kPa.

effects). The data for the volume of water are given in Figure 12-1. The slope of the line gives the partial molar volume

$$\bar{V} = \left(\frac{\partial V}{\partial n}\right)_{T,P} = \frac{V}{n} = \mathrm{V}$$

in which V is the molar volume of the pure phase. Similarly, for any extensive thermodynamic property J of a pure phase

$$\bar{J} = \frac{J}{n} = \mathrm{J} \tag{12-25}$$

12-5 ESCAPING TENDENCY

Partial Molar Free Energy and Escaping Tendency

G. N. Lewis proposed the term "escaping tendency" to give a strong kinetic-molecular flavor to the concept of partial molar free energy, or chemical potential. Let us consider two solutions of iodine, in water and carbon tetrachloride, which have reached equilibrium with each other at a fixed pressure and temperature (Figure 12-2). In this system at equilibrium let us carry out a transfer of an infinitesimal quantity of iodine from the water to the carbon tetrachloride phase. In view of Equation 12-12 we can say that

$$\mu_{I_2(H_2O)}\, dn_{I_2(H_2O)} + \mu_{I_2(CCl_4)}\, dn_{I_2(CCl_4)} = 0$$

In this closed system any loss of iodine from the water phase is accompanied by an equivalent gain in the carbon tetrachloride, thus

$$-dn_{I_2(H_2O)} = dn_{I_2(CCl_4)} \qquad (12\text{-}26)$$

Hence

$$\mu_{I_2(H_2O)}\, dn_{I_2(H_2O)} + \mu_{I_2(CCl_4)}[-dn_{I_2(H_2O)}] = 0$$

It follows that

$$\mu_{I_2(H_2O)} = \mu_{I_2(CCl_4)} \qquad (12\text{-}27)$$

for this system in equilibrium at constant pressure and temperature. Thus at equilibrium the chemical potential, or partial molar free energy, of the iodine is the same in all phases in which it is present, or the escaping tendency of the iodine in the water is the same as that of the iodine in the carbon tetrachloride.

It may be helpful to consider also the situation in which the iodine will diffuse spontaneously (at constant pressure and temperature) from the water into the carbon tetrachloride—a case in which the concentration in the water phase is greater than that which would exist in equilibrium with the carbon tetrachloride phase. From Equation 12-12 we can write

$$\mu_{I_2(H_2O)}\, dn_{I_2(H_2O)} + \mu_{I_2(CCl_4)}\, dn_{I_2(CCl_4)} < 0$$

For the spontaneous diffusion of iodine Equation 12-26 is valid in this closed system. Hence

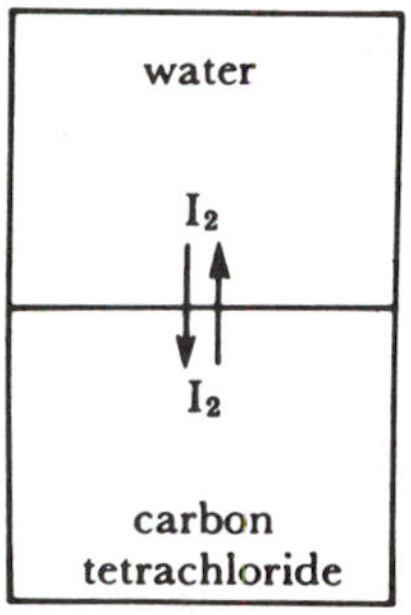

Figure 12-2. Schematic diagram of equilibrium distribution of iodine between water and carbon tetrachloride at fixed temperature and pressure.

$$\mu_{I_2(H_2O)}\, dn_{I_2(H_2O)} + \mu_{I_2(CCl_4)}[-dn_{I_2(H_2O)}] < 0$$

or

$$[\mu_{I_2(H_2O)} - \mu_{I_2(CCl_4)}]\, dn_{I_2(H_2O)} < 0 \tag{12-28}$$

Since the water loses iodine

$$dn_{I_2(H_2O)} < 0$$

that is, dn is a negative number. In such a case Equation 12-28 is valid only if the difference in chemical potentials is a positive number. Therefore

$$\mu_{I_2(H_2O)} > \mu_{I_2(CCl_4)} \tag{12-29}$$

Thus we may say that the escaping tendency of the iodine is greater in the water than in the carbon tetrachloride phase.

In general, when the partial molar free energy, or chemical potential, of a given species is greater in one phase than in a second, we also shall say that the escaping tendency is greater in the former case than in the latter. The escaping tendency thus is a qualitative phrase, corresponding to the property given precisely by the chemical potential. Therefore the escaping tendency can be used for comparative purposes in isothermal transformations.

12-6 CHEMICAL EQUILIBRIUM IN SYSTEMS OF VARIABLE COMPOSITION

We can apply the criterion of equilibrium expressed in Equation 12-12 to chemically reacting systems. Consider the reaction

$$a\text{A} + b\text{B} + \cdots = r\text{R} + s\text{S} + \cdots \tag{12-30}$$

If this chemical reaction is at equilibrium at a fixed pressure and temperature it follows from Equation 12-12 that

$$\mu_A\, dn_A + \mu_B\, dn_B + \cdots + \mu_R\, dn_R + \mu_S\, dn_S + \cdots = 0 \tag{12-31}$$

However, the various dn's in Equation 12-31 are not independent but, in view of the stoichiometry of the reaction of Equation 12-30, must be related as follows:

$$-\frac{dn_A}{a} = -\frac{dn_B}{b} = \cdots = \frac{dn_R}{r} = \frac{dn_S}{s} = \cdots \tag{12-32}$$

Since reactants disappear and products appear in the reaction, the corresponding dn's in Equation 12-32 have opposite signs. In view of the series of equalities in this equation let us define a quantity $d\xi$ such that

$$d\xi = -\frac{dn_A}{a} = \cdots = \frac{dn_R}{r} = \cdots = \frac{dn_i}{\nu_i} \tag{12-33}$$

in which ν_i is merely a generalized notation for the *stoichiometric coefficients*, $-a$, $-b$, r, s, and so on. The quantity ξ is called the *extent of reaction* or the *progress variable*. From the relationships of Equation 12-33, Equation 12-31 can be rewritten as

$$-a\mu_A d\xi - b\mu_B d\xi - \cdots + r\mu_R d\xi + s\mu_S d\xi + \cdots = 0 \qquad (12\text{-}34)$$

If Equation 12-34 is integrated with respect to ξ from $\xi = 0$ to $\xi = 1$ at constant values of the chemical potentials (fixed composition of the reacting mixture) then we obtain, at equilibrium,

$$\sum \nu_i \mu_i = 0 \qquad (12\text{-}35)$$

in which it is understood that ν_i is a negative number for the stoichiometric coefficients of the reactants and a positive number for the products. Another way of writing Equation 12-35 is

$$\sum (|\nu_i|\,\mu_i)_{\text{reactants}} = \sum (|\nu_i|\,\mu_i)_{\text{products}} \qquad (12\text{-}36)$$

The concept of escaping tendency also can be applied to the chemical reaction in Equation 12-30. At equilibrium, from Equation 12-36,

$$\sum (|\nu_i|\,\mu_i)_{\text{reactants}} = \sum (|\nu_i|\,\mu_i)_{\text{products}}$$

For a chemical transformation capable of undergoing a spontaneous change it follows from Equations 12-12 and 12-33 that

$$\sum \nu_i \mu_i < 0$$

or that

$$\sum (|\nu_i|\,\mu_i)_{\text{reactants}} > \sum (|\nu_i|\,\mu_i)_{\text{products}}$$

Thus we can compare the *sums* of $\nu_i\mu_i$ for reactants and products to arrive at a decision as to whether a transformation is at equilibrium or capable of a spontaneous change. Although we can compare escaping tendencies or μ's of a given substance under different conditions at constant temperature, it is meaningless to compare individual escaping tendencies of different substances because we have no way of determining absolute values of $\bar{G}$ or μ. For similar reasons we cannot compare escaping tendencies of a single substance at different temperatures.

EXERCISES

1. Show that μ_i is equal to the three partial derives in Equation 12-9.
2. Show that $\Sigma_i\ \mu_i\, dn_i \leqslant 0$ is a criterion of spontaneity and equilibrium when only PdV work is done at (a) constant T, V; (b) constant S, P; (c) constant S, V; (d) constant E, V.
3. Show that

$$\left(\frac{\partial \bar{S}}{\partial P}\right)_T = -\left(\frac{\partial \bar{V}}{\partial T}\right)_P$$

4. Starting with the relationship for the corresponding extensive quantities show that

a. $\left(\frac{\partial \bar{G}_i}{\partial P}\right)_{T,X_i,X_j} = \bar{V}_i$ b. $\left(\frac{\partial \bar{G}_i}{\partial T}\right)_{P,X_i,X_j} = -\bar{S}_i$

c. $\bar{G}_i = \bar{H}_i + T\left(\frac{\partial \bar{G}_i}{\partial T}\right)_{P,X_i,X_j}$ d. $\left(\frac{\partial(\mu/T)}{\partial T}\right)_{P,X_i,X_j} = -\frac{\bar{H}_i}{T^2}$ (12-37)

5. For a general chemical transformation

$$a\text{A} + b\text{B} + \cdots = r\text{R} + s\text{S} + \cdots$$

at a fixed temperature, an alternative statement to Equation 12-35 is that equilibrium exists if

$$\left(\frac{\partial G}{\partial \xi}\right)_{P,T} = 0$$

in which G is the total free energy of all the components and ξ is the extent of reaction. Verify this statement, and also that

$$\left(\frac{\partial A}{\partial \xi}\right)_{V,T} = 0$$

is valid if the system is at equilibrium.

6. For many biochemical reactions, such as the hydrolysis of adenosine triphosphate,

adenosine triphosphate (ATP) + H_2O = adenosine diphosphate (ADP)

+ inorganic phosphate (P_i) (12-38)

the equilibrium constant is written as

$$K_{\text{observed}} = \frac{[\text{ADP}][\text{P}_i]}{[\text{ATP}]} \tag{12-39}$$

in which the concentrations are total concentrations of the species shown and their products of ionization. Since the species whose concentrations are given in Equation 12-39 ionize as weak acids and also form complexes with Mg^{2+}, the value of $K_{observed}$ and the value of $\Delta G^\circ_{observed}$ are functions of T, pH, and pMg at constant pressure (pMg = $-\log\,[Mg^{2+}]$). An equation for the total differential of ΔG° can be written as

$$d\,\Delta G^\circ = \left(\frac{\partial\,\Delta G^\circ}{\partial T}\right)_{pH,pMg} dT + \left(\frac{\partial\,\Delta G^\circ}{\partial\,pH}\right)_{T,pMg} d\,pH + \left(\frac{\partial\,\Delta G^\circ}{\partial\,pMg}\right)_{T,pH} d\,pMg \tag{12-40}$$

It also can be shown [R. A. Alberty, *J. Am. Chem. Soc.* **91**, 3899 (1969)] that

$$\left(\frac{\partial\,\Delta G^\circ}{\partial\,pH}\right)_{T,pMg} = -2.3RTn_H \tag{12-41}$$

and

$$\left(\frac{\partial\,\Delta G^\circ}{\partial\,pMg}\right)_{T,pH} = -2.3RTn_{Mg} \tag{12-42}$$

in which n_H is the number of H^+ ions produced in the reaction in Equation 12-38 and n_{Mg} is the number of Mg^{2+} ions produced.

a. Show that

$$\left(\frac{\partial\,\Delta S^\circ}{\partial\,pMg}\right)_{T,pH} = 2.3R\left[n_{Mg} + T\left(\frac{\partial n_{Mg}}{\partial T}\right)_{pH,pMg}\right]$$

b. Show that

$$\left(\frac{\partial n_H}{\partial\,pMg}\right)_{T,pH} = \left(\frac{\partial n_{Mg}}{\partial\,pH}\right)_{T,pMg}$$

c. Starting with a total differential for ΔH° in terms of T, pH, and pMg and assuming that

$$\left(\frac{\partial\,\Delta H^\circ}{\partial\,pH}\right)_{T,pMg} = 2.3RT^2\left(\frac{\partial n_H}{\partial T}\right)_{pH,pMg}$$

and

$$\left(\frac{\partial\,\Delta H^\circ}{\partial\,pMg}\right)_{T,pH} = 2.3RT^2\left(\frac{\partial n_{Mg}}{\partial T}\right)_{pH,pMg}$$

show that

$$\left(\frac{\partial\, \Delta c_p^\circ}{\partial\, \mathrm{pH}}\right)_{T,\mathrm{pMg}} = 2.3RT\left[2\left(\frac{\partial n_\mathrm{H}}{\partial T}\right)_{\mathrm{pH,pMg}} + T\left(\frac{\partial^2 n_\mathrm{H}}{\partial T^2}\right)_{\mathrm{pH,pMg}}\right]$$

7. Usually, chemical reactions are carried out in vessels that fit readily on a desk top so that the difference in gravitational energy at the top and bottom of the vessel is negligible. However, in some situations a very high cylindrical vessel may be used, in which case the gravitational energy of a portion of the material of fixed total mass m at the top is significantly different from that at the bottom. In this situation the Gibbs free energy, G, may be a function of h, the height above ground level, as well as of T, P, and n_i.

   a. Write an equation for dG for this situation.

   b. It can be shown that

$$\left(\frac{\partial G}{\partial h}\right)_{T,P,n_i} = mg$$

   Rewrite the equation in (a) accordingly.

   c. Starting with the fundamental definition of G, consider a system of mass m undergoing a reversible change in height in a gravitational field at constant pressure and temperature and show that $dG = mgdh$.

   d. Using the results in (c) and (b) find an equation for $\Sigma\, \mu_i\, dn_i$.

   e. For a system of many components i at constant pressure and temperature and at any height h

$$G = \sum n_i \mu_i$$

   Assuming this relationship plus the result in (d), find an equation for $\Sigma n_i\, d\mu_i$.

8. For a system in which equilibrium between pure solid solute and solute in solution is maintained as the temperature is changed at constant pressure show that

$$\left(\frac{\partial \mu_{1(\text{sat soln})}}{\partial T}\right)_P = -\bar{s}_1 - \frac{X_2}{X_1}(\bar{s}_2 - s_2)$$

in which X_2 and X_1 are mole fractions of solute and solvent, respectively, $\bar{s}_2$ and $\bar{s}_1$ are partial molar entropies, and s_2 is the molar entropy of pure solid solute.

CHAPTER 13

Mixture of Gases

In this chapter we will apply the concepts developed in Chapter 12 to gaseous systems, first to mixtures of ideal gases, then to pure real gases, and finally to mixtures of real gases.

13-1 MIXTURES OF IDEAL GASES

In Chapter 6 we defined an ideal gas on the basis of two properties:

$$PV = nRT \tag{6-1}$$

and

$$\left(\frac{\partial E}{\partial V}\right)_T = 0 \tag{6-2}$$

We define an ideal gas mixture as one that follows Dalton's law,

$$P = n_1 \frac{RT}{V} + n_2 \frac{RT}{V} + \cdots = \frac{RT}{V} \sum_i n_i \tag{13-1}$$

and for which the partial molar energy, $\bar{E}_i$, of each component is dependent only on the temperature and is independent of the pressure and the composition.

We will see that the relationships that are derived for mixtures of ideal gases will form convenient bases for the treatment of nonideal gases and solutions.

The Entropy and Free Energy of Mixing Ideal Gases

The entropy and free energy of mixing ideal gases can be calculated on the basis of a thought experiment with a van't Hoff equilibrium box (Figure 13-1). Consider a cylinder in equilibrium with a thermal reservoir at temperature T so that the experiment is isothermal. The two gases in the cylinder are separated by two semipermeable pistons, the one on the right permeable only to A and the one on the left permeable only to B. To carry out the mixing process in a reversible manner the external pressure, P', on the right piston is kept infinitesimally less than the pressure of B in the mixture; and the external pressure, P'', on the left piston is kept infinitesimally less than the pressure of A in the mixture. Initially, both A and B are present in the separate compartments at the same pressure, P.

The work done by the gases in the mixing process is the sum of the work done by A in expanding against the left piston and the work done by B in expanding against the right piston. That is,

$$\begin{aligned} W_{rev} &= W_A + W_B \\ &= -\int_{V_A}^{V_A+V_B} P_A dV - \int_{V_B}^{V_A+V_B} P_B dV \\ &= -\int_{V_A}^{V_A+V_B} n_A RT \frac{dV}{V} - \int_{V_B}^{V_A+V_B} n_B RT \frac{dV}{V} \end{aligned}$$

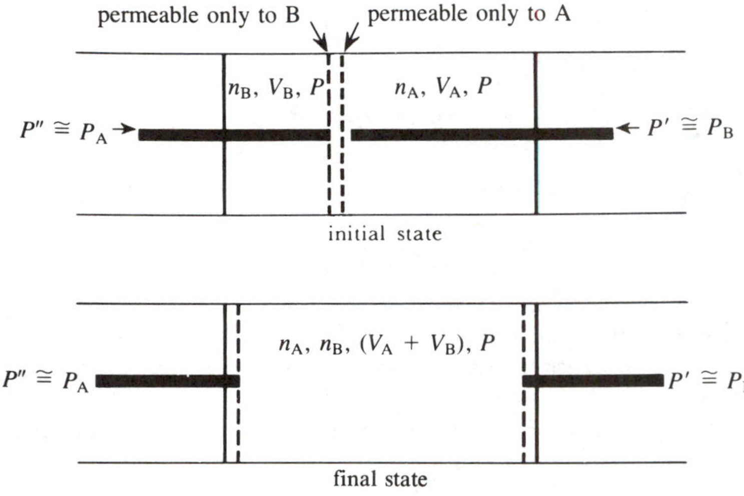

Figure 13-1. van't Hoff equilibrium box.

$$= -n_A RT \ln \frac{V_A + V_B}{V_A} - n_B RT \ln \frac{V_A + V_B}{V_B} \tag{13-2}$$

Since A and B initially were at the same temperature and pressure, and since the final mixture is at the original temperature, and the total pressure of the mixture is equal to the initial pressures of the individual gases, the volumes V_A, V_B, and $V_A + V_B$ are in the same proportion as the respective numbers of moles of gas. Thus Equation 13-2 can be written

$$W_{rev} = -n_A RT \ln \frac{n_A + n_B}{n_A} - n_B RT \ln \frac{n_A + n_B}{n_B}$$

$$= n_A RT \ln \frac{n_A}{n_A + n_B} + n_B RT \ln \frac{n_B}{n_A + n_B}$$

$$= n_A RT \ln X_A + n_B RT \ln X_B \tag{13-3}$$

in which X_A and X_B are the mole fractions. Since the mixing process is isothermal and the gases form an ideal mixture, $\Delta E = 0$, and

$$Q_{rev} = -W_{rev} = -n_A RT \ln X_A - n_B RT \ln X_B \tag{13-4}$$

Then from the entropy change in a reversible, isothermal process (Equation 7-70)

$$\Delta S_{mixing} = \frac{Q_{rev}}{T} = -n_A R \ln X_A - n_B R \ln X_B \tag{13-5}$$

Since both X_A and X_B are less than 1, ΔS_{mixing} is a positive quantity.

For the reversible mixing the entropy change in the surroundings is equal, but opposite in sign, and the total entropy change is zero. If the mixing process were allowed to proceed irreversibly by puncturing the two pistons, ΔS for the system would be the same, but ΔS for the surroundings would be zero since no work would be done and no heat exchanged. Thus the total change in entropy for the irreversible process would be positive.

For the isothermal process involving ideal gases, ΔH is zero and

$$\Delta G_{mixing} = -T\Delta S_{mixing}$$

so that

$$\Delta G_{mixing} = n_A RT \ln X_A + n_B RT \ln X_B \tag{13-6}$$

The Chemical Potential of a Component of an Ideal Gas Mixture

ΔG_{mixing} also is equal to the difference between the free energy of the mixture and the free energies of the unmixed gases. That is,

$$\Delta G_{\text{mixing}} = G_{\text{mixture}} - G_{\text{pure gases}} = [n_A\mu_A + n_B\mu_B]_{\text{mixed}} - [n_A G_A + n_B G_B]_{\text{unmixed}} \quad (13\text{-}7)$$

Equation 8-56 for the isothermal expansion of an ideal gas is

$$\Delta G = nRT \ln \frac{P_2}{P_1}$$

If the change in state is the expansion of one mole of ideal gas from a standard pressure $P^\circ = 0.1$ MPa to a pressure P, Equation 8-56 can be written

$$\Delta G = G - G^\circ = RT \ln P/P^\circ \quad (13\text{-}8)$$

Substituting into Equation 13-7 we obtain

$$\Delta G_{\text{mixing}} = n_A\mu_A + n_B\mu_B - n_A(G_A^\circ + RT \ln P/P^\circ) - n_B(G_B^\circ + RT \ln P/P^\circ) \quad (13\text{-}9)$$

From Equations 13-6 and 13-9 we have

$$n_A RT \ln X_A + n_B RT \ln X_B = n_A\mu_A + n_B\mu_B \\ -n_A G_A^\circ - n_A RT \ln P/P^\circ - n_B G_B^\circ - n_B RT \ln P/P^\circ \quad (13\text{-}10)$$

The coefficients of n_A and n_B on the two sides of the equation must be equal; thus

$$RT \ln X_A = \mu_A - G_A^\circ - RT \ln P/P^\circ$$

Therefore

$$\mu_A = G_A^\circ + RT \ln X_A + RT \ln P/P^\circ \\ = G_A^\circ + RT \ln PX_A/P^\circ$$

and similarly for μ_B. We will define the partial pressure of an ideal gas as the product of its mole fraction and the total pressure. Thus we can write

$$\mu_A = G_A^\circ + RT \ln p_A/P^\circ \quad (13\text{-}11)$$

Equation 13-8, which gives the molar free energy of a pure ideal gas, is of the same form as Equation 13-11, which gives the chemical potential of a component of an ideal gas mixture, except that for the latter, partial pressure is substituted for total pressure. If the standard state of a component of the mixture is defined as one in which the *partial pressure* of that component is 0.1 MPa, then

$$\mu_A^\circ = G_A^\circ$$

and we can write

$$\mu_A = \mu_A^\circ + RT \ln p_A/P^\circ \quad (13\text{-}12)$$

Chemical Equilibrium in Ideal Gas Mixtures

For the reaction (Equation 12-30 applied to ideal gases)

$$a\text{A}(p_\text{A}) + b\text{B}(p_\text{B}) + \cdots = r\text{R}(p_\text{R}) + s\text{S}(p_\text{S}) + \cdots$$

we can substitute the expression in Equation 13-12 for the chemical potentials in Equation 12-35, that is

$$-a(\mu_\text{A}^\circ + RT\ln p_\text{A}/P^\circ) - b(\mu_\text{B}^\circ + RT\ln p_\text{B}/P^\circ) + r(\mu_\text{R}^\circ + RT\ln p_\text{R}/P^\circ) + s(\mu_\text{S}^\circ + RT\ln p_\text{S}/P^\circ) = 0$$

or, rearranging,

$$-(r\mu_\text{R}^\circ + s\mu_\text{S}^\circ - a\mu_\text{A}^\circ - b\mu_\text{B}^\circ)/RT = \ln\left[\frac{(p_\text{R}/P^\circ)^r\,(p_\text{S}/P^\circ)^s}{(p_\text{A}/P^\circ)^a\,(p_\text{B}/P^\circ)^b}\right]_\text{equil} \qquad (13\text{-}13)$$

Since all the quantities on the left-hand side are constant at constant temperature, the quantity in brackets is also a constant at constant temperature, and, in particular, independent of the total pressure and the initial composition of the system. We therefore designate the quantity in brackets as K_p, the equilibrium constant in terms of partial pressures for a mixture of ideal gases. Thus,

$$\Delta G^\circ = -RT\ln K_p \qquad (13\text{-}14)$$

The subscript p distinguishes the ideal gas equilibrium constant in terms of partial pressures from other forms for the constants that will be derived for real systems.

13-2 THE FUGACITY FUNCTION OF A PURE GAS

We expressed the molar free energy of a pure ideal gas as (Equation 13-8)

$$G_\text{A} = G_\text{A}^\circ + RT\ln P/P^\circ$$

and the chemical potential of a component of an ideal gas mixture as (Equation 13-12)

$$\mu_\text{A} = \mu_\text{A}^\circ + RT\ln p_\text{A}/P^\circ$$

as a result of substituting $V = nRT/P$ in the integral

$$\Delta G = \int_{P^\circ}^{P} V\,dP \qquad (13\text{-}15)$$

It would be possible to apply Equation 13-15 to real gases by substituting a different empirical expression for V as a function of P for each gas, but no simple closed form is applicable to all gases. A simple form of the equation for the chemical potential and a simple form of the equation for an equilibrium constant that is independent of the gases involved is so convenient, however, that G. N. Lewis suggested an alternative procedure. He defined a new function, the *fugacity, f*, with a universal relationship to the chemical potential, and let the dependence of f on P vary for different gases. The fugacity is defined to have the dimensions of pressure.

An advantage of the fugacity over the chemical potential as a measure of escaping tendency is that an absolute value of the fugacity can be calculated, whereas an absolute value of the chemical potential cannot be calculated.

One part of the definition of fugacity can be stated as

$$\mu = \mu^\circ + RT \ln (f/f^\circ) \tag{13-16}$$

in which μ° is a function only of the temperature. The standard chemical potential is characteristic of each gas, and the standard state chosen. For a pure gas, the value of f° is chosen equal to P°, 0.1 MPa.

Since all gases approach ideality as their pressure is decreased, and since Equation 13-16 is of the same form as Equation 13-12 for an ideal gas, it is convenient to complete the definition of f by stating

$$\lim_{P \to 0} \frac{f}{P} = 1 \tag{13-17}$$

that is, as the pressure approaches zero the fugacity approaches the pressure. Figure 13-2 indicates the relationship between P and f for ideal and real gases. The standard state for a real gas is chosen as the state at which the fugacity is equal to 0.1 MPa, 1 bar, along a line extrapolated from values of f at low pressure, as indicated in Figure 13-2.

From Equation 13-16 we can see that the free energy change for the isothermal expansion of a real gas is

$$\Delta G = nRT \ln \frac{f_2}{f_1} \tag{13-18}$$

As the pressure approaches zero Equation 13-17 applies and ΔG approaches the value calculated from Equation 8-56.

Change of Fugacity with Pressure

The dependence of fugacity on pressure can be derived by differentiating Equation 13-16:

$$\left(\frac{\partial \mu}{\partial P}\right)_T = RT\left[\frac{\partial \ln (f/f^\circ)}{\partial P}\right]_T \tag{13-19}$$

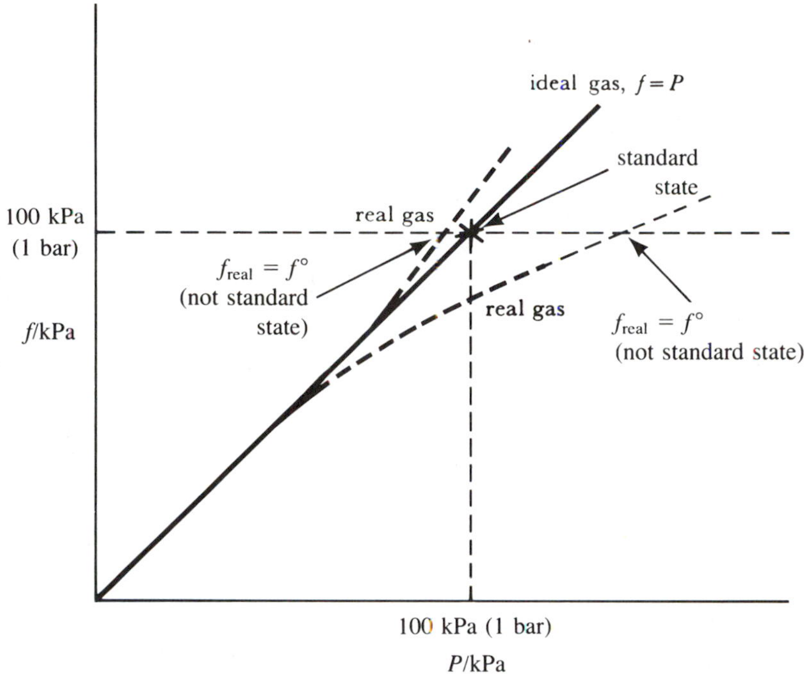

Figure 13-2. Characteristics of the fugacity for ideal and real gases.

Combining Equation 13-19 with Equation 12-17,

$$\left(\frac{\partial \mu}{\partial P}\right)_T = \bar{V} = V$$

we obtain

$$\left[\frac{\partial \ln f}{\partial P}\right]_T = \frac{V}{RT} \tag{13-20}$$

This equation can be integrated to find a fugacity at one pressure from that at another:

$$RT \ln \frac{f_2}{f_1} = \int_{P_1}^{P_2} V\, dP \tag{13-21}$$

Change of Fugacity with Temperature

Let us consider an isothermal process in which a gas is transformed from one state, A, at a pressure P, to another, A*, at a different pressure, P^*. Such a transformation can be represented as follows:

$$\mathrm{A}(P) = \mathrm{A}^*(P^*) \tag{13-22}$$

The free energy change for such a transformation is given by the expression

$$\begin{aligned} \Delta G &= \mu^* - \mu \\ &= RT \ln (f^*/f^\circ) - RT \ln (f/f^\circ) \end{aligned} \tag{13-23}$$

and

$$\frac{\mu^*}{T} - \frac{\mu}{T} = R \ln (f^*/f) \tag{13-24}$$

The partial derivative of the fugacity with respect to temperature is given by

$$\left[\frac{\partial(\mu^*/T)}{\partial T}\right]_{P^*} - \left[\frac{\partial(\mu/T)}{\partial T}\right]_P = R\left(\frac{\partial \ln f^*}{\partial T}\right)_{P^*} - R\left(\frac{\partial \ln f}{\partial T}\right)_P$$

From Equation 12-37 we have

$$\left[\frac{\partial(\mu^*/T)}{\partial T}\right]_{P^*} - \left[\frac{\partial(\mu/T)}{\partial T}\right]_P = -\frac{\bar{H}^*}{T^2} + \frac{\bar{H}}{T^2} \tag{13-25}$$

So

$$\left(\frac{\partial \ln f^*}{\partial T}\right)_{P^*} - \left(\frac{\partial \ln f}{\partial T}\right)_P = -\frac{\bar{H}^*}{RT^2} + \frac{\bar{H}}{RT^2} \tag{13-26}$$

If the pressure P^* approaches zero, the ratio of the fugacity to the pressure approaches one, and we can write

$$\left(\frac{\partial \ln f^*}{\partial T}\right)_{P^*} = \left(\frac{\partial \ln P^*}{\partial T}\right)_{P^*} = 0$$

And from Equation 13-26 we obtain

$$\left(\frac{\partial \ln f}{\partial T}\right)_P = \frac{\bar{H}^* - \bar{H}}{RT^2} \tag{13-27}$$

in which $\bar{H}^*$ is the partial molar enthalpy of the substance in State A*, that is, the state of zero pressure. Therefore the difference $(\bar{H}^* - \bar{H})$ is the change in molar enthalpy when the gas goes from State A to its state of zero pressure, that is, at infinite volume.

The pressure dependence of this enthalpy change is given by the expression

$$\left[\frac{\partial(\bar{H}^* - \bar{H})}{\partial P}\right]_T = -\left(\frac{\partial \bar{H}}{\partial P}\right)_T = -\left(\frac{\partial H}{\partial P}\right)_T \tag{13-28}$$

because $(\partial \bar{H}^*/\partial P)_T$ is zero, since $\bar{H}^*$ is the partial molar enthalpy at a fixed (zero) pressure. In Equation 13-28 we replace the partial molar enthalpy, $\bar{H}$, by the molar enthalpy, H, since we are dealing with a pure gas.

From Equation 6-66 we know that the pressure coefficient of the molar enthalpy of a gas is related to the Joule–Thomson coefficient, $\mu_{J.T.}$, by the equation

$$\left(\frac{\partial H}{\partial P}\right)_T = -C_p \mu_{J.T.}$$

Combining Equations 13-28 and 6-66 we find

$$\left[\frac{\partial(\bar{H}^* - \bar{H})}{\partial P}\right]_T = C_p \mu_{J.T.} \tag{13-29}$$

Because of this relationship between $(\bar{H}^* - \bar{H})$ and $\mu_{J.T.}$, the former quantity frequently is referred to as the "Joule–Thomson heat." The pressure coefficient of this Joule–Thomson enthalpy change can be calculated from the known values of the Joule–Thomson coefficient and the heat capacity of the gas. Similarly, since $(\bar{H}^* - \bar{H})$ is a derived function of the fugacity, knowledge of the temperature dependence of the latter can be used to calculate the Joule–Thomson coefficient.

13-3 CALCULATION OF THE FUGACITY OF A REAL GAS

Several methods have been developed for calculating fugacities from measurements of pressures and molar volumes of real gases.

Graphical or Numerical Methods

Using the α Function. A typical molar volume–pressure graph for a real gas is illustrated in Figure 13-3, together with the corresponding graph for an ideal gas. From Equation 13-21 we can write

$$RT \ln \frac{f_2}{f_1} = \int_{P_1}^{P_2} v dP$$

The ratio of the fugacity f_2 at the pressure P_2 to the fugacity f_1 at the pressure P_1 can be obtained by graphical or numerical integration, as indicated by the shaded area in Figure 13-3. However, as P_1 approaches zero, the area becomes infinite. Hence this direct method is not suitable for determining absolute values of the fugacity of a real gas.

Equation 13-21 is based only on part of the definition of fugacity. The second part of the definition states that while f approaches zero as P approaches zero, the ratio f/P approaches one. Hence this ratio might be integrable to zero pressure.

If we take the pressure coefficient of the ratio f/P we obtain

$$\left[\frac{\partial \ln (f/P)}{\partial P}\right]_T = \left(\frac{\partial \ln f}{\partial P}\right)_T - \left(\frac{\partial \ln P}{\partial P}\right)_T \tag{13-30}$$

The pressure coefficient of $\ln f$ is given by Equation 13-20,

$$\left(\frac{\partial \ln f}{\partial P}\right)_T = \frac{v}{RT}$$

in which v is the molar volume of the gas. Thus Equation 13-30 becomes

$$\left[\frac{\partial \ln (f/P)}{\partial P}\right]_T = \frac{v}{RT} - \frac{\partial \ln P}{\partial P}$$

$$= \frac{v}{RT} - \frac{1}{P}$$

$$= \frac{1}{RT}\left(v - \frac{RT}{P}\right) \tag{13-31}$$

If we call the quantity within the parentheses $-\alpha$, that is, if

$$\alpha = \left(\frac{RT}{P} - v\right) \tag{13-32}$$

we obtain

$$\left[\frac{\partial \ln (f/P)}{\partial P}\right]_T = -\frac{\alpha}{RT} \tag{13-33}$$

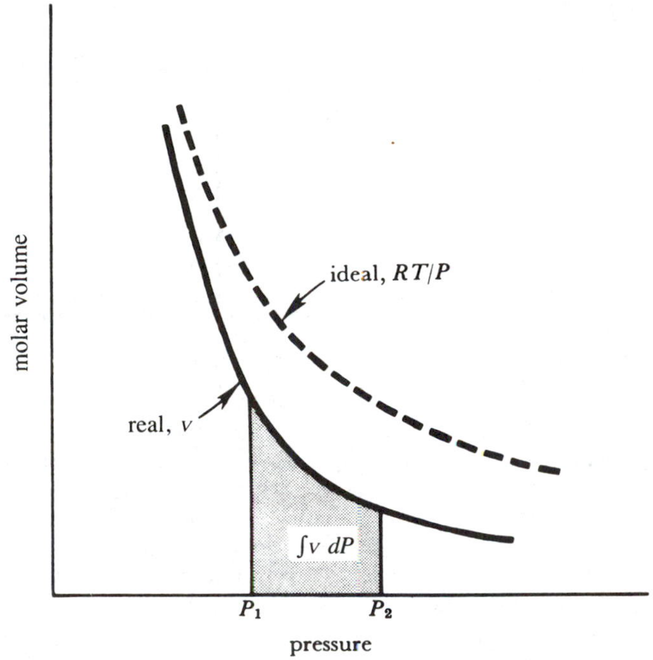

Figure 13-3. Comparison of molar volume–pressure isotherms for a possible real gas and an ideal gas.

Integration of this equation for isothermal conditions from zero pressure to some pressure P gives

$$\int_0^{\ln f/P} d\ln\frac{f}{P} = -\frac{1}{RT}\int_0^P \alpha\,dP$$

or

$$\ln\frac{f}{P} - \ln\left(\frac{f}{P}\right)_{P=0} = -\frac{1}{RT}\int_0^P \alpha\,dP \qquad (13\text{-}34)$$

Since f/P approaches one as P approaches zero, the second term on the left side of Equation 13-34 goes to zero. Hence

$$\ln\left(\frac{f}{P}\right) = -\frac{1}{RT}\int_0^P \alpha\,dP \qquad (13\text{-}35)$$

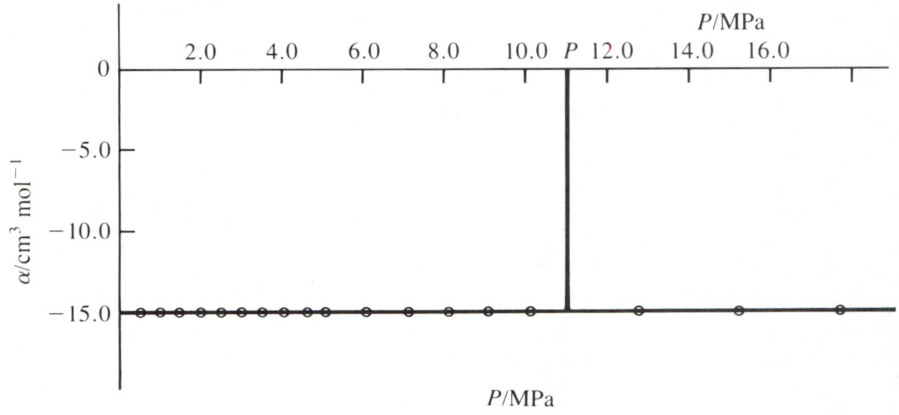

Figure 13-4. The α function for hydrogen gas at 300 K. Data from [1].

Thus to evaluate f it is necessary to integrate $\alpha\, dP$. Both v and RT/P approach infinity as the pressure goes to zero. Nevertheless, the difference between them generally does not approach zero. Usually, α can be measured for several pressures and an extrapolation made to zero pressure. A typical graph for α (for hydrogen gas) is illustrated in Figure 13-4. The shaded area indicates the graphical evaluation of the integral in Equation 13-35 [2].

This procedure of using a finite difference between two quantities, both of which become infinite, is of general usefulness. We also will use it in Chapter 21 to obtain the standard potential of a cell.

Based on Compressibility Factor. The behavior of most pure gases can be represented adequately by a single chart of the compressibility factor, Z, which has been defined as (Equation 6-56)

$$Z = \frac{Pv}{RT}$$

If Z is plotted as a function of the reduced pressure, $P_r = P/P_c$, then at a given reduced temperature, $T_r = T/T_c$, all gases fit a single curve. At another reduced temperature, T_r', a new curve is obtained for Z versus P_r, but it too fits all gases.

If Z could be related to α, it would be possible to plot some function of the fugacity against P_r at a given value of T_r for which all gases would fit the same curve. At another reduced temperature, T_r', a new curve would be obtained.

Let us derive a relationship between α and Z. It follows from Equation 13-32 that

$$\alpha = \frac{RT}{P} - v = \frac{RT}{P}\left(1 - \frac{Pv}{RT}\right) = \frac{RT}{P}(1 - Z) \qquad (13\text{-}36)$$

From Equation 13-36 and Equation 13-33 we obtain

$$d \ln \frac{f}{P} = -\frac{1}{RT}\frac{RT}{P}(1 - Z)\, dP$$

which by integration yields

$$\ln \frac{f}{P} = -\int_0^P \frac{1 - Z}{P}\, dP \qquad (13\text{-}37)$$

The integration of Equation 13-37 can be carried out, graphically or numerically, to provide a chart of f/P (or γ, the *fugacity coefficient* [3]) as a function of P_r and T_r. A single chart of "universal fugacity coefficients" is applicable to all pure gases within the precision to which the compressibility factor chart is valid. Several investigators have prepared such charts, a typical example of which is illustrated in Figure 13-5. The values of T_r and P_r can be calculated from the critical constants of the gas, γ can be read from the chart, and the fugacity can be calculated from the expression

$$f = \gamma P \qquad (13\text{-}38)$$

Analytical Methods

Based on the Virial Equation. If the pressure–volume behavior of a gas is represented by a virial equation of the form (Equation 6-54)

$$Pv = RT + BP + CP^2 + \cdots$$

Equation 13-35 becomes

$$\ln (f/P) = (1/RT)\int_0^P (B + CP + \cdots)\, dP$$

$$= (1/RT)(BP + CP^2/2 + \cdots) \qquad (13\text{-}39)$$

and the fugacity can be evaluated at any pressure from values of the virial coefficients. The limiting value of α at zero pressure can be seen to be equal to the value of $-B$, where B is the second virial coefficient.

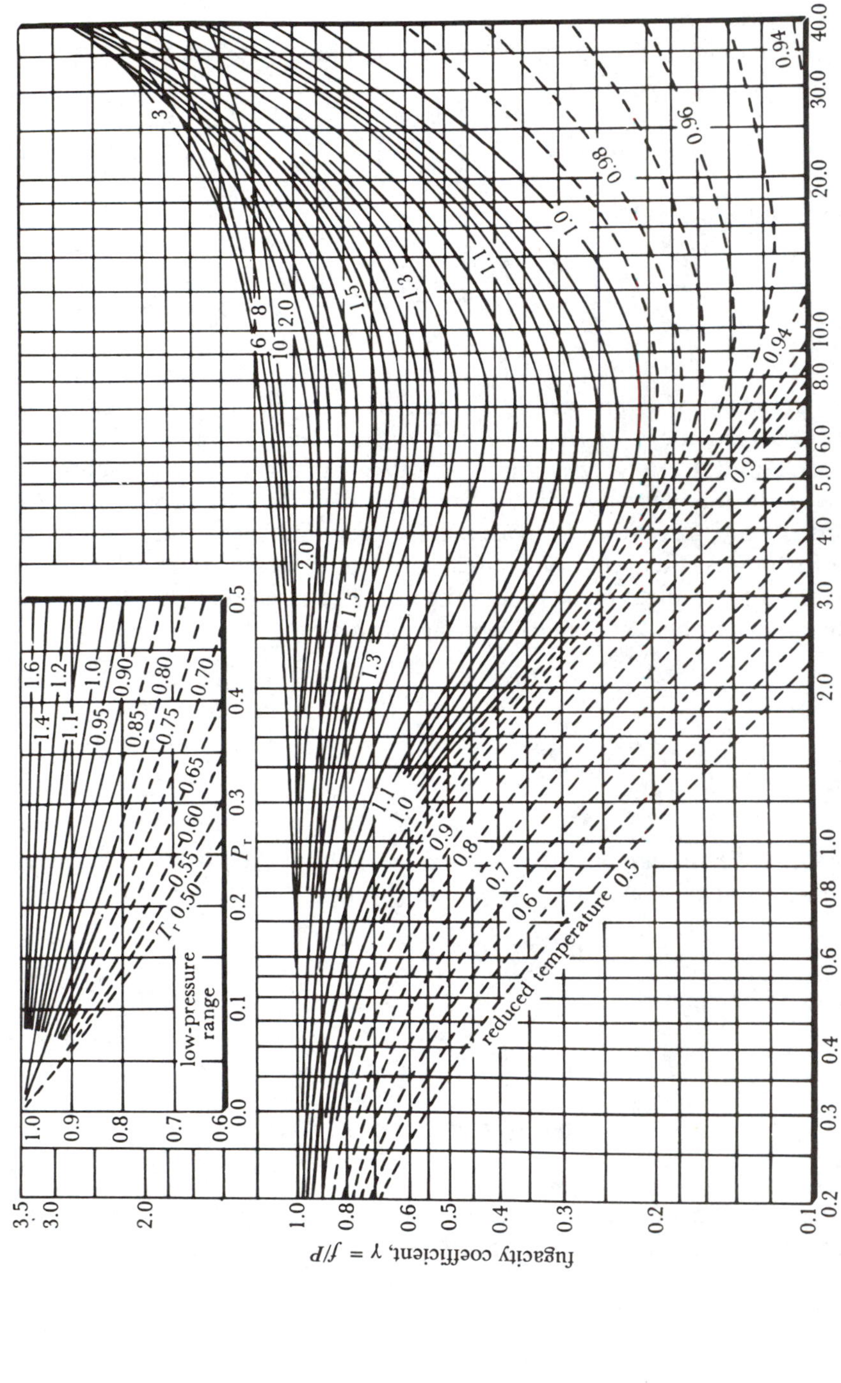

Figure 13-5. Fugacity coefficients of gases. Based on data taken from B. W. Gamson and K. M. Watson, *Natl. Petrol. News, Tech. Sec.* **36**, R623 (Sept. 6, 1944).

Based on the van der Waals Equation of State. We can integrate $d \ln (f/P)$ by using an equation of state such as the van der Waals equation. Integrating as in Equation 13-34 we obtain

$$RT \ln \frac{f}{P} = -\int_0^P \alpha \, dP = \int_0^P \left(v - \frac{RT}{P}\right) dP$$

$$= \int_0^P v \, dP - \int_0^P RT \, d \ln P \tag{13-40}$$

To evaluate the first integral it is necessary to substitute for v or dP. A trial will show that it is simpler to substitute for dP. Thus, solving the van der Waals equation of state for P we obtain

$$P = \frac{RT}{v - b} - \frac{a}{v^2}$$

and

$$dP = -\frac{RT}{(v - b)^2} \, dv + \frac{2a}{v^3} \, dv \tag{13-41}$$

If we insert Equation 13-41 into Equation 13-40 we obtain

$$RT \ln \frac{f}{P} = \int_\infty^V v\left[-\frac{RT}{(v - b)^2} + \frac{2a}{v^3}\right] dv - RT \ln P \Big]_0^P$$

$$= -\int_\infty^V \frac{RTv}{(v - b)^2} \, dv + \int_\infty^V \frac{2a}{v^2} \, dv - RT \ln P \Big]_0^P$$

$$= -\int_\infty^V \frac{RT(v - b + b)}{(v - b)^2} \, dv + \int_\infty^V \frac{2a}{v^2} \, dv - RT \ln P \Big]_0^P$$

$$= -\int_\infty^V \frac{RT(v - b)}{(v - b)^2} \, dv - \int_\infty^V \frac{RTb}{(v - b)^2} \, dv + \int_\infty^V \frac{2a}{v^2} \, dv - RT \ln P \Big]_0^P$$

$$= -RT \ln (v - b) \Big]_\infty^V + \frac{RTb}{v - b} \Big]_\infty^V - \frac{2a}{v} \Big]_\infty^V - RT \ln P \Big]_0^P \tag{13-42}$$

If we combine the first and fourth terms on the right side of Equation 13-42 before substituting limits we obtain

$$RT \ln \frac{f}{P} = -RT \ln \{P(v-b)\}\Bigg]_{\substack{P=0\\ V=\infty}}^{P,V} + \frac{RTb}{v-b}\Bigg]_{\infty}^{V} - \frac{2a}{v}\Bigg]_{\infty}^{V} \qquad (13\text{-}43)$$

Since

$$\lim_{\substack{P\to 0\\ V\to\infty}} P(v-b) = Pv = RT$$

$$RT \ln \frac{f}{P} = -RT \ln \{P(v-b)\} + \frac{RTb}{v-b} - \frac{2a}{v} + RT \ln RT + 0 + 0$$

Hence

$$\ln \frac{f}{P} = \ln \frac{RT}{P(v-b)} + \frac{b}{v-b} - \frac{2a}{RTv} \qquad (13\text{-}44)$$

Thus the fugacity of a gas that obeys the van der Waals equation can be evaluated from the constants a and b at any given pressure P and corresponding molar volume, v.

An Approximate Method. When the third virial coefficient is sufficiently small, it frequently happens that α is roughly constant, particularly at relatively low pressures. A good example is hydrogen gas (Figure 13-4). When this is the case, we can integrate Equation 13-35 analytically and obtain

$$RT \ln \frac{f}{P} = -\alpha P = BP \qquad (13\text{-}45)$$

where B is the second virial coefficient.

This equation can be converted into several other approximate forms. For example,

$$\ln \frac{f}{P} = -\frac{\alpha P}{RT}$$

Therefore

$$\frac{f}{P} = e^{-\alpha P/RT} \qquad (13\text{-}46)$$

The exponential in Equation 13-46 can be expanded as a Taylor series to give

$$\frac{f}{P} = 1 - \frac{\alpha P}{RT} + \frac{1}{2!}\left(\frac{\alpha P}{RT}\right)^2 - \cdots \tag{13-47}$$

If $\alpha P \ll RT$, we can neglect all terms of higher power than (αP) and we obtain

$$\frac{f}{P} = 1 - \frac{\alpha P}{RT} = \frac{RT - [(RT/P) - v]P}{RT}$$

Therefore

$$\frac{f}{P} = \frac{Pv}{RT} = Z \tag{13-48}$$

Another relationship can be obtained by defining an *ideal* pressure, P_i:

$$P_i = \frac{RT}{v}$$

With this relationship Equation 13-48 becomes

$$\frac{f}{P} = \frac{P}{P_i} \tag{13-49}$$

Thus the fugacity can be estimated from the observed pressure, P, and the ideal pressure calculated from the observed volume. The error [4] in Equation 13-49 is less than 1% for oxygen up to a pressure of 10 MPa. For carbon dioxide the error is 1% at 2.5 MPa and 4% at 5 MPa. If $\alpha P \ll RT$, the numerical value of f/P is relatively insensitive to variations in the value of α as large as 30%.

13-4 JOULE–THOMSON EFFECT FOR A VAN DER WAALS GAS

We have pointed out that the temperature coefficient of the fugacity function is related to the Joule–Thomson coefficient, $\mu_{J.T.}$. Let us consider an example of the calculation of $\mu_{J.T.}$ from a fugacity equation. We will restrict our discussion to relatively low pressures and to a gas that obeys the van der Waals equation.

Approximate Value of α for a van der Waals Gas

From the van der Waals equation (6-53)

$$\left(P + \frac{a}{v^2}\right)(v - b) = RT$$

we obtain

$$Pv + \frac{a}{v} - Pb - \frac{ab}{v^2} = RT$$

and

$$v + \frac{a}{Pv} - b - \frac{ab}{Pv^2} = \frac{RT}{P}$$

so that

$$\alpha = \frac{RT}{P} - v = \frac{a}{Pv} - b - \frac{ab}{Pv^2}$$

When the pressure is low enough that

$$Pv \cong RT$$

then

$$\alpha \cong \frac{a}{RT} - b - \frac{abP}{(RT)^2} \tag{13-50}$$

At low pressures $bP/RT \ll Pv/RT \cong 1$; hence, the third term on the right in Equation 13-50 is negligible in comparison to the first term, and

$$\alpha \cong \frac{a}{RT} - b \tag{13-51}$$

Fugacity at Low Pressures

According to Equation 13-35

$$RT \ln \frac{f}{P} = -\int_0^P \alpha \, dP$$

Substituting α from Equation 13-50 into Equation 13-35 we obtain

$$RT \ln \frac{f}{P} = -\int_0^P \frac{a}{RT} \, dP + \int_0^P b \, dP + \int_0^P \frac{abP}{(RT)^2} \, dP$$

$$= -\frac{aP}{RT} + bP + \frac{abP^2}{2(RT)^2}$$

and

$$\ln \frac{f}{P} = -\frac{aP}{(RT)^2} + \frac{bP}{RT} + \frac{abP^2}{2(RT)^3} \tag{13-52}$$

Enthalpy of a van der Waals Gas

According to Equation 13-27

$$\frac{\bar{H}^* - \bar{H}}{RT^2} = \left(\frac{\partial \ln f}{\partial T}\right)_P$$

Differentiating Equation 13-52 with respect to T we obtain

$$\left(\frac{\partial \ln f/P}{\partial T}\right)_P = \left(\frac{\partial \ln f}{\partial T}\right)_P = \frac{\bar{H}^* - \bar{H}}{RT^2} = \frac{2aP}{R^2T^3} - \frac{bP}{RT^2} - \frac{3}{2}\frac{abP^2}{R^3T^4}$$

or

$$\bar{H}^* - \bar{H} = \frac{2aP}{RT} - bP - \frac{3abP^2}{2R^2T^2} \tag{13-53}$$

Joule–Thomson Coefficient

From Equation 13-29 we see that

$$\mu_{\text{J.T.}} = \frac{1}{c_p}\left[\frac{\partial(\bar{H}^* - \bar{H})}{\partial P}\right]_T$$

Applied to Equation 13-53, Equation 13-29 becomes

$$\mu_{\text{J.T.}} = \frac{1}{c_p}\left(\frac{2a}{RT} - b - \frac{3abP}{R^2T^2}\right) \tag{13-54}$$

and the more approximate result (starting with Equation 13-51 for α) is

$$\mu_{\text{J.T.}} = \frac{1}{c_p}\left(\frac{2a}{RT} - b\right) \tag{13-55}$$

Since c_p is positive, the sign of the Joule–Thomson coefficient depends on the sign of the expression in parentheses in Equations 13-54 and 13-55. The expression in Equation 13-54 is a quadratic in T and there are two values of T at any value of P for which $\mu_{\text{J.T.}} = 0$. Thus, Equation 13-54 predicts two values of the Joule–Thomson inversion temperature, T_i, for any pressure low enough for Equation 13-50 to be a good approximation for α. As we saw in Section 6-2 and Figure 6-6, this prediction fits, at least qualitatively, the experimental data for the Joule–Thomson experiment for N_2.

At sufficiently high temperatures and low pressures Equation 13-55 applies. This equation predicts a single value of T_i that is independent of pressure. It can be seen from Figure 6-6 that this clearly is an approximate relationship for the upper inversion temperature.

13-5 MIXTURES OF REAL GASES

Now that we have obtained expressions for the fugacity of a real gas and its temperature and pressure coefficients, let us consider the application of the concept of fugacity to components of a mixture of real gases.

Fugacity of a Component of a Gaseous Solution

The equation for the fugacity of component *i* of a mixture has the same form as Equation 13-16,

$$\mu_i = \mu_i^\circ + RT \ln (f_i/f^\circ) \tag{13-56}$$

in which μ_i° and f° have the same values as for the pure component. On the basis of this definition the fugacity of a component of a gaseous mixture is equal to that of the pure gas in equilibrium with the mixture across a membrane permeable only to that component.

Since the mixture approaches ideality as the total pressure approaches zero, Equation 13-56 should approach Equation 13-12. The second part of the definition of fugacity for a gaseous component, analogous to Equation 13-17, is

$$\lim_{P \to 0} \frac{f_i}{p_i} = 1 \tag{13-57}$$

in which p_i is the partial pressure of the component in the mixture, defined as $p_i = X_i P$.

The form of Equation 13-20 for a component of a solution is

$$\left(\frac{\partial \ln f_i}{\partial P} \right)_{T,X_i} = \frac{\bar{V}_i}{RT} \tag{13-58}$$

The integration of Equation 13-58 for a component of a mixture leads to a problem of nonconvergence at $P = 0$, just as for a single gas. To circumvent this difficulty we shall consider the ratio of the fugacity to the partial pressure of a component, just as we considered the ratio of the fugacity to the pressure of a single gas.

By steps analogous to Equations 13-30 and 13-31 we can show that

$$\left[\frac{\partial \ln (f_i/p_i)}{\partial P}\right]_{T,X_i} = \frac{\bar{V}_i}{RT} - \frac{1}{P} \tag{13-59}$$

Integrating Equation 13-59 between $P = 0$ and P we obtain

$$\ln \frac{f_i}{p_i} - \ln \left(\frac{f_i}{p_i}\right)_{P=0} = \int_0^P \left(\frac{\bar{V}_i}{RT} - \frac{1}{P}\right) dP \tag{13-60}$$

From Equation 13-57 we can see that $\ln f_i/p_i$ approaches zero as P approaches zero, thus

$$\ln \frac{f_i}{p_i} = \int_0^P \left(\frac{V_i}{RT} - \frac{1}{P}\right) dP \tag{13-61}$$

If sufficient data are available on the dependence of the volume of the mixture on composition and pressure, the fugacities of the components can be calculated by means of Equation 13-61.

Approximate Rule for Solutions of Real Gases [5] [6]

Since the evaluation of Equation 13-61 requires a great deal of data, and since adequate data are available for only a few mixtures of gases, it is useful to have approximate relationships that can be used to estimate the fugacity of components in a solution of gases.

The most common approximate relationship is the Lewis and Randall rule. Lewis and Randall suggested that, even though the gases in the solution are not ideal, the mixture behaves as an ideal solution in that the fugacity of each component obeys the equation

$$f_i = X_i f_i^{\bullet} \tag{13-62}$$

in which $f_i^{\bullet}$ is the fugacity of the pure gas at the same temperature and at the same total pressure, P, and X_i is the mole fraction of the particular component. Thus the fugacity of a component can be estimated from the fugacity of the pure gas and the composition of the mixture.

Fugacity Coefficients in Gaseous Solution

The *fugacity coefficient*, γ_i, of a constituent of a gaseous solution is defined by the expression

$$\gamma_i = \frac{f_i}{p_i} \tag{13-63}$$

To the level of approximation provided by the Lewis and Randall rule γ_i is given by the equation

$$\begin{aligned} \gamma_i &= \frac{X_i f_i^\bullet}{p_i} \\ &= \frac{X_i f_i^\bullet}{X_i P} \\ &= \gamma_i^\bullet \end{aligned}$$

in which $\gamma_i^\bullet$ is the fugacity coefficient of the pure constituent at the same pressure and temperature. Thus the fugacity coefficients obtained from Figure 13-5 for a pure gas could be used to estimate the fugacity of a component in a mixture.

Equilibrium Constant and Free Energy Change for Reactions Involving Real Gases

For the reaction

$$a\mathrm{A(g)} + b\mathrm{B(g)} = c\mathrm{C(g)} + d\mathrm{D(g)} \tag{13-64}$$

we can show, by a procedure analogous to that used for ideal gases, that the standard free energy change is related to the equilibrium constant in terms of fugacities by the equation

$$\Delta G^\circ = -RT \ln K_f \tag{13-65}$$

In this equation K_f is given by the ratio

$$K_f = \frac{(f_\mathrm{C}/f^\circ)^c (f_\mathrm{D}/f^\circ)^d}{(f_\mathrm{A}/f^\circ)^a (f_\mathrm{B}/f^\circ)^b} \tag{13-66}$$

and is the thermodynamic equilibrium constant, K. To obtain ΔG° from equilibrium data it is necessary to calculate the equilibrium constant in terms of fugacities rather than partial pressures.

Since $f_i = \gamma_i p_i$ (Equation 13-63) K also can be expressed as

$$\begin{aligned} K = K_f &= \frac{(p_\mathrm{C}\gamma_\mathrm{C}/p^\circ)^c (p_\mathrm{D}\gamma_\mathrm{D}/p^\circ)^d}{(p_\mathrm{A}\gamma_\mathrm{A}/p^\circ)^a (p_\mathrm{B}\gamma_\mathrm{B}/p^\circ)^b} \\ &= \left(\frac{\gamma_\mathrm{C}^c \gamma_\mathrm{D}^d}{\gamma_\mathrm{A}^a \gamma_\mathrm{B}^b}\right) \frac{(p_\mathrm{C}/p^\circ)^c (p_\mathrm{D}/p^\circ)^d}{(p_\mathrm{A}/p^\circ)^a (p_\mathrm{B}/p^\circ)^b} \end{aligned} \tag{13-67}$$

If we apply the Lewis and Randall rule $\gamma_i^\bullet$ can be used to replace γ_i and the result is

$$K = \left(\frac{\gamma_C^{\bullet c} \gamma_D^{\bullet d}}{\gamma_A^{\bullet a} \gamma_B^{\bullet b}} \right) \frac{(p_C/p^\circ)^c (p_D/p^\circ)^d}{(p_A/p^\circ)^a (p_B/p^\circ)^b}$$

$$= K_p K_\gamma \tag{13-68}$$

in which K_p represents the partial pressure equilibrium constant, and K_γ is the corresponding ratio of the fugacity coefficients for the respective pure gases at the specified *total* pressure. Approximate values of K_γ can be obtained from tables and graphs of fugacity coefficients of *pure* gases, as in Figure 13-5, and K_p can be calculated by the methods described in Chapter 8.

EXERCISES

1. Consider a gas with the equation of state

$$Pv = RT + BP \tag{13-69}$$

   for which B is a small *negative* number.

   a. Draw a rough sketch of a graph of Pv versus P for this gas. Include a dotted line for the corresponding graph of an ideal gas.

   b. Draw a dotted curve for a graph of v versus P for an ideal gas. On this same graph draw a curve for v versus P for a gas with the equation of state given by Equation 13-69.

   c. As P approaches zero what happens to the two curves in the graph in (b)?

   d. Rearrange Equation 13-69 to read explicitly for v. As P approaches zero what does v approach?

   e. Rearrange Equation 13-69 into one for $[v - (RT/P)]$. As P approaches zero what does the quantity in brackets approach?

   f. Draw a graph of $[v - (RT/P)]$ versus P for the gas with the equation of state given by Equation 13-69.

   Derive the following:

   g. An expression for $\ln f$.

   h. An expression for γ.

   i. An expression for the Joule–Thomson heat.

   j. An expression for the Joule–Thomson coefficient.

2. For helium, B in the equation of state (13-69) is essentially constant and equals 11.5 cm^3 mol^{-1} in the temperature range 30°C to 90°C [T. L. Cottrell and R. A. Hamilton, *Trans. Faraday Soc.* **52**, 156 (1956); A. Michels and H. Wouters, *Physica* **8**, 923 (1941)]. Find explicit answers to questions (g–j) of Exercise 1 for He at 0.1 MPa and 60°C.

3. If the fugacity function is defined by Equation 13-56, show that for a solution of two components

$$X_1\left(\frac{\partial \ln f_1}{\partial X_1}\right)_{P,T} = X_2\left(\frac{\partial \ln f_2}{\partial X_2}\right)_{P,T}$$

4. For hydrogen at 0°C, Amagat has prepared the data given in Table 13-1. Find the fugacity of hydrogen at 100 MPa by the graphical method using the α function.

Table 13-1

P/MPa	Pv/RT	P/MPa	Pv/RT
10.13	1.069	60.80	1.431
20.27	1.138	70.93	1.504
30.40	1.209	81.06	1.577
40.53	1.283	91.19	1.649
50.66	1.356	101.33	1.720

5. Proceeding in a manner analogous to that used for ideal gases prove the following equation for real gases:

$$\Delta G° = -RT \ln K_f \tag{13-65}$$

6. R. H. Ewell [*Ind. Eng. Chem.* **32**, 147 (1940)] has suggested the following reaction as a method for the production of hydrogen cyanide:

$$N_2(g) + C_2H_2(g) = 2HCN(g) \tag{13-70}$$

a. From data in tables of the National Bureau of Standards calculate $\Delta G°_{298}$ for this reaction.

b. By methods discussed previously calculate $\Delta G°$ at 300°C. The value obtained should be 25,882 J mol^{-1}.

c. Calculate K_f for Reaction 13-70 at 300°C.

d. Refer to the International Critical Tables to find the critical temperatures and pressures for the gases in Reaction 13-70. Tabulate these values. Also tabulate the reduced temperatures for 300°C and the reduced pressures for a total pressure of 0.5 MPa and 20.0 MPa.

e. Referring to Figure 13-5, find γ's for the gases in Reaction 13-70 at total pressures of 0.5 MPa and 20.0 MPa. Tabulate these values.

f. Calculate K_γ at 0.5 MPa and 20.0 MPa and add these values to the table in (e).

g. Calculate K_p at 0.5 MPa and 20.0 MPa and add these values to the table in (e).

h. If we start with an equimolar mixture of N_2 and C_2H_2, what fraction of C_2H_2 is converted to HCN at 0.5 MPa total pressure? At 20.0 MPa total pressure?

i. What is the effect of increasing total pressure on the yield of HCN?

j. According to Le Chatelier's principle what should be the effect of increasing total pressure on the yield of HCN?

7. The thermodynamic equilibrium constant, K, for the formation of ammonia according to the equation

$$\tfrac{1}{2}N_2(g) + \tfrac{3}{2}H_2(g) = NH_3(g)$$

is 0.0067 at 450°C.

a. Calculate the degree of dissociation of ammonia at 450°C and 30.40 MPa total pressure. At this temperature and pressure the fugacities of the pure gases, in MPa, are: H_2, 33.13; N_2, 34.65; NH_3, 27.66.

b. Make a corresponding calculation assuming that the fugacity coefficients each are unity.

8. Table 13-2 shows values of v for O_2 at 300 K taken from L. A. Weber, *J. Res. Natl. Bur. Std. (U.S.)* **74A**, 93(1970).

Table 13-2

P/MPa	v/cm^3 mol^{-1}
0.101325	24602.02
1.00000	2478.85
5.0000	484.94
10.000	237.81
20.000	118.38
30.00	82.28

a. Calculate α as a function of P for O_2 at 300 K and draw a smooth curve from your calculated points.

b. Calculate f/P at several values of P with Equation 13-35, using values read from your smooth curve in a numerical integration.

c. Calculate f/P at the same values of P with Equation 13-48, which is based on the assumption that α is constant.

d. Why do the values in (c) agree with the values in (b) for $P \leqslant 10.000$ MPa, despite the clear variation of α of 25% between 0 and 10 MPa? *Hint:* consider the numerical properties of the ln function and its argument when the argument is less than 1.

REFERENCES

1. H. L. Johnston and D. White, *Trans. Am. Soc. Mech. Eng.* **22**, 7855 (1950).
2. Several investigators report values of α that deviate from $-15\ \text{cm}^3\ \text{mol}^{-1}$ at pressures below 1 MPa, but some values are greater than $-15\ \text{cm}^3\ \text{mol}^{-1}$ and some are less. The limiting value of α at zero pressure must be equal to $-B$ (Equation 13-39), which is given as 14.8 ± 0.5, in J. H. Dymond and E. B. Smith, "The Virial Coefficients of Gases," Clarendon Press, Oxford, 1969, p. 57. It would seem, therefore, that the disagreement in the data at low pressure is due to the difficulty in measuring a small difference between two very large numbers.
3. Values of γ for specific gases are available in *TRC Thermodynamic Tables–Hydrocarbons*, Thermodynamics Research Center: The Texas A&M University System, College Station, Texas.
4. G. N. Lewis and M. Randall, *Thermodynamics*, McGraw-Hill, New York, 1923, p. 198.
5. Lewis and Randall, *Thermodynamics*, p. 226.
6. More detailed treatments of the thermodynamic properties of gaseous solutions have been described by J. A. Beattie, *Chem. Rev.* **44**, 141 (1949); O. Redlich and J. N. S. Kwong, *Chem. Rev.* **44**, 233 (1949); K. S. Pitzer and G. O. Hultgren, *J. Am. Chem. Soc.* **80**, 4793 (1958); C. M. Knobler, *Chem. Thermodynamics* **2**, 199 (1978).

CHAPTER 14

The Phase Rule

The basis of classical thermodynamics was developed by J. Willard Gibbs in his essay [1], "On the Equilibrium of Heterogeneous Substances." In particular, he derived the phase rule, which describes the conditions of equilibrium for multiphase, multicomponent systems, which are so important to the geologist and the materials scientist. In this chapter we will present a derivation of the phase rule and apply the result to several examples.

14-1 DERIVATION OF THE PHASE RULE

The phase rule is expressed in terms of P, the number of phases in the system; C, the number of components; and F, the number of degrees of freedom, or the variance of the system.

The number of phases is the number of different homogeneous regions in the system. Thus in a system containing liquid water and a number of chunks of ice there are only two phases. The number of degrees of freedom is the number of intensive variables that can be altered without the appearance or disappearance of a phase. First we will discuss a system that does not react chemically, that is, one in which the number of components is simply the number of chemical species.

Nonreacting Systems

If we express the composition of a phase in terms of mole fractions, then $(C - 1)$ intensive variables are needed to describe the composition if every component appears in the phase. In a system of P phases, $P(C - 1)$ intensive variables are used to describe the system. As was pointed out in Chapter 6, a one-phase, one-component system can be described by a large number of intensive variables, yet the specification of the values of any two such variables is sufficient to fix the state of such a system. Thus, $2P$ variables are needed to describe the temperature and pressure of each phase, or any convenient choice of two intensive variables. Therefore,

$$P(C - 1) + 2P = P(C + 1) \tag{14-1}$$

variables describe the state of a system.

Mechanical Equilibrium. For a system of fixed total volume and of uniform temperature throughout, the condition of equilibrium is given by Equation 8-14 as

$$dA = 0$$

If phase I of the system changes its volume, with a concurrent change in the volume of phase II, then, at constant temperature, it follows from Equation 8-35 that

$$\begin{aligned} dA &= dA_{\text{I}} + dA_{\text{II}} \\ &= -P_{\text{I}}dV_{\text{I}} - P_{\text{II}}dV_{\text{II}} = 0 \end{aligned}$$

Since the total volume is fixed,

$$dV_{\text{I}} = -dV_{\text{II}}$$

and

$$P_{\text{II}}dV_{\text{I}} - P_{\text{I}}dV_{\text{I}} = 0 \tag{14-2}$$

The constraint of Equation 14-2 can be met only if $P_{\text{I}} = P_{\text{II}}$, which is the condition for *mechanical equilibrium*. (We will discuss later several cases to which this requirement does not apply.) Or, to put the argument differently, if the pressures of two phases are different, the phase with the higher pressure will spontaneously expand; and the phase with the lower pressure will spontaneously contract, with a decrease in A, until the pressures are equal. Thus, for P phases, $P - 1$ independent relationships among the pressures of the phases can be written.

Thermal Equilibrium. For a system at constant total energy and constant total volume (an isolated system), the condition of equilibrium is given by Equation 7-105 as

$$dS = 0$$

If an infinitesimal amount of heat, DQ, is transferred reversibly from phase I to phase II, it follows from Equation 7-48 that

$$\begin{aligned} dS &= dS_{\mathrm{I}} + dS_{\mathrm{II}} \\ &= -\frac{DQ}{T_{\mathrm{I}}} + \frac{DQ}{T_{\mathrm{II}}} = 0 \end{aligned} \qquad (14\text{-}3)$$

The constraint of Equation 14-3 can be met only if $T_{\mathrm{I}} = T_{\mathrm{II}}$, which is the condition for *thermal equilibrium*. Or, to put the argument differently, if the temperatures of two phases differ, heat will flow irreversibly from the phase at higher temperature to the phase at lower temperature, with an increase in entropy, until the temperatures are equal. Thus, for P phases, $P - 1$ independent relationships among the temperatures of the phases can be written.

Transfer Equilibrium. For a system at constant temperature and pressure, the condition of equilibrium is given by Equation 8-19 as

$$dG = 0$$

If dn moles of a substance are transferred from phase I to phase II, then it follows from Equations 12-5 and 12-12 that

$$dG = -\mu_{\mathrm{I}} dn + \mu_{\mathrm{II}} dn = 0 \qquad (14\text{-}4)$$

The condition of Equation 14-4 can be met only if $\mu_{\mathrm{I}} = \mu_{\mathrm{II}}$, which is the condition of *transfer equilibrium* between phases. Or, to put the argument differently, if the chemical potentials (escaping tendencies) of a substance in two phases differ, there will be spontaneous transfer from the phase of higher chemical potential to the phase of lower chemical potential, with a decrease in the free energy of the system, until the chemical potentials are equal (see Section 12-5). For each component present in all P phases, there are $(P - 1)$ equations of the form of Equation 14-4 that provide constraints at transfer equilibrium. Furthermore, an equation of the form of Equation 14-4 can be written for each one of the C components in the system in transfer equilibrium between any two phases. Thus, $C(P - 1)$ independent relationships among the chemical potentials can be written. Since chemical potentials are functions of the mole fractions at constant temperature and pressure, there are $C(P - 1)$ relationships among the mole fractions. Thus summing the independent relationships for temperature, pressure, and composition in the system, we find

$$2(P - 1) + C(P - 1) = (C + 2)(P - 1) \qquad (14\text{-}5)$$

independent relationships exist among the variables.

The Phase Rule. The number of degrees of freedom is the difference between the number of variables needed to describe the system and the number of independent relationships among those variables:

$$F = P(C + 1) - (P - 1)(C + 2)$$

$$= C - P + 2 \quad (14\text{-}6)$$

In a system in which one component is absent from a phase, the number of variables needed to describe the system decreases by one. Since the number of independent relationships also decreases by one the number of degrees of freedom remains the same.

Reacting Systems

For a system undergoing R independent chemical reactions among N chemical species there are R equilibrium expressions to be added to the relationships among the intensive variables. The total number of intensive variables becomes (from Equation 14-1)

$$P(N - 1) + 2P = P(N + 1)$$

and the number of independent relationships becomes

$$(N + 2)(P - 1) + R$$

Thus the number of degrees of freedom for a reacting system is

$$F = P(N + 1) - (N + 2)(P - 1) - R$$

$$= N - R - P + 2 \quad (14\text{-}7)$$

The general definition for the number of components in a system is the minimum number of chemical species from which all phases in the system can be prepared. This number is $(N - R)$ because each equilibrium relationship decreases by one the number of species required to prepare a phase. The quantity $(N - R)$ is equivalent in Equation 14-7 to C in Equation 14-6. For example, water in equilibrium with its vapor at room temperature and atmospheric pressure is a one-component system. Water in equilibrium with H_2 and O_2 at a temperature and pressure at which dissociation takes place is a two-component system unless the mole ratio of H_2/O_2 is exactly 2; then $C = 1$. Water in equilibrium with OH^- and H^+ ions at room temperature and atmospheric pressure is a one-component system because the requirement for electrical neutrality in ionic solutions imposes an additional relationship on the system.

14-2 ONE-COMPONENT SYSTEMS

The number of degrees of freedom for a one-component system,

$$F = 3 - P \quad (14\text{-}8)$$

is at most two since the minimum value for *P* is one. Thus the temperature and pressure can be varied independently for a one-component, one-phase system and the system can be represented as an area on a temperature versus pressure diagram.

If two phases of one component are present, only one degree of freedom remains, either temperature or pressure. Two phases in equilibrium are represented by a curve on a T–P diagram. When either temperature or pressure is fixed, the other is determined by the Clapeyron equation (9-9). If three phases of one component are present, no degrees of freedom remain. Three phases in equilibrium are represented on a T–P diagram by a point called the triple point. Varying either temperature or pressure will cause the disappearance of a phase.

An interesting example of a one-component system is SiO_2, which can exist in five different crystalline forms, as a liquid, and as a vapor. Since $c = 1$, the maximum number of phases that can coexist at equilibrium is three. Each phase occupies an area on the T–P diagram; the two-phase equilibria are represented by curves, and the three-phase equilibria by points. Figure 14-1 [2], which displays the equilibrium relationships among the solid forms of SiO_2, was obtained from calculations of the temperature and pressure dependence of ΔG (as described in Section 8-3) and from experimental determination of equilibrium temperature as a function of equilibrium pressure.

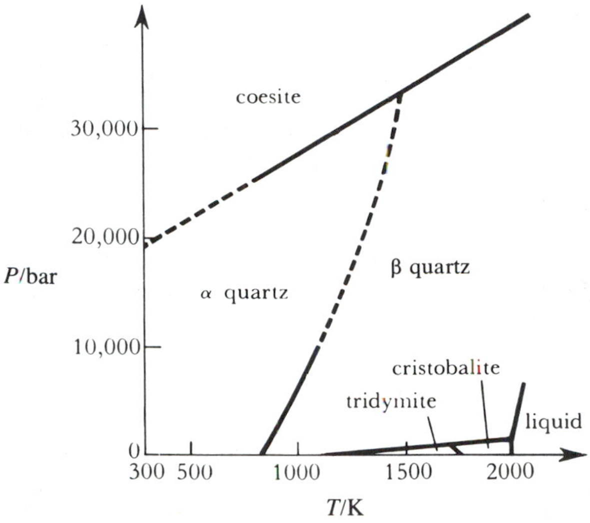

Figure 14-1. Phase diagram for SiO_2. From R. Kern and A. Weisbrod, *Thermodynamics for Geologists*, Freeman, Cooper and Co., San Francisco, 1967, p. 123.

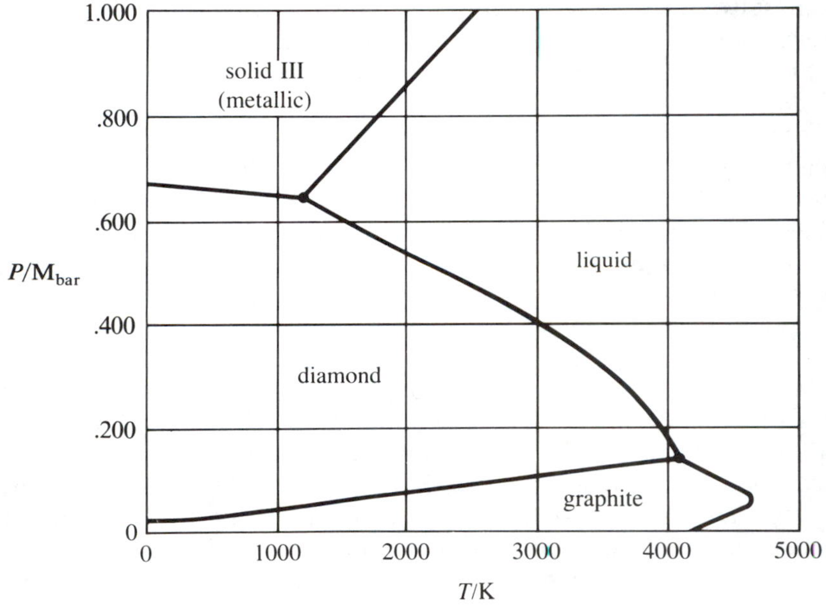

Figure 14-2. Phase diagram for carbon. From C. G. Suits, *Am. Scientist* **52**, 395 (1964).

A one-phase system that is important in understanding the geology of diamonds as well as the industrial production of diamonds is that of carbon, shown in Figure 14-2. The phase diagram shows clearly that graphite is the stable solid phase at low pressure. Thus, diamond can spontaneously change to graphite at atmospheric pressure ($\sim 10^5$ Pa); diamond owners obviously need not worry, however; the transition in the solid state is infinitely slow at ordinary temperatures.

Although phase diagrams such as Figure 14-1 and Figure 14-2 describe the conditions of T and P at which different phases are stable, they do not describe the properties of the system. Since the specification of two intensive variables is sufficient to fix all other intensive variables, the variation of any other intensive variable can be described in terms of a surface above the T–P plane, and the height of any point in the surface above the T–P plane represents the value of the intensive variable. Figure 14-3 shows such a surface for the molar volume v as a function of T and P. When two or more phases are present at equilibrium, v is a multivalued function of T and P. Similar surfaces can be constructed to describe other thermodynamic properties, such as G, H, and S, relative to some standard value.

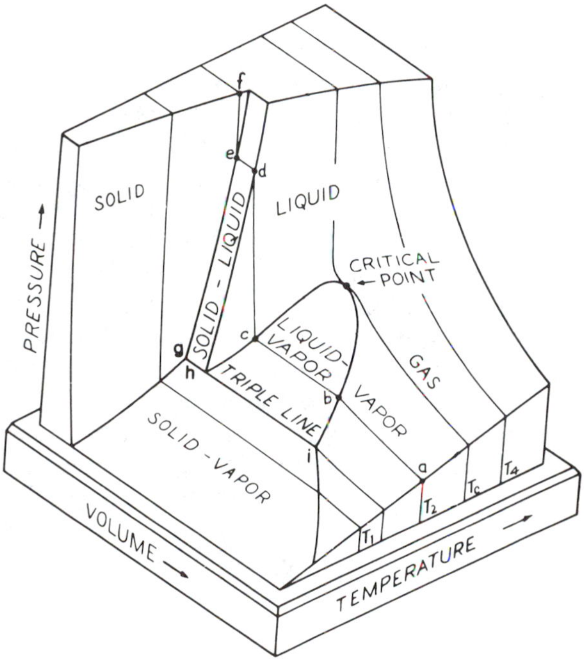

Figure 14-3. A *P-v-T* surface for a substance that contracts on freezing. Here T_1 represents an isotherm below the triple-point temperature, T_2 represents an isotherm between the triple point and the critical point, T_c is the critical temperature, and T_4 represents an isotherm above the critical temperature. Points *g, h,* and *i* represent the molar volumes of solid, liquid, and vapor, respectively, in equilibrium at the triple point. Points *e* and *d* represent the molar volumes of solid and liquid, respectively, in equilibrium at temperature T_2 and the corresponding equilibrium pressure. Points *c* and *b* represent the molar volumes of liquid and vapor, respectively, in equilibrium at temperature T_2 and the corresponding equilibrium pressure. (From Sears-Salinger, *Thermodynamics, Kinetic Theory, and Statistical Thermodynamics*, Addison-Wesley, Reading, Mass., 1975.)

14-3 TWO-COMPONENT SYSTEMS

The number of degrees of freedom for a two-component system,

$$F = 4 - P \tag{14-9}$$

has a maximum value of three. Since a complete representation of such a system requires three coordinates, we can decrease the variance by fixing the temperature and leaving pressure and composition as the variables of the system, or by fixing the pressure and leaving temperature and composition as the variables of the system. In a reduced-phase diagram for a two-component system a single phase is represented by an area, and the two phases in equilibrium by a curve relating the two variables. Since the composition of the two phases generally is different, two conjugate curves are required.

Figure 14-4 [3] is a reduced two-component diagram for the mineral olivine, which is a solid solution of fayalite, Fe_2SiO_4, and forsterite, Mg_2SiO_4. Above the liquidus curve the system exists as a single-liquid phase and below the solidus curve the system exists as a single-solid phase. Between the two curves liquid and solid phases are in equilibrium, and their compositions are given by the intersections of a constant-temperature line with the liquidus and solidus curves. A point in the area between the solidus and liquidus curves represents the overall composition of the system of two phases. This value, since it depends on the relative amounts of the two phases, is not described by the phase rule, which is concerned only with the number of phases and their composition. Although the area between the solidus and liquidus curves is frequently called a "two-phase area," only the curves correspond geometrically to a value of $F = 1$.

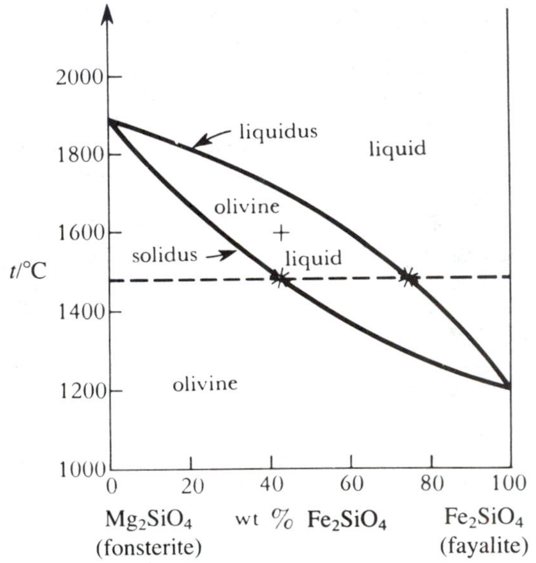

Figure 14-4. Phase diagram for olivine at constant pressure. From R. Kern and A. Weisbrod, *Thermodynamics for Geologists*, Freeman, Cooper and Co., San Francisco, 1967, p. 197.

Other two-component systems may exhibit either limited solubility or complete insolubility in the solid state. An example with limited solubility is the silver–copper system, of which the reduced-phase diagram is shown in Figure 14-5. Area *L* represents a single-liquid phase, with $F = 2$, and S_1 and S_3 represent solid-solution phases rich in Ag and Cu, respectively. Area S_2 is a "two-phase area," and the curves *AB* and *DE* represent the compositions of the two solid-solution phases that are in equilibrium at any temperature. At 1052 K, liquid of composition *C* is in equilibrium with solid solutions of composition *B* and *D*. This "triple-point," with $F = 0$, is represented by a line because the three phases have different compositions. Between 1052 K and 1234 K, lines *FB* and *FC* represent the compositions of solid solution and liquid solution, respectively, in equilibrium, with overall system composition richer in Ag than *C*. Between 1052 K and 1356.55 K, lines *CG* and *GD* represent the compositions of liquid solution and solid solution, respectively, in equilibrium with overall system composition richer in Cu than *C*.

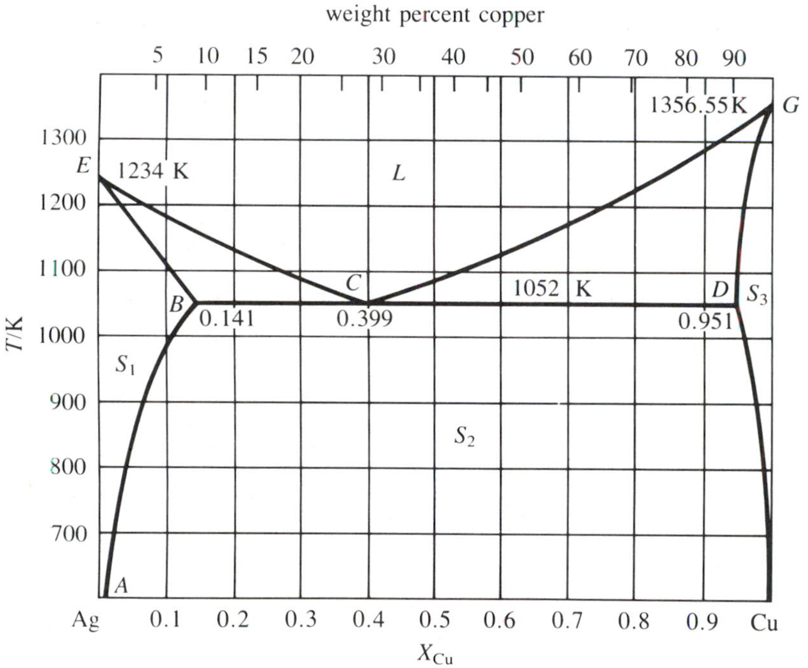

Figure 14-5. Phase diagram for the system Ag-Cu at constant pressure. With permission from R. Hultgren, P. D. Desai, D. T. Hawkins, M. Gleiser, and K. K. Kelley, "Selected Values of Thermodynamic Properties of Binary Alloys," American Society for Metals, Metals Park, Ohio, 1973, p. 46.

Figure 14-6 is the phase diagram of the system Na–K at constant pressure, a system that exhibits limited solubility and the occurrence of a solid compound Na_2K (β), which melts at 280.06 K to form a solid solution rich in Na and liquid of composition *H*.

Although the phase rule is concerned with the existence of the relationships among system variables that are represented by the equilibrium curves, it provides no information about the nature of those relationships. We will consider the dependence of the chemical potential on the system variables for various systems in later chapters.

Two Phases at Different Pressures

In deriving the phase rule we assumed that all phases are at the same pressure. In mineral systems, fluid phases can be at a pressure different from the solid phases if the rock column above them is permeable to the fluid. Under these circumstances the system has an additional degree of freedom and the equilibrium at any depth depends both on the fluid pressure, P_F, and the pressure on the solid, P_S, at that level. Each pressure is determined by ρ, the

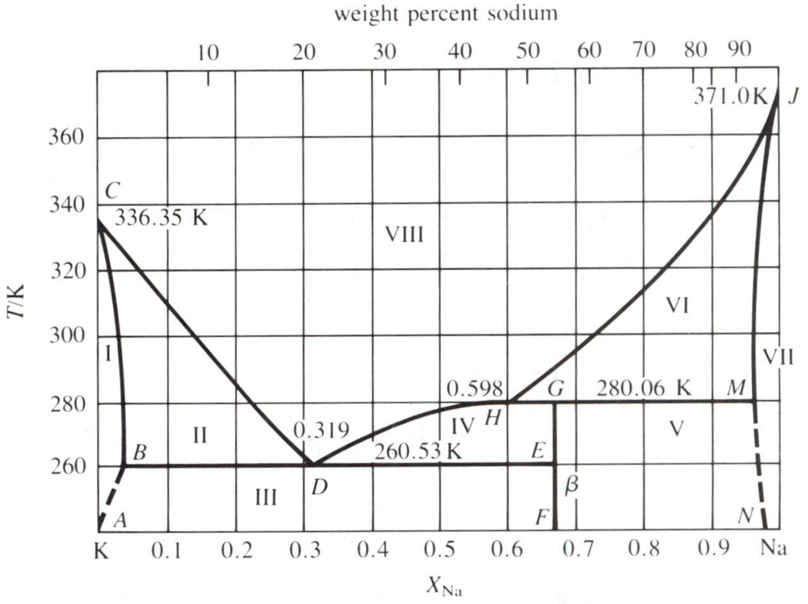

Figure 14-6. The phase diagram at constant pressure of the system Na-K. With permission from R. Hultgren, P. D. Desai, D. T. Hawkins, M. Gleiser, and K. K. Kelley, "Selected Values of Thermodynamic Properties of Binary Alloys," American Society of Metals, Metal Park, Ohio, 1973, p. 1057.

density of the phase, and h, the height of the column between the surface and the level being studied.

The equations required to calculate the effect of pressure and temperature on ΔG are modified from Equation 8-95 to include a term for each pressure at any temperature T. For example, for the gypsum–anhydrite equilibrium,

$$\underset{\text{gypsum}}{CaSO_4 \cdot 2H_2O(s, P_S, T)} = \underset{\text{anhydrite}}{CaSO_4(s, P_S, T)} + 2H_2O(l, P_F, T) \qquad (14\text{-}10)$$

the free energy change is given by

$$\Delta G(P_F, P_S, T) = \Delta G(P = P^\circ, T) + P_S(\Delta v_S) + P_F(\Delta v_F) \qquad (14\text{-}11)$$

in which Δv_S represents the molar volume change of the solid phases in the transformation and Δv_F represents the molar volume change of the fluid phase in the transformation. (See Table 2-1 for the definition of one mole for a chemical transformation.) That is,

$$\Delta v_S = v_{CaSO_4(s)} - v_{CaSO_4 \cdot 2H_2O(s)} = -29.48 \text{ cm}^3 \qquad (14\text{-}12)$$

and

$$\Delta v_F = 2v_{H_2O(l)} = 36.14 \text{ cm}^3 \qquad (14\text{-}13)$$

The validity of Equation 14-11 can be seen from the addition of the following steps:

1. $CaSO_4 \cdot 2H_2O(s, P = P^\circ, T) = CaSO_4(s, P = P^\circ, T) + 2H_2O(l, P = P^\circ, T)$,

$$\Delta G = \Delta G(P = P^\circ, T)$$

2. $CaSO_4 \cdot 2H_2O(s, P_S, T) = CaSO_4 \cdot 2H_2O(s, P = P^\circ, T)$,

$$\Delta G = v_{CaSO_4 \cdot 2H_2O(s)}(P^\circ - P_S) \cong -P_S v_{CaSO_4 \cdot 2H_2O(s)}$$

3. $CaSO_4(s, P = P^\circ, T) = CaSO_4(s, P_S, T)$,

$$\Delta G = v_{CaSO_4(s)}(P_S - P^\circ) \cong P_S v_{CaSO_4(s)}$$

4. $2H_2O(l, P = P^\circ, T) = 2H_2O(l, P_F, T)$,

$$\Delta G = 2v_{H_2O(l)}(P_F - P^\circ) \cong 2P_F v_{H_2O(l)}$$

The approximations for Equations 2 through 4 are valid because P°, atmospheric pressure, is small compared to P_F and P_S, the high pressures found in geologic formations.

Since the sum of Equations 1 through 4 is the same as Equation 14-10, ΔG

for Equation 14-10 is the sum of the ΔG's for Equations 1 through 4, and Equation 14-11 is obtained.

The equilibrium diagram [4] for Equation 14-10 is shown in Figure 14-7. If gypsum and anhydrite are both under liquid water at 1 bar, then equilibrium can be attained only at 40°C (see Figure 14-7). If the liquid pressure is increased, then the temperature at which the two solids, both subject to this liquid pressure, are in equilibrium is given by the curve with positive slope on the right side of Figure 14-7. This curve also describes the behavior of the gypsum–anhydrite equilibrium in a rock formation that is completely impermeable to the fluid phase, so that the pressure on the fluid phase is equal to the pressure on the solid phase. Thus the right curve applies to any situation in which P_F is equal to P_S. Under these conditions the net ΔV for the transformation of Equation 14-10 is $36.14 - 29.48 = 6.66$ cm^3 (see Equations 14-12 and 14-13), and $6.66 \times P_F$ makes a positive contribution to the ΔG of Equation 14-11. Thus an increase in pressure should shift the equilibrium from anhydrite to gypsum.

In contrast, if the rock above the layer being studied is completely permeable to the fluid, the pressure on the solid phases is that of the overlying

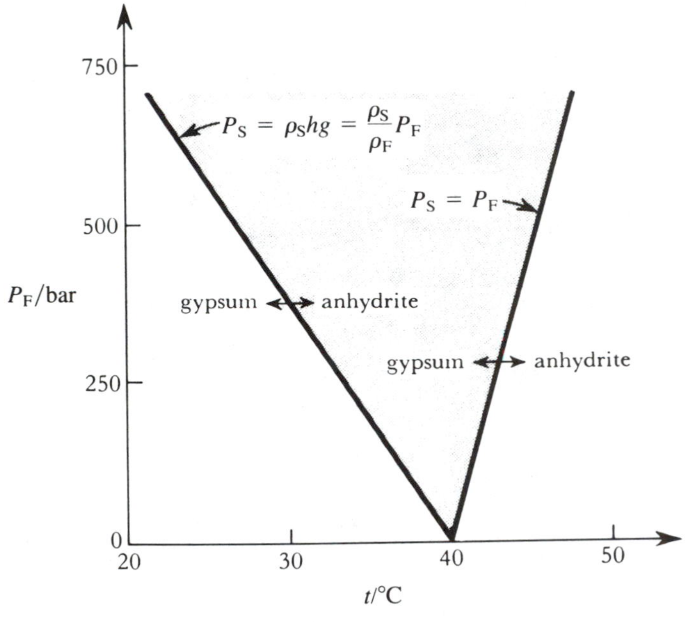

Figure 14-7. Equilibrium diagram for Equation 14-10. From R. Kern and A. Weisbrod, *Thermodynamics for Geologists*, Freeman, Cooper and Co., San Francisco, 1967, p. 274.

rock, $\rho_S hg$, whereas the pressure on the fluid phase is $\rho_F hg$, in which g is the acceleration due to gravity. Under these circumstances the equilibrium temperature for the transformation in Equation 14-10 varies with pressure according to the curve with negative slope at the left side of Figure 14-7. In this case $P_S\ \Delta v_S\ (-29.48P_S)$ exceeds $P_F\ \Delta v_F\ (36.14P_F)$ in magnitude, and the net contribution to ΔG of the $P\ \Delta v$ terms in Equation 14-11 is negative. Hence, increased pressure shifts the equilibrium from gypsum to anhydrite. If the rock is partially permeable, the equilibrium curve falls between the two curves shown, the exact position depending on the ratio of P_F to P_S. At some value of the ratio between unity and ρ_F/ρ_S, the equilibrium temperature becomes independent of the pressure. Whatever the position of the equilibrium curve, gypsum is the stable solid phase at low temperatures (to the left of the curve) and anhydrite is the stable solid phase at high temperatures (to the right of the curve).

Phase Rule Criterion of Purity

Equation 14-6 is written as if the number of degrees of freedom of a system were calculated from known values of the number of phases and the number of

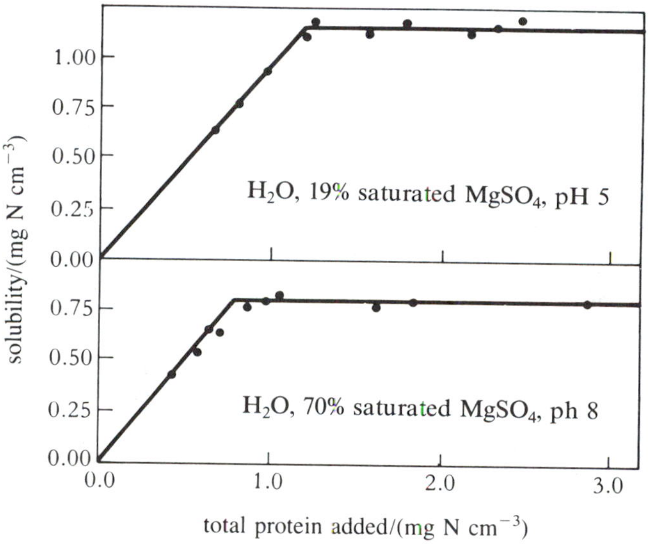

Figure 14-8. Solubility curves of chymotrypsinogen A in various solvents. Adapted from J. H. Northrop, M. Kunitz, and R. M. Herriott, *Crystalline Enzymes*, 2d ed., Columbia University Press, New York, 1948.

components. In practice, an experimentalist often determines F and P from his or her observations and then calculates C, the number of components.

The determination of the purity of a homogeneous solid from solubility measurements is an example of this application of the phase rule. The experimental procedure is to measure the concentration of dissolved material in equilibrium with excess solid, at a fixed temperature and pressure, as a function of the amount of solid added. If the solid is pure, the solid and solvent constitute a two-component system. At constant temperature and pressure, $F = 2 - P$ and a two-phase system of solid and saturated solution has zero degrees of freedom. If the solid contains more than one species, the system has three components and $F = 1$ when two phases are present. In the former case the solubility is independent of the amount of excess solid; in the latter case the solubility increases with the amount of added solid. Figure 14-8 shows the data of Butler [5] on the solubility of chymotrypsinogen A, the precursor of the pancreatic enzyme chymotrypsin. These data represent one of the earliest rigorous demonstrations of the purity of a protein.

EXERCISES

1. What would be the number of degrees of freedom in a system in which pure H_2O was raised to a temperature sufficiently high to allow dissociation into H_2 and O_2?

2. At atmospheric pressure (101.325 kPa) α quartz is in equilibrium with β quartz at 847 ± 1.5 K. [D. R. Stull and H. Prophet, "JANAF Thermochemical Tables," NSRDS-NBS37, 2d ed., 1971.] The enthalpy change of the transition is 728 ± 167 J mol^{-1}. C. Berger, L. Eyraud, M. Richard, and R. Riviere [*Bull. Soc. chim. France* 628, (1966)] measured the volume change by an X-ray diffraction method, and they reported a value of 0.154 ± 0.015 cm^3. Use these values to calculate the slope at atmospheric pressure of the equilibrium curve between α quartz and β quartz in Figure 14-1. Compare your results with the value of the pressure derivative of the equilibrium temperature, dT/dP, equal to 0.21 K (TPa)$^{-1}$ [R. E. Gibson, *J. Phys. Chem.* **32**, 1197 (1928)].

3. For Equation 14-10 [4]

$$\Delta G\ (P = P°,\ T) = -10{,}430 \text{ J mol}^{-1} + (685.8 \text{ J mol}^{-1} \text{ K}^{-1})T$$

The value of $\Delta V_S = -29.48$ cm^3 mol^{-1} and the value of $\Delta V_F =$ 36.14 cm^3 mol^{-1} ($2H_2O$). Calculate the ratio of P_S to P_F at which $\Delta G\ (P_F, P_S, T)$ is independent of pressure.

4. In Figure 14-6, identify the phases in equilibrium and the curves that describe the composition of each phase in Areas I, II, III, IV, V, VI, VII, and VIII. Identify the phases in equilibrium and the composition of those phases at 260.53 K along line *BDE*, and at 280.06 K along line *HGM*.

REFERENCES

1. J. Willard Gibbs, *Trans. Conn. Acad. Sci.* **3**, 108–248 (1876); **3**, 343–542 (1878). Reprinted in *The Collected Works of J. Willard Gibbs*, Vol. I, Yale University Press, New Haven, 1957.
2. R. Kern and A. Weisbrod, *Thermodynamics for Geologists*, Freeman, Cooper and Co., San Francisco, 1967, p. 123.
3. Kern and Weisbrod, *Thermodynamics for Geologists*, p. 197; N. L. Bowen and J. R. Schairer, *Amer. J. Sci.* **29**, 151–217 (1935).
4. Kern and Weisbrod, *Thermodynamics for Geologists*, p. 274; G. J. F. McDonald, *Amer. J. Sci.* **251**, 884–898 (1953).
5. J. A. V. Butler, *J. Gen. Physiol.* **24**, 189 (1940); M. Kunitz and J. H. Northrop, *Compt. rend. Trav. Lab. Carlsberg, Ser. chim.* **22**, 288 (1938).

CHAPTER 15

The Ideal Solution

In Chapter 14 we discussed multiphase-multicomponent systems in terms of the phase rule and its graphical representation. We begin our analytical description of such systems in this chapter by considering the ideal solution. Although not many pairs of substances form ideal solutions, we shall find that the relationships that describe an ideal solution provide limiting rules for real solutions and thereby provide a framework for the discussion of real solutions.

15-1 DEFINITION

We define an ideal solution by visualizing a liquid–vapor or solid–vapor equilibrium in which each component in the vapor phase obeys a generalized form of Raoult's law, a form expressed in fugacities instead of partial pressures. Specifically, the fugacity, $f_{i(\text{gas})}$, of each component in the gas phase is equal to $f^{\bullet}_{i(\text{gas})}$, the fugacity of the pure vapor in equilibrium with the pure condensed phase, multiplied by $X_{i(\text{cond})}$, the mole fraction of the component in the condensed phase.

$$f_{i(\text{gas})} = f^{\bullet}_{i(\text{gas})} X_{i(\text{cond})} \qquad (15\text{-}1)$$

For an ideal two-component system, the dependence of the fugacities in the gas phase on the mole fractions in the condensed phase is illustrated in Figure 15-1. Equation 15-1 reduces to the historical form of Raoult's law, $p_i = p^{\bullet}_i X_i$, when the vapors are an ideal mixture of ideal gases.

From Equation 13-56,

$$\mu_{i(\text{gas})} = \mu^{\circ}_{i(\text{gas})} + RT \ln (f_{i(\text{gas})}/f^{\circ}_{(\text{gas})})$$

and

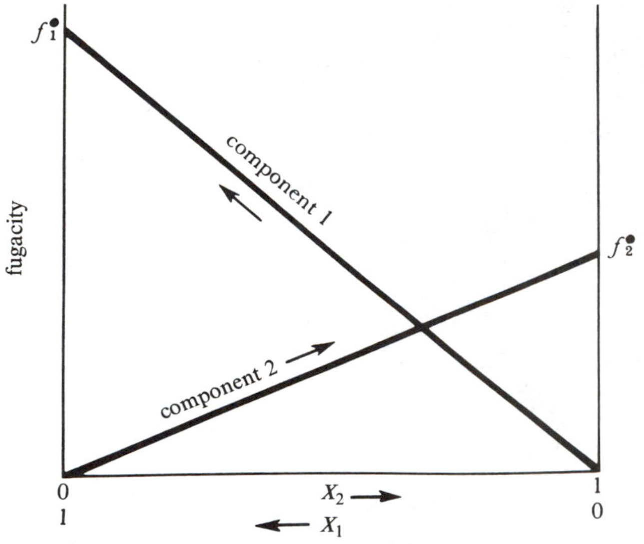

Figure 15-1. Dependence of fugacities in gas phase on composition in an ideal solution.

$$\mu^{\bullet}_{i(\text{gas})} = \mu^{\circ}_{i(\text{gas})} + RT \ln (f^{\bullet}_{i(\text{gas})}/f^{\circ}_{(\text{gas})})$$

where $\mu^{\bullet}_{i(\text{gas})}$ is determined at the same temperature and pressure as that of the solution. Thus, for a vapor in equilibrium with a condensed ideal solution

$$\begin{aligned} \mu_{i(\text{gas})} &= \mu^{\bullet}_{i(\text{gas})} + RT \ln (f_{i(\text{gas})}/f^{\bullet}_{i(\text{gas})}) \\ &= \mu^{\bullet}_{i(\text{gas})} + RT \ln X_{i(\text{cond})} \end{aligned} \tag{15-2}$$

But, at equilibrium between the phases,

$$\begin{aligned} \mu_{i(\text{cond})} &= \mu_{i(\text{gas})} \\ &= \mu^{\bullet}_{i(\text{gas})} + RT \ln X_{i(\text{cond})} \\ &= \mu^{\bullet}_{i(\text{cond})} + RT \ln X_{i(\text{cond})} \end{aligned} \tag{15-3}$$

Consequently we can also write

$$f_{i(\text{cond})} = f^{\circ}_{i(\text{cond})} X_{i(\text{cond})}$$

if we define $f_{i(\text{cond})}$ by the equation

$$\mu_{i(\text{cond})} = \mu^{\circ}_{i(\text{cond})} + RT \ln (f_{i(\text{cond})}/f^{\circ}_{i(\text{cond})})$$

It is convenient to choose the pure condensed phase at the temperature and pressure of the solution as the standard state for the component in the solution. Thus, Equation 15-3 can be written

$$\mu_{\text{i(cond)}} = \mu^{\circ}_{i(\text{cond})} + RT \ln X_{i(\text{cond})} \tag{15-4}$$

15-2 SOME CONSEQUENCES OF THE DEFINITION

If Equation 15-1 or Equation 15-3 is used to define an ideal solution of two components, values for the changes in thermodynamic properties resulting from the formation of such a solution follow directly.

Volume Changes

There is no change in volume when mixing pure components that form an ideal solution. That this statement is valid for an ideal solution can be shown as follows. At any fixed mole fraction, differentiation of Equation 15-3 with respect to pressure yields

$$\left(\frac{\partial \mu_i}{\partial P}\right)_{T,X_i} = \left(\frac{\partial \mu_i^\circ}{\partial P}\right)_T = \left(\frac{\partial \mu_i^\bullet}{\partial P}\right)_T \tag{15-5}$$

But from Equation 12-17 we have [1]

$$\left(\frac{\partial \mu_i}{\partial P}\right)_{T,X_i} = \bar{V}_i$$

Hence Equation 15-5 becomes

$$\bar{V}_i = V_i^\bullet \tag{15-6}$$

in which $V_i^\bullet$ represents the molar volume of pure component *i*. Thus the partial molar volume of each component in solution is equal to the molar volume of the corresponding pure substance.

Before the two pure components are mixed, the total volume, V_{initial}, is

$$V_{\text{initial}} = n_1 V_1^\bullet + n_2 V_2^\bullet$$

When the solution is formed, the total volume, V_{final}, is, from Equation 12-19,

$$V_{\text{final}} = n_1 \bar{V}_1 + n_2 \bar{V}_2$$

The volume change on mixing is

$$\Delta V = V_{\text{final}} - V_{\text{initial}} = n_1 \bar{V}_1 + n_2 \bar{V}_2 - (n_1 V_1^\bullet + n_2 V_2^\bullet) = 0 \tag{15-7}$$

Heat Effects

No heat is evolved when pure components that form an ideal solution are mixed. The validity of this statement can be shown from consideration of the temperature coefficient of the chemical potential. Again, from Equation 15-3 at fixed mole fraction,

$$\left[\frac{\partial(\mu_i/T)}{\partial T}\right]_{P,X_i} = \left[\frac{\partial(\mu_i^\circ/T)}{\partial T}\right]_P = \left[\frac{\partial(\mu_i^\bullet/T)}{\partial T}\right]_P \tag{15-8}$$

From Section 12-3, we see that thermodynamic relationships for extensive thermodynamic properties also apply to partial molar properties. From Equation 8-75

$$\left[\frac{\partial(\mu_i/T)}{\partial T}\right]_{P,X_i} = -\frac{\bar{H}_i}{T^2} \tag{15-9}$$

Thus Equation 15-8 becomes

$$\bar{H}_i = H_i^\bullet \tag{15-10}$$

in which $H_i^\bullet$ is the molar enthalpy of pure component i. Before mixing the two components, the enthalpy, H_{initial}, is given by

$$H_{\text{initial}} = n_1 H_1^\bullet + n_2 H_2^\bullet$$

and after the formation of the solution the enthalpy is

$$H_{\text{final}} = n_1 \bar{H}_1 + n_2 \bar{H}_2$$

Thus the enthalpy change on mixing is

$$\Delta H = H_{\text{final}} - H_{\text{initial}} = n_1\bar{H}_1 + n_2\bar{H}_2 - (n_1 H_1^\bullet + n_2 H_2^\bullet) = 0 \tag{15-11}$$

Since ΔH is a measure of the heat exchanged in a constant-pressure process, no heat is evolved or absorbed on mixing an ideal solution.

From Equation 12-15

$$\left(\frac{\partial \bar{H}_i}{\partial T}\right)_{P,X_i} = \bar{C}_{pi} \quad \text{and} \quad \left(\frac{\partial H_i^\bullet}{\partial T}\right)_P = C_{pi}^\bullet$$

Therefore

$$\bar{C}_{pi} = C_{pi}^\bullet \tag{15-12}$$

for each component of an ideal solution.

15-3 THERMODYNAMICS OF TRANSFER OF A COMPONENT FROM ONE IDEAL SOLUTION TO ANOTHER

Figure 15-2 illustrates the transfer process and gives the values of the partial molar thermodynamic properties in the two solutions. We can represent the transfer process by the equation

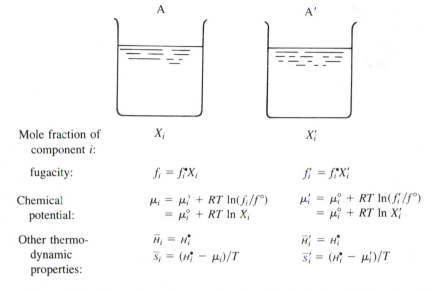

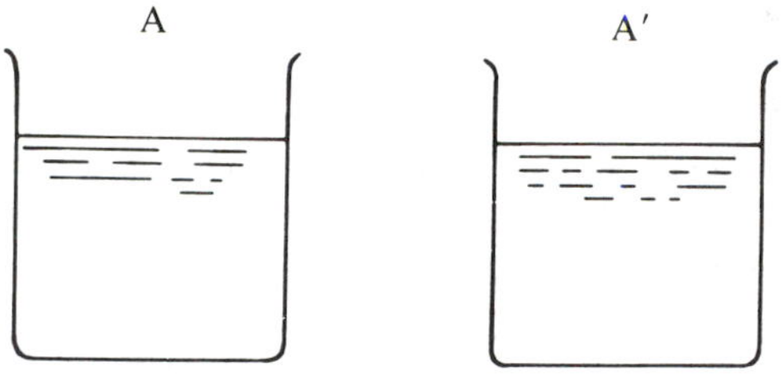

Mole fraction of component i:	X_i	X_i'
fugacity:	$f_i = f_i^\bullet X_i$	$f_i' = f_i^\bullet X_i'$
Chemical potential:	$\mu_i = \mu_i^\circ + RT \ln(f_i/f^\circ)$ $= \mu_i^\circ + RT \ln X_i$	$\mu_i' = \mu_i^\circ + RT \ln(f_i'/f^\circ)$ $= \mu_i^\circ + RT \ln X_i'$
Other thermodynamic properties:	$\bar{H}_i = H_i^\bullet$ $\bar{S}_i = (H_i^\bullet - \mu_i)/T$	$\bar{H}_i' = H_i^\bullet$ $\bar{S}_i' = (H_i^\bullet - \mu_i')/T$

Figure 15-2. Thermodynamic properties of two ideal solutions of different mole fraction, X_i and X_i', prepared from the same components.

$$\text{component } i(\text{A}) = \text{component } i(\text{A}') \tag{15-13}$$

and for the change in any property J we can write

$$\Delta J = J_{\text{final}} - J_{\text{initial}}$$

Thus the change in chemical potential, and the change in free energy, is

$$\begin{aligned} \Delta G &= \Delta \mu_i \\ &= \mu_i' - \mu_i \\ &= \mu_i^\circ + RT \ln X_i' - [\mu_i^\circ + RT \ln X_i] \\ &= RT \ln \frac{X_i'}{X_i} \end{aligned} \tag{15-14}$$

Similarly, the enthalpy change is

$$\begin{aligned} \Delta H &= \bar{H}_i' - \bar{H}_i \\ &= H_i^\bullet - H_i^\bullet = 0 \end{aligned} \tag{15-15}$$

Finally, the entropy change is

$$\begin{aligned} \Delta S &= \bar{S}_i' - \bar{S}_i \\ &= \frac{H_i^\bullet - \mu_i'}{T} - \frac{H_i^\bullet - \mu_i}{T} \end{aligned}$$

$$= -\frac{\mu_i' - \mu_i}{T} \tag{15-16}$$

and from Equation 15-14 and Equation 15-16

$$\Delta S = -R \ln \frac{X_i'}{X_i} \tag{15-17}$$

15-4 THERMODYNAMICS OF MIXING

In a similar way we can consider an integral mixing process for the formation of an ideal solution from the components, as illustrated in Figure 15-3. The mixing process can be represented by the equation

$$\begin{aligned} &n_1 \text{ moles component 1(pure)} + n_2 \text{ moles component 2(pure)} \\ &\quad = \text{solution containing } n_1 \text{ moles component } 1(X_1) \\ &\qquad + n_2 \text{ moles component } 2(X_2) \end{aligned} \tag{15-18}$$

Thus

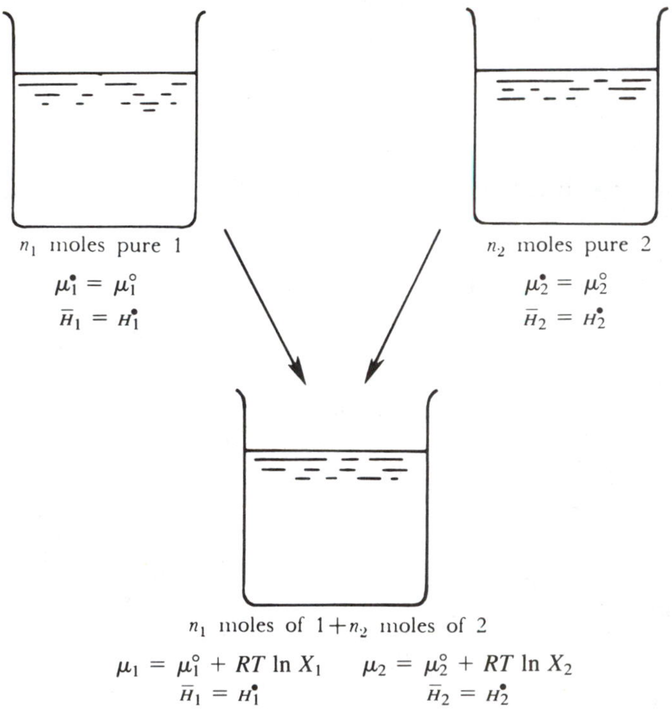

Figure 15-3. Thermodynamics of formation of an ideal solution from pure components.

$$\Delta J = J_{\text{final}} - J_{\text{initial}}$$
$$= n_1\bar{J}_1 + n_2\bar{J}_2 - n_1J_1^\bullet - n_2J_2^\bullet$$

Then the change in free energy is

$$\Delta G = n_1\mu_1 + n_2\mu_2 - n_1\mu_1^\bullet - n_2\mu_2^\bullet$$
$$= n_1[\mu_1^\circ + RT \ln X_1 - \mu_1^\circ] + n_2[\mu_2^\circ + RT \ln X_2 - \mu_2^\circ]$$
$$= n_1 RT \ln X_1 + n_2 RT \ln X_2 \qquad (15\text{-}19)$$

For the enthalpy change

$$\Delta H = n_1\bar{H}_1 + n_2\bar{H}_2 - n_1H_1^\bullet - n_2H_2^\bullet$$
$$= 0 \qquad (15\text{-}20)$$

Since, for an isothermal change,

$$\Delta S = \frac{\Delta H - \Delta G}{T}$$

then

$$\Delta S = -\frac{\Delta G}{T}$$
$$= -n_1R \ln X_1 - n_2R \ln X_2 \qquad (15\text{-}21)$$

The values for the formation of an ideal solution are identical with those we derived for mixing ideal gases in Chapter 13. A mixture of ideal gases is a special case of an ideal solution. The equations that we have derived are equally applicable to solid, liquid, and gaseous solutions as long as there is no phase change in the mixing process. For the special case of one mole of solution the thermodynamic changes are

$$\Delta \mathrm{G} = \frac{\Delta G}{n_1 + n_2}$$
$$= X_1RT \ln X_1 + X_2RT \ln X_2 \qquad (15\text{-}22)$$
$$\Delta \mathrm{H} = 0 \qquad (15\text{-}23)$$

and

$$\Delta \mathrm{S} = -X_1R \ln X_1 - X_2R \ln X_2 \qquad (15\text{-}24)$$

15-5 EQUILIBRIUM BETWEEN A PURE SOLID AND AN IDEAL LIQUID SOLUTION

For some ideal solutions the range of composition that can be attained is limited because of the limited solubility of one or both components. As an example, let us consider the solution of naphthalene in benzene.

When ΔH is measured for the change

naphthalene(solid) = naphthalene(solution)

it is found that ΔH is equal to ΔH_{fusion} of naphthalene. Therefore it is reasonable to consider that dissolved naphthalene can be regarded as being in the liquid state rather than in the solid state. If we examine the process

naphthalene(supercooled liquid) = naphthalene(solution)

we find the accompanying heat effect is zero. Clearly this is consistent with the observation that dissolved naphthalene behaves as if it were a liquid. The liquid is called supercooled because the temperature is below the melting point of naphthalene.

If the solution is ideal, the chemical potential, μ_2, of the dissolved solute at a fixed temperature and pressure is given by the expression

$$\mu_2 = \mu_2^\circ + RT \ln X_2 \quad (15\text{-}25)$$

$$= \mu_2^\bullet + RT \ln X_2 \quad (15\text{-}26)$$

in which $\mu_2^\bullet$ is the chemical potential of the *pure, supercooled, liquid naphthalene.* This supercooled, liquid naphthalene is characterized also by the molar volume and enthalpy, $\bar{V}_2$ and $\bar{H}_2$, respectively, that are equal to the corresponding quantities for the dissolved solute.

Above a specified concentration (at a given temperature and pressure) no more naphthalene will dissolve at equilibrium; that is, the solution becomes saturated. When solid naphthalene is in equilibrium with the solution,

naphthalene(solid) = naphthalene(satd soln) (15-27)

it follows that

$$\mu_{2(\text{solid})} = \mu_{2(\text{satd soln})} \quad (15\text{-}28)$$

$$= \mu_2^\bullet + RT \ln X_{2(\text{satd soln})} \quad (15\text{-}29)$$

Clearly,

$$\mu_{2(\text{solid})} < \mu_2^\bullet \quad (15\text{-}30)$$

since we know that supercooled, liquid naphthalene can be transformed spontaneously into the solid, and since $X_{2(\text{satd soln})}$ in Equation 15-29 is always less than one; in other words, the escaping tendency of the supercooled liquid is greater than that of the solid. By rearranging Equation 15-29 we obtain

$$\ln X_{2(\text{satd soln})} = \frac{\mu_{2(\text{solid})} - \mu_2^\bullet}{RT} \quad (15\text{-}31)$$

or

$$X_{2(\text{satd soln})} = \exp\left\{\frac{\mu_{2(\text{solid})} - \mu_2^\bullet}{RT}\right\} \quad (15\text{-}32)$$

In this equation X_2 represents the mole fraction of naphthalene *in the saturated solution in benzene*. It is determined only by the chemical potential of solid naphthalene and pure, supercooled liquid naphthalene. No property of the solvent (benzene) appears in Equation 15-32. Thus we arrive at the conclusion that the solubility of naphthalene (in terms of mole fraction) is the same in all solvents with which it forms an ideal solution. Furthermore, there is nothing in the derivation of Equation 15-32 that restricts its application to naphthalene. Hence the *solubility* (in terms of mole fraction) of any specified solid is *the same in all solvents with which it forms an ideal solution.*

Change of Solubility with Pressure at a Fixed Temperature

When a solid is in equilibrium with the solute in an ideal solution under isothermal conditions:

$$\text{solid} = \text{solute in solution} \tag{15-33}$$

then

$$\mu_{2(\text{solid})} = \mu_{2(\text{satd})}$$

If the pressure is changed, the solubility can change, but if equilibrium is maintained, then

$$d\mu_{2(\text{solid})} = d\mu_{2(\text{satd})}$$

The chemical potential of the solid is a function only of the pressure at constant temperature, whereas the chemical potential of the solute in the saturated solution is a function of both pressure and mole fraction. Thus

$$d\mu_{2(\text{solid})} = \left(\frac{\partial \mu_{2(\text{solid})}}{\partial P}\right)_T dP$$

$$= d\mu_{2(\text{satd})} = \left(\frac{\partial \mu_{2(\text{satd})}}{\partial P}\right)_{T,X_2} dP + \left(\frac{\partial \mu_{2(\text{satd})}}{\partial X_2}\right)_{T,P} dX_2 \tag{15-34}$$

From Equation 12-17 and Equation 15-6

$$\left(\frac{\partial \mu_{2(\text{solid})}}{\partial P}\right)_T = V_{2(\text{solid})} \quad \text{and} \quad \left(\frac{\partial \mu_{2(\text{satd})}}{\partial P}\right)_{T,X_2} = \bar{V}_{2(\text{satd})} = V_2^{\bullet} \tag{15-35}$$

and, from Equation 15-3,

$$\left(\frac{\partial \mu_{2(\text{satd})}}{\partial X_2}\right)_{T,P} = \frac{RT}{X_2} \tag{15-36}$$

Substituting from Equations 15-35 and 15-36 into Equation 15-34, we obtain

$$v_{2(\text{solid})}\, dP = v_2^\bullet\, dP + RT\,\frac{dX_2}{X_2}$$

or

$$\frac{dX_2}{X_2} = d\ln X_2 = \frac{v_{2(\text{solid})} - v_2^\bullet}{RT}\, dP \qquad (15\text{-}37)$$

Rearranging we obtain

$$\left[\frac{\partial \ln X_{2(\text{satd})}}{\partial P}\right]_T = \frac{v_{(\text{solid})} - v_2^\bullet}{RT} \qquad (15\text{-}38)$$

in which we recognize explicitly that X_2 is the mole fraction of solute in the saturated solution.

Since $v_2^\bullet$ is the molar volume of pure, supercooled solute, Equation 15-38 can be written

$$\left[\frac{\partial \ln X_{2(\text{satd})}}{\partial P}\right]_T = -\frac{\Delta v_{\text{fusion}}}{RT} \qquad (15\text{-}39)$$

In this case Δv is the molar volume change for the transition from the pure, solid solute to the supercooled, liquid solute.

Change of Solubility with Temperature

The procedure for deriving the temperature coefficient of the solubility of a solute in an ideal solution parallels that just used for the pressure coefficient. The condition for maintenance of equilibrium with a change in temperature is

$$d\mu_{2(\text{solid})} = d\mu_{2(\text{satd})} \qquad (15\text{-}40)$$

Since at constant pressure the chemical potential of the pure solid is a function only of the temperature, and the chemical potential of the solute is a function of the temperature and mole fraction, we can express Equation 15-40 as

$$\left(\frac{\partial \mu_{2(\text{solid})}}{\partial T}\right)_P dT = \left(\frac{\partial \mu_{2(\text{satd})}}{\partial T}\right)_{P,X_2} dT + \left(\frac{\partial \mu_{2(\text{satd})}}{\partial X_2}\right)_{P,T} dX_2 \qquad (15\text{-}41)$$

From Equation 12-16 we have

$$\left(\frac{\partial \mu_{2(\text{solid})}}{\partial T}\right)_P = -S_{2(\text{solid})} \quad \text{and} \quad \left(\frac{\partial \mu_{2(\text{satd})}}{\partial T}\right)_{P,X_2} = -\bar{S}_{2(\text{satd})}$$

and from Equation 15-3

$$\left(\frac{\partial \mu_{2(\text{satd})}}{\partial X_2}\right)_{T,P} = \frac{RT}{X_2}$$

Consequently, Equation 15-41 becomes

$$-S_{2(\text{solid})}\, dT = -\bar{S}_{2(\text{satd})}\, dT + \frac{RT}{X_2}\, dX_2$$

or

$$d \ln X_2 = \frac{\bar{S}_{2(\text{satd})} - S_{2(\text{solid})}}{RT}\, dT \tag{15-42}$$

and

$$\left[\frac{d \ln X_{2(\text{satd})}}{dT}\right]_P = \frac{\bar{S}_{2(\text{satd})} - S_{2(\text{solid})}}{RT} \tag{15-43}$$

From Figure 15-2, since $\mu_{2(\text{satd})} = \mu_{2(\text{solid})}$ at equilibrium,

$$\bar{S}_{2(\text{satd})} - S_{2(\text{solid})} = \frac{\bar{H}_{2(\text{satd})} - H_{2(\text{solid})}}{T}$$

$$= \frac{H_2^\bullet - H_{2(\text{solid})}}{T} \tag{15-44}$$

Thus Equation 15-43 can be written

$$\left[\frac{\partial \ln X_{2(\text{satd})}}{\partial T}\right]_P = \frac{\Delta H_{\text{fusion}(2)}}{RT^2} \tag{15-45}$$

and $\Delta H_{\text{fusion}(2)}$ is the molar enthalpy change for the transition from pure, solid solute to pure, supercooled, liquid solute.

Since T is the temperature at which a liquid solution is in equilibrium with a pure, solid solute, one also can interpret Equation 15-45 as describing the way in which the freezing point of the solution depends on concentration; from this point of view, component 2 is the solvent. This is shown explicitly by inverting the derivative:

$$\left[\frac{\partial\, T_{\mathrm{m}}}{\partial \ln X_2}\right]_P = \frac{RT_{\mathrm{m}}^2}{\Delta H_{\mathrm{fusion(2)}}} \tag{15-46}$$

The interpretation of Equations 15-45 and 15-46 can be illustrated graphically by the reduced-phase diagram for a two-component system at constant pressure, as shown for the system diopside–anorthite in Figure 15-4.

As we pointed out in Chapter 14, each of the two-phase regions has one degree of freedom, and the equilibrium relationship between T and X is given by curves AB or BC when component 2 or component 1, respectively, is in higher concentration. In the region below the horizontal line only the temperature can vary since both phases are pure solids.

The equation of curve AB can be obtained by integrating Equation 15-46, and the equation of curve BC can be obtained by integrating the corresponding

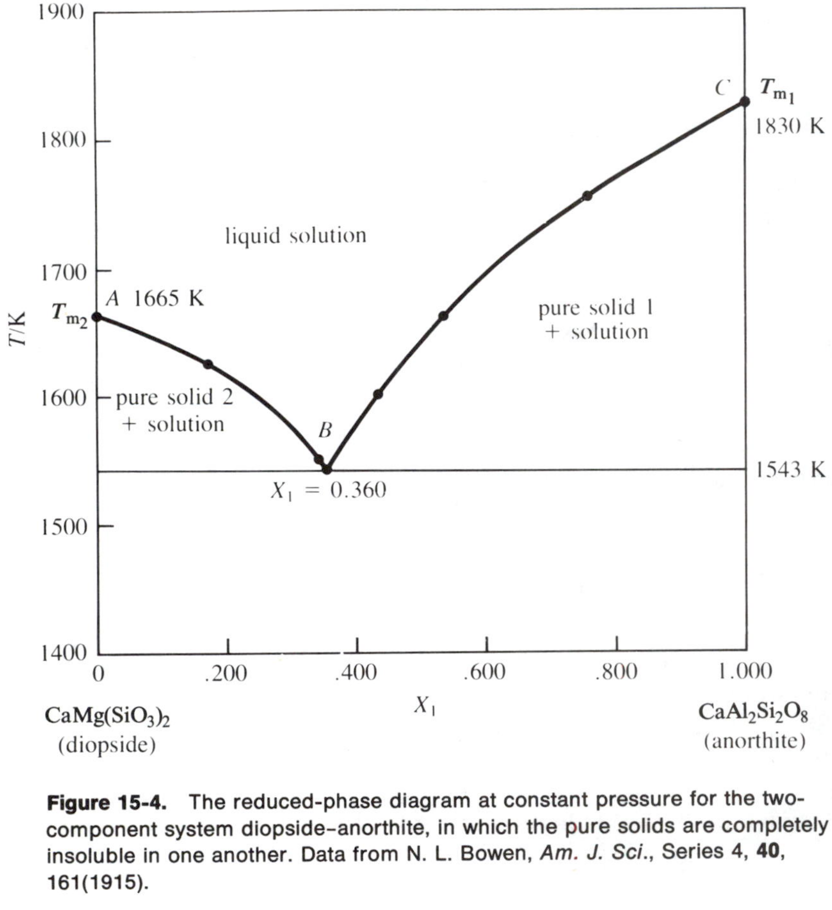

Figure 15-4. The reduced-phase diagram at constant pressure for the two-component system diopside–anorthite, in which the pure solids are completely insoluble in one another. Data from N. L. Bowen, *Am. J. Sci.*, Series 4, **40**, 161(1915).

equation:

$$\left(\frac{\partial T_{\mathrm{m}}}{\partial \ln X_1}\right)_P = \frac{RT_{\mathrm{m}}^2}{\Delta H_{\mathrm{fusion}(1)}} \tag{15-47}$$

in which $\Delta H_{\mathrm{fusion}(1)}$ is the molar enthalpy for the transition from pure, solid component 1 to pure, supercooled, liquid component 1.

15-6 EQUILIBRIUM BETWEEN AN IDEAL SOLID SOLUTION AND AN IDEAL LIQUID SOLUTION

In Chapter 14 we discussed briefly the solid–liquid equilibrium diagram of the mineral olivine. Olivine is an ideal, solid solution of forsterite (Mg_2SiO_4) and fayalite (Fe_2SiO_4) in the solid state, and an ideal, liquid solution of the same components in the molten state. The relationships that we have developed in this chapter permit us to interpret the olivine phase diagram (Figure 14-4) in a quantitative way.

Composition of the Two Phases in Equilibrium

Let us designate forsterite as component 1 and fayalite as component 2. According to Equations 15-3 and 15-4 we can write for each component in each phase

$$\mu_{i(\mathrm{solid})} = \mu^{\circ}_{i(\mathrm{solid})} + RT \ln X_{i(\mathrm{solid})} = \mu^{\bullet}_{i(\mathrm{solid})} + RT \ln X_{i(\mathrm{solid})} \tag{15-48}$$

$$\mu_{i(\mathrm{liquid})} = \mu^{\circ}_{i(\mathrm{liquid})} + RT \ln X_{i(\mathrm{liquid})} = \mu^{\bullet}_{i(\mathrm{liquid})} + RT \ln X_{i(\mathrm{liquid})} \tag{15-49}$$

At equilibrium at constant temperature and pressure, the chemical potential of each component must be the same in both phases. That is

$$\mu_{i(\mathrm{liquid})} = \mu_{i(\mathrm{solid})} \tag{15-50}$$

Substituting from Equations 15-48 and 15-49 into Equation 15-50 we obtain

$$\ln \frac{X_{i(\mathrm{liquid})}}{X_{i(\mathrm{solid})}} = \frac{\mu^{\bullet}_{i(\mathrm{solid})} - \mu^{\bullet}_{i(\mathrm{liquid})}}{RT} \tag{15-51}$$

The two equations of the form of Equation 15-51, together with the restrictions that $X_{1(\mathrm{solid})} + X_{2(\mathrm{solid})} = 1$ and $X_{1(\mathrm{liquid})} + X_{2(\mathrm{liquid})} = 1$, uniquely determine the compositions of the two phases in equilibrium at any temperature and pressure.

Since the temperature interval in which solid and liquid phases can be in equilibrium is between the melting points of the pure components (see Figure 14-4), one component is above its melting point and one is below its melting point. For the component above its melting point

$$\mu^\bullet_{\text{(solid)}} > \mu^\bullet_{\text{(liquid)}}$$

and therefore

$$X_{\left(\substack{\text{solid}\\ \text{solution}}\right)} < X_{\left(\substack{\text{liquid}\\ \text{solution}}\right)}$$

The opposite is true for the other component.

Temperature Dependence of the Equilibrium Compositions

If the equilibrium is to be maintained as the temperature is changed at constant pressure

$$d\mu_{i\text{(liquid)}} = d\mu_{i\text{(solid)}} \tag{15-52}$$

for both components. For each component μ_i is a function of temperature and composition, so Equation 15-52 becomes

$$\left[\frac{\partial \mu_{i\text{(solid)}}}{\partial T}\right]_{P,X_i} dT + \left[\frac{\partial \mu_{i\text{(solid)}}}{\partial X_{i\text{(solid)}}}\right]_{T,P} dX_{i\text{(solid)}}$$

$$= \left[\frac{\partial \mu_{i\text{(liquid)}}}{\partial T}\right]_{P,X_i} dT + \left[\frac{\partial \mu_{i\text{(liquid)}}}{\partial X_{i\text{(liquid)}}}\right]_{T,P} dX_{i\text{(liquid)}} \tag{15-53}$$

Substituting the appropriate expressions for the partial derivatives in Equation 15-53, we write

$$-\bar{S}_{i\text{(solid)}}\, dT + \frac{RT}{X_{i\text{(solid)}}}\, dX_{i\text{(solid)}} = -\bar{S}_{i\text{(liquid)}}\, dT + \frac{RT}{X_{i\text{(liquid)}}}\, dX_{i\text{(liquid)}} \tag{15-54}$$

Rearranging terms, we obtain

$$\frac{dX_{i\text{(solid)}}}{X_{i\text{(solid)}}} - \frac{dX_{i\text{(liquid)}}}{X_{i\text{(liquid)}}} = \frac{\bar{S}_{i\text{(solid)}} - \bar{S}_{i\text{(liquid)}}}{RT}\, dT \tag{15-55}$$

As in Equation 15-43, from Figure 14-4 we can conclude that

$$\bar{S}_{i\text{(solid)}} - \bar{S}_{i\text{(liquid)}} = \frac{\bar{H}_{i\text{(solid)}} - \bar{H}_{i\text{(liquid)}}}{T}$$

$$= \frac{H^\bullet_{i\text{(solid)}} - H^\bullet_{i\text{(liquid)}}}{T}$$

and Equation 15-55 becomes

$$d\ln\frac{X_{i\text{(solid)}}}{X_{i\text{(liquid)}}} = -\frac{\Delta H_{\text{fusion}(i)}}{RT^2}\, dT \tag{15-56}$$

The integration of Equation 15-56 for the two components leads to the temperature–composition curves of the solid and liquid phases.

An analysis of vapor–liquid equilibrium for an ideal solution has been published [2].

EXERCISES

1. a. Calculate ΔG, ΔH, and ΔS (per mole of benzene) at 298 K for the addition of an infinitesimal quantity of pure benzene to 1 mole of an ideal solution of benzene and toluene in which the mole fraction of the latter is 0.6.

   b. Calculate ΔG, ΔH, and ΔS (per mole of solution) at 298 K for the mixing of 0.4 mole of pure benzene with 0.6 mole of pure toluene.

2. Calculate the entropy (per mole of mixture) of "unmixing" ^{235}U and ^{238}U from a sample of pure uranium from natural sources. The former isotope occurs to the extent of 0.7 mole % in the natural (ideal) mixture.

3. For the mixing of two pure components to form 1 mole of an ideal solution, ΔG_{mixing} is given by Equation 15-22.

   a. Plot ΔG_{mixing} as a function of X_2.

   b. Prove analytically, from Equation 15-22, that the curve in (a) has a minimum at $X_2 = 0.5$.

4. Suppose that a pure gas dissolves in some liquid solvent to produce ideal solutions. Show that the solubility of this gas must fit the following relationships:

   a. $$\left(\frac{\partial \ln X_{2(\text{satd})}}{\partial P}\right)_T = \frac{\bar{V}_{2(g)} - \bar{V}_2}{RT} \qquad (15\text{-}57)$$

   b. $$\left(\frac{\partial \ln X_{2(\text{satd})}}{\partial T}\right)_P = -\frac{\Delta H_{\text{vaporization solute}}}{RT^2} \qquad (15\text{-}58)$$

   c. $$\ln X_{2(\text{satd})} = \frac{\Delta H_{\text{vap}}}{R}\left(\frac{1}{T} - \frac{1}{T_{\text{bp}}}\right) \qquad (15\text{-}59)$$

   in which T_{bp} is the boiling point of the pure solute.

5. If two lead amalgam electrodes of different compositions (X_2 and X_2') are prepared and immersed in a suitable electrolyte, an electrical cell is obtained in which the following transfer process occurs when the cell is discharged:

   $$\text{Pb (in amalgam; } X_2') = \text{Pb (in amalgam; } X_2)$$

   For a particular cell, operated at 27°C and 1 bar, $X_2' = 0.000625$ and $X_2 = 0.0165$. Assume that Pb and Hg form ideal solutions in this concentration range.

a. Calculate ΔG for the reaction if the cell is discharged reversibly.

b. Calculate ΔG for the reaction if the cell is short-circuited so that no electrical work is done.

c. What is ΔH for the reaction carried out under the conditions in (a)?

d. What is ΔS for the reaction carried out under the conditions in (a)?

e. What is Q for the reaction carried out under the conditions in (a)?

f. What is Q for the reaction carried out under the conditions in (b)?

6. L. G. Maury, R. L. Burwell, Jr., and R. H. Tuxworth have published data [*J. Am. Chem. Soc.* **76**, 5831 (1954)] for the equilibrium

3-methylhexane = 2-methylhexane

ΔG° (pure liquid is the standard state for each substance) is -1194 J mol^{-1} at 0°C. In a solution containing only the two isomers, equilibrium is attained when the mole fraction of the 3-methylhexane is 0.372. Is the equilibrium solution ideal? Show the computations on which your answer is based.

7. Integrate Equations 15-46 and 15-47 from some point X, T to $X = 1$, $T_m = T_m$(pure), assuming that ΔH_{fusion} is constant throughout the temperature range. The melting points of pure naphthalene and pure benzene are 80.2°C and 5.4°C, respectively. The average enthalpies of fusion of naphthalene and benzene in the temperature range are 10,040 J mol^{-1} and 19,200 J mol^{-1}, respectively. Calculate the temperature and composition for the naphthalene-benzene system that correspond to point B, the eutectic point, in Figure 15-4.

REFERENCES

1. Although we have defined the standard state at a fixed pressure, the pressure of the solution, it is possible to change the equilibrium vapor pressure by changing the pressure applied to the condensed phase, as described in Section 9-2.

2. M. P. Silverman, *J. Chem. Educ.* **62**, 112 (1985).

CHAPTER 16

Dilute Solutions of Nonelectrolytes

We will proceed in our discussion of solutions from ideal to nonideal solutions, limiting ourselves at first to nonelectrolytes. For dilute solutions of nonelectrolytes, several limiting laws have been found to describe the behavior of these systems with increasing precision as infinite dilution is approached. If we take any one of them as an empirical rule, we can derive the others from it on the basis of thermodynamic principles.

16-1 HENRY'S LAW

The empirical description of dilute solutions that we take as the starting point of our discussion is Henry's law. Recognizing that when the vapor phase is in equilibrium with the solution, μ_2 in the condensed phase is equal to $\mu_{2(\text{gas})}$, we can state this law as follows: For dilute solutions of a nondissociating solute at constant temperature, the fugacity of the solute in the gas phase is proportional to its mole fraction in the condensed phase: that is

$$f_{2(\text{gas})} = k_2 X_{2(\text{cond})} \tag{16-1}$$

This is a generalized statement of Henry's law, which originally was expressed in terms of vapor pressure instead of fugacity. As solutions become more and more dilute, this law becomes increasingly more accurate. We can indicate the limiting nature of Henry's law explicitly by writing it as

$$\lim_{X_2 \to 0} \frac{f_{2(\text{gas})}}{X_{2(\text{cond})}} = k_2 \tag{16-2}$$

The difference between Raoult's law (Equation 15-1) and Henry's law lies in the proportionality constant relating the fugacity to the mole fraction. For

Raoult's law this constant is $f^\bullet_{2(gas)}$, the fugacity of the vapor in equilibrium with the pure solute. Generally, however, for Henry's law

$$k_2 \neq f^\bullet_{2(gas)} \tag{16-3}$$

This distinction can be made clearer by a graphical illustration (Figure 16-1). A typical fugacity–mole fraction curve is shown by the solid line. If the solute formed an ideal solution with the solvent, the fugacity of the solute would be represented by the broken line (Raoult's law). The actual behavior of the solute does not approach Raoult's law, except when its mole fraction approaches 1 (that is, under circumstances when it no longer would be called the solute). When the solute is present in small quantities, its fugacity deviates widely from Raoult's law. However, as X_2 approaches 0, the fugacity does approach a linear dependence on X_2. This limiting linear relationship, which is a graphical illustration of Henry's law, is represented by the dotted line in Figure 16-1.

When a solution obeys Henry's law, the expression for the chemical potential (from Equation 13-16 and Equation 16-1) is

$$\mu_2 = \mu_{2(gas)} = \mu^\circ_{2(gas)} + RT \ln \left(\frac{f_{2(gas)}}{f^\circ_{gas}} \right)$$

$$= \mu^\circ_{2(gas)} + RT \ln \left(\frac{k_2 X_2}{f^\circ_{gas}} \right)$$

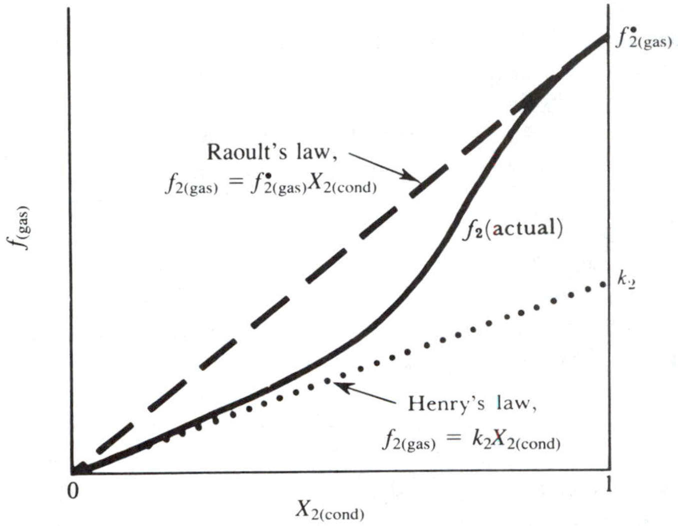

Figure 16-1. Distinction between Henry's law and Raoult's law.

$$= \mu^{\circ}_{2(\text{gas})} + RT \ln \left(\frac{k_2}{f^{\circ}_{\text{gas}}} \right) + RT \ln X_2 \tag{16-4}$$

If we define the first two terms on the right-hand side of Equation 16-4 as μ_2°, the standard chemical potential of the solute in solution, then

$$\mu_2 = \mu_2^{\circ} + RT \ln X_2 \tag{16-5}$$

and the standard state is a *hypothetical* standard state of unit mole fraction of solute, one at the point of *extrapolation* of Henry's-law behavior to $X_2 = 1$ (see Figure 16-1). In this case, if we wish to use an expression of the form of Equation 13-16 to describe the chemical potential of the solute, then we must set $f^{\circ}_{2(\text{gas})} = k_2$, and not use the $f^{\circ}_{\text{gas}} = f^{\bullet}_{2(\text{gas})}$ generally assigned by Raoult's law. This change in assignment of the standard state can also be seen from Figure 16-1.

If mole fraction is not a convenient unit of composition, Henry's law can be stated in other units. Since the law applies primarily to very dilute solutions

$$X_2 = \frac{n_2}{n_1 + n_2} \cong \frac{n_2}{n_1} \tag{16-6}$$

in which n_2 is the number of moles of solute and n_1 is the number of moles of solvent. Consequently Equation 16-1 can be revised to the form

$$f_{2(\text{gas})} = k_2' \left(\frac{n_2}{n_1} \right) = k_2'' \left(\frac{m_2}{m_2^{\circ}} \right) \tag{16-7}$$

where the ratio is used to keep the dimensions of k_2'' the same as those of k_2', and m_2, the molality, is the number of moles of solute per unit mass of solvent, usually the kilogram. Even when Equation 16-1 is valid throughout a wide range of composition, Equations 16-6 and 16-7 are still approximate. In limiting-law form Equation 16-7 becomes

$$\lim_{m_2 \to 0} \frac{f_2}{m_2/m_2^{\circ}} = k_2'' \tag{16-8}$$

From Equation 13-16 and Equation 16-7, the chemical potential can be expressed as a function of molality.

$$\mu_2 = \mu^{\circ}_{2(\text{gas})} + RT \ln \left(\frac{f_{2(\text{gas})}}{f^{\circ}_{\text{gas}}} \right)$$

$$= \mu^{\circ}_{2(\text{gas})} + RT \ln \left[\frac{k_2'' m_2/m_2^{\circ}}{f^{\circ}_{\text{gas}}} \right] \tag{16-9}$$

We can define the standard chemical potential of the solute in solution as

$$\mu_2^\circ = \mu_{2(\text{gas})}^\circ + RT \ln \left(\frac{k_2''}{f_{\text{gas}}^\circ} \right) \tag{16-10}$$

Thus

$$\mu_2 = \mu_2^\circ + RT \ln \left(\frac{m_2}{m_2^\circ} \right) \tag{16-11}$$

In Equation 16-11 the choice of m_2° is entirely arbitrary. However, it is conventional to choose $m_2^\circ = 1$ mol kg^{-1}; that is, the standard state of the solute is a *hypothetical* one molal standard state that is the point of extrapolation of Henry's-law behavior to a molality of 1 mol kg^{-1}. In a figure analogous to Figure 16-1, but with m_2 along the horizontal axis, the standard state would be a point on the Henry's law dotted line directly above $m_2 = 1$ mol kg^{-1}.

For solutions obeying Henry's law, as for ideal solutions, and for solutions of ideal gases, the chemical potential is a linear function of the logarithm of the composition variable, and the standard chemical potential depends on the choice of composition variable.

16-2 NERNST'S DISTRIBUTION LAW

If a quantity of a solute, A, is distributed between two immiscible solvents, for example I_2 between carbon tetrachloride and water, then at equilibrium the partial molar free energies or escaping tendencies are the same in both phases; thus for

$$\text{A(in solvent a)} = \text{A(in solvent b)}$$

$$\mu_2 = \mu_2'$$

If the chemical potential is expressed in terms of mole fraction

$$\mu_2^\circ + RT \ln X_2 = \mu_2^{\circ\prime} + RT \ln X_2' \tag{16-12}$$

Equation 16-12 can describe either an ideal solution (see Equation 15-4) or a solution sufficiently dilute that Henry's law is followed (see Equation 16-5). In either case it follows that

$$\ln \frac{X_2}{X_2'} = -\frac{\mu_2^\circ - \mu_2^{\circ\prime}}{RT}$$

and

$$\frac{X_2}{X_2'} = \exp \left[-\frac{\mu_2^\circ - \mu_2^{\circ\prime}}{RT} \right]$$

$$= \kappa \tag{16-13}$$

The value of κ is constant since the standard chemical potentials in the two solvents are constants at a fixed temperature. Nernst's distribution law also can be stated in terms of molality:

$$\frac{m_2}{m_2'} = \kappa' \tag{16-14}$$

if $m_2^\circ = m_2^{\circ\prime}$; that is, if the same standard state is chosen for both solvents. Again κ' is a constant, but it differs in magnitude from κ. Equation 16-14 is valid only for low molalities, even for ideal solutions, as can be seen from Equation 16-6.

16-3 RAOULT'S LAW

We can show that if the solute obeys Henry's law in very dilute solutions, the solvent follows Raoult's law in the same solutions. Let us start from the Gibbs–Duhem equation (12-23), which relates changes in the chemical potential of the solute to changes in the chemical potential of the solvent; that is, for a two-component system

$$n_1\, d\mu_1 + n_2\, d\mu_2 = 0$$

If Equation 12-23 is divided by $n_1 + n_2$, we obtain

$$X_1 d\mu_1 + X_2 d\mu_2 = 0$$

or

$$X_1 \frac{\partial \mu_1}{\partial X_1} + X_2 \frac{\partial \mu_2}{\partial X_1} = 0 \tag{16-15}$$

Since, for a two-component system

$$X_1 + X_2 = 1$$

and

$$dX_1 = -dX_2$$

Equation 16-15 can be rewritten as

$$X_1 \frac{\partial \mu_1}{\partial X_1} - X_2 \frac{\partial \mu_2}{\partial X_2} = 0$$

or alternatively, with the constancy of temperature and pressure explicitly indicated, as

$$\left(\frac{\partial \mu_1}{\partial \ln X_1}\right)_{T,P} = \left(\frac{\partial \mu_2}{\partial \ln X_2}\right)_{T,P} \tag{16-16}$$

Let us apply Equation 16-16, which is a general relationship for any two-component solution, to a solution for which Henry's law describes the behavior of the solute. From Equation 16-5,

$$\left(\frac{\partial \mu_2}{\partial \ln X_2}\right)_{T,P} = RT \tag{16-17}$$

Thus, from Equation 16-16 and Equation 16-17

$$\left(\frac{\partial \mu_1}{\partial \ln X_1}\right)_{T,P} = RT \tag{16-18}$$

Integration of Equation 16-18 at constant T,P leads to

$$\int d\mu_1 = RT \int d \ln X_1$$

or

$$\mu_1 = RT \ln X_1 + C \tag{16-19}$$

in which C is a constant of integration. For the solvent at $X_1 = 1$, $\mu_1 = \mu_1^\bullet = \mu_1^\circ = C$, and hence

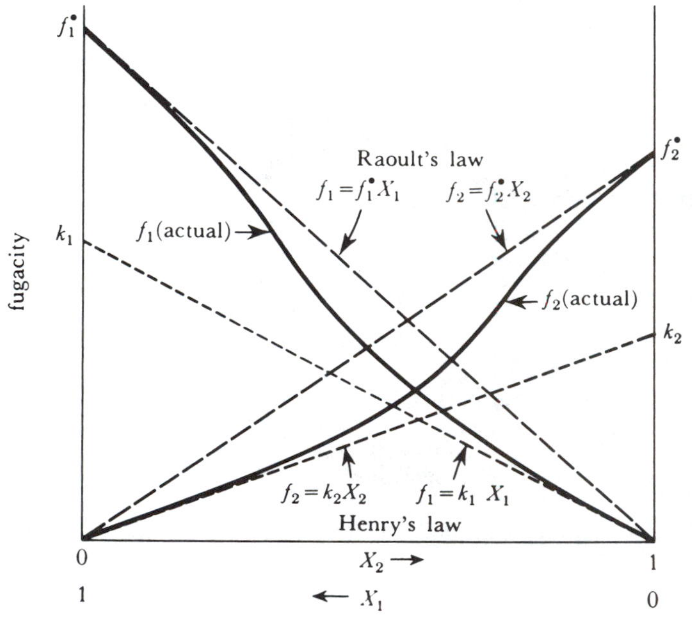

Figure 16-2. Limiting-law validity of Henry's law and Raoult's law.

$$\mu_1 = \mu_1^{\bullet} + RT \ln X_1$$
$$= \mu_1^{\circ} + RT \ln X_1$$

These equations are the same as Equation 15-3 and Equation 15-4, statements of Raoult's law; thus, the solvent obeys Raoult's law when the solute obeys Henry's law. Since Henry's law is a limiting law for the solute in dilute solution, Raoult's law for the solvent in the same solution is also a limiting law. We can express this limiting law in terms of the fugacity by rearranging Equation 15-1 to read

$$\lim_{X_1 \to 1} \frac{f_1}{X_1} = f_1^{\bullet} \tag{16-20}$$

This behavior of a two-component mixture is illustrated in Figure 16-2, which shows the actual fugacity, the values calculated from Henry's law, and those calculated from Raoult's law, as a function of mole fraction.

16-4 VAN'T HOFF'S LAW OF OSMOTIC PRESSURE [1]

As we indicated in Chapter 14, the requirement that all phases be at the same pressure at equilibrium does not apply in all cases, and, in particular, it does not apply if the phases are separated by a rigid membrane. If the membrane is permeable to only one component, we can show that the pressure on the two phases must be different if equilibrium is maintained at a *fixed temperature*.

Consider the apparatus illustrated schematically in Figure 16-3, in which two portions of pure solvent are separated by a membrane, M, that is

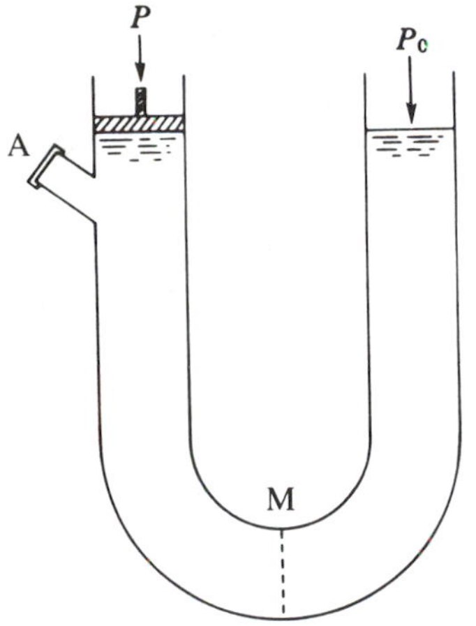

Figure 16-3. Schematic diagram of apparatus for measurement of osmotic pressure.

permeable to the solvent. The liquid levels are equal, the pressures P and P_0 are equal at equilibrium, and

$$\mu_1^\bullet(\text{left}) = \mu_1^\bullet(\text{right}) \tag{16-21}$$

If we add through the sidearm, A, some solute to which the membrane is impermeable, then with adequate mixing the solute is distributed uniformly throughout the left chamber but is absent from the right chamber. Solvent then will be observed to move from the right to the left side because

$$\mu_1(\text{left}) < \mu_1^\bullet(\text{right})$$

The chemical potential of the pure solvent is always greater than the chemical potential of the solvent in the solution.

The movement of solvent can be prevented and equilibrium restored if the pressure P is made sufficiently greater than P_0. In the new equilibrium state

$$\mu_1(\text{left}) = \mu_1^\bullet(\text{right}) \tag{16-22}$$

and

$$d\mu_1(\text{left}) = d\mu_1^\bullet(\text{right}) = 0 \tag{16-23}$$

since $\mu_1^\bullet(\text{right})$ was not affected by either the addition of solute to the left chamber or the increase of pressure on the left chamber.

Since μ_1 is a function of the pressure, P, and the mole fraction of solute, X_2, we can express $d\mu_1$ in Equation 16-23 as

$$d\mu_1 = \left(\frac{\partial \mu_1}{\partial P}\right)_{T,X_2} dP + \left(\frac{\partial \mu_1}{\partial \ln X_2}\right)_{T,P} d \ln X_2 = 0 \tag{16-24}$$

From the Gibbs–Duhem equation at constant temperature and pressure (Equation 12-23), we can write

$$\left(\frac{\partial \mu_1}{\partial \mu_2}\right)_{T,P} = -\frac{n_2}{n_1} = -\frac{X_2}{X_1}$$

By the chain rule of differential calculus,

$$\left(\frac{\partial \mu_1}{\partial \ln X_2}\right)_{T,P} = \left(\frac{\partial \mu_1}{\partial \mu_2}\right)_{T,P} \left(\frac{\partial \mu_2}{\partial \ln X_2}\right)_{T,P} = -\frac{X_2}{X_1} \left(\frac{\partial \mu_2}{\partial \ln X_2}\right)_{T,P} \tag{16-25}$$

Substituting from Equation 16-25 in Equation 16-24, and substituting $\bar{v}_1$ for $(\partial \mu_1/\partial P)_{T,X_2}$, from Equation 12-17, we have

$$\bar{v}_1 \, dP - \frac{X_2}{X_1} \left(\frac{\partial \mu_2}{\partial \ln X_2}\right)_{T,P} \frac{dX_2}{X_2} = 0$$

With the insertion of Equation 16-17

$$\left(\frac{\partial \mu_2}{\partial \ln X_2}\right)_{T,P} = RT$$

we find that

$$\bar{V}_1\, dP - \frac{RT}{X_1}\, dX_2 = 0$$

or

$$\left(\frac{\partial P}{\partial X_2}\right)_T = \frac{RT}{X_1\, \bar{V}_1} \tag{16-26}$$

Since Henry's law is valid only in the limit of very dilute solutions, we can write a limiting law for Equation 16-26 as

$$\lim_{X_2 \to 0} \left(\frac{\partial P}{\partial X_2}\right)_T = \lim_{X_2 \to 0} \frac{RT}{X_1\, \bar{V}_1}$$

$$= \frac{RT}{V_1^{\bullet}}$$

or

$$dP = \frac{RT}{V_1^{\bullet}}\, dX_2 \tag{16-27}$$

If we assume that $V_1^{\bullet}$ is independent of P over the range for which Equation 16-27 is valid, we can integrate Equation 16-27 as follows:

$$\int_{P_0}^{P} dP = \frac{RT}{V_1^{\bullet}} \int_{X_2=0}^{X_2} dX_2$$

or

$$P - P_0 = \frac{RT}{V_1^{\bullet}}\, X_2 \tag{16-28}$$

The difference in pressures, $P - P_0$, required to maintain osmotic equilibrium is defined as the *osmotic pressure* and is denoted by Π. Equation 16-28 thus becomes

$$\Pi = \frac{RT}{V_1^\bullet} X_2 \tag{16-29}$$

Since the solution is dilute, so that $X_2 \cong n_2/n_1$,

$$\Pi = \frac{RT}{V_1^\bullet} \frac{n_2}{n_1}$$

$$= \frac{n_2 RT}{V_1} \tag{16-30}$$

in which V_1 is the total volume of solvent. Because the solution is dilute, $V_1 \cong V$, the total volume of the solution, and

$$\Pi = \frac{n_2}{V} RT$$

$$= c_2 RT = \frac{w_2 RT}{M_2} \tag{16-31}$$

in which c_2 is the concentration of solute in moles per unit volume, usually 1 dm^3; w_2 is the mass concentration (mass per unit volume); and M_2 is the molar mass. Equation 16-31 is clearly a limiting law. A more accurate expression is

$$\lim_{c_2 \to 0} \frac{\Pi}{c_2} = RT \tag{16-32}$$

or

$$\lim_{w_2 \to 0} \frac{\Pi}{w_2} = \frac{RT}{M_2} \tag{16-33}$$

Values for Π can be determined experimentally for varying mass concentrations. The ratio Π/w_2, plotted against w_2 and extrapolated to $w_2 = 0$, gives a value of RT/M_2. The unknown molecular weight then is calculated.

In this form van't Hoff's law of osmotic pressure is also used to determine the molecular weights of biological and synthetic macromolecules.

When the osmotic pressure is measured for a solution of macromolecules that contains more than one species of macromolecule (for example, a synthetic polymer with a distribution of molar masses, or a protein molecule that undergoes association or dissociation), the osmotic pressures of the various species are additive. That is, in sufficiently dilute solution

$$\Pi = \sum_i \Pi_i$$

$$= RT \sum_i \frac{w_i}{M_i} \tag{16-34}$$

If we divide each summation by $w = \Sigma w_i$, Equation 16-34 becomes

$$\frac{\Pi}{w} = \frac{RT \sum_i (w_i/M_i)}{\sum_i w_i}$$

$$= \frac{RT}{\left(\sum_i w_i\right)\Big/\sum_i (w_i/M_i)} \tag{16-35}$$

The number-average molecular weight, M_n, is defined as

$$M_n = \frac{\sum_i c_i M_i}{\sum_i c_i}$$

where c_i is the molar concentration of each species i. Consequently,

$$M_n = \frac{\sum_i (w_i/M_i)\, M_i}{\sum_i (w_i/M_i)}$$

$$= \frac{\sum_i w_i}{\sum_i (w_i/M_i)} \tag{16-36}$$

Therefore, we can rewrite Equation 16-35 in the form of a limiting law as

$$\lim_{w \to 0} \frac{\Pi}{w} = \frac{RT}{M_n} \tag{16-37}$$

Thus, the result of an osmotic pressure experiment with a mixture of macromolecules yields the number-average molecular weight, M_n.

Osmotic Work in Biological Systems

As we mentioned at the end of Chapter 8, an important application of the free energy function to biological systems is the calculation of the minimum work required to maintain a nonequilibrium concentration gradient across a membrane. In living cells these gradients are not maintained by pressure differences, as in the osmotic pressure experiment described here, but by active transport processes whose mechanisms are just beginning to be understood. The function of a thermodynamic analysis is to verify that the energy-producing part of the mechanism is adequate to maintain the observed gradient.

If we assume that solutes in biological systems are at low enough concentrations to obey Henry's law, their chemical potential is given as (Equation 16-11)

$$\mu_2 = \mu_2^\circ + RT \ln \left(\frac{m_2}{m_2^\circ}\right)$$

Thus the free energy change for the transfer of one mole of the solute from a molality m_2 to a molality m_2' is

$$\begin{aligned}\Delta G &= \mu_2' - \mu_2 \\ &= RT \ln \frac{m_2'}{m_2} \qquad (16\text{-}38)\end{aligned}$$

If m_2' is greater than m_2, then ΔG is positive and work must be done on the system by the surroundings in order to carry out the transfer. According to Equation 8-104

$$\Delta G < W_{\text{net,irrev}}$$

in which W_{net} is positive when work is done on the system. Thus

$$W_{\text{net,irrev}} > RT \ln \frac{m_2'}{m_2}$$

and, since W_{net} and ΔG are positive,

$$|W_{\text{net,irrev}}| > RT \ln \frac{m_2'}{m_2} \qquad (16\text{-}39)$$

For example, if the molality of glucose in blood is 5.5×10^{-3} mol kg^{-1} and the molality of glucose in urine is 5.5×10^{-5} mol kg^{-1}, then the kidney must do work at least equal to

$$\begin{aligned}RT \ln \frac{5.5 \times 10^{-3} \text{ mol kg}^{-1}}{5.5 \times 10^{-5} \text{ mol kg}^{-1}} &= (8.314 \text{ J K}^{-1} \text{ mol}^{-1})(310 \text{ K}) \ln 100 \\ &= 11{,}870 \text{ J mol}^{-1}\end{aligned}$$

per mole of glucose transported from the kidney to the blood against this gradient. Any mechanism suggested for carrying out the active transport of glucose in the kidney must provide at least this much work.

16-5 VAN'T HOFF'S LAW OF FREEZING POINT DEPRESSION AND BOILING POINT ELEVATION

Let us consider a pure solid phase, such as ice, in equilibrium with a pure liquid phase, such as water, at some specified temperature and pressure. If the two phases are in equilibrium

$$\mu^{\bullet}_{1(\text{solid})} = \mu^{\bullet}_1 \qquad (16\text{-}40)$$

in which $\mu^{\bullet}_{1(\text{solid})}$ represents the chemical potential of the pure solid. If solute is added to the system, and if it dissolves only in the liquid phase, then the chemical potential of the solvent will be decreased:

$$\mu_1 < \mu^{\bullet}_1$$

To reestablish equilibrium, $\mu^{\bullet}_{1(\text{solid})}$ must be decreased also. This can be accomplished by decreasing the temperature. The chemical potential of the liquid solvent is decreased by the drop in temperature as well as by the addition of solute. Equilibrium is reestablished if

$$d\mu^{\bullet}_{1(\text{solid})} = d\mu_1 \qquad (16\text{-}41)$$

Since the chemical potential of the solid phase depends only on the temperature, whereas that of the solvent in the solution depends on both temperature and concentration of added solute, the total differentials of Equation 16-41 can be expressed in terms of the appropriate partial derivatives, as follows:

$$\left(\frac{\partial \mu^{\bullet}_{1(\text{solid})}}{T}\right)_P dT = \left(\frac{\partial \mu_1}{\partial T}\right)_{P,X_2} dT + \left(\frac{\partial \mu_1}{\partial \ln X_2}\right)_{P,T} d\ln X_2 \qquad (16\text{-}42)$$

From Equation 12-16,

$$\left(\frac{\partial \mu^{\bullet}_{1(\text{solid})}}{\partial T}\right)_P = -s^{\bullet}_{\text{solid}} \quad \text{and} \quad \left(\frac{\partial \mu_1}{\partial T}\right)_{P,X_2} = -\bar{s}_1 \qquad (16\text{-}43)$$

and from Equation 16-25 and Equation 16-17

$$\left(\frac{\partial \mu_1}{\partial \ln X_2}\right)_{P,T} d\ln X_2 = -\frac{RT}{X_1}\, dX_2 \qquad (16\text{-}44)$$

Substituting from Equations 16-43 and 16-44 into Equation 16-42, we have

$$-s^{\bullet}_{\text{solid}}\, dT = -\bar{s}_1\, dT - \frac{RT}{X_1}\, dX_2$$

or

$$\left(\frac{\partial T}{\partial X_2}\right)_P = -\frac{RT}{X_1(\bar{s}_1 - s^{\bullet}_{\text{solid}})} \qquad (16\text{-}45)$$

If we express Equation 16-45 as a limiting law, consistent with the observation that Henry's law is valid only in very dilute solutions.

$$\lim_{X_2 \to 0} \left(\frac{\partial T}{\partial X_2} \right)_P = - \frac{RT}{s_1^\bullet - s_{\text{solid}}^\bullet}$$

$$= - \frac{RT}{\Delta s_{\text{fusion(1)}}} \quad (16\text{-}46)$$

At equilibrium at constant temperature and pressure (Equations 9-8 and 9-9)

$$\Delta s = \frac{\Delta H}{T}$$

so that

$$\lim_{X_2 \to 0} \left(\frac{\partial T}{\partial X_2} \right)_P = - \frac{RT^2}{\Delta H_{\text{fusion(1)}}} \quad (16\text{-}47)$$

in which $\Delta H_{\text{fusion(1)}}$ is the molar enthalpy of the transition from pure solid component 1 to pure, supercooled, liquid component 1. This result is analogous to that in Equation 15-47, to which Equation 16-47 is a limiting-law equivalent. Whether we choose to use the equations to describe the temperature dependence of the solubility of a component or the concentration dependence of the freezing point is a matter of point of view.

If $\Delta H_{\text{fusion(1)}}$ is assumed to be constant in the small-temperature range from the freezing point of the pure solvent, T_0, to the freezing point of solution, T, Equation 16-47 is integrated as

$$\frac{\Delta H_{\text{fusion(1)}}}{R} \int_{T_0}^{T} \frac{dT}{T^2} = - \int_0^{X_2} dX_2$$

or

$$\frac{\Delta H_{\text{fusion(1)}}}{R} \left[\frac{1}{T_0} - \frac{1}{T} \right] = -X_2 \quad (16\text{-}48)$$

If $T_0 \cong T$, a further approximation is

$$\Delta T = - \frac{RT_0^2}{\Delta H_{\text{fusion(1)}}} X_2 \quad (16\text{-}49)$$

The preceding expression, like the other laws of the dilute solution, is a limiting law. It is expressed more accurately as

$$\lim_{X_2 \to 0} \frac{\Delta T}{X_2} = - \frac{RT_0^2}{\Delta H_{\text{fusion(1)}}} \quad (16\text{-}50)$$

By a similar set of arguments it can be demonstrated that the boiling point elevation for dilute solutions containing a nonvolatile solute is given by the expression

$$\lim_{X_2 \to 0} \frac{\Delta T}{X_2} = \frac{RT_0^2}{\Delta H_{\text{vaporization}(1)}} \tag{16-51}$$

According to Equation 16-6, in a solution sufficiently dilute that the limiting form of Equation 16-50 applies,

$$X_2 = \frac{n_2}{n_1}$$

$$= n_2 \frac{M_1}{w_1}$$

where w_1 is the mass of solvent, usually expressed in kilograms. But n_2/w_1 is equal to m_2, so that in dilute solution,

$$X_2 = m_2 M_1 \tag{16-52}$$

If we substitute for X_2 from Equation 16-52 in Equation 16-50, written for dilute solutions, then

$$\Delta T = \left(\frac{RT^2 M_1}{\Delta H_{\text{fusion}(1)}} \right) m_2 \tag{16-53}$$

$$= K_f m_2 \tag{16-54}$$

where K_f is the molal freezing point depression constant of the solvent, equal to $RT^2 M_1/\Delta H_{\text{fusion}(1)}$.

Since the laws of dilute solution are limiting laws, they may not provide an adequate approximation at finite concentrations. For a more satisfactory treatment of solutions of finite concentrations, for which deviations from the limiting laws become appreciable, the use of a new function is needed. This new function, the activity, is described in the following chapters.

EXERCISES

1. Derive Equation 16-51, the van't Hoff expression for the elevation of the boiling point.
2. If a molecule of solute dissociates into two particles in dilute solution, Henry's law can be expressed by the relationship

$$f_2 = K(X_2)^2 \tag{16-55}$$

in which X_2 is the mole fraction of solute without regard to dissociation.

a. Derive the Nernst law for the distribution of this solute between two solvents; in one solvent the solute dissociates and in the other it does not.

b. Derive the other laws of the dilute solution for such a dissociating solute.

3. Derive the laws of the dilute solution for a solute, one molecule of which dissociates into ν particles.

4. a. The standard free energy of formation at 298 K of α-D-glucose(s) is −902,900 J mol^{-1}. The solubility of this sugar in 80% ethanol at this temperature is 20 g kg^{-1} solvent, and the solute obeys Henry's law up to saturation in this solvent. Compute $\Delta G f^\circ$ for this sugar at 298 K in 80% ethanol. The standard state for this dissolved solute is a hypothetical 1-molal solution.

   b. The standard free energy of formation at 298 K of β-D-glucose(s) is −901,200 J mol^{-1}. The solubility of this sugar in 80% ethanol at this temperature is 49 g kg^{-1} solvent, and this solute also obeys Henry's law up to saturation. Calculate the equilibrium constant for the reaction (in 80% ethanol)

$$\alpha\text{-D-glucose} = \beta\text{-D-glucose}$$

5. A solution of α-D-glucose in water obeys Henry's law up to high concentrations and throughout a wide range of temperature and pressure. (*Note:* This does not mean that the Henry's law constant k is independent of pressure or temperature; it is not.)

   a. Calculate the heat absorbed when a solution containing 0.01 mole of glucose and 1000 g of solvent is mixed with one containing 0.05 mole of glucose and 1000 g of solvent.

   b. Calculate the volume change in the mixing described in (a).

6. Compute $\Delta G f^\circ$ of O_2(aq) at 25°C, that is, ΔG° of formation of oxygen dissolved in water at a hypothetical molality of 1 mol kg^{-1}. The solubility of oxygen in water exposed to air is 0.00023 mol kg^{-1}. The saturated solution may be assumed to obey Henry's law.

7. According to W. M. Latimer [*Oxidation Potentials*, 2d ed., Prentice-Hall, New York, 1952, p. 54], $\Delta G f^\circ$ of Cl_2(aq), that is, ΔG° of formation of chlorine (hypothetically) dissolved in water at a molality of 1 mol kg^{-1}, is 6900 J mol^{-1} at 25°C. The enthalpy of solution of gaseous chlorine at 1 atm (101.3 kPa) into a saturated aqueous solution is −25,000 J mol^{-1}.

   a. Neglecting hydrolysis reactions of Cl_2 in water, and assuming that dissolved chlorine obeys Henry's law, calculate the solubility of chlorine (in moles per kilogram of water) when the pressure of the pure gas is 101.3 kPa. Assume also that the pure gas behaves ideally.

   b. If dissolved chlorine obeys Henry's law, calculate $\Delta H f^\circ$ of Cl_2(aq).

8. The solubility of nitrogen in water exposed to air at 0°C is 0.84×10^{-3} mol kg^{-1}. Dissolved N_2 obeys Henry's law. Calculate $\Delta G f^\circ_{273}$ of N_2(aq).

REFERENCE

1. The use of the Gibbs–Duhem equation to derive the limiting laws for colligative properties is based on the work of W. Bloch.

CHAPTER 17

Activities and Standard States for Nonelectrolytes

In the preceding chapters we considered Raoult's law and Henry's law, which are laws that describe the thermodynamic behavior of dilute solutions of nonelectrolytes; these laws are strictly valid only in the limit of infinite dilution. These laws led to a simple linear dependence of the chemical potential on the logarithm of the mole fraction of solvent and solute, as in Equations 15-4 (Raoult's law) and 16-5 (Henry's law), or on the logarithm of the molality of the solute, as in Equation 16-11 (Henry's law). These equations are of the same form as the equation derived for the dependence of the chemical potential of an ideal gas on the pressure, Equation 13-11.

When we try to describe the behavior of solutions over the entire range of composition, we find there is no intrinsic relationship between chemical potential and a composition variable. When faced with a similar situation with real gases, G. N. Lewis invented the *fugacity* function so that he could define a linear dependence of the chemical potential of a gas on the logarithm of the fugacity. The characteristics of specific gases were then expressed implicitly by the dependence of the fugacity on the pressure. Similarly, Lewis invented the (dimensionless) *activity* function so that he could define a linear dependence of the chemical potential on the logarithm of the activity. The characteristics of specific solvent–solute combinations were then expressed by the dependence of the activity on the mole fraction or the molality.

17-1 DEFINITIONS

Activity

The *activity*, a_i, of a component of a solution is defined by the equation

$$\mu_i = \mu_i^\circ + RT \ln a_i \qquad (17\text{-}1)$$

together with one of the following limiting conditions:
for solvent,

$$\lim_{X_1 \to 1} \frac{a_1}{X_1} = 1 \tag{17-2}$$

for solute on the mole fraction scale,

$$\lim_{X_2 \to 0} \frac{a_2}{X_2} = 1 \tag{17-3}$$

and for solute on the molality scale,

$$\lim_{m_2 \to 0} \frac{a_2}{m_2/m_2^\circ} = 1 \tag{17-4}$$

In Equation 17-4, m_2° is the molality corresponding to μ_2°—that is, the molality of the standard state. The latter three constraints must be imposed because we want Equation 17-1 to approach Equation 15-4, Equation 16-5, or Equation 16-11 in the appropriate limit.

Activity Coefficient

The deviation of a real solution from the limiting-law behavior of Raoult's law is described conveniently by a function called the *activity coefficient*, defined (on a mole fraction scale) as

$$\gamma_i = \frac{a_i}{X_i} \tag{17-5}$$

Equation 17-1 can then be written

$$\mu_i = \mu_i^\circ + RT \ln X_i + RT \ln \gamma_i \tag{17-6}$$

The deviation of the behavior of a solute from Henry's law, on the mole fraction scale, is described conveniently by the activity coefficient, defined as

$$\gamma_2 = \frac{a_2}{X_2} \tag{17-7}$$

Equation 17-1 can then be written

$$\mu_2 = \mu_2^\circ + RT \ln X_2 + RT \ln \gamma_2 \tag{17-8}$$

If the molality is a more convenient composition measure than the mole fraction, the activity coefficient of the solute is defined as

$$\gamma_2 = \frac{a_2}{m_2/m_2^\circ} \tag{17-9}$$

and Equation 17-1 can be written

$$\mu_2 = \mu_2^\circ + RT \ln \left(\frac{m_2}{m_2^\circ} \right) + RT \ln \gamma_2 \tag{17-10}$$

Although we cannot determine its absolute value, the chemical potential of a component of a solution has a value that is independent of the choice of concentration scale and standard state. The standard chemical potential, the activity, and the activity coefficient have values that do depend on the choice of concentration scale and standard state. To complete the definitions we have given, we must define the standard states we wish to use.

Whether μ_i, γ_i, and a_i refer to a mole fraction composition scale or a molality composition scale will be clear from the context in which they are used. We will not attempt to use different symbols for each scale.

17-2 CHOICE OF STANDARD STATES

From the nature of the definition (Equation 17-1) it is clear that the activity of a given component may have any numerical value, depending on the state chosen for reference, but a_i° must be equal to 1. There is no reason other than convenience for one state to be chosen as the standard in preference to any other. It frequently will be convenient to change standard states as we proceed from one type of problem to another. Nevertheless, certain choices generally have been adopted. Unless a clear statement is made to the contrary, we will assume the following conventional standard states in all of our discussions.

Gases

If Equation 17-1 is to be consistent with Equation 13-55, it is clear that, for a real gas

$$a_i = \frac{f_i}{f^\circ} \tag{17-11}$$

and that the standard state of a gas is that state, along a line extrapolated from values of f at low pressure, at which $f = 1$ bar (0.1 MPa), as indicated in Figure 13-2. For an ideal gas, since $f = P$,

$$a_i = \frac{p_i}{P^\circ} \tag{17-12}$$

and $P^\circ = 1$ bar (0.1 MPa).

Liquids and Solids

Pure Substances. In most problems involving pure substances, it is convenient to choose the pure solid or the pure liquid at each temperature and at a pressure of 1 bar (0.1 MPa) as the standard state. According to this convention, the activity of a pure solid or pure liquid at 1 bar is equal to 1 at any temperature.

Solvent in Solution. The reference state we shall use for the component of a solution designated as the solvent is the pure substance at the same temperature and pressure as the solution. The activity of the pure solvent is equal to 1 at any temperature.

The activity of the solvent in solution cannot be greater than unity at any finite concentration. If it were, a portion of pure solvent would separate spontaneously from the solution, because the chemical potential of the solvent in the solution, and hence its escaping tendency, would be greater than that of the pure substance.

The solid curve in Figure 17-1 shows the activity of the solvent in a solution as a function of the mole fraction of solvent. If the solution were ideal, Equations 15-4 and 17-1 would both be applicable over the whole range of mole fraction, so that $a_1 = X_1$, a relationship indicated by the broken line. Also, since Equation 17-1 approaches Equation 15-4 in the limit as $X_1 \rightarrow 1$ for the

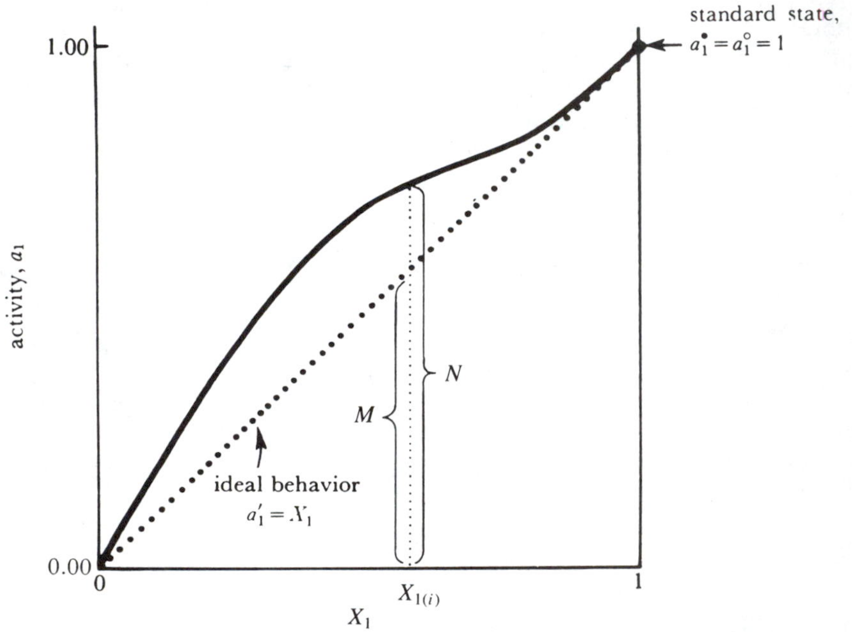

Figure 17-1. Activity and standard state for the solvent.

real solution, the solid curve approaches the ideal line asymptotically as $X_1 \to 1$.

The broken line in Figure 17-1 has a slope of 1. In Figure 17-1 when X_1, the abscissa, is equal to 1, a_1, the ordinate, is equal to 1. The activity coefficient, a_1/X_1, at any concentration of solvent, X_i, is given by the ratio N/M because N is a_1 and, since the slope of the broken line is 1, M is equal to X_1. In the example illustrated by Figure 17-1, the activity coefficient of the solvent is always greater than or equal to 1. If the solid line lay below the (broken) line for ideal behavior, the activity coefficient always would be less than or equal to 1. Generally, in a real solution

$$\gamma_1 = \frac{a_1}{X_1} \neq 1 \tag{17-13}$$

Since, empirically, the solvent approaches Raoult's-law behavior in the limit of infinite dilution, we also can state that

$$\lim_{X_1 \to 1} \frac{a_1}{X_1} = \lim_{X_1 \to 1} \gamma_1 = 1 \tag{17-14}$$

The activity coefficient of the solvent is 1 at all concentrations in an ideal solution because

$$\gamma_1 = \frac{a_1}{X_1} = \frac{X_1}{X_1} = 1 \tag{17-15}$$

Thus, the ideal solution is a reference for the solvent in a real solution, and the activity coefficient of the solvent measures the deviation from ideality.

Solute in Solution. If the mole fraction scale of composition is used, it is convenient to choose a standard state such that the activity would approach the mole fraction in the limit of infinite dilution in which Henry's law is valid. That is,

$$\lim_{X_2 \to 0} \frac{a_2}{X_2} = 1 \tag{17-16}$$

since Equation 17-1 approaches Equation 16-5 in the limit as $X_2 \to 0$. The solid curve in Figure 17-2 represents the activity of the solute as a function of the mole fraction, X_2, of the solute, when the standard state is chosen to be the hypothetical state of unit mole fraction extrapolated along the Henry's law line. In the example in Figure 17-2, there is *no real state* of the solution in which the activity of the solute is equal to 1.

This choice of a standard state for the solute may appear strange at first glance. It might seem that the choice of pure solute as standard state would be a simpler one. The latter procedure would require either experimental information on the pure solute in the same physical state (for example, liquid or solid) as the solution, or data for solutions of sufficiently high concentration

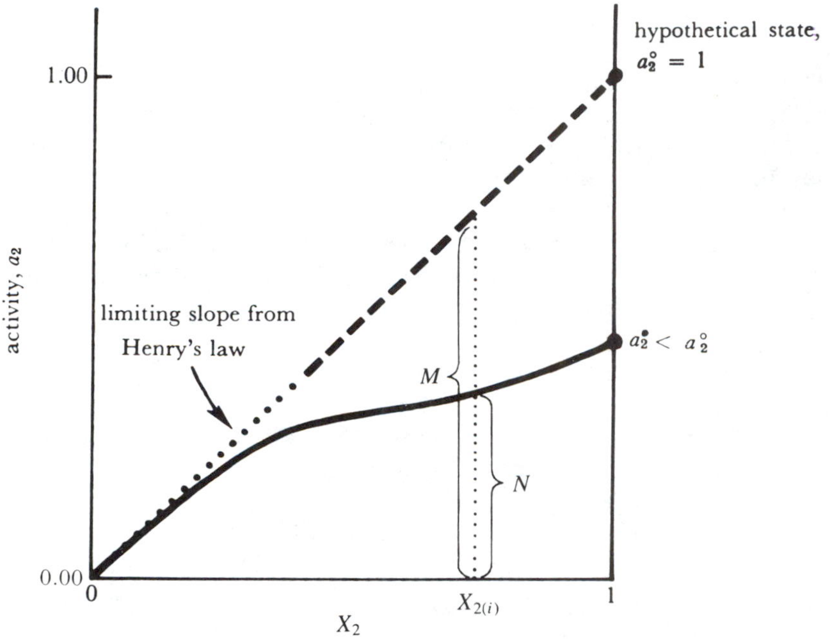

Figure 17-2. Activity and activity coefficient when standard state is set by extrapolation of Henry's law.

of solute so that Raoult's law might be approached and might be used for the extrapolation to obtain a_2 at $X_2 = 1$. Such information for most solutes is not available. The details of the solid curve in Figure 17-2 may be known only for small values of X_2 and not throughout the entire concentration range. Thus there are usually no means for determining the value that a_2 would approach for pure, supercooled, liquid solute. However, with data available only at low concentrations of solute, it is feasible to find the limiting slope and the constant in Henry's law and hence to determine a_2° for the *hypothetical* state selected as the standard state.

When data are available for the solute in substantially the entire concentration range, from mole fraction 0 to 1, the choice of standard state, either the hypothetical unit mole fraction (Henry's law) or the actual unit mole fraction (Raoult's law), is arbitrary. Figure 17-2 shows the relationships for activity and activity coefficient when Henry's law is used to define the standard state, and Figure 17-3 shows the same relationships when pure solute is chosen as the standard state.

It can be seen from Figures 17-2 and 17-3 that the numerical values of the activity and activity coefficient of the solute are different for the two choices of standard state. The scale of activities, for example, is necessarily different. The activity coefficient at mole fraction $X_{2(i)}$ is given by the ratio N/M in both figures. Thus when the standard state is chosen on the basis of Henry's law, the

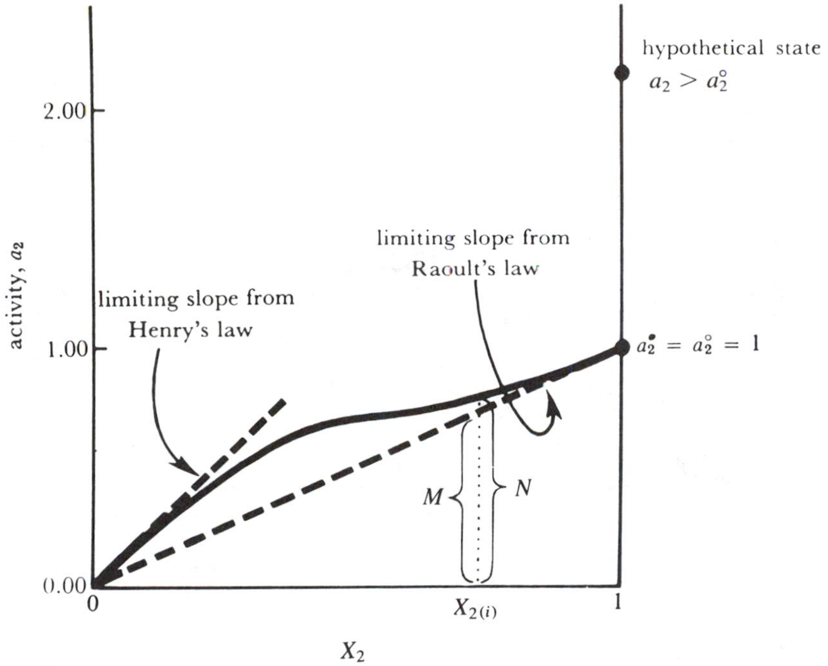

Figure 17-3. Activity and activity coefficient when pure solute is made the standard state.

activity coefficients are less than 1, whereas when the pure solute is chosen as standard state, the activity coefficients all are greater than 1. If the actual curve for the activity were below the Raoult's-law curve, as in Figure 16-1, the opposite relationship would be true.

However, the choice of standard states makes no difference in the value of ΔG for the transfer of solute between two solutions of different concentrations in the same solvent. We can see this by applying the definition of activity to the equation for the free energy change in a transfer process:

$$\begin{aligned}\Delta G &= \mu_2' - \mu_2 \\ &= \mu_2^\circ + RT \ln a_2' - [\mu_2^\circ + RT \ln a_2] \\ &= RT \ln a_2' - RT \ln a_2 \\ &= RT \ln \frac{a_2'}{a_2} \qquad (17\text{-}17)\end{aligned}$$

The values μ_2' and μ_2 are characteristic of the states of the system and do not depend on the choice of a standard state. It can be seen that the curves in Figure 17-2 and Figure 17-3 differ only by a scale factor that is determined by the choice of the point at which $a_2 = a_2^\circ = 1$. If both a_2' and a_2 are based on the same standard state, their ratio is independent of that choice.

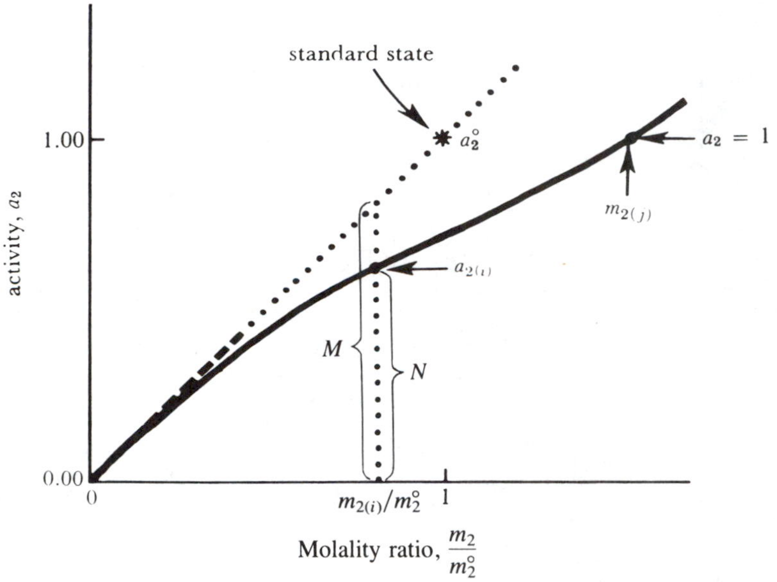

Figure 17-4. Activity and activity coefficient when the standard state is established on a molality basis.

The molality is used as a concentration scale primarily for solutions for which data are not available for concentrations approaching pure solute, so the Henry's-law basis for choosing the standard state is the only practical choice. The behavior of an example of such a solution is shown in Figure 17-4.

For such solutions the definition of activity is completed by the requirement that the activity approach the molality ratio in the limit of infinite dilution. That is,

$$\lim_{m_2 \to 0} \frac{a_2}{m_2/m_2^\circ} = \lim_{m_2 \to 0} \gamma_2 = 1 \tag{17-18}$$

The standard state chosen on this basis is indicated by the asterisk in Figure 17-4. This reference state is a "*hypothetical 1-molal solution*," that is, a state that has the activity that a 1-molal solution would have *if it obeyed the limiting law*. It is misleading to say that the standard state of the solute is the infinitely dilute solution because a_2 equals zero in an infinitely dilute solution. It is clear from the graph that the activity of the solute may be greater than 1 at high m_2. The activity coefficient at molality $m_{2(i)}$ [that is, $\gamma_{2(i)} = a_{2(i)}/(m_{2(i)}/m_2^\circ)$]is represented by the ratio *N/M* (since the height M is numerically equal to the distance $m_{2(i)}/m_2^\circ$ along the abscissa) and represents the deviation of the behavior of the solute from Henry's-law behavior. The activity coefficient, $\gamma_{2(i)}$, also is equal to the slope of a line from the origin to the point $a_{2(i)}$.

We can see from Figure 17-4 that for the particular system depicted there is a solution of some finite concentration $m_{2(j)}$ for which the activity is 1. Nevertheless, it would be misleading to call this solution of molality $m_{2(j)}$ the standard state. The standard state has the properties that a 1-molal solution would have *if it obeyed the limiting law*; the solution of molality $m_{2(j)}$ generally does not obey the limiting law.

17-3 GENERAL RELATIONSHIP BETWEEN EQUILIBRIUM CONSTANT AND FREE ENERGY

With the definition of the activity function we are able to derive a general expression relating $\Delta G°$ of a reaction to the equilibrium constant and hence to eliminate the restrictions imposed upon previous relationships.

Let us consider the chemical reaction

$$aA(a_A) + bB(a_B) = cC(a_C) + dD(a_D)$$

From Equation 12-11

$$dG = \mu_A \, dn_A + \mu_B \, dn_B + \mu_C \, dn_C + \mu_D \, dn_D \tag{17-19}$$

Applying the procedure used in Equations 12-32 through 12-34 and integrating from $\xi = 0$ to $\xi = 1$, we obtain

$$\Delta G = c\mu_C + d\mu_D - a\mu_A - b\mu_B \tag{17-20}$$

Substitution for the chemical potentials from Equation 17-1 gives

$$\Delta G = c(\mu_C^\circ + RT \ln a_C) + d(\mu_D^\circ + RT \ln a_D) \\ - a(\mu_A^\circ + RT \ln a_A) - b(\mu_B^\circ + RT \ln a_B)$$

This may be written

$$\Delta G = (c\mu_C^\circ + d\mu_D^\circ - a\mu_A^\circ - b\mu_B^\circ) + RT \ln \frac{a_C^c a_D^d}{a_A^a a_B^b} \tag{17-21}$$

The expression in parentheses in Equation 17-21 is equal to $\Delta G°$, so we can write

$$\Delta G = \Delta G^\circ + RT \ln \frac{a_C^c a_D^d}{a_A^a a_B^b} \tag{17-22}$$

Equation 17-22 is a general relationship for the calculation of ΔG for any reaction from the value for $\Delta G°$ and from the activities a_A, a_B, a_C, and a_D.

At equilibrium at constant temperature and pressure, $\Delta G = 0$ and

$$\frac{\Delta G^\circ}{RT} = -\ln \left[\frac{a_C^c a_D^d}{a_A^a a_B^b} \right]_{\text{equil}} \tag{17-23}$$

From the definitions of standard states for components of solutions, it is clear that ΔG° is a function of temperature and pressure, but it is a constant for a particular reaction at a fixed temperature and pressure. Thus we can write

$$\frac{\Delta G^\circ}{RT} = -\ln K_a \tag{17-24}$$

in which K_a is the equilibrium constant in terms of activities. Thus

$$K_a = \left[\frac{a_C^c a_D^d}{a_A^a a_B^b}\right]_{\text{equil}} \tag{17-25}$$

If the mole fraction is a convenient variable, Equation 17-3 can be used to write Equation 17-25 as

$$K_a = \frac{X_C^c \gamma_C^c \, X_D^d \gamma_D^d}{X_A^a \gamma_A^a \, X_B^b \gamma_B^b} \tag{17-26}$$

$$= \frac{X_C^c \, X_D^d \gamma_C^c \gamma_D^d}{X_A^a \, X_B^b \gamma_A^a \gamma_B^b} \tag{17-27}$$

$$= K_X K_\gamma \tag{17-28}$$

Similarly, for reactants and products for which molality is the convenient composition variable, we can write

$$K_a = K_m K_\gamma$$

where

$$K_m = \frac{(m_C/m^\circ)^c \, (m_D/m^\circ)^d}{(m_A/m^\circ)^a \, (m_B/m^\circ)^b} \tag{17-29}$$

For some reactions a constant can be expressed in terms of mole fractions for some components and molalities for other components.

If the value of K_a for a reaction is calculated from the value of ΔG°, to obtain equilibrium yields in terms of m_i or X_i we must have values of the γ_i to substitute into Equation 17-26 or Equation 17-29. The determination of these quantities from experimental data will be discussed in Chapters 19 and 20.

17-4 DEPENDENCE OF ACTIVITY ON PRESSURE

Since $\mu_i = \mu_i^\circ + RT \ln a_i$

$$\left(\frac{\partial \mu_i}{\partial P}\right)_{T,X} = \left(\frac{\partial \mu_i^\circ}{\partial P}\right)_{T,X} + RT\left(\frac{\partial \ln a_i}{\partial P}\right)_{T,X} \tag{17-30}$$

From Equation 12-17

$$\left(\frac{\partial \mu_i}{\partial P}\right)_{T,X} = \bar{V}_i$$

If the standard state is defined to change with pressure, then

$$\left(\frac{\partial \ln a_i}{\partial P}\right)_{T,X} = \frac{\bar{V}_i - \bar{V}_i^\circ}{RT} \tag{17-31}$$

For solvents, since the standard state is the pure substance,

$$\bar{V}_1^\circ = V_1^\bullet \tag{17-32}$$

in which $V_1^\bullet$ is the molar volume for pure solvent. For solutes for which the standard state is obtained from Henry's law (Figures 17-2 and 17-4), the value of V_2° can be obtained from Equations 16-4 and 16-5, or Equation 16-10, by applying Equation 12-17. When the concentration of solute is expressed on the mole fraction scale (Equations 16-4 and 16-5),

$$\mu_2^\circ = \mu_{2(\text{gas})}^\circ + RT \ln\left(\frac{k_2}{f_{\text{gas}}^\circ}\right) \tag{17-33}$$

Then

$$\bar{V}_2^\circ = \left(\frac{\partial \mu_2^\circ}{\partial P}\right)_{T,X}$$

$$= RT\left(\frac{\partial \ln k_2}{\partial P}\right)_{T,X} \tag{17-34}$$

Similarly, when the concentration of solute is expressed on the molality scale (Equation 16-10),

$$\bar{V}_2^\circ = \left(\frac{\partial \mu_2^\circ}{\partial P}\right)_{T,m_2}$$

$$= RT\left(\frac{\partial \ln k_2''}{\partial P}\right)_{T,m_2} \tag{17-35}$$

Only one term appears in the derivative in each case because both $\mu^\circ_{2(\text{gas})}$ and f°_{gas} are defined at a fixed pressure and are not functions of the pressure of the system.

Thus $\bar{V}^\circ_2$ can be calculated from determinations of the Henry's-law constant as a function of pressure. It can be shown that $\bar{V}^\circ_2$ is equal to $\bar{V}^\infty_2$, the partial molar volume of the solute at infinite dilution.

17-5 DEPENDENCE OF ACTIVITY ON TEMPERATURE

From previous discussions of temperature coefficients of the free energy functions, we expect the expression for the temperature dependence of the activity to involve an enthalpy function. We need to develop relationships for the enthalpies of the standard states.

Standard Partial Molar Enthalpies

For pure solids and liquids the standard enthalpy is the enthalpy of the substance at the specified temperature and at 1 bar.

Solvent. Pure solvent at the same temperature and pressure (generally 1 bar) as the solution usually is chosen as the reference state for the solvent. It follows that

$$\bar{H}_{1(\text{standard state})} = \bar{H}^\circ_1 = H^\bullet_1 \tag{17-36}$$

Solute. The standard state for the solute is the *hypothetical* unit mole fraction state (Figure 17-2), or the *hypothetical* 1-molal solution (Figure 17-4). In both cases, the standard state is obtained by extrapolation from the Henry's-law line that describes behavior at infinite dilution. Thus, the partial molar enthalpy of the standard state is *not* that of the *actual* pure solute or the *actual* 1-molal solution.

The chemical potential of the solute that follows Henry's law is given by Equation 16-11

$$\mu_2 = \mu^\circ_2 + RT \ln \left(\frac{m_2}{m^\circ_2} \right)$$

If we divide each term in Equation 16-11 by T and differentiate with respect to T at constant P and m_2, the result is

$$\left[\frac{\partial(\mu/T)}{\partial T} \right]_{P,m_2} = \left[\frac{\partial(\mu^\circ_2/T)}{T} \right]_{P,m_2} \tag{17-37}$$

From Equation 12-37

$$\left[\frac{\partial(\mu/T)}{\partial T}\right]_{P,m_2} = -\frac{H}{T^2}$$

Substituting from Equation 12-37 into Equation 17-37, we have

$$-\frac{\bar{H}_2}{T^2} = -\frac{\bar{H}_2^\circ}{T^2}$$

or

$$\bar{H}_2 = \bar{H}_2^\circ \tag{17-38}$$

Thus, the partial molar enthalpy along the Henry's-law line is constant and equal to the standard partial molar enthalpy. The only real solution along the Henry's-law line is the infinitely dilute solution, so

$$\bar{H}_2^\circ = \bar{H}_2^\infty \tag{17-39}$$

For this reason, the infinitely dilute solution frequently is called the reference state for the partial molar enthalpy of both solvent and solute.

Equation for Temperature Coefficient

From Equation 17-1

$$\mu_i = \mu_i^\circ + RT \ln a_i$$

or

$$\frac{\mu_i}{T} = \frac{\mu_i^\circ}{T} + R \ln a_i \tag{17-40}$$

From Equation 12-37

$$\left[\frac{\partial(\mu/T)}{\partial T}\right]_{P,m_2} = -\frac{\bar{H}}{T^2}$$

If we differentiate Equation 17-40 with respect to temperature at constant molality and pressure, and substitute from Equation 12-37, the result is

$$-\frac{\bar{H}_i}{T^2} = -\frac{\bar{H}_i^\circ}{T^2} + R\left(\frac{\partial \ln a_i}{\partial T}\right)_{P,m_2}$$

or

$$\left(\frac{\partial \ln a_i}{\partial T}\right)_{P,m_2} = -\frac{\bar{H}_i - \bar{H}_i^\circ}{RT^2} \tag{17-41}$$

From Equations 17-36, 17-38, and 17-39, we see that $\bar{H}_i^\circ$ for both solute and solvent are equal to $\bar{H}_i^\infty$. Therefore, we can write Equation 17-41 for either solvent or solute, as

$$\left(\frac{\partial \ln a_i}{\partial T}\right)_{P,m_2} = -\frac{\bar{H}_i - \bar{H}_i^\infty}{RT^2} \tag{17-42}$$

in which $\bar{H}_i^\infty - \bar{H}_i$ is the change in partial molar enthalpy on dilution to an infinitely dilute solution.

From the definition of the activity coefficient (Equations 17-5 and 17-9),

$$\gamma_i = \frac{a_i}{m_i/m_i^\circ} \qquad or \qquad \gamma_i = \frac{a_i}{X_i}$$

we can show that for the solute

$$\left(\frac{\partial \ln \gamma_2}{\partial T}\right)_{P,m_2} = -\frac{\bar{H}_2 - \bar{H}_2^\infty}{RT^2} \tag{17-43}$$

and for the solvent

$$\left(\frac{\partial \ln \gamma_1}{\partial T}\right)_{P,X_1} = -\frac{\bar{H}_1 - \bar{H}_1^\infty}{RT^2} \tag{17-44}$$

17-6 STANDARD ENTROPY

We have pointed out that there may be a concentration $m_{2(j)}$ of the solute in the real solution that has an activity of 1, which is equal to the activity of the hypothetical 1-molal standard state. Also, $\bar{H}_2$ of the solute in the standard state equals the partial molar enthalpy of the solute at infinite dilution. We might inquire whether the *partial molar entropy* of the solute in the standard state, $\bar{S}_2^\circ$, corresponds to the partial molar entropy in either of these two solutions.

Let us compare $\bar{S}_2$ for a real solution with $\bar{S}_2^\circ$ of the hypothetical 1-molal solution. For any component of a solution, from Equation 12-14, we can write

$$\mu_2 = \bar{H}_2 - T\bar{S}_2$$

Hence

$$-T\bar{S}_2 = \mu_2 - \bar{H}_2 \qquad \text{and} \qquad -T\bar{S}_2^\circ = \mu_2^\circ - \bar{H}_2^\circ$$

and

$$-T(\bar{S}_2 - \bar{S}_2^\circ) = (\bar{H}_2^\circ - \bar{H}_2) + (\mu_2 - \mu_2^\circ) \tag{17-45}$$

At infinite dilution, that is, when $m_2 = 0$

$$\bar{S}_2 \neq \bar{S}_2^\circ \qquad (\text{at } m_2 = 0) \tag{17-46}$$

since

$$\mu_2 \neq \mu_2^\circ \qquad (\text{at } m_2 = 0) \tag{17-47}$$

even though

$$\bar{H}_2 = \bar{H}_2^\circ \qquad (\text{at } m_2 = 0)$$

Hence the partial molar entropy of the solute in the standard state is not that of the solute at infinite dilution. Similarly, at the molality $m_{2(j)}$ (Figure 17-4) where a_2 is unity

$$\bar{S}_2 \neq \bar{S}_2^\circ \qquad (\text{at molality where } a_2 = 1) \tag{17-48}$$

because

$$\bar{H}_2 \neq \bar{H}_2^\circ \qquad (\text{at molality where } a_2 = 1) \tag{17-49}$$

even though

$$\mu_2 = \mu_2^\circ \qquad (\text{at molality where } a_2 = 1) \tag{17-50}$$

Thus $\bar{S}_2$ can be equal to $\bar{S}_2^\circ$ only for a solution with some molality $m_{2(k)}$ at which

$$(\bar{H}_2^\circ - \bar{H}_2) = (\bar{G}_2^\circ - \bar{G}_2) = \mu_2^\circ - \mu_2 \tag{17-51}$$

The particular value of the molality $m_{2(k)}$ at which $\bar{S}_{2(k)} = \bar{S}_2^\circ$ differs from solute to solute and for different solvents with the same solute.

We can summarize our conclusions about the thermodynamic properties of the solute in the hypothetical 1-molal standard state as follows. Such a solute is characterized by values of the thermodynamic functions that are represented by μ_2°, $\bar{H}_2^\circ$, and $\bar{S}_2^\circ$. There frequently is also a real solution at some molality $m_{2(j)}$ (Figure 17-4) for which $\mu_2 = \mu_2^\circ$, that is, for which the activity has a value of 1. The real solution for which $\bar{H}_2$ is equal to $\bar{H}_2^\circ$ is the one at infinite dilution. Furthermore, $\bar{S}_2$ has a value equal to $\bar{S}_2^\circ$ for some real solution only at a molality $m_{2(k)}$ that is neither zero nor $m_{2(j)}$. Thus there are three different real concentrations of the solute for which the thermodynamic quantities μ_2, $\bar{H}_2$, and $\bar{S}_2$, respectively, have the same values as in the hypothetical standard state.

The standard thermodynamic properties for the solvent are

$$\mu_1^\circ = \mu_1^\bullet = \mu_1^\infty \tag{17-52}$$

Table 17-1 Standard States for Thermodynamic Calculations (For every case it is assumed that the temperature has been specified.)

Physical State	Chemical Potential	Enthalpy[a]	Entropy	Volume[b]
Pure gas	Hypothetical ideal gas at 1 bar (0.1 MPa). There also will be a pressure, usually near 1 bar, at which the real gas has a fugacity of unity.	Hypothetical ideal gas at 1 bar; also real gas at zero pressure. (See Exercise 1, p. 363).	Hypothetical ideal gas at 1 bar (0.1 MPa). There also will be a pressure of the real gas, not zero and not that of unit fugacity, with an entropy equal to that in the standard state.	
Pure liquid or pure solid	1 bar (0.1 MPa)	1 bar (0.1 MPa)	1 bar (0.1 MPa)	
Solvent in a solution[c]	Pure solvent at same pressure as solution.	Pure solvent at same pressure as solution.	Pure solvent at same pressure as solution.	Pure solvent at same pressure as solution.
Solute in a solution[c,d]	Hypothetical 1-molal solution obeying limiting law corresponding to Henry's law at same pressure as solution. There also may be a finite concentration, not equal to zero, at which the activity of the solute is unity. This may be considered at the standard state *only in free energy calculations.*	Hypothetical 1-molal solution obeying limiting law corresponding to Henry's law at same pressure as solution. The value of the partial molar enthalpy in the standard state, $\bar{H}_2^\circ$, is always equal to that at infinite dilution. Hence the *infinitely dilute solution* can be used as *reference state* in all *enthalpy calculations*, but not in free energy or entropy calculations.	Hypothetical 1-molal solution obeying limiting law corresponding to Henry's law at same pressure as solution. There also may be a solution of finite concentration, not equal to zero but also not having an activity of unity, with a partial molar entropy of solute equal to that in the reference state.	Hypothetical 1-molal solution obeying limiting law corresponding to Henry's law at same pressure as solution. The value of $\bar{V}_2^\circ$ is always equal to that at infinite dilution.

[a]The standard state for the heat capacity is the same as that for the enthalpy. For a proof of this statement for the solute in a solution, see Exercise 2, p. 363.

[b]Since the standard state for gases, pure solids, and pure liquids is defined at a fixed pressure, $\bar{V}_2^\circ = (\partial \mu^\circ/\partial P)_T \equiv 0$.

[c]This choice of standard state for components of a solution is different from that used by many American thermodynamicists. It seems preferable to the choice of a 1-bar standard state, however, because it is more consistent with the extrapolation procedure by which the standard state is determined experimentally, and it leads to a value of the activity coefficient equal to 1 when the solution is ideal or very dilute whatever the pressure. A more detailed rationale for this choice can be found in the references [1].

[d]A more elegant (though more difficult to visualize) formulation of the procedure for the selection of the standard state for a solute may be made as follows. From (Equation 17-1)

$$\mu_2 = \mu_2^\circ + RT \ln a_2$$

and (Equation 17-9)

$$a_2 = \frac{m_2 \gamma_2}{m_2^\circ}$$

it follows that

$$\mu_2^\circ = \mu_2 - RT \ln \frac{m_2}{m_2^\circ} - RT \ln \gamma_2$$

Therefore the state in which the solute has a partial molar free energy of μ_2° can be found from the following limit, since γ_2 approaches unity as m_2 approaches zero:

$$\lim_{m_2 \to 0} \left(\mu_2 - RT \ln \frac{m_2}{m_2^\circ} \right) = \mu_2^\circ$$

With this method of formulation it also is possible to show that frequently there is a real solution at some molality m_j for which $\mu_2 = \mu_2^\circ$, that $\bar{H}_2^\circ$ corresponds to $\bar{H}_2$ for a real solution at infinite dilution, and that $\bar{S}_2^\circ$ equals $\bar{S}_2$ for a real solution at a molality m_k, which is neither zero nor m_j.

and

$$\bar{H}_1^{\circ} = H_1^{\bullet} = \bar{H}_1^{\infty} \tag{17-53}$$

so

$$\bar{S}_1^{\circ} = S_1^{\bullet} = \bar{S}_1^{\infty} \tag{17-54}$$

Table 17-1 summarizes the information on the standard states of pure phases as well as those of solvents and solutes.

17-7 DEVIATIONS FROM IDEALITY IN TERMS OF EXCESS THERMODYNAMIC QUANTITIES

Various functions have been used to express the deviation of observed behavior of solutions from that expected for ideal systems. Some, such as the activity coefficient, are most convenient for measuring deviations from ideality for a particular component of a solution. However, the most convenient measure for the solution as a whole is the series of *excess* functions [2], which are defined in the following way.

We have derived an expression for the free energy of mixing two pure substances to form one mole of an ideal solution (Equation 15-22),

$$\Delta G_{\text{mix}}^{\text{I}} = X_1 RT \ln X_1 + X_2 RT \ln X_2 \tag{17-55}$$

In actual systems the observed value for the free energy of mixing, ΔG_{mix}, may differ from $\Delta G_{\text{mix}}^{\text{I}}$. We define this difference, the excess free energy of mixing, $\Delta G_{\text{mix}}^{\text{E}}$, as

$$\Delta G_{\text{mix}}^{\text{E}} = \Delta G_{\text{mix}} - \Delta G_{\text{mix}}^{\text{I}} = \Delta G_{\text{mix}} - X_1 RT \ln X_1 - X_2 RT \ln X_2 \tag{17-56}$$

Similarly, we define the excess entropy of mixing, $\Delta S_{\text{mix}}^{\text{E}}$, as

$$\Delta S_{\text{mix}}^{\text{E}} = \Delta S_{\text{mix}} - \Delta S_{\text{mix}}^{\text{I}} = \Delta S_{\text{mix}} + X_1 R \ln X_1 + X_2 R \ln X_2 \tag{17-57}$$

For the excess enthalpy of mixing, $\Delta H_{\text{mix}}^{\text{E}}$, since $\Delta H_{\text{mix}}^{\text{I}}$ is zero

$$\Delta H_{\text{mix}}^{\text{E}} = \Delta H_{\text{mix}} \tag{17-58}$$

It can be shown that the usual relationships between temperature coefficients of the free energy and entropy or enthalpy, respectively, also apply if stated for excess functions. Thus

$$\Delta S_{\text{mix}}^{\text{E}} = -\frac{\partial \Delta G_{\text{mix}}^{\text{E}}}{\partial T} \tag{17-59}$$

and

$$\Delta H_{\text{mix}}^{\text{E}} = \Delta G_{\text{mix}}^{\text{E}} - T\frac{\partial \Delta G_{\text{mix}}^{\text{E}}}{\partial T} \tag{17-60}$$

Excess thermodynamic functions can be evaluated most readily when the vapor pressures of both solute and solvent in a solution can be measured.

EXERCISES

1. For a pure gas the fugacity is defined so that (Equation 13-17)

$$\lim_{P \to 0} \frac{f}{P} = 1$$

Show that, on this basis, the enthalpy of the gas in the standard state must be equal to that at zero pressure. (*Hint:* If the gas were ideal, $f = P$ at all pressures. For the real gas, $f^\circ = k = 0.1$ MPa. Proceed by analogy with the discussion in Section 13-2.)

2. a. Rearrange Equation 17-41 so that RT^2 occurs as a factor on the left side.
   b. Differentiate both sides of the equation obtained in (a) with respect to T.
   c. Prove that $\bar{C}^\circ_{p(2)}$ is equal to $\bar{C}_{p(2)}$ for the infinitely dilute solution. Keep in mind that the temperature coefficient of ln γ_2 is zero in an infinitely dilute solution.

3. Frequently it is necessary to convert solute activity coefficients based on mole fraction to a molality basis, or vice versa. The equation for making this conversion can be derived in the following way.
   a. Starting from the fundamental definitions of activity on each concentration basis, prove that for a solute

$$\frac{a_X}{a_m} = \frac{k_2''}{k_2}$$

in which the subscript X refers to a mole fraction basis and m to a molality basis, and k_2 and k_2'' are the Henry's-law constants on a mole fraction and molality basis, respectively.

   b. Show further that

$$\frac{\gamma_X}{\gamma m} = \frac{m_2/m_2^\circ}{X_2} \frac{k_2''}{k_2}$$

(k_2 and k_2'' are defined by Equations 16-1 and 16-7.)

   c. In very dilute solutions Henry's law is valid in the form of either Equation 16-7 or Equation 16-1. Prove that

$$k_2'' = k_2 M_1 m_2^\circ$$

in which M_1 is the molar mass of the solvent, when the mass and the molar mass of the solvent are both expressed in kilograms.

   d. Show that for any concentration

$$\frac{\gamma_m}{\gamma m} = 1 + m_2 M_1$$

(A factor of 1000 appears in these equations if the molar mass of the solvent is expressed in grams; that is, $k_2'' = k_2 M_1 m_2^\circ/1000$, and $\gamma_X/\gamma_m = 1 + m_2 M_1/1000$.)

4. Molality, m_2, and molarity, c_2, are related by the expression

$$m_2 = \frac{c_2}{\rho - c_2 M_2}$$

in which ρ is the density of the solution, with mass in kilograms and volume in cubic decimeters (liters). If γ_c, the activity coefficient on a molarity basis, is defined by

$$\gamma_c = \frac{a_c}{c}$$

show that

$$\frac{\gamma_m}{\gamma_c} = \left[\frac{\rho}{\rho_1} - \frac{c_2 M_2}{\rho_1}\right]$$

in which ρ_1 is the density of pure solvent.

5. Theoretical molecular-statistical analyses of certain types of solutions (regular solutions, with solvent and solute molecules of the same size) indicate that the free energy of mixing per mole of solution is given by

$$\Delta G_{\text{mix}} = RT(X_1 \ln X_1 + X_2 \ln X_2) + X_1 X_2 \omega$$

in which ω is a parameter related to the interaction energies between molecules. Show that

$$\Delta S^{\text{E}}_{\text{mix}} = -X_1 X_2 \frac{\partial \omega}{\partial T}$$

and

$$\Delta H^{\text{E}}_{\text{mix}} = X_1 X_2 \left(\omega - T\frac{\partial \omega}{\partial T}\right)$$

6. Starting with Equation 17-56 for $\Delta G^{\text{E}}_{\text{mix}}$, derive Equations 17-59 and 17-60 for the excess entropy and enthalpy of mixing, respectively.

7. Show that $\bar{V}_i^\circ = \bar{V}_i^\infty$ by an argument similar to the one used to demonstrate that $\bar{H}_i^\circ = \bar{H}_i^\infty$.

8. At 42°C the heat of mixing of 1 mole of water and 1 mole of ethanol is −343.1 J. The vapor pressure of water above the solution is 0.821 p_1° and that of ethanol is 0.509 $p_2^\bullet$ in which $p^\bullet$ is the vapor pressure of the corresponding pure liquid. Assume vapors behave as ideal gases. Compute the excess entropy of mixing.

9. Rhombic sulfur is soluble in CS_2, and so is monoclinic. For the transition

$$\text{S(rhombic)} = \text{S(monoclinic)}, \qquad \Delta G^\circ_{298} = 96 \text{ J mol}^{-1}$$

The solubility of monoclinic sulfur in CS_2 is 22-molal. What is the solubility of rhombic sulfur in the same solvent? Assume that the activity coefficient for both forms of dissolved sulfur is 1.

REFERENCES

1. K. Denbigh, *The Principles of Chemical Equilibrium*, 3d ed., Cambridge University Press, Cambridge, 1971, pp. 287–288. B. Perlmutter-Hayman, *J. Chem. Educ.* **61**, 782 (1984).
2. G. Scatchard and C. L. Raymond, *J. Am. Chem. Soc.* **60**, 1278 (1938).

CHAPTER 18

Calculation of Partial Molar Quantities from Experimental Data: Volume and Enthalpy

In this chapter we will consider the methods by which values of partial molar quantities can be obtained from experimental data. Most of the methods are applicable to any thermodynamic property, J, but special emphasis will be placed on the partial molar volume and the partial molar enthalpy, since these are needed to determine the pressure and temperature coefficients of the chemical potential. Furthermore, the volume function, being easy to visualize, serves well in an initial exposition of partial molar quantities.

18-1 GRAPHICAL DIFFERENTIATION OF J AS A FUNCTION OF COMPOSITION

A common procedure for the calculation of partial molar quantities, defined (as in Equation 12-10) as

$$\bar{J}_i = \left(\frac{\partial J}{\partial n_i}\right)_{T,P,n_j} \tag{18-1}$$

is the graphical differentiation of curves of J as a function of composition. A particularly simple case is shown in Figure 18-1, in which the volume is a linear function of the mole number of glycolamide in a fixed volume of water. In this case the partial molar volume of solute, $\bar{V}_2$, is constant and is equal to the slope of the line. The partial molar volume, $\bar{V}_2$, represents the effective volume of the solute in solution, that is, the increase in volume per mole of solute added. Since, from Equation 12-19,

$$V = n_1\bar{V}_1 + n_2\bar{V}_2$$

the linear dependence of V on n_2 at constant n_1 also indicates that $\bar{V}_1$ is constant

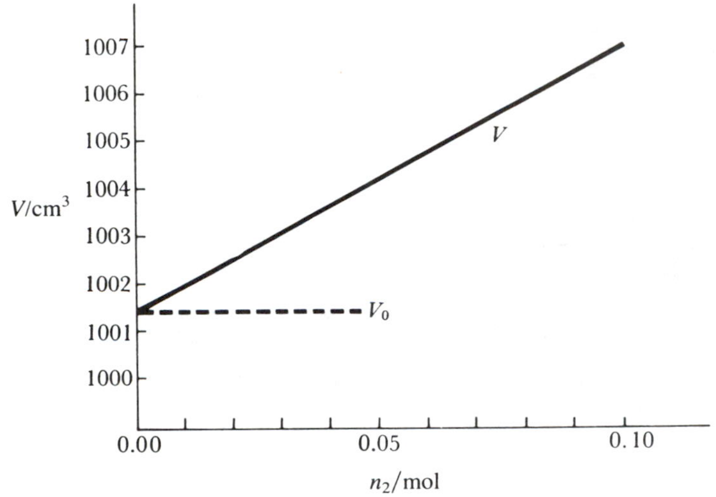

Figure 18-1. Linear dependence of volume on concentration for dilute solutions of glycolamide in water. Based on data of F. T. Gucker, Jr., W. L. Ford, and C. E. Moser, *J. Phys. Chem.* **43**, 153 (1939). Precise measurements at higher concentrations indicate that this line does have some curvature, which is not detectable in dilute solution.

and is equal to $V_1^\bullet$, the molar volume of pure solvent. In contrast, the value of $\bar{V}_2$ generally is not equal to the molar volume of pure solute. For example, glycolamide (Figure 18-1) has an effective volume, $\bar{V}_2$, in dilute aqueous solution of 56.2 cm^3 mol^{-1}; the pure solid has a molar volume of 54.0 cm^3 mol^{-1}. For iodine, $\bar{V}_2$ as a solute in liquid perfluoro-*n*-heptane is 100 cm^3 mol^{-1}, whereas the solid has a molar volume of 51 cm^3 mol^{-1}, and the supercooled liquid has a molar volume of 59 cm^3 mol^{-1}. Hydrogen has an effective volume in aqueous solution, at 101.32 kPa (1 atm) and 25°C, of 26 cm^3 mol^{-1}, in contrast to approximately 25,000 cm^3 mol^{-1} for the pure gas. Furthermore, $\bar{V}_2$ for hydrogen (as well as for many other solutes) varies greatly with solvent; it is 50 cm^3 mol^{-1} in ether and 38 cm^3 mol^{-1} in acetone. Even more surprising are the effective volumes of some salts in water. For NaCl the molar volume of the crystal is 27 cm^3 mol^{-1} compared to 16.4 cm^3 mol^{-1} for $\bar{V}_2$. For Na_2CO_3 the molar volume of the pure solid is about 42 cm^3 mol^{-1} compared to -6.7 cm^3 mol^{-1} for $\bar{V}_2$; that is, Na_2CO_3 dissolved in water has a *negative* effective volume. These observations show very clearly that $\bar{V}_2$ reflects not only the volume of the solute molecule, but also the effect that the solvent–solute interaction has on the volume of the solvent.

Partial Molar Volumes

For most solutions, the volume is not a linear function of the composition; the slope of the volume–concentration curve and the value of $\bar{V}_2$ are functions of

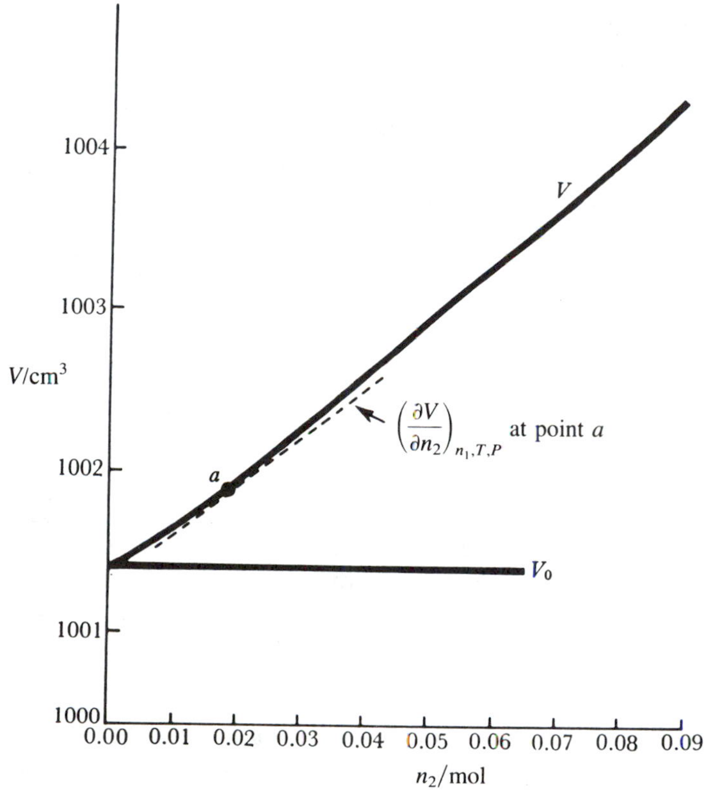

Figure 18-2. Nonlinear dependence of volume on concentration for dilute solutions of sulfuric acid in water. Based on data of I. M. Klotz and C. F. Eckert, *J. Am. Chem. Soc.* **64**, 1878 (1942).

the composition, as illustrated in Figure 18-2. In this case it can be seen that the volume does not change by equal increments when equal quantities of solute are added successively to a fixed quantity of solvent. Thus let us consider what must be the significance of the slope, $(\partial V/\partial n_2)_{n_1,T,P}$, represented by the dotted line in Figure 18-2. According to the principles of calculus, this slope represents the change in volume per mole of added solute, n_2 (temperature, pressure, and moles of solvent, n_1, being maintained constant), at a fixed point on the curve—in other words, at some specified value of n_2. The value of n_2 must be specified since the slope depends on the position on the curve at which it is measured. In practice, this slope, which we represent by $\bar{V}_2$, as in Equation 18-1, refers to either one of the following two experiments:

1. Measure the change in total volume, V, of the solution when one mole of solute is added to a very large quantity (strictly speaking, an infinite quantity) of the solution at the desired concentration. Because very large

quantities of solution are used, the addition of one mole of solute does not change the concentration of the solution appreciably.

2. Measure the change in total volume, V, of the solution when a small quantity of the solute is added to the solution. Then calculate the change for one mole (that is, divide ΔV by Δn_2) as if there were no change in composition when a whole mole of solute is added. Repeat this procedure but add a smaller quantity, $\Delta n_2'$, of solute and compute $\Delta V'/\Delta n_2'$. Repeat again with a still smaller quantity of added solute. The limiting value,

$$\lim_{\Delta n_2 \to 0} \frac{\Delta V}{\Delta n_2} = \left(\frac{\partial V}{\partial n_2}\right)_{T,P,n_1}$$

gives $\bar{V}_2$.

These are two equivalent points of view concerning the meaning of $\bar{V}_2$. Clearly, $\bar{V}_2$ does not correspond to the actual change in volume when a whole mole of solute is added to a limited quantity of solution since $\bar{V}_2$ is the increase in V upward along the dotted line in Figure 18-2, whereas the actual volume follows the solid line. To calculate $\bar{V}_2$ from the data represented in Figure 18-2, numerical or graphical differentiation is necessary, and the procedure described in Chapter 2 is used.

Most frequently, volume data for solutions are tabulated as density, ρ, versus composition. The procedure for obtaining $\bar{V}$ is illustrated by reference to the densities and weight percent concentrations of ethanol–water mixtures (Table 18-1, Columns 1 and 4) at 25°C.

To obtain $\bar{V}_2$, that is, $(\partial V/\partial n_2)_{n_1}$, we need values of V for a fixed quantity, n_1, of water, but for variable quantities, n_2, of ethanol. For this purpose we convert the relative weights given in Column 1 to relative numbers of moles, that is, to n_2/n_1 in Column 2. The numbers in Column 2 also are the moles of ethanol accompanying one mole of water in each of the solutions listed in Column 1.

From this information and the density (Column 4), we can calculate the volume in cubic centimeters that contains one mole of water. The mass, M, of a solution containing n_2 moles of ethanol and one mole of water is

$$M = n_2 \times (\text{molecular weight of } C_2H_5OH) + 1 \times (\text{molecular weight of } H_2O) \qquad (18\text{-}2)$$

Hence the volume per mole of water is

$$V = \frac{M}{\rho} \text{ cm}^3 \,(\text{mole } H_2O)^{-1} \qquad (18\text{-}3)$$

Numerical values for these volumes of solution containing a fixed quantity (one mole) of water are listed in Column 5. The partial molar volumes can be

Table 18-1 Densities and Partial Volumes of Ethanol–Water Mixtures

1	2	3	4	5	6	7	8
Wt % Ethanol	$\frac{C_2H_5OH}{H_2O} = \frac{n_2}{n_1}$	$\frac{H_2O}{C_2H_5OH} = \frac{n_1}{n_2}$	$\rho/(\text{g cm}^{-3})$	$V/[\text{cm}^3(\text{mol } H_2O)^{-1}]$	$\frac{\Delta V}{\Delta n_2}$	$V/[\text{cm}^3(\text{mol } C_2H_5OH)^{-1}]$	$\frac{\Delta V}{\Delta n_1}$
20	0.097769	10.228241	0.96639	23.3032		238.3510	
					53.59		18.064
25	0.130357	7.671181	0.95895	25.0496		192.1602	
					54.32		17.969
30	0.167603	5.966474	0.95067	27.0726		161.5282	
					55.10		17.839
35	0.210578	4.748826	0.94146	29.4403		139.8072	
					55.75		17.700
40	0.260716	3.835590	0.93148	32.2354		123.6419	
					56.31		17.555
45	0.319970	3.125296	0.92085	35.5719		111.1726	
					56.68		17.436
50	0.391074	2.556394	0.90985	39.6021		101.2650	
					57.03		17.300
55	0.477979	2.092140	0.89850	44.5582		93.2220	
					57.26		17.189
60	0.586611	1.704707	0.88699	50.7785		86.5624	
					57.50		17.048
65	0.726280	1.376878	0.87527	58.8096		80.9737	
					57.70		16.905
70	0.912564	1.095883	0.86340	69.5576		76.2236	

determined graphically from a chord-area plot of the ratio of increments listed in Column 6 versus n_2/n_1 since

$$\left(\frac{\Delta V}{\Delta n_2}\right)_{n_1} = \left(\frac{\Delta V}{\Delta(n_2/n_1)}\right)_{n_1=1} \tag{18-4}$$

Such a graph is illustrated in Figure 18-3. The partial derivative $(\partial V/\partial n_2)_{n_1}$ can be determined from a smooth curve drawn through the chords to balance the areas, as in the figure.

The partial molar volumes for water in the ethanol solutions can be calculated by an analogous procedure. In this case we wish to find V for a fixed quantity of solute, let us say one mole, but variable quantities of water. The first step consists of calculating the moles of water per mole of ethanol, n_1/n_2, from the data on percentage composition. Such values are listed in Column 3 of

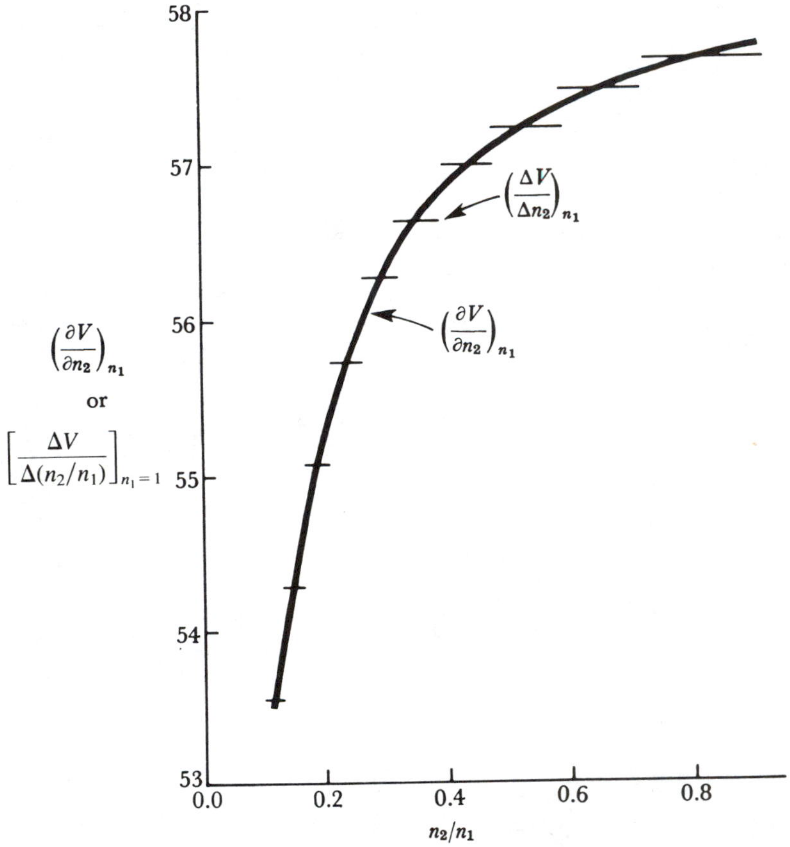

Figure 18-3. Graphical differentiation to obtain $\bar{V}_2$ in ethanol–water solutions.

Table 18-1. From this information and the density, we can calculate the volumes of solutions containing one mole of ethanol but different quantities of water. These volumes are listed in Column 7. The partial molar volumes of the water, $\bar{V}_1$, then can be determined from a chord-area plot of the ratio of increments $\Delta V/\Delta n_1$ (listed in Column 8) versus n_1/n_2 because

$$\left(\frac{\Delta V}{\Delta n_1}\right)_{n_2} = \left(\frac{\Delta V}{\Delta(n_1/n_2)}\right)_{n_2=1} \tag{18-5}$$

If we wish to use the technique of numerical differentiation discussed in Section 2-3 to calculate $\bar{V}_1$, we must plot $V/[\text{cm}^3(\text{mol } C_2H_5OH)^{-1}]$ against n_1/n_2 accurately, and read from the smooth curve values of V at equal intervals of n_1/n_2. The derivative can then be calculated as in Section 2-3. With data of the precision illustrated in Table 18-1, a very large scale graph is required to take advantage of the precision of the data.

Alternatively, we can fit the values of $V/[\text{cm}^3 (\text{mol } C_2H_5OH)^{-1}]$ as a function of n_1/n_2 to an empirical polynomial and use the resulting function to calculate values of V at equal intervals of n_1/n_2.

An interesting alternative procedure is the tangent method. Since (Equation 12-19)

$$J = n_1\bar{J}_1 + n_2\bar{J}_2$$

the molar value of J is

$$\begin{aligned} \mathrm{J} &= \frac{J}{n_1 + n_2} = \frac{n_1}{n_1 + n_2}\bar{J}_1 + \frac{n_2}{n_1 + n_2}\bar{J}_2 \\ &= X_1\bar{J}_1 + X_2\bar{J}_2 && (18\text{-}6) \\ &= X_1\bar{J}_1 + (1 - X_1)\bar{J}_2 && (18\text{-}7) \end{aligned}$$

If J is plotted against X (as done for V in Figure 18-4), the slope of the tangent line can be obtained by differentiating Equation 18-7 as

$$\frac{d\mathrm{J}}{dX_1} = \bar{J}_1 - \bar{J}_2 \tag{18-8}$$

If the ordinate values of points on the tangent line (see Figure 18-4) are represented by y, then the equation for this line can be written as

$$y = (\bar{J}_1 - \bar{J}_2)X_1 + \text{constant} \tag{18-9}$$

since the slope is given by Equation 18-8. The constant can be evaluated as follows. The tangent line passes through the point $[X_1, \mathrm{J} \text{ at } X_1]$ of the curve for J. At this point y, being equal to J, can be expressed as (Equation 18-7)

$$y = X_1\bar{J}_1 + (1 - X_1)\bar{J}_2 \tag{18-10}$$

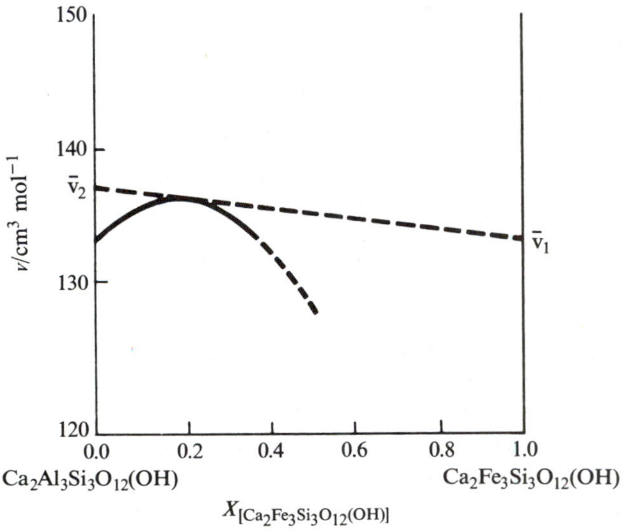

Figure 18-4. Molar volume as a function of mole fraction in a solid solution of $Ca_2Al_3Si_3O_{12}(OH)$ and $Ca_2Fe_3Si_3O_{12}(OH)$. The straight dashed line illustrates the determination of $\bar{v}_1$ and $\bar{v}_2$ from the tangent line. From R. Kern and A. Weisbrod, *Thermodynamics for Geologists*, Freeman, Cooper, San Francisco, 1967, p. 168.

If we equate the values of y in Equations 18-9 and 18-10, we find that the constant equals $\bar{J}_2$. Thus the equation for the tangent line becomes

$$y = (\bar{J}_1 - \bar{J}_2)X_1 + \bar{J}_2 \tag{18-11}$$

It follows from Equation 18-11 that the intercept of the tangent line on the left ordinate, corresponding to $X_1 = 0$, is $\bar{J}_2$, whereas the intercept on the right ordinate, corresponding to $X_1 = 1$, is $\bar{J}_1$.

A graph for the molar volumes of the solid solution of $Ca_2Al_3Si_3O_{12}(OH)$ and $Ca_2Fe_3Si_3O_{12}(OH)$ is shown in Figure 18-4 [1].

Partial Molar Enthalpies

Absolute values of partial molar enthalpies cannot be determined, just as absolute values of enthalpies cannot be determined. Thus, it is necessary to choose some state as a reference and to express the partial molar enthalpy relative to that reference state. The most convenient choice for the reference state usually is the infinitely dilute solution. Without committing ourselves to this choice exclusively, we will nevertheless use it in most of our problems.

The relative values of partial molar enthalpies are used so frequently that it has become customary to use the special symbol $\bar{L}_i$ to represent them. Thus $\bar{L}_i$, *the relative partial molar enthalpy*, is defined by the equation

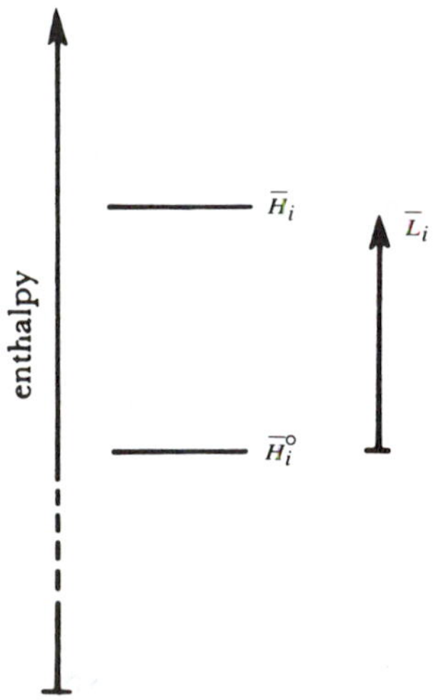

Figure 18-5. Relative partial molar enthalpy.

$$\bar{L}_i = \bar{H}_i - \bar{H}_i^\circ \tag{18-12}$$

in which $\bar{H}_i^\circ$ is the partial molar enthalpy of component i in the standard state, that is, the infinitely dilute solution. For the solvent, $\bar{H}_1^\circ$ equals $H_1^\bullet$, the molar enthalpy of the pure solvent. For the solute, $\bar{H}_2^\circ$ generally does not equal $H_2^\bullet$, the molar enthalpy of the pure solute.

Relative partial molar enthalpies also can be visualized in terms of a diagram, such as Figure 18-5. Although the absolute position of $\bar{H}_i$ or $\bar{H}_i^\circ$ on the enthalpy scale cannot be specified, the difference between them can be determined.

Experimental data from which relative partial molar enthalpies are calculated consist of enthalpy changes for mixing processes, commonly those that give *integral heats of solution*. An example of an integral heat of solution is the enthalpy change for the process of dissolving one mole of NaCl in 1000 g (55.51 moles) of pure H_2O to give a 1-molal solution, as shown by Equation 18-13:

$$\text{NaCl(s)} + 55.51\ \text{H}_2\text{O(l)} = \text{solution of 1 NaCl and 55.51 H}_2\text{O} \tag{18-13}$$

For such a process

$$\Delta H = H_{\text{final}} - H_{\text{initial}} \tag{18-14}$$

For any extensive thermodynamic property of a solution (Equation 12-19)

$$J = n_1\bar{J}_1 + n_2\bar{J}_2$$

Thus

$$\Delta H = H_{\text{final}} - H_{\text{initial}} = n_1\bar{H}_1 + n_2\bar{H}_2 - n_1H_1^\bullet - n_2H_2^\bullet \tag{18-15}$$

in which $H_1^\bullet$ and $H_2^\bullet$ are the molar enthalpies of pure solvent and pure solute, respectively. Since $\bar{H}_1^\circ = H_1^\bullet$

$$\begin{aligned}\Delta H &= n_1(\bar{H}_1 - \bar{H}_1^\circ) + n_2(\bar{H}_2 - H_2^\bullet) \\ &= n_1\bar{L}_1 + n_2(\bar{H}_2 - H_2^\bullet)\end{aligned} \tag{18-16}$$

If we add $(\bar{H}_2^\circ - \bar{H}_2^\circ)$ to the preceding equation we obtain

$$\begin{aligned}\Delta H &= n_1\bar{L}_1 + n_2\,[\bar{H}_2 - \bar{H}_2^\circ - (H_2^\bullet - \bar{H}_2^\circ)] \\ &= n_1\bar{L}_1 + n_2(\bar{H}_2 - \bar{H}_2^\circ) - n_2(H_2^\bullet - \bar{H}_2^\circ)\end{aligned}$$

or

$$\Delta H = n_1\bar{L}_1 + n_2\bar{L}_2 - n_2L_2^\bullet \tag{18-17}$$

Differentiation of Equation 18-17 with respect to n_1 at constant n_2 yields

$$\left(\frac{\partial\,\Delta H}{\partial n_1}\right)_{n_2} = \bar{L}_1 \tag{18-18}$$

If ΔH is plotted against n_1 at constant n_2, a graphical differentiation by the chord-area method will yield $\bar{L}_1$ as a function of composition.

Differentiation of Equation 18-17 with respect to n_2 at constant n_1 yields

$$\left(\frac{\partial\,\Delta H}{\partial n_2}\right)_{n_1} = \bar{L}_2 - L_2^\bullet \tag{18-19}$$

To obtain $\bar{L}_2$ from the experimental data, it is necessary to have a value of $L_2^\bullet$, the relative molar enthalpy of the pure solute. Since

$$\lim_{n_2\to 0} \bar{L}_2 = 0$$

we can see that

$$\lim_{n_2\to 0}\left(\frac{\partial\,\Delta H}{\partial n_2}\right)_{n_1} = -L_2^\bullet \tag{18-20}$$

From the limiting value of the slope of a ΔH versus n_2 plot, it is possible to calculate $L_2^\bullet$. It follows that $\bar{L}_2$ at any concentration can be evaluated from

$$\bar{L}_2 = \left(\frac{\partial\,\Delta H}{\partial n_2}\right)_{n_1} - \lim_{n_2\to 0}\left(\frac{\partial\,\Delta \mathrm{H}}{\partial n_2}\right)_{n_1} \tag{18-21}$$

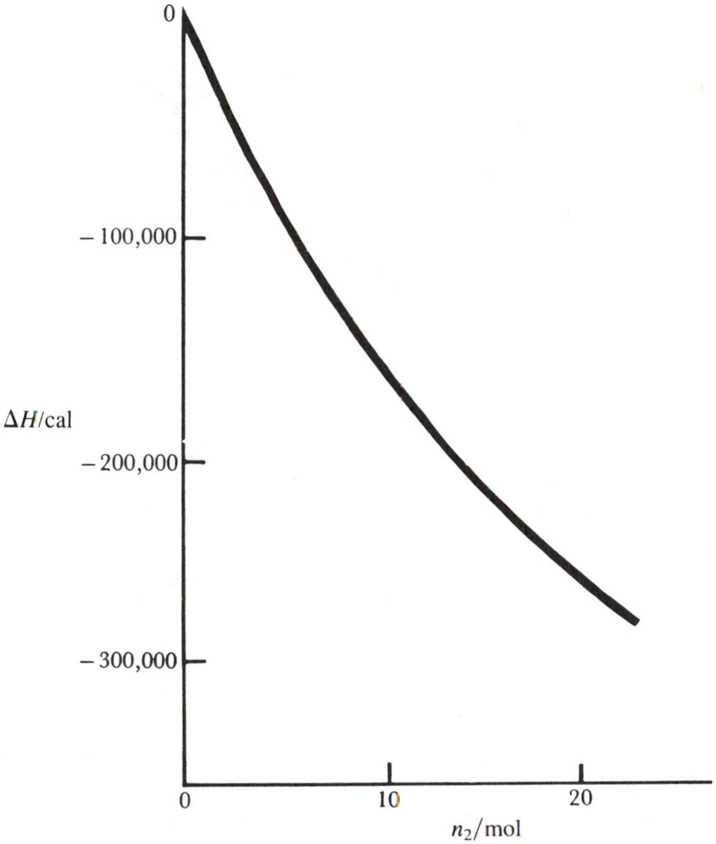

Figure 18-6. Heat exchanged upon the dissolution of n_2 moles of gaseous HCl in 1000 g of water. Based on data from G. N. Lewis and M. Randall, *Thermodynamics*, McGraw-Hill, New York, 1923, p. 96.

Lewis and Randall [2] have recalculated some of Thomsen's data on the heat absorbed when n_2 moles of gaseous HCl is dissolved in 1000 g of H_2O. A plot of ΔH as a function of n_2 at constant n_1 is shown in Figure 18-6.

To find $\bar{L}_2$ it is necessary to calculate $(\partial\, \Delta H/\partial n_2)_{n_1}$ at various molalities of HCl. These slopes have been evaluated by the chord-area method and are assembled in Figure 18-7. The extrapolation required by Equation 18-21 also has been carried out, and the value obtained for $-L_2^\circ$ is $-17{,}300$ cal mol^{-1}. Hence the relative partial molar enthalpies of HCl in aqueous solutions can be expressed by the equation

$$\bar{L}_2 = \left(\frac{\partial\, \Delta H}{\partial n_2}\right)_{n_1} + 17{,}300 \qquad (18\text{-}22)$$

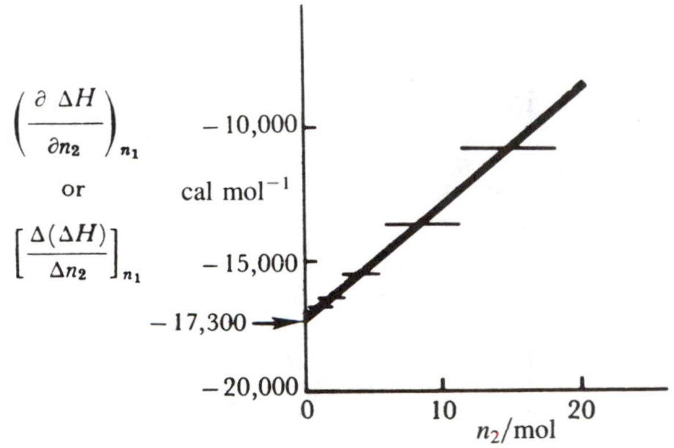

Figure 18-7. Chord-area plot of the slopes of heat of solution curve in Figure 18-6.

Thus for a 10-molal solution, the slope can be read from Figure 18-7 and $\bar{L}_2$ can be calculated:

$$\bar{L}_2 = -13{,}000 + 17{,}300 = 4300 \text{ cal mol}^{-1}$$

In this particular problem it is simple to put Equation 18-22 into a more explicit analytical form since the derivative $(\partial\ \Delta H/\partial n_2)_{n_1}$ in Figure 18-7 apparently is linear. This linear graph can be represented analytically by the equation

$$\left(\frac{\partial\ \Delta H}{\partial n_2}\right)_{n_1} = -17{,}300 + 430 n_2 \qquad (18\text{-}23)$$

Hence, for 1000 g of solvent, Equation 18-22 can be reduced to

$$\bar{L}_2 = 430 n_2 (\text{cal mol}^{-1})$$

or

$$\bar{L}_2 = 430 m_2 (\text{cal mol}^{-1}) \qquad (18\text{-}24)$$

in which m_2 is the molality of the solute. Although Thomsen's data provide a basis for illustrating this method of calculation, they imply a relationship between ΔH and m_2 at low molalities that is probably not an accurate description of the system. A better functional relationship, based on more modern data, will be discussed in Chapter 20.

18-2 PARTIAL MOLAR QUANTITIES OF ONE COMPONENT FROM THOSE OF ANOTHER COMPONENT BY NUMERICAL INTEGRATION

Rearrangement of Equation 12-22 leads to

$$d\bar{J}_1 = -\frac{n_2}{n_1}\,d\bar{J}_2 \tag{18-25}$$

This equation can be integrated from the infinitely dilute solution ($n_2 = 0$) to any finite concentration to give

$$\int_{\bar{J}_1^\circ}^{\bar{J}_1} d\bar{J}_1 = \int_{\bar{J}_2^\circ}^{\bar{J}_2} -\frac{n_2}{n_1}\,d\bar{J}_2$$

or

$$\bar{J}_1 - \bar{J}_1^\circ = -\int_{\bar{J}_2^\circ}^{\bar{J}_2} \frac{n_2}{n_1}\,d\bar{J}_2 \tag{18-26}$$

in which $\bar{J}_1^\circ$ represents the partial molar quantity of the solvent at infinite dilution of the solute, that is, pure solvent in its standard state. (For example, for water at room temperature $\bar{V}_1^\circ = V^\bullet = 18$ cm^3 mol^{-1}.)

Partial Molar Volumes

When $\bar{V}_2$ is available as a function of the composition (that is, as a function of n_2 at fixed n_1), it is possible to calculate $\bar{V}_1$ by graphical integration of Equation 18-26, as applied to volumes:

$$\bar{V}_1 - \bar{V}_1^\circ = \bar{V}_1 - V_1^\bullet = -\int_{\bar{V}_2^\circ}^{\bar{V}_2} \frac{n_2}{n_1}\,d\bar{V}_2 \tag{18-27}$$

Values of n_2/n_1 are plotted against $\bar{V}_2$, as in Figure 18-8, and a numerical integration is carried out from tabulated values of n_2/n_1 and $\bar{V}_2$. If the intervals, $\Delta\bar{V}_2$, are sufficiently small

$$\sum^{\cdots}\left(\frac{n_2}{n_1}\right)\Delta\bar{V}_2 \cong \int_{\bar{V}_2^\circ}^{\bar{V}_2} \frac{n_2}{n_1}\,d\bar{V}_2 \tag{18-28}$$

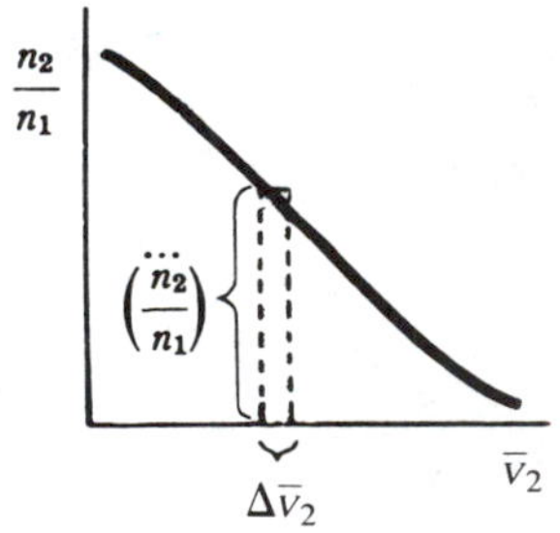

Figure 18-8. Graph of data for numerical integration to determine $\bar{V}_1$.

in which $(\dddot{n_2/n_1})$ represents the average value of this ratio in the interval $\Delta\bar{V}_2$. Since $V_1^{\bullet}$ is the molar volume of pure solvent, it can be calculated from the known molecular weight and density of the pure solvent.

Partial Molar Enthalpies

If Equation 18-26 is applied to relative partial molar enthalpies

$$\bar{L}_1 - \bar{L}_1^{\circ} = \bar{L}_1 = -\int_{\bar{L}_2^{\circ}}^{\bar{L}_2} \frac{n_2}{n_1}\, d\bar{L}_2 \qquad (18\text{-}29)$$

By the same graphical procedure described for partial molar volumes, the partial molar enthalpies of solvent can be evaluated from the values for the solute.

18-3 ANALYTIC METHODS

When the value of an intensive property, J, can be expressed as an algebraic function of the composition, the partial molar quantities can be determined analytically.

Partial Molar Volumes

In the case of sodium chloride in water at 25°C and 1 atm, for example, V (in cubic centimeters) for 1000 g of water can be expressed in terms of the following series in the molality, m_2:

$$V = 1001.38 + 16.6253m_2 + 1.7738m_2^{3/2} + 0.1194m_2^2 \qquad (18\text{-}30)$$

or,

$$V = 1001.38 + 16.6253n_2 + 1.7738n_2^{3/2} + 0.1194n_2^2 \qquad (18\text{-}31)$$

The value of $\bar{V}_2$ can be obtained by differentiation, since the quantity of solvent is fixed:

$$\bar{V}_2 = \left(\frac{\partial V}{\partial n_2}\right)_{n_1}$$

$$= 16.6253 + 2.6607m_2^{1/2} + 0.2388m_2 \qquad (18\text{-}32)$$

$$= 16.6253 + 2.6607n_2^{1/2} + 0.2388n_2 \text{ (for 1000 g of water)} \qquad (18\text{-}33)$$

The partial molar volume of the solvent, $\bar{V}_1$, can be obtained by the integration illustrated in Equation 18-27. To evaluate $d\bar{V}_2$ we merely need to differentiate Equation 18-33:

$$d\bar{V}_2 = (1.3304n_2^{-1/2} + 0.2388)\, dn_2 \qquad (18\text{-}34)$$

Substituting from Equation 18-34 into Equation 18-27 we have

$$\bar{V}_1 - V_1^{\bullet} = -\int_{\bar{V}_2^{\circ}}^{\bar{V}_2} \frac{n_2}{n_1}\, d\bar{V}_2 = -\int_0^{n_2} \frac{n_2}{n_1} (1.3304n_2^{-1/2} + 0.2388)\, dn_2 \qquad (18\text{-}35)$$

Since $n_1 = 1000/18.02 = 55.51$ and since $V_1^{\bullet} = 18.08 \text{ cm}^3 \text{ mol}^{-1}$

$$\bar{V}_1 = 18.08 - \frac{1}{n_1}\int_0^{n_2} (1.3304n_2^{1/2} + 0.2388n_2)\, dn_2$$

$$= 18.08 - \frac{0.8869}{55.51} n_2^{3/2} - \frac{0.1194}{55.51} n_2^2$$

$$= 18.08 - 0.015977n_2^{3/2} - 0.002151n_2^2 \text{ (for 1000 g of solvent)} \qquad (18\text{-}36)$$

$$= 18.08 - 0.015977m_2^{3/2} - 0.002151m_2^2 \qquad (18\text{-}37)$$

Partial Molar Enthalpies

Let us consider the case of an imaginary solute, B, in which the heat accompanying the dissolution of n_2 moles of B in 1000 g of water can be expressed by the relationship

$$\Delta H = 40m_2 + 30m_2^2 \qquad (18\text{-}38)$$

To calculate $\bar{L}_2$ we find the limiting value of the partial derivative of ΔH, as suggested by Equation 18-21:

$$\left(\frac{\partial\ \Delta H}{\partial n_2}\right)_{n_1=55.51} = \left(\frac{\partial\ \Delta H}{\partial m_2}\right) = 40 + 60m_2$$

and

$$\lim_{m_2\to 0}\left(\frac{\partial\ \Delta H}{\partial m_2}\right) = 40$$

Hence, by substitution into Equation 18-21 we obtain

$$\bar{L}_2 = 40 + 60m_2 - 40$$

or

$$\bar{L}_2 = 60m_2 \tag{18-39}$$

The relative partial molar enthalpy of the solvent is obtained by appropriate substitutions into Equation 18-29:

$$\bar{L}_1 = -\frac{1}{55.51}\int_0^{m_2} m_2(60dm_2) = -\frac{30}{55.51}m_2^2$$

$$= -0.54m_2^2 \tag{18-40}$$

Values of $\bar{L}_1$ for the HCl solutions for which $\bar{L}_2$ are given in Equation 18-24 also can be obtained by analytical integration of Equation 18-29:

$$\bar{L}_1 = -\frac{1}{55.51}\int_0^{m_2} m_2(430dm_2)$$

$$= -3.87m_2^2\ (\text{cal mol}^{-1}) \tag{18-41}$$

18-4 PARTIAL MOLAR QUANTITIES FROM APPARENT MOLAR QUANTITIES

The apparent molar quantity is a convenient function for calculating partial molar quantities; hence, we shall examine some of its properties.

Definition of Apparent Molar Quantity

As an example, we note in Figure 18-9 that the value of J for a solution increases with added solute, as indicated by the solid curve. In actual practice it

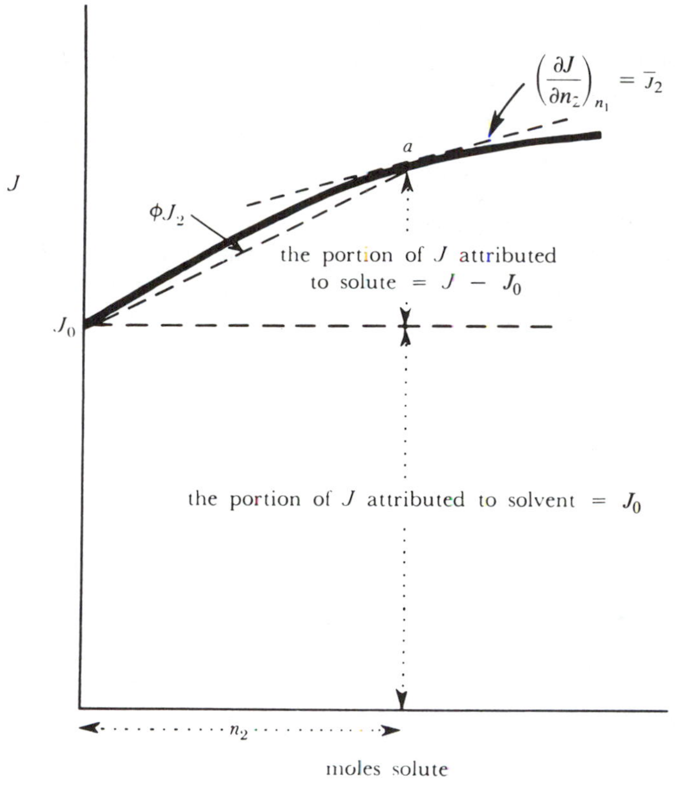

Figure 18-9. Significance of the apparent molar property, ϕJ_2.

is impossible to say which part of J for a solution should be attributed to the solvent and which to the solute. Nevertheless, J at the point a may be considered arbitrarily to consist of two portions. The first, J_0, that of the pure solvent, also can be attributed to the solvent in the solution at every finite concentration of solute. Then the difference $J - J_0$ can be attributed to the solute, as indicated in Figure 18-9. This *apparent* contribution to J by the solute, $J - J_0$, is converted to a molar value at point a if $J - J_0$ is divided by the moles of solute, n_2, present in the solution at this point. Hence ϕJ_2, the *apparent molar property* of the solute, is defined by the expression

$$\phi J_2 = \frac{\text{total } J - J \text{ of pure solvent}}{\text{moles of solute}}$$

$$= \frac{J - n_1 J_1^\bullet}{n_2} \tag{18-42}$$

The corresponding quantity, the apparent molar property of the solvent, ϕJ_1, generally is not considered.

It can be shown that, at infinite dilution, the apparent molar property is equal to the partial molar property. By substituting Equation 12-19 for J into Equation 18-42 we obtain

$$\phi J_2 = \frac{J - n_1 J_1^\bullet}{n_2} = \frac{n_1 \bar{J}_1 + n_2 \bar{J}_2 - n_1 J_1^\bullet}{n_2} \tag{18-43}$$

Since $\bar{J}_1$ at infinite dilution is equal to $J_1^\bullet$, the limit of ϕJ_2 at $n_2 = 0$ is an indeterminate form (0/0). The limit can be evaluated by applying L'Hopital's rule and differentiating the numerator and denominator of Equation 18-43 with respect to n_2 at constant n_1; that is,

$$\lim_{n_2 \to 0} \phi J_2 = \lim_{n_2 \to 0} \frac{\bar{J}_2}{1} = \bar{J}_2^\circ \tag{18-44}$$

It is also clear from inspection of Figure 18-9 that ϕJ_2 approaches $\bar{J}_2$ as n_2 approaches zero.

Relationship of ϕJ_2 to J

Equation 18-42 can be rearranged to give J explicitly:

$$J = n_2 \phi J_2 + n_1 J_1^\bullet \tag{18-45}$$

By differentiating Equation 18-45 with respect to n_2 we have

$$\bar{J}_2 = \left(\frac{\partial J}{\partial n_2}\right)_{n_1}$$

$$= \phi J_2 + n_2 \left(\frac{\partial \phi J_2}{\partial n_2}\right)_{n_1} \tag{18-46}$$

since both n_1 and $J_1^\bullet$ are independent of n_2. Similarly,

$$\bar{J}_1 = \left(\frac{\partial J}{\partial n_1}\right)_{n_2}$$

$$= n_2 \left(\frac{\partial \phi J_2}{\partial n_1}\right)_{n_2} + J_1^\bullet \tag{18-47}$$

The calculation of partial molar quantities from apparent molar quantities generally is used when ϕJ_2 is expressible as an analytic function such as an equation of the form

$$\phi J_2 = a + bm_2 + cm_2^2 \tag{18-48}$$

in which m_2 is the molality and a, b, and c are constants. Let us consider a series of solutions containing 1000 g of solvent, so that

$$n_2 = m_2$$

and

$$dn_2 = dm_2$$

It follows directly from Equation 18-46 that

$$\bar{J}_2 = m_2 \left(\frac{\partial \phi J_2}{\partial m_2} \right)_{n_1} + \phi J_2$$

$$= m_2[b + 2cm_2] + a + bm_2 + cm_2^2$$

Therefore

$$\bar{J}_2 = a + 2bm_2 + 3cm_2^2 \tag{18-49}$$

There are several possible procedures for calculating $\bar{J}_1$. We could use Equation 18-47, carrying out the required differentiation with the aid of the substitution

$$m_2 = \frac{\text{moles solute}}{\text{kilograms solvent}} = \frac{n_2}{n_1 M_1} \tag{18-50}$$

where M_1 is expressed in kilograms.

Probably a simpler method, however, is to use the general relationship (Equation 12-19)

$$J = n_1\bar{J}_1 + n_2\bar{J}_2$$

as a substitution into Equation 18-45 to obtain

$$n_1\bar{J}_1 + n_2\bar{J}_2 = n_2\phi J_2 + n_1 J_1^\bullet \tag{18-51}$$

Considering again solutions containing 1000 g of solvent, and using Equations 18-48 and 18-49 for ϕJ_2 and $\bar{J}_2$, respectively, we obtain

$$n_1\bar{J}_1 + m_2(a + 2bm_2 + 3cm_2^2) = m_2(a + bm_2 + cm_2^2) + n_1 J_1^\bullet$$

Hence

$$n_1(\bar{J}_1 - J_1^\bullet) = -bm_2^2 - 2cm_2^3 \tag{18-52}$$

Since

$$n_1 = \frac{1}{M_1}$$

it follows that

$$\bar{J}_1 = J_1^{\bullet} - bM_1m_2^2 - 2cM_1m_2^3 \tag{18-53}$$

Equations 18-49 and 18-53, although derived for solutions containing 1000 g of solvent, are applicable also to any quantity of solution containing a different amount of solvent, since they are expressed in terms of the molality, which is an intensive quantity.

Apparent Molar Enthalpies

Sometimes it is convenient, particularly when measuring the heat of dilution (addition of solvent to solution) rather than the heat of solution of solute in solvent, to analyze calorimetric data in terms of the function known as the *apparent molar enthalpy*, which is given by the equation

$$\phi H_2 = \frac{H - n_1H_1^{\bullet}}{n_2} \tag{18-54}$$

Rearranging Equation 18-54, we have

$$\begin{aligned} n_2\phi H_2 &= H - n_1H_1^{\bullet} \\ &= n_1\bar{H}_1 + n_2\bar{H}_2 - n_1H_1^{\bullet} \end{aligned}$$

Since

$$\lim_{m_2 \to 0} \phi H_2 = \phi H_2^{\circ} = \bar{H}_2^{\circ}$$

we can write

$$\begin{aligned} n_2\phi H_2 - n_2\phi H_2^{\circ} &= n_1\bar{H}_1 + n_2\bar{H}_2 - n_1H_1^{\bullet} - n_2\bar{H}_2^{\circ} \\ &= n_1\bar{L}_1 + n_2\bar{L}_2 \end{aligned} \tag{18-55}$$

Let us define the relative apparent molar enthalpy by the expression

$$\phi L_2 = \phi H_2 - \phi H_2^{\circ} \tag{18-56}$$

It follows that

$$n_2\phi L_2 = n_1\bar{L}_1 + n_2\bar{L}_2 \tag{18-57}$$

or

$$n_2\phi L_2 = L \tag{18-58}$$

The validity of Equation 18-58 can be seen from a graphical representation of the quantities involved. Although the absolute value of H° may not be known, it can be represented by an arbitrary point on a graph such as Figure 18-10. Since L is given by the difference between the enthalpy of the specified solution and the enthalpy at zero molality

$$\phi L_2 = \frac{L - L^{\circ}}{n_2} = \frac{L}{n_2} \tag{18-59}$$

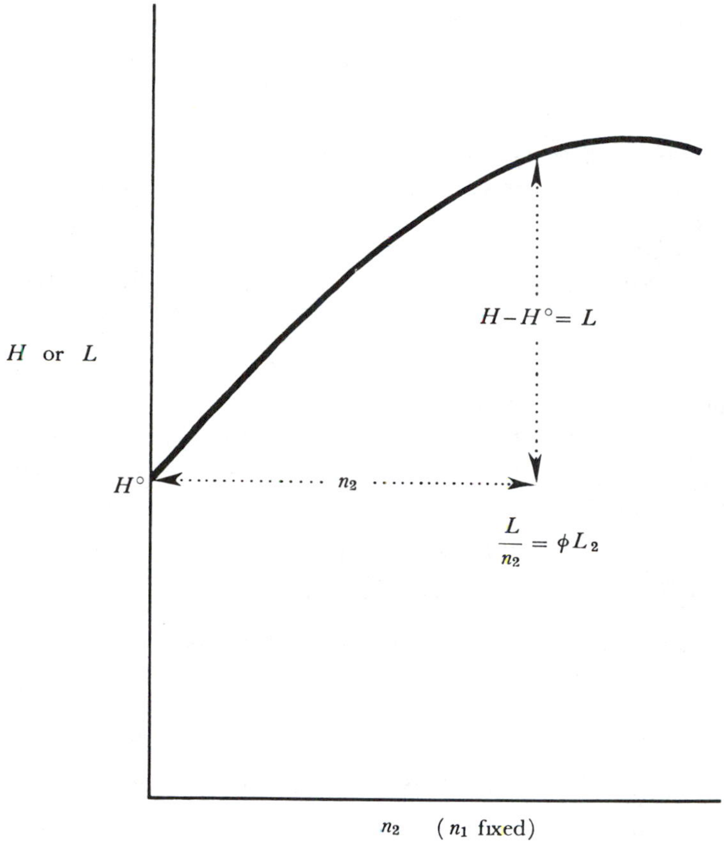

Figure 18-10. Relative apparent molar enthalpy.

Differentiation of Equation 18-58 with respect to n_2 gives

$$\bar{L}_2 = \left(\frac{\partial L}{\partial n_2}\right)_{n_1} = \frac{\partial}{\partial n_2}(n_2 \phi L_2) = \phi L_2 + n_2 \left(\frac{\partial \phi L_2}{\partial n_2}\right)_{n_1} \tag{18-60}$$

Substitution of this expression into Equation 18-57 gives

$$n_1 \bar{L}_1 = n_2 \phi L_2 - n_2 \bar{L}_2$$

$$= n_2 \phi L_2 - n_2 \left[\phi L_2 + n_2 \left(\frac{\partial \phi L_2}{\partial n_2}\right)_{n_1}\right]$$

and

$$\bar{L}_1 = -\frac{n_2^2}{n_1}\left(\frac{\partial \phi L_2}{\partial n_2}\right)_{n_1} \tag{18-61}$$

For Equations 18-60 and 18-61 to be useful, we need to consider how to obtain ϕL_2 from experimental data for heats of dilution. For the dilution process

$$\left[\begin{array}{c}\text{solution of } n_1 \text{ moles } H_2O \\ \text{and } n_2 \text{ moles HCl}\end{array}\right] + n_1' \text{ moles } H_2O = \left[\begin{array}{c}\text{solution of } n_1'' \text{ moles} \\ H_2O \text{ and } n_2 \text{ moles HCl}\end{array}\right] \tag{18-62}$$

in which

$$n_1'' = n_1 + n_1'$$

the *integral heat of dilution* is

$$\begin{aligned}\Delta H &= H_{\text{final}} - H_{\text{initial}} \\ &= n_1''\bar{H}_1'' + n_2\bar{H}_2'' - (n_1'H_1^\bullet + n_1\bar{H}_1 + n_2\bar{H}_2)\end{aligned} \tag{18-63}$$

By adding $n_1H_1^\bullet + n_2\bar{H}_2^\circ - n_1H_1^\bullet - n_2\bar{H}_2^\circ$ to Equation 18-63, we obtain

$$\begin{aligned}\Delta H &= [n_1''\bar{H}_1'' - (n_1 + n_1')H_1^\bullet + n_2\bar{H}_2'' - n_2\bar{H}_2^\circ] - [n_1\bar{H}_1 - n_1H_1^\bullet + n_2\bar{H}_2 - n_2\bar{H}_2^\circ] \\ &= L'' - L\end{aligned} \tag{18-64}$$

Substituting from Equation 18-58 into Equation 18-64 we have

$$\Delta H = n_2\phi L_2'' - n_2\phi L_2 \tag{18-65}$$

If the dilution process is carried out on a quantity of solution containing one mole of solute, then

$$\Delta H_{n_2=1} = \phi L_2'' - \phi L_2 \tag{18-66}$$

To obtain a value of ϕL_2 rather than a difference, we must repeat the experiment with successively greater quantities of solvent until the resulting solution is sufficiently dilute that we can use the limiting relationship

$$\lim_{m_2''\to 0} \phi L_2'' = 0 \tag{18-67}$$

From Equation 18-67 it follows that

$$\begin{aligned}\lim_{m_2''\to 0}(\Delta H_{n_2=1}) &= \lim_{m_2''\to 0}(\phi L_2'' - \phi L_2) \\ &= -\phi L_2\end{aligned} \tag{18-68}$$

As an example of the use of this method let us consider some data on the heats of dilution of aqueous solutions of hydrochloric acid [3]. The dilution experiments, and accompanying heats, can be summarized by the following equations:

1. solution of 1HCl and $3H_2O + 2H_2O$
   = solution of 1HCl and $5H_2O$, $\Delta H = -7196$ J mol^{-1}
2. solution of 1HCl and $3H_2O + 9H_2O$
   = solution of 1HCl and $12H_2O$, $\Delta H = -13{,}514$ J mol^{-1}
3. solution of 1HCl and $3H_2O + 22H_2O$
   = solution of 1HCl and $25H_2O$, $\Delta H = -15{,}690$ J mol^{-1}
4. solution of 1HCl and $25H_2O + 25H_2O$
   = solution of 1HCl and $50H_2O$, $\Delta H = -1033$ J mol^{-1}
5. solution of 1HCl and $25H_2O + 75H_2O$
   = solution of 1HCl and $100H_2O$, $\Delta H = -1619$ J mol^{-1}
6. solution of 1HCl and $25H_2O + 375H_2O$
   = solution of 1HCl and $400H_2O$, $\Delta H = -2297$ J mol^{-1}
7. solution of 1HCl and $400H_2O + 1200H_2O$
   = solution of 1HCl and $1600H_2O$, $\Delta H = -381$ J mol^{-1}

From these data it is possible to tabulate values for the heats of dilution of the solution with one mole of HCl and three moles of H_2O to solutions with various final values of n_1. For example, the ΔH of dilution of the following process:

8. solution of 1HCl and $3H_2O + 397H_2O$
   = solution of 1HCl and $400H_2O$, $\Delta H = -17{,}987$ J mol^{-1}

is obtained by adding Equations 3 and 6. The resulting values of ΔH are in Table 18-2.

The extrapolation of the heats of dilution to infinite dilution is illustrated in Figure 18-11. The square root of the molality, $m_2^{1/2}$, has been chosen as the abscissa in this case instead of the molality because the extrapolation is more convenient for solutions of electrolytes. The reasons for this choice will be made explicit in Chapter 20.

The value of ϕL_2 obtained from the extrapolation is +18,671 J mol^{-1} for the solution of one mole of HCl and three moles H_2O. From this value and those of ΔH in the last column of Table 18-2, $\phi L_2''$ for the other solutions can be obtained. The values calculated are listed in Table 18-3 together with the relative partial molar enthalpies, which can be derived from the calculated values by applying Equations 18-60 and 18-61.

18-5 CHANGES IN J FOR SOME PROCESSES INVOLVING SOLUTIONS

There are two kinds of processes involving solutions, other than chemical changes, that we need to consider. One is a transfer or differential process and the other is a mixing or an integral process.

Table 18-2 Enthalpies of Dilution of Hydrochloric Acid Solutions at 25°C

$$HCl \cdot 3H_2O + n_1H_2O(l) = HCl \cdot (3 + n_1)H_2O$$

n_1/mol	$\frac{n_1+3}{n_2}$	m_2''/(mol kg^{-1})	$[m_2''$/(mol kg^{-1})$]^{1/2}$	$\Delta H = \phi L_2'' - \phi L_2(HCl \cdot 3H_2O)$ /(J mol^{-1})
6397	6400	0.0087	0.093	−18,552
3197	3200	0.0173	0.132	−18,472
1597	1600	0.0347	0.186	−18,368
797	800	0.0694	0.263	−18,209
397	400	0.139	0.374	−17,987
197	200	0.278	0.527	−17,703
97	100	0.555	0.745	−17,309
47	50	1.110	1.054	−16,723
22	25	2.220	1.490	−15,690
17	20	2.775	1.660	−15,188
15	18	3.083	1.756	−14,895
12	15	3.700	1.924	−14,351
9	12	4.625	2.151	−13,514
7	10	5.550	2.356	−12,636
5	8	6.938	2.634	−11,381
3	6	9.250	3.041	−9,037
2	5	11.10	3.33	−7,196
1	4	13.88	3.72	−4,226
0	3	18.50	4.30	0

$$\lim_{m_2'' \to 0} \Delta H = -18{,}671 \text{ J mol}^{-1}$$

Differential Process

Consider the equation

$$\text{glycine(s)} = \text{glycine}(m_2 = 1,\text{aq}) \tag{18-69}$$

An infinitesimal transfer of glycine from the solid phase to the solution at constant temperature and pressure results in a corresponding change, dJ, in the thermodynamic property, J, of the system comprised of crystalline glycine and a 1-molal aqueous solution of glycine. The application of Equation 12-21 leads to the expression

$$dJ = J^{\bullet}_{2(s)}\, dn_s + \bar{J}_{2(m_2=1)}\, dn_2 \tag{18-70}$$

Since mass is conserved in the transfer

$$-dn_{2(s)} = dn_{2(m_2=1)} = dn$$

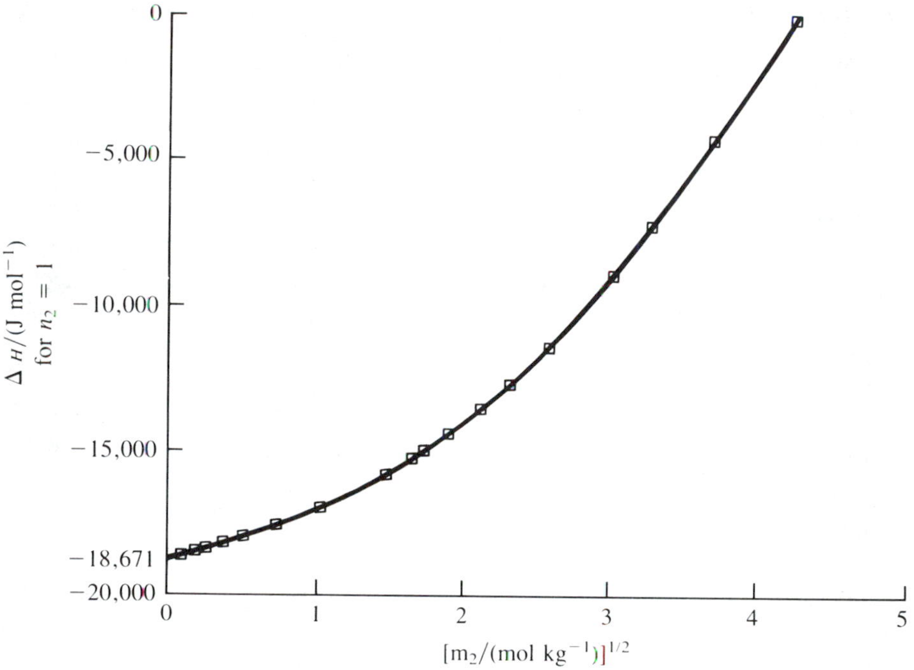

Figure 18-11. Extrapolation of data for heats of dilution of hydrochloric acid, with initial slope constrained to value required by Debye–Hückel limiting law (see Chapter 20).

and Equation 18-70 can be written

$$dJ = -J_{2(s)}\,dn + \bar{J}_{2(m_2=1)}\,dn$$
$$= [\bar{J}_{2(m_2=1)} - J^{\bullet}_{2(s)}]\,dn \qquad (18\text{-}71)$$

Equation 18-71 can be integrated to obtain the change in J for a finite change of state:

$$\int_{J_1}^{J_2} dJ = \int_{n_2=0}^{n_2=1} [\bar{J}_{2(m_2=1)} - J^{\bullet}_{2(s)}]\,dn \qquad (18\text{-}72)$$

or

$$\Delta J = J_2 - J_1 = \int_{n_2=0}^{n_2=1} [\bar{J}_{2(m_2=1)} - J^{\bullet}_{2(s)}]\,dn \qquad (18\text{-}73)$$

It is characteristic of a differential process that the transfer occurs without a change in the composition of any phase. If a finite change of state is to occur without a change of composition, the aqueous solution of Equation 18-69 must have a volume sufficiently large that the addition of one mole of solid glycine

Table 18-3 Relative Enthalpies of Aqueous Solutions of Hydrochloric Acid at 25°C

m''_2/(mol/kg^{-1})	$\phi L''_2$/(J mol^{-1})	$\bar{L}_2$/(J mol^{-1})	$\bar{L}_1$/(J mol^{-1})
0.0087	119	259	−0.0133
0.0173	199	364	−0.0373
0.0347	303	506	−0.103
0.0694	462	704	−0.284
0.139	684	986	−0.804
0.278	968	1,380	−2.261
0.555	1,362	1,985	−6.740
1.110	1,948	3,011	−22.07
2.220	2,981	4,988	−81.41
2.775	3,483	5,967	−125.3
3.083	3,776	6,578	−157.4
3.700	4,320	7,751	−229.1
4.625	5,157	9,550	−364.1
5.550	6,035	11,390	−532.7
6.938	7,290	14,200	−848.5
9.250	9,634	18,900	−1534
11.10	11,475	22,600	−2210
13.88	14,445	27,880	−3395
18.50	18,671	35,820	−5695

does not change the composition, so that $\bar{J}_{2(m_2=1)}$, like $J^\bullet_{2(s)}$, is a constant. Then the integral on the right side of Equation 18-73 can be evaluated as

$$\Delta J = J_2 - J_1 = \bar{J}_{2(m_2=1)} - J^\bullet_{2(s)} \tag{18-74}$$

For the case in which J represents the volume of the system, we can use the data (Table 18-4) of Gucker, Ford, and Moser on the partial molar volumes in aqueous glycine solutions [4]. Then we calculate ΔV for the change of state in Equation 18-69 as

$$\begin{aligned} \Delta V &= \bar{V}_{2(m_2=1)} - V^\bullet_{2(s)} \\ &= 44.88\ \text{cm}^3\ \text{mol}^{-1} - 46.71\ \text{cm}^3\ \text{mol}^{-1} \\ &= -1.83\ \text{cm}^3\ \text{mol}^{-1} \end{aligned} \tag{18-75}$$

Thus the volume change for Equation 18-69 is the sum of the volume change for the disappearance of one mole of solid glycine, $-V^\bullet_{2(s)}$, and the volume change for the addition of one mole of solid glycine to a large volume of solution with $m_2 = 1$, $\bar{V}_2$.

Table 18-4 Partial Molar Volumes in Aqueous Solutions of Glycine

$m/(\text{mol kg}^{-1})$	$\bar{V}_2/(\text{cm}^3\ \text{mol}^{-1})$	$\bar{V}_1/(\text{cm}^3\ \text{mol}^{-1})$
0	43.20	18.07
1	44.88	18.05
Pure solid	46.71	

A process analogous to that of Equation 18-69 is

$$H_2O(l) = H_2O(\text{solution, } m_2 \text{ of glycine is } 1) \tag{18-76}$$

for which

$$\Delta J = \bar{J}_{1(m_2=1)} - J_1^{\bullet}$$

or, for the volume,

$$\begin{aligned}\Delta V &= \bar{V}_{1(m_2=1)} - V_1^{\bullet}\\ &= 18.05\ \text{cm}^3\ \text{mol}^{-1} - 18.07\ \text{cm}^3\ \text{mol}^{-1}\\ &= -0.02\ \text{cm}^3\ \text{mol}^{-1}\end{aligned} \tag{18-77}$$

Integral Process

More typical of common experience is a mixing process such as

$$\text{glycine(s)} + 55.51H_2O(l) = \text{solution}\begin{bmatrix}55.51H_2O;\\ 1\ \text{glycine};\\ m_2 = 1\end{bmatrix} \tag{18-78}$$

for which the expression for ΔJ is

$$\Delta J = J_{\text{final}} - J_{\text{initial}}$$

But the expression for J_{final} and J_{initial} includes terms for all the components of the initial and final phases, since the composition of the phases changes during the mixing process. Thus

$$\Delta J = n_1\bar{J}_1 + n_2\bar{J}_2 - n_1J_1^{\bullet} - n_2J_{2(s)}^{\bullet}$$

and for Equation 18-78

$$\Delta J = 55.51\ (\bar{J}_1 - J_1^{\bullet}) + \bar{J}_2 - J_{2(s)}^{\bullet} \tag{18-79}$$

EXERCISES

1. Show that $\partial J/\partial X_1$ is not identical with $\partial J/\partial n_1$ if n_2 is held constant in both cases.

2. a. If V is the volume of a two-component solution containing 1 kg of solvent show that

$$\bar{V}_2 = \frac{M_2 - V(\partial \rho/\partial m_2)_{T,P}}{\rho}$$

in which M_2 is the molar mass of the solute, ρ is the density of the solution, and m_2 is the molality of solute.

b. Show that

$$\bar{V}_1 = \frac{M_1\left[1 + m_2 V\left(\frac{\partial \rho}{\partial m_2}\right)_{T,P}\right]}{\rho}$$

3. If c is the concentration in moles of solute per dm^3 of solution, ρ is the density of the solution, and ρ_0 is the density of pure solvent, show that

a. $\phi V_2 = \frac{1}{c} - \frac{1}{\rho_0}\left[\frac{\rho}{c} - M_2\right]$

b. $\bar{V}_2 = \frac{M_2 - (\partial \rho/\partial c)_{T,P}}{\rho - c(\partial \rho/\partial c)_{T,P}}$

c. $\bar{V}_1 = \frac{M_1}{\rho - c(\partial \rho/\partial c)_{T,P}}$

d. For a particular two-component solution, the density can be expressed as a linear function of the molar concentration of the solute. Prove that (i) $\bar{V}_1 = V_1^\bullet$ and (ii) $\bar{V}_2$ = constant.

4. a. Verify the calculations in Table 18-1 for the solutions that are 20 and 25 weight percent ethanol.

b. Plot the volume per mole of water (Column 5) versus n_2/n_1.

c. Make a careful graph of $\Delta V/\Delta n_2$ versus n_2/n_1. Draw a smooth curve through the chord plot.

d. Make a careful graph of $\Delta V/\Delta n_1$ versus n_1/n_2. Draw a smooth curve through the chord plot.

e. Determine the partial molar volumes of water and of ethanol, respectively, for a 45% ethanol solution.

f. As a method of checking the calculations in Table 18-1, calculate the volume of 500 g of a 45% ethanol solution from the partial molar volumes, and compare the value obtained with that which can be calculated directly by using the density.

g. Make a graph of $\bar{v}_1$ versus n_1/n_2. Carry out an appropriate integration to find the difference between $\bar{v}_2$ in a 65% solution and $\bar{v}_2$ in a 25% solution.

h. Find $\bar{v}_2$ for a 25% solution from the chord-area graph in (c), and add it to the difference calculated in (g). Compare the sum with the value of $\bar{v}_2$ in a 65% solution, which can be read from the graph in (c).

5. Using the data in Exercise 4 and data for the density of pure H_2O (0.99708 g cm^{-3} at 25°C) and of pure ethanol (0.78506 g cm^{-3} at 25°C), compute the volume changes for the following processes:

a. $C_2H_6O(l) = C_2H_6O$(45% aqueous ethanol)

b. $H_2O(l) = H_2O$(45% aqueous ethanol)

6. Compute $v = V/(n_1 + n_2)$ for each of the ethanol–water mixtures in Table 18-1 and plot the value of v as a function of X_{H_2O}. Use the tangent method to calculate the values of $\bar{v}_1$ and $\bar{v}_2$ in a 45% ethanol solution and compare them with the results in Exercise 4(e).

7. a. Compute the volume changes for the process

$$4.884C_2H_6O(l) + 15.264H_2O(l) = \begin{bmatrix} 4.884C_2H_6O & \text{(in large quantity of} \\ & \text{aqueous solution of} \\ 15.264H_2O & \text{45\% ethanol)} \end{bmatrix}$$

b. Compute the volume change for the process

$$4.884C_2H_6O(l) + 15.264H_2O(l) = 500 \text{ g of aqueous solution of 45\% ethanol}$$

c. Compare the answers in (a) and (b). They should be the same, within computational error. Why?

d. Write word statements that emphasize the differences in meaning between the equations in Exercises 5(a), 5(b), 7(a), and 7(b).

8. The specific heats and apparent molar heat capacities of aqueous solutions of glycolamide at 25°C [F. T. Gucker, Jr., W. L. Ford, and C. E. Moser, *J. Phys. Chem.* **43**, 153 (1939); *J. Phys. Chem.* **45**, 309 (1941)] are listed in Table 18-5.

Table 18-5

m_2/(mol kg^{-1})	C_p/(cal g^{-1})	$\phi C_{p(2)}$/(cal mol^{-1})
0.000	(1.00000)	
0.2014	0.99223	35.9
0.4107	0.98444	36.0
0.7905	0.97109	36.33
1.2890	0.95467	
1.7632	0.94048	36.84
2.6537	0.91666	37.41
4.3696	0.87899	38.29
4.3697	0.87900	38.29
6.1124	0.84891	39.01

a. Calculate the value of $\phi C_{p(2)}$ of a 1.2890-molal solution.

b. Plot $\phi C_{p(2)}$ as a function of the molality.

c. Using the method of least squares, derive an equation for $\phi C_{p(2)}$ as a function of m_2.

d. Derive an equation for the partial molar heat capacity of glycolamide based on the least-squares equation for $\phi C_{p(2)}$.

e. Determine a few numerical values of $\bar{C}_{p(2)}$ and plot them on the graph on which you have $\phi C_{p(2)}$.

f. Derive the equation for the partial molar heat capacity of water in aqueous solutions of glycolamide.

9. According to D. A. MacInnes and M. O. Dayhoff [*J. Am. Chem. Soc.* **75**, 5219 (1953)] the apparent molar volume of KI in CH_3OH can be expressed by the equation

$$\phi V_2 = 21.45 + 11.5\sqrt{m_2}$$

when V is expressed in cm^3. The density of pure methanol is 0.7865 g cm^{-3}. Compute ΔV for the process

$$CH_3OH(l) = CH_3OH(soln, m_2 = 1)$$

10. For iodine in methanol solution, the apparent molar volume of the solute is essentially constant, irrespective of concentration, and is equal to 62.3 cm^3 mol^{-1} at 25°C [MacInnes and Dayhoff, *J. Am. Chem. Soc.* **75**, 5219 (1953)]. Solid I_2 has a density of 4.93 g cm^{-3}, and pure methanol has a density of 0.7865 g cm^{-3} at 25°C.

a. Compute ΔV for $I_2(s) = I_2(m_2 = 1)$

b. Compute ΔV for $I_2(s) = I_2(m_2 = 0)$

c. Compute ΔV for $I_2(m_2 = 1) = I_2(m_2 = 0)$

d. Compute ΔV for $CH_3OH(l) = CH_3OH(soln, m_2 = 1)$

e. Compute ΔV for $I_2(s) + 31.2\ CH_3OH(l) = soln(m_2 = 1)$.

11. Gucker, Planck, and Pickard [*J. Am. Chem. Soc.* **61**, 459 (1939)] have found that the equation

$$\phi L_2 = 128.9 m_2$$

expresses the relative apparent molar enthalpy of aqueous solutions of sucrose at 20°C when ϕL_2 is in cal mol^{-1}. Derive expressions for $\bar{L}_2$ and $\bar{L}_1$, respectively, as a function of the molality.

12. If the heat of solution of HCl(g) in 25 moles of H_2O is $-71{,}756$ J mol^{-1}, calculate L_2(g) for HCl (see pp. 386–391).

13. a. Calculate the differential heat of solution of 1 mole of HCl(g) in $HCl \cdot 25H_2O$, that is, in a solution of 1 mole of HCl and 25 moles of H_2O.

b. Calculate the differential heat of solution of 1 mole of H_2O(l) in $HCl \cdot 25H_2O$.

c. From (a) and (b) calculate the integral heat of solution of HCl(g) in 25 moles of H_2O(l).

14. Prove that $\partial \bar{L}_1/\partial T = \bar{C}_{p(1)} - C^{\bullet}_{p(1)}$.
15. With the aid of the data in Exercise 11 and the following equation for the relative apparent molar enthalpy at 30°C (in units of cal mol^{-1}),

$$\phi L_2 = 140.2 m_2$$

derive expressions for $\bar{C}_{p(2)}$ and $\bar{C}_{p(1)}$ for aqueous sucrose solutions at 25°C. $C^{\circ}_{p(2)}$ is 151.50 cal mol^{-1} K^{-1}.
16. The heat absorbed when m_2 moles of NaCl is dissolved in 1000 g of H_2O is given by the expression (for ΔH in cal mol^{-1})

$$\Delta H = 923 m_2 + 476.1 m_2^{3/2} - 726.1 m_2^2 + 243.5 m_2^{5/2}$$

   a. Derive an expression for $\bar{L}_2$ and compute values of the relative partial molar enthalpy of NaCl in 0.01- and 0.1-molal solutions.
   b. Derive an expression for $\bar{L}_1$ and compute its value for 0.01- and 0.1-molal solutions.
   c. If the expression in this exercise for ΔH per 1000 g of H_2O is divided by m_2, the number of moles of solute in 1000 g of H_2O, we obtain the heat absorbed when one mole of NaCl is dissolved in enough water to give a molality equal to m_2. Show that this equation is

   $$\Delta H \text{ (for 1 mole of NaCl)} = 923 + 476.1 m_2^{1/2} - 726.1 m_2 + 243.5 m_2^{3/2}$$

   d. Calculate $(\partial\ \Delta H/\partial n_1)_{n_2}$ from the preceding expression. Derive an equation for $\bar{L}_1$.
17. The following information for partial molar enthalpies [Gucker, Pickard, and Ford, *J. Am. Chem. Soc.* **62**, 2698 (1940)] is for glycine and its aqueous solutions at 25°C:

m_2/(mol kg^{-1})	$\bar{L}_1$/(cal mol^{-1})	$\bar{L}_2$/(cal mol^{-1})
1.000	1.537	−165.5
3.33(saturated)		−354
Glycine(pure)		−3765

   a. The heat of solution of an infinitesimal quantity of pure solid glycine in a saturated aqueous solution is 3411 cal mol^{-1}. Show that $L^{\bullet}_{2(s)}$ is −3765 cal mol^{-1}.
   b. Calculate ΔH per mole for the addition of an infinitesimal quantity of solid glycine to a 1-molal aqueous solution.
   c. Calculate ΔH per mole for the addition of an infinitesimal quantity of solid glycine to an infinitely dilute aqueous solution.
   d. Calculate ΔH for the addition of one mole of solid glycine to 1000 g of pure water to form a 1-molal aqueous solution.
18. At 50°C the apparent molal volume of urea, $CO(NH_2)_2$, in water solutions is given by the following equation up to 4-molal concentration (with ϕV_2 in cm^3):

$$\phi V_2 = 45.60 + 0.07m_2$$

The density of solid urea is 1.335 g cm^{-3} and of pure liquid water is 0.988 g cm^{-3}. Calculate ΔV for each of the following processes:

a. urea(s) = urea(aq soln, m_2 = 1)

b. $CO(NH_2)_2$(s) + 55.51 H_2O(l) = solution(m_2 = 1)

19. J. C. G. Calado and E. J. S. Gomes de Azevedo [*J. Chem. Soc., Faraday Trans.* **1, 79**, 2657 (1983)] have measured the molar volume of liquid mixtures of ethane and ethene at 161.39 K. Their values are

X(ethene)	v/cm^3 mol^{-1}	Δv/cm^3 mol^{-1}
0.0000	52.548	0.000
0.2444	51.684	0.131
0.3253	51.365	0.141
0.4214	50.988	0.156
0.5357	50.519	0.153
0.6407	50.086	0.148
0.7115	49.793	0.142
0.7898	49.445	0.114
1.0000	48.475	0.000

Calculate the partial molar volumes of each component at the given mole fractions.

REFERENCES

1. R. Kern and A. Weisbrod, *Thermodynamics for Geologists*, Freeman, Cooper, San Francisco, 1967, p. 168.
2. G. N. Lewis and M. Randall, *Thermodynamics*, McGraw-Hill, New York, 1923, p. 96.
3. F. D. Rossini, *J. Res. Natl. Bur. Std.* **9**, 679 (1932).
4. F. T. Gucker, W. L. Ford, and C. E. Moser, *J. Phys. Chem.* **43**, 153 (1939).

CHAPTER 19

Determination of Nonelectrolyte Activities

Having established the definitions and conventions for the activity function in Chapter 17, we are in a position to understand the experimental methods that have been used to determine numerical values of this quantity.

19-1 MEASUREMENTS OF VAPOR PRESSURE

If the vapor pressure of a substance in solution is sufficiently great to be determined experimentally, the activity can be calculated directly. Since the standard state differs for the solvent and solute, each must be considered separately.

Solvent

The activity of the solvent is related directly to the vapor pressure when the vapor is an ideal gas. Since the pure solvent is the reference state, the chemical potential, μ_1, of the solvent in any solution is given by the expression (Equation 17-1),

$$\mu_1 = \mu_1^{\bullet} + RT \ln a_1$$

The chemical potential of the solvent in the liquid phase in equilibrium with vapor is equal to the chemical potential of the solvent vapor. Thus

$$\mu_1^{\bullet} + RT \ln a_1 = \mu_{1(\text{gas})}^{\circ} + RT \ln \frac{p_1}{p_{\text{gas}}^{\circ}} \qquad (19\text{-}1)$$

Similarly, for the pure solvent in equilibrium with its vapor,

$$\mu_1^\bullet = \mu_{1(\text{gas})}^\circ + RT \ln \frac{p_1^\bullet}{p_{\text{gas}}^\circ} \tag{19-2}$$

Thus, from Equations 19-1 and 19-2, we conclude that

$$a_1 = \frac{p_1}{p_1^\bullet} \tag{19-3}$$

in which p_1 represents the partial pressure of the vapor of the solvent over the solution. If the vapor is not ideal, fugacities must be used in place of partial pressures (see Chapter 13).

Solute

If the solute is sufficiently volatile to allow a determination of its vapor pressure over the solution, its activity also can be calculated from its partial pressure in the vapor. Since the chemical potential in the vapor is equal to the chemical potential of the same component in the liquid

$$\mu_2 = \mu_2^\circ + RT \ln a_2 = \mu_{2(\text{gas})} = \mu_{2(\text{gas})}^\circ + RT \ln \frac{p_2}{p_{\text{gas}}^\circ} \tag{19-4}$$

provided that the vapor behaves as an ideal gas. If data are available throughout the entire range of composition, and pure solute is chosen as the standard state, then μ_2° is equal to $\mu_2^\bullet$. If a hypothetical standard state is chosen, then μ_2° is equal to the chemical potential of the solute in that standard state—that is, the chemical potential the solute would have, either at $X_2 = 1$ or at $m_2 = 1$, if Henry's law described its behavior at that concentration. For the solute in its standard state

$$\mu_2^\circ = \mu_{2(\text{gas})}^\circ + RT \ln \frac{p_2^\circ}{p_{\text{gas}}^\circ} \tag{19-5}$$

where p_2° is the vapor pressure of the solute vapor in equilibrium with the solution in which the solute is in its standard state. Thus, from Equations 19-4 and 19-5, we conclude that

$$a_2 = \frac{p_2}{p_2^\circ} \tag{19-6}$$

Since the standard states are, by definition, hypothetical, we must find a way to determine p_2° from experimental data.

If Henry's law is followed, it is expressed on the molality scale, if the vapor is ideal, as

$$p_2 = k_2'' \left(\frac{m_2}{m_2^\circ} \right) \tag{19-7}$$

(see Equation 16-7). Since p_2° is the value of p_2 when $m_2 = m_2^\circ$,

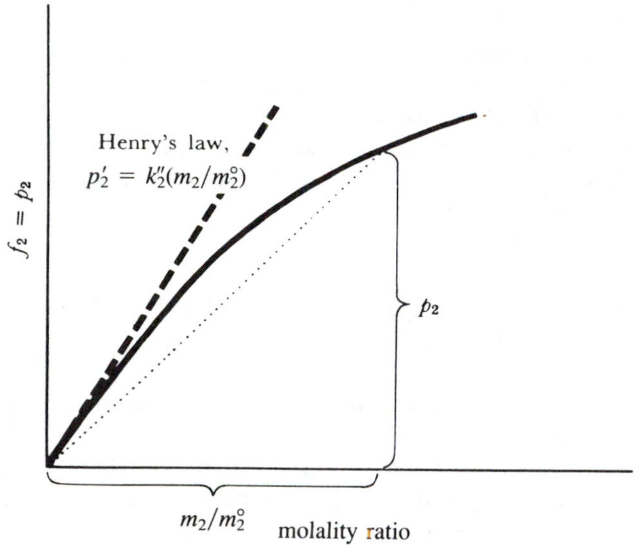

Figure 19-1. Partial pressure of solute as a function of molality.

$$p_2^\circ = k_2'' \tag{19-8}$$

We can see from Figure 19-1 that k_2'' is the limiting slope of the curve of p_2 against m_2/m_2° as $m_2 \to 0$. Thus,

$$\lim_{m_2 \to 0} \left(\frac{p_2}{m_2/m_2^\circ} \right) = k_2'' = p_2^\circ \tag{19-9}$$

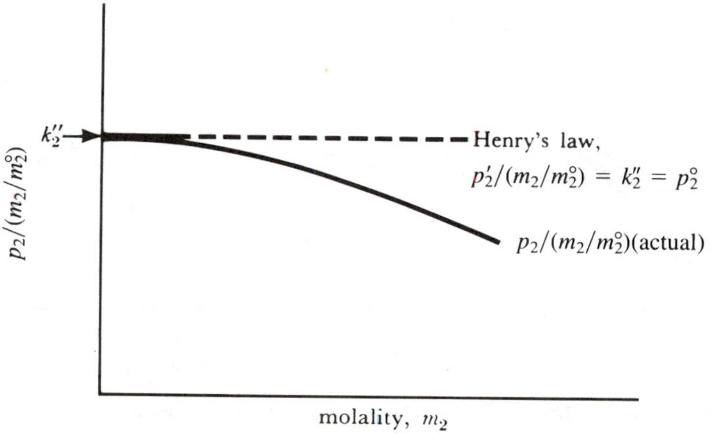

Figure 19-2. Determination of partial pressure of solute in its standard state.

The extrapolation of this ratio to $m_2 = 0$ is shown in Figure 19-2. The curve in Figure 19-2 should have a horizontal limiting slope. If an experimental curve does not have a horizontal limiting slope, the data have not been obtained in sufficiently dilute solution to show Henry's-law behavior.

Using these equations and the measured partial pressure of the solute in the vapor, we have enough information to calculate the activity of the solute, if the vapor behaves as an ideal gas.

19-2 DISTRIBUTION OF SOLUTE BETWEEN TWO IMMISCIBLE SOLVENTS

If the activity of a solute is known in one solvent, then its activity in another solvent immiscible with the first can be determined from concentration-distribution measurements. As an example, let us consider a rather extreme situation, such as that illustrated in Figure 19-3, in which the shapes of the fugacity curves are different in two different solvents. The limiting behavior at infinite dilution, Henry's law, is indicated for each solution. The graphs reveal that the standard states are different in the two solvents since the hypothetical 1-molal solutions have different fugacities.

If the solute in solution A is in equilibrium with that in solution B, its escaping tendency is the same in both solvents. Consequently, its chemical potential, μ_2, at equilibrium also must be identical in both solvents. Nevertheless, the solute will have different activities in solutions A and B since (Equation 17-1)

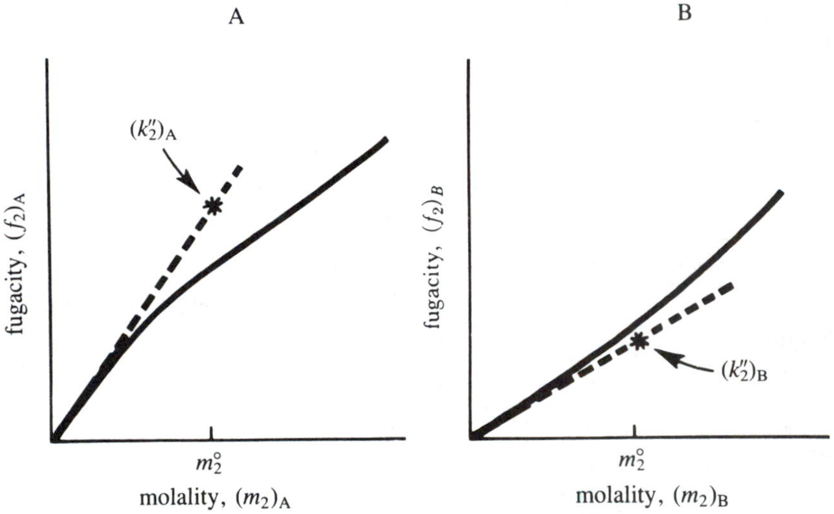

Figure 19-3. Comparison of fugacity-molality curves for solute in two immiscible solvents; used to determine Henry's-law constants.

$$\mu_2 = \mu_2^\circ + RT \ln a_2$$

and $(\mu_2^\circ)_A$ differs from $(\mu_2^\circ)_B$. If the activity of the solute is known in one of the solvents, then the activity in the other solvent can be calculated as follows. At equilibrium

$$(\mu_2)_A = (\mu_2)_B \tag{19-10}$$

Since

$$(\mu_2)_A = (\mu_2^\circ)_A + RT \ln (a_2)_A$$

and

$$(\mu_2)_B = (\mu_2^\circ)_B + RT \ln (a_2)_B$$

it follows from Equation 19-10 that

$$(\mu_2^\circ)_A + RT \ln (a_2)_A = (\mu_2^\circ)_B + RT \ln (a_2)_B$$

or

$$RT \ln \frac{(a_2)_B}{(a_2)_A} = (\mu_2^\circ)_A - (\mu_2^\circ)_B \tag{19-11}$$

Therefore, to calculate $(a_2)_B$ from $(a_2)_A$ we must find the difference between the chemical potentials in the respective standard states. From Equation 16-10

$$\mu_2^\circ = \mu_{2(\text{gas})}^\circ + RT \ln \left(\frac{k_2''}{f_{\text{gas}}^\circ} \right)$$

Thus

$$(\mu_2^\circ)_B - (\mu_2^\circ)_A = RT \ln \left[\frac{(k_2'')_B}{(k_2'')_A} \right] \tag{19-12}$$

and

$$\frac{(a_2)_B}{(a_2)_A} = \frac{(k_2'')_A}{(k_2'')_B} \tag{19-13}$$

To determine the ratio of the activities in the two solvents, we need to determine the ratio of the Henry's-law constants for the solute in the two solutions.

For these solutions, from Equation 16-8,

$$\lim_{(m_2)_A \to 0} \frac{(f_2)_A}{(m_2)_A/m_2^\circ} = (k_2'')_A$$

and

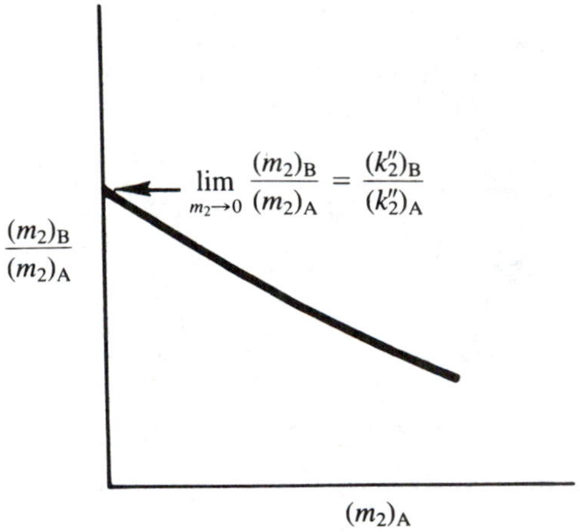

Figure 19-4. Extrapolation of distribution data to obtain conversion factor for activities.

$$\lim_{(m_2)_B \to 0} \frac{(f_2)_B}{(m_2)_B/m_2^\circ} = (k_2'')_B$$

Since, at equilibrium, $(f_2)_A = (f_2)_B$

$$\lim_{m_2 \to 0} \frac{(m_2)_A}{(m_2)_B} = \frac{(k_2'')_B}{(k_2'')_A} \tag{19-14}$$

Thus, Equation 19-13 can be written

$$(a_2)_B = (a_2)_A \lim_{m_2 \to 0} \frac{(m_2)_B}{(m_2)_A} \tag{19-15}$$

If the activity in one solvent is known as a function of molality, and if the equilibrium molalities in both solvents can be determined for a range of molalities, Equation 19-15 can be used to calculate the activity in the other solvent. A value for

$$\lim_{m_2 \to 0} \frac{(m_2)_B}{(m_2)_A}$$

is obtained by plotting $(m_2)_B/(m_2)_A$ against $(m_2)_A$ and extrapolating to $(m_2)_A = 0$, as in Figure 19-4.

19-3 MEASUREMENTS OF CELL POTENTIALS

Although potential measurements are used primarily to determine activities of electrolytes, such measurements can also be used to obtain information on activities of nonelectrolytes. In particular, the activities of components of alloys can be calculated from the potentials of cells such as the following for lead amalgam:

$$\text{Pb(amalgam}, X_2'); \text{Pb(CH}_3\text{COO)}_2, \text{CH}_3\text{COOH}; \text{Pb(amalgam}, X_2) \quad (19\text{-}16)$$

In this cell two amalgams with different mole fractions of lead are dipped into a common electrolyte solution containing a lead salt. The activities of lead in these amalgams can be calculated from emf measurements with this cell.

If we adopt the convention of writing the chemical reaction in the cell as occurring so that electrons will move in the outside conductor from left to right, the reaction at the left electrode is

$$\text{Pb(amalgam}, X_2') = \text{Pb}^{2+} + 2e^- \quad (19\text{-}17)$$

and that at the right electrode is

$$\text{Pb}^{2+} + 2e^- = \text{Pb(amalgam}, X_2) \quad (19\text{-}18)$$

Since the electrolyte containing Pb^{2+} is common to both electrodes, the cell reaction, that is, the sum of Reactions 19-17 and 19-18, is

$$\text{Pb(amalgam}, X_2') = \text{Pb(amalgam}, X_2) \quad (19\text{-}19)$$

Since the two amalgams have the same solvent (mercury), we may choose the same standard state for Pb in each amalgam. Equation 17-17 then will represent the free energy change for Equation 19-19:

$$\Delta G = RT \ln \frac{a_2}{a_2'}$$

If concentrations in the alloy are expressed in mole fraction units, then (Equation 17-5)

$$a_2 = X_2\gamma_2$$

Since, from Equation 8-108,

$$\Delta G = -n\mathscr{F}\mathscr{E}$$

it follows that

$$-n\mathscr{F}\mathscr{E} = RT \ln \frac{a_2}{a_2'} \quad (19\text{-}20)$$

$$= RT \ln \frac{X_2\gamma_2}{a_2'} \quad (19\text{-}21)$$

so

$$\mathscr{E} = -\frac{RT}{n\mathscr{F}} \ln X_2 - \frac{RT}{n\mathscr{F}} \ln \gamma_2 + \frac{RT}{n\mathscr{F}} \ln a_2' \quad (19\text{-}22)$$

Table 19-1 [1] Electromotive Force of Pb (amalgam, $X_2' = 6.253 \times 10^{-4}$); $Pb(CH_3COO)_2$, CH_3COOH; Pb (amalgam, X_2) at 25°C[a]

X_2	$-\mathscr{E}$/Volts	$n\mathscr{F}\mathscr{E}/RT + \ln X_2$	$1000a^2$	γ
0.0006253	0.000000	−7.3773	0.6099	0.975
0.0006302	0.000204	−7.3854	0.6197	0.983
0.0009036	0.0004636	−7.3700	0.8750	0.968
0.001268	0.008911	−7.3640	1.220	0.962
0.001349	0.009659	−7.3603	1.294	0.959
0.001792	0.013114	−7.3453	1.693	0.945
0.002055	0.014711	−7.3327	1.917	0.933
0.002744	0.018205	−7.3155	2.516	0.917
0.002900	0.018886	−7.3132	2.653	0.915
0.003086	0.019656	−7.3110	2.817	0.913
0.003203	0.020068	−7.3059	2.909	0.908
0.003729	0.021827	−7.2908	3.335	0.894
0.003824	0.022111	−7.2877	3.410	0.892
0.004056	0.022802	−7.2826	3.598	0.887
0.004516	0.023954	−7.2649	3.936	0.872
0.005006	0.025160	−7.2557	4.323	0.864
0.005259	0.025692	−7.2478	4.506	0.857
0.005670	0.026497	−7.2353	4.798	0.846
0.006085	0.027256	−7.2237	5.090	0.836
0.006719	0.028340	−7.2090	5.538	0.824
0.007858	0.029951	−7.1778	6.278	0.799
0.007903	0.030010	−7.1767	6.306	0.798
0.008510	0.030771	−7.1619	6.691	0.786
0.009737	0.032062	−7.1277	7.399	0.760
0.01125	0.033437	−7.0903	8.234	0.732
0.01201	0.033974	−7.0668	8.586	0.715
0.01388	0.035226	−7.0195	9.465	0.682
0.01406	0.035323	−7.0142	9.537	0.678
0.01456	0.035609	−7.0015	9.751	0.670
0.01615	0.036375	−6.9575	10.35	0.641
0.01650(satd)	0.036394	−6.9375	10.37	0.628

[a]M. M. Haring, M. R. Hatfield, and P. P. Zapponi, *Trans. Electrochem. Soc.* **75**, 473 (1939).

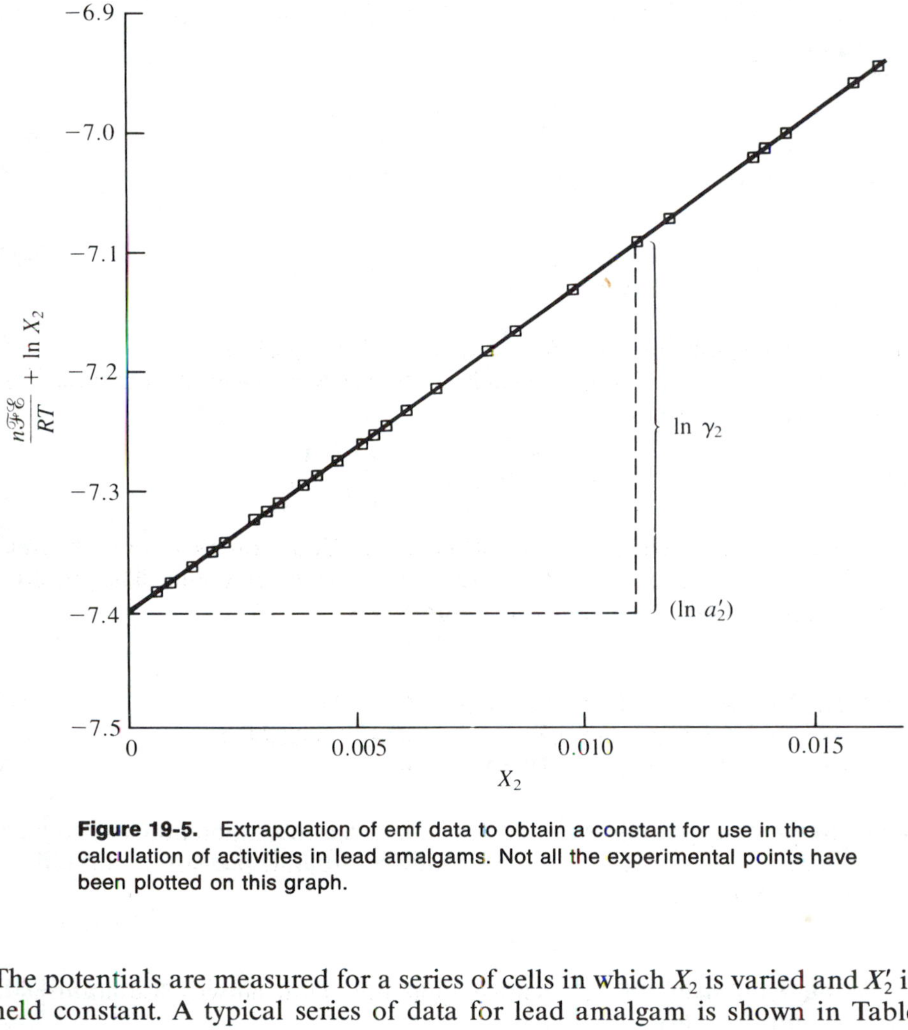

Figure 19-5. Extrapolation of emf data to obtain a constant for use in the calculation of activities in lead amalgams. Not all the experimental points have been plotted on this graph.

The potentials are measured for a series of cells in which X_2 is varied and X_2' is held constant. A typical series of data for lead amalgam is shown in Table 19-1.

To obtain values of a_2 or of γ_2 from these data, we rearrange Equation 19-22 to the form

$$\mathcal{E} + \frac{RT}{n\mathcal{F}} \ln X_2 = \frac{RT}{n\mathcal{F}} \ln a_2' - \frac{RT}{n\mathcal{F}} \ln \gamma_2 \tag{19-23}$$

or

$$\frac{n\mathcal{F}\mathcal{E}}{RT} + \ln X_2 = \ln a_2' - \ln \gamma_2 \tag{19-24}$$

The left side of Equation 19-24 is evaluated from the data in Table 19-1, with $n = 2$, since two electrons are transferred per Pb transferred. Then this quantity is plotted against X_2 and extrapolated to $X_2 = 0$, as in Figure 19-5. From Equations 17-7 and 17-16 we can write

$$\lim_{X_2 \to 0} \gamma_2 = \lim_{X_2 \to 0} \frac{a_2}{X_2} = 1$$

Thus, we can write a limiting form of Equation 19-24,

$$\lim_{X_2 \to 0} \left[\frac{n\mathcal{F}\mathcal{E}}{RT} + \ln X_2 \right] = \ln a_2' \qquad (19\text{-}25)$$

Then the extrapolated value from Figure 19-5 is equal to $\ln a_2'$.

Once a_2' is known, a_2 can be determined at various mole fractions from Equation 19-20 in the form

$$\mathcal{E} - \frac{RT}{n\mathcal{F}} \ln a_2' = -\frac{RT}{n\mathcal{F}} \ln a_2$$

The results for the lead amalgams of the cell in Equation 19-16 are assembled in Table 19-1. The activity coefficients also have been calculated. The graphical representation of $\ln \gamma_2$ is shown in Figure 19-5.

19-4 DETERMINATION OF THE ACTIVITY OF ONE COMPONENT FROM KNOWN VALUES OF THE ACTIVITY OF THE OTHER

The fundamental relationship between the chemical potentials of the two components of a solution at a fixed temperature and pressure is the Gibbs–Duhem equation (12-23),

$$n_1 d\mu_1 + n_2 d\mu_2 = 0$$

From the relationship of the chemical potential to the activity (Equation 17-1), we can write Equation 12-23 as

$$n_1 d \ln a_1 + n_2 d \ln a_2 = 0 \qquad (19\text{-}26)$$

If Equation 19-26 is divided by $(n_1 + n_2)$, the result is

$$X_1 d \ln a_1 + X_2 d \ln a_2 = 0 \qquad (19\text{-}27)$$

Calculation of Activity of Solvent from That of Solute

If adequate data are available for the activity of the solute, the activity of the solvent can be obtained by rearranging Equation 19-27 as

$$d \ln a_1 = -\frac{X_2}{X_1} d \ln a_2 \tag{19-28}$$

and integrating. Since a_2 approaches zero as X_2 approaches zero, $\ln a_2$ is an indeterminate quantity at one of the limits of integration. Although both a_2 and X_2 approach zero, their ratio, $a_2/X_2 = \gamma_2$, approaches one (Equation 17-3). Thus it is necessary to convert Equation 19-28 into a corresponding equation for the activity coefficients.

Since

$$X_1 + X_2 = 1$$

then

$$dX_1 = -dX_2$$

and

$$\frac{dX_1}{X_1} = -\frac{dX_2}{X_1} = -\frac{X_2}{X_1}\frac{dX_2}{X_2}$$

or

$$d \ln X_1 = -\frac{X_2}{X_1} d \ln X_2 \tag{19-29}$$

The subtraction of Equation 19-29 from Equation 19-28 gives the expression

$$d \ln \frac{a_1}{X_1} = -\frac{X_2}{X_1} d \ln \frac{a_2}{X_2} \tag{19-30}$$

or

$$d \ln \gamma_1 = -\frac{X_2}{X_1} d \ln \gamma_2 \tag{19-31}$$

Integrating Equation 19-31 from the infinitely dilute solution to some finite concentration, we obtain

$$\int_0^{\ln \gamma_1} d \ln \gamma_1 = -\int_0^{\ln \gamma_2} \frac{X_2}{X_1} d \ln \gamma_2$$

and

$$\ln \gamma_1 = -\int_0^{\ln \gamma_2} \frac{X_2}{X_1} d \ln \gamma_2 \tag{19-32}$$

If X_2/X_1 is plotted against $\ln \gamma_2$, the integration of Equation 19-32 can be carried out graphically, or a numerical integration can be done directly from tabulated values of X_2/X_1 and $\ln \gamma_2$.

Calculation of Activity of Solute from That of Solvent

To calculate the activity of the solute from that of the solvent, it is useful to rearrange Equation 19-31 to the form

$$d \ln \gamma_2 = -\frac{X_1}{X_2} d \ln \gamma_1 \tag{19-33}$$

which, on integration, gives

$$\ln \gamma_2 = -\int_0^{\ln \gamma_1} \frac{X_1}{X_2} d \ln \gamma_1 \tag{19-34}$$

However, the evaluation of the integral in Equation 19-34 is difficult because X_1/X_2 approaches infinity in the limit of infinitely dilute solutions.

One method of overcoming this difficulty is as follows. Instead of setting the lower limit in the integration of Equation 19-33 at infinite dilution, let us use a temporary lower limit at a finite concentration, X_2'. Thus, in place of Equation 19-34 we obtain

$$\ln \frac{\gamma_2}{\gamma_2'} = -\int_{\ln \gamma_1'}^{\ln \gamma_1} \frac{X_1}{X_2} d \ln \gamma_1 \tag{19-35}$$

The evaluation of the integral in Equation 19-35 offers no difficulties since X_1/X_2 is finite at the lower limit. Using Equation 19-35 we can obtain precise values of the ratio γ_2/γ_2' as a function of X_2. If γ_2/γ_2' is plotted against X_2, as shown in Figure 19-6, the values can be extrapolated to a finite limiting value. From Equation 17-3 we know that

$$\lim_{X_2 \to 0} \frac{a_2}{X_2} = \lim_{X_2 \to 0} \gamma_2 = 1$$

Thus

$$\lim_{X_2 \to 0} \frac{\gamma_2}{\gamma_2'} = \frac{1}{\gamma_2'} \tag{19-36}$$

Once a value of γ_2' is obtained by extrapolation, values of γ_2 in each of the other solutions can be calculated from the values of γ_2/γ_2' from Equation 19-35.

The activity of a solute also can be computed from the activity of the solvent with an experimental technique known as the "isopiestic method."

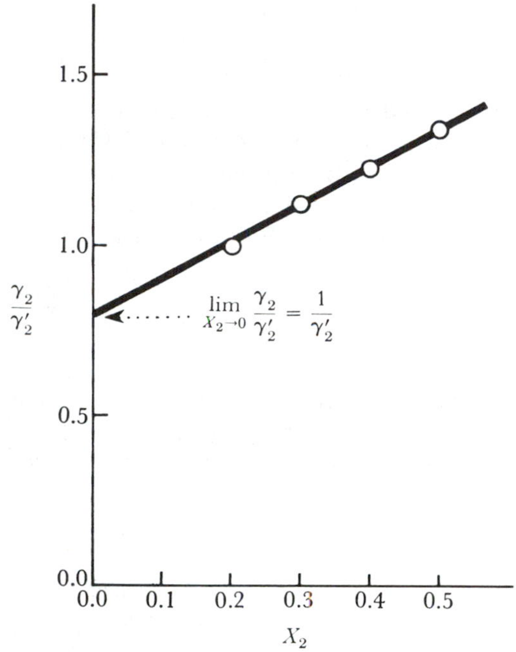

Figure 19-6. Extrapolation of relative activity coefficients to obtain $1/\gamma_2'$ for the calculation of activity coefficients. In this particular problem, the reference state used is for $X_2 = 0.2$.

With this technique, we compare solutions of two different nonvolatile solutes, for one of which, the reference solution, the activity of the solvent has been determined previously with high precision. If both solutions are placed in an evacuated container, solvent will evaporate from the solution with higher vapor pressure and condense into the solution with lower vapor pressure until equilibrium is attained. The solute concentration for each solution then is determined by analysis. Once the molality of the reference solution is known, the activity of the solvent in the reference solution can be read from records of previous experiments with reference solutions. Since the standard state of the solvent is the same for all solutes, the activity of the solvent is the same in both solutions at equilibrium. Once the activity of the solvent is known as a function of m_2 for the new solution, the activity of the new solute can be calculated by the methods [2] discussed previously in this section.

19-5 MEASUREMENTS OF FREEZING POINTS

Perhaps the method of most general applicability for determining activities of nonelectrolytes in solutions is the one based on measurements of the lowering of the freezing point of a solution. Since measurements are made of the properties of the solvent, activities of the solute are calculated by methods described in the preceding section.

Elaborate procedures have been developed for obtaining activity coefficients from freezing-point and thermochemical data. However, to avoid duplication, the details will not be outlined here, since a completely general discussion, applicable to solutions of electrolytes as well as nonelectrolytes, is presented in Chapter 21 of the Third Edition of this work [3].

EXERCISES

1. The data in Table 19-2 on the partial pressures (in mm Hg) of toluene and of acetic acid at 69.94°C have been taken from the International Critical Tables, Vol. III, p. 217, 223, and 288. For the purposes of this exercise, assume that the partial pressure of each component is identical with its fugacity.

Table 19-2

X_1 (Toluene)	X_2 (Acetic Acid)	p_1/(mm Hg) (Toluene)	p_2/(mm Hg) (Acetic Acid)
0.0000	1.0000	0	136
0.1250	0.8750	54.8	120.5
0.2310	0.7690	84.8	110.8
0.3121	0.6879	101.9	103.0
0.4019	0.5981	117.8	95.7
0.4860	0.5140	130.7	88.2
0.5349	0.4651	137.6	83.7
0.5912	0.4088	154.2[a]	78.2
0.6620	0.3380	155.7	69.3
0.7597	0.2403	167.3	57.8
0.8289	0.1711	176.2	46.5
0.9058	0.0942	186.1	30.5
0.9565	0.0435	193.5	17.2
1.0000	0.0000	202	0

[a]This evidently is a typographical error. The correct figure is near 145.2. It also is likely (D. H. Volman, private correspondence) that the International Critical Tables did not interpret correctly the mole fraction scale used by the original investigator [J. von Zawidski, *Z. physik. Chem.* **35**, 129 (1900)].

a. Draw a graph of f_1 versus X_1. Indicate Raoult's law by a dotted line.

b. Draw a graph of f_2 versus X_2. Indicate Raoult's law and Henry's law, each by a dotted line.

c. Calculate the constant in Henry's law for acetic acid in toluene solutions by extrapolating a graph of f_2/X_2 versus X_2 to infinite dilution.

d. Calculate the activities and activity coefficients of acetic acid on the basis of a k_2'' established from Henry's law. Plot these values against X_2.

e. Calculate the activities and activity coefficients of acetic acid when the pure liquid is taken as the standard state. Plot these values on the same graph as in (d).

f. Calculate the activities and activity coefficients of toluene, the solvent in these solutions.

2. From vapor pressure data for water–propanol solutions [J. A. V. Butler, D. W. Thompson, and W. H. Maclennan, *J. Chem. Soc.* 674 (1933)], activities and activity coefficients can be calculated on the basis of different choices of standard state. Some results of such calculations are summarized in Table 19-3; compositions in all cases are expressed in terms of mole fraction.

Table 19-3

	γ_2 for Propanol If		γ_1 for Water If	
X_2 (Propanol)	Hypothetical Unit Mole Fraction Standard State	Pure Component Standard State	Hypothetical Unit Mole Fraction Standard State	Pure Component Standard State
0.00	1.000	14.4	0.270	1.000
0.01	0.854	12.3	0.268	0.994
0.10	0.420	6.05	0.286	1.060
0.40	0.113	1.63	0.410	1.52
0.60	0.083	1.19	0.564	2.09
0.90	0.0695	0.986	0.92	3.40
1.00	0.0694	1.000	1.000	3.7

a. Using pure water and pure propanol as the standard states, calculate the free energy change for mixing nine moles of pure water with one mole of pure propanol to form a solution of mole fraction $X_2 = 0.1$.

b. Using Henry's law to establish the standard state of the solute, propanol, and pure water as the standard state of the solvent, water, calculate ΔG for the process in (a).

c. Compare the results in (a) and (b) with the ideal free energy of mixing for the solution in (a).

3. The partial pressures (in mm Hg) listed in Table 19-4 have been measured for the acetone component of an acetone–chloroform solution [J. von Zawidzki, *Z. physik. Chem.* **35**, 129 (1900)].

a. Calculate the activity coefficients, $a_1/X_1 = \gamma_1$, for acetone.

b. Make a graph of X_1/X_2 versus $\log \gamma_1$.

c. Compute γ_2/γ_2' at values of X_2 of 0.2, 0.3, 0.4, and 0.5. Let $X_2' = 0.2$.

d. Compute values of γ_2 at the same mole fractions, X_2, as in (c).

Table 19-4

X_1(Acetone)	p_1(Acetone)/(mm Hg)
1.000	344.5
0.9405	322.9
0.8783	299.7
0.8165	275.8
0.7103	230.7
0.6387	200.3
0.5750	173.7

4. Potentials [G. Spiegel and H. Ulrich, *Z. physik. Chem.* **A178**, 187 (1937)] of lithium amalgam electrodes in the cell

   Li(amalgam, X_2), LiCl in acetonitrile, Li(amalgam, $X_2' = 0.0003239$)

   at 25°C are given in Table 19-5.

   a. Plot $\mathscr{E}$ versus X_2. Note the limit as X_2 approaches zero.

   b. Calculate the activities and activity coefficients of lithium in the amalgams.

Table 19-5

$X_{2(\text{lithium})}$	$\mathscr{E}$/Volts
0.0003239	0.0000
0.001846	0.0458
0.002345	0.0517
0.004218	0.0684
0.008779	0.0894
0.01300	0.1006
0.02265(satd)	0.1189

5. Calculate the emf of the following cells at 25°C. In each case consider the activity coefficient as 1.

   a. H_2(76 cm of Hg), HCl(m_2 = Y), H_2(70 cm of Hg).

   b. H_2(70 cm of Hg), HCl(m_2 = Y), H_2(76 cm of Hg).

   c. Tl(in amalgam, $X_{Tl} = 0.0001$), TlCl(m_2 = Y), Tl(in amalgam, $X_{Tl} = 0.001$).

   d. Tl(in amalgam, $X_{Tl} = 0.001$), Tl_2SO_4(m_2 = Y), Tl(in amalgam, $X_{Tl} = 0.0001$).

   e. Cd(in amalgam, $X_{Cd} = 0.001$), $CdCl_2$(m_2 = Y), Cd(in amalgam, X_{Cd} = 0.0001).

   f. Cd(in amalgam, $X_{Cd} = 0.0001$), $CdSO_4$(m_2 = Y), Cd(in amalgam, X_{Cd} = 0.001).

6. At 42.05°C the heat of mixing of one mole of water and one mole of ethanol is −82.0 cal. The vapor pressure of water above the solution is 0.821 $p_1^\bullet$ and that of ethanol is 0.509 $p_2^\bullet$.
   a. Calculate the entropy of mixing for this solution.
   b. Compute the excess entropy above that for the mixing of two components that form an ideal solution.

REFERENCES

1. The use of a spreadsheet to produce Table 19-1 has been described by R. M. Rosenberg and E. V. Hobbs, *J. Chem. Educ.* **62**, 140 (1985).
2. See, for example, G. N. Lewis and M. Randall, *Thermodynamics*, 2d ed. (revised by K. S. Pitzer and L. Brewer), McGraw-Hill, New York, 1961, pp. 320–323.
3. I. M. Klotz and R. M. Rosenberg, *Chemical Thermodynamics*, 3d ed., Benjamin/Cummings, Menlo Park, 1972, pp. 374–383.

CHAPTER 20

Activity, Activity Coefficients, and Osmotic Coefficients of Strong Electrolytes

In Chapters 17 and 19 we developed procedures for defining standard states for nonelectrolyte solutes and for determining the numerical values of the corresponding activities and activity coefficients from experimental measurements. The activity of the solute is defined by Equation 17-1 and either Equation 17-3 or Equation 17-4 for the hypothetical unit mole fraction standard state or the hypothetical 1-molal standard state, respectively. The activity of the solute is obtained from the activity of the solvent by use of the Gibbs–Duhem equation, as in Section 19-4. When the solute activity is plotted against the appropriate composition variable, the portion of the resulting curve in the dilute region in which the solute follows Henry's law is extrapolated to $X_2 = 1$ or $(m_2/m^\circ) = 1$ to find the standard state.

When activity data for a strong electrolyte such as HCl is plotted against (m_2/m°), as illustrated in Figure 20-1, the initial slope is equal to zero. Thus, an extrapolation to the standard state yields a value of the activity in the standard state equal to zero, contrary to the definition of activity in Equation 17-1. Therefore, it is clear that the procedure for determining standard states must be modified for electrolytes.

20-1 DEFINITIONS AND STANDARD STATES FOR DISSOLVED ELECTROLYTES

Uni-univalent Electrolytes

Since a plot of the activity of an electrolyte such as aqueous HCl against the first power of m_2/m° gives a limiting slope of zero, we might examine graphs in which the activity is plotted against other powers of the molality ratio. Such a plot is shown in Figure 20-2, in which the activity of aqueous HCl is plotted

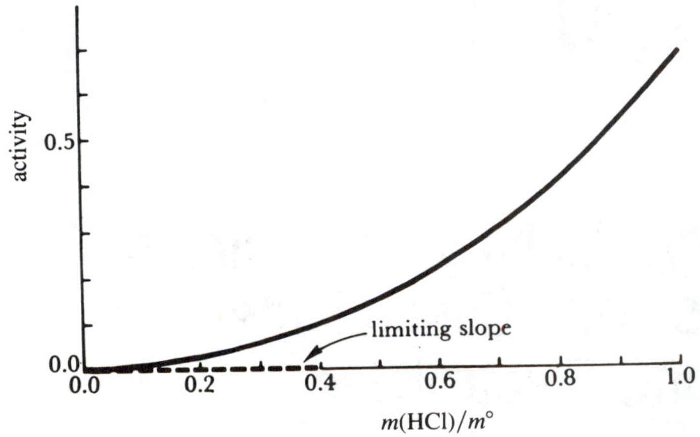

Figure 20-1. Activity versus molality ratio for hydrochloric acid. Based on data from G. N. Lewis and M. Randall, *Thermodynamics*, McGraw-Hill, New York, 1923, p. 336.

against the square of the molality ratio. The curve has a finite, nonzero limiting slope.

This result suggests that the appropriate form of the limiting law for uni-univalent electrolytes (such as HCl) is

$$\lim_{m_{\mathrm{HCl}} \to 0} \frac{a_{\mathrm{HCl}}}{(m_{\mathrm{HCl}}/m^\circ)^2} = 1 \tag{20-1}$$

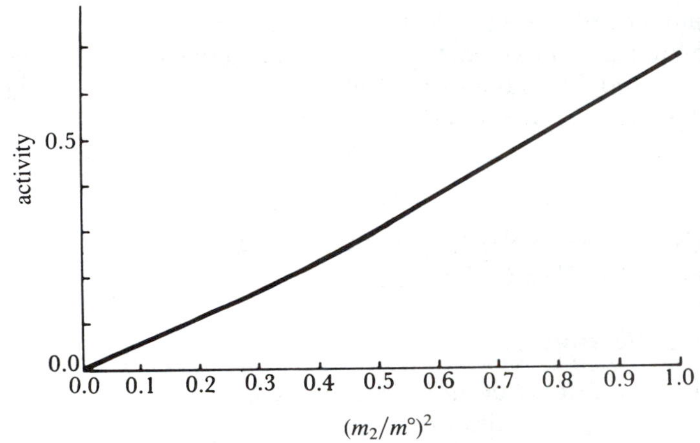

Figure 20-2. Activity versus the square of the molality ratio for aqueous hydrochloric acid. Data are the same as for Figure 20-1.

The extrapolation of the curve in Figure 20-2 from the linear portion near zero concentration in which Henry's-law behavior is observed to a molality ratio equal to 1 leads to a hypothetical unit molality ratio standard state. No real state of the system corresponds to the extrapolated point, but the properties of the standard state can be calculated from the properties of dilute solutions in which Henry's law is followed.

Thus far, we have not introduced any assumptions about the dissociation of electrolytes in order to describe their experimental behavior. As far as thermodynamics is concerned, such details need not be considered. We can take the limiting law in the form of Equation 20-1 as an experimental fact and derive thermodynamic relationships from it. Nevertheless, in view of the general applicability of the ionic theory, it is desirable to relate our results to that theory.

For example, the empirical relation between the activity and the molality ratio can be understood on the assumption that the chemical potential of the electrolyte is the sum of the chemical potentials of the constituent ions. That is, for HCl as the solute,

$$\mu_{\mathrm{HCl}} = \mu_{\mathrm{H}^+} + \mu_{\mathrm{Cl}^-} \tag{20-2}$$

If we apply Equation 17-1, the definition of the activity, to Equation 20-2, the result is

$$\mu^\circ_{\mathrm{HCl}} + RT \ln a_{\mathrm{HCl}} = \mu^\circ_{\mathrm{H}^+} + RT \ln a_{\mathrm{H}^+} + \mu^\circ_{\mathrm{Cl}^-} + RT \ln a_{\mathrm{Cl}^-} \tag{20-3}$$

We can also assume that

$$\mu^\circ_{\mathrm{HCl}} = \mu^\circ_{\mathrm{H}^+} + \mu^\circ_{\mathrm{Cl}^-} \tag{20-4}$$

so that

$$a_{\mathrm{HCl}} = (a_{\mathrm{H}^+})(a_{\mathrm{Cl}^-}) \tag{20-5}$$

The individual ion activities should follow the limiting relations

$$\lim_{m_{\mathrm{HCl}}\to 0} \frac{a_{\mathrm{H}^+}}{m_{\mathrm{H}^+}/m^\circ} = 1 \tag{20-6}$$

and

$$\lim_{m_{\mathrm{HCl}}\to 0} \frac{a_{\mathrm{Cl}^-}}{m_{\mathrm{Cl}^-}/m^\circ} = 1 \tag{20-7}$$

so the product of the limits is

$$\lim_{m_{\mathrm{HCl}}\to 0} \frac{(a_{\mathrm{H}^+})(a_{\mathrm{Cl}^-})}{(m_{\mathrm{H}^+}/m^\circ)(m_{\mathrm{Cl}^-}/m^\circ)} = \lim_{m_{\mathrm{HCl}}\to 0} \frac{a_{\mathrm{HCl}}}{(m_{\mathrm{HCl}}/m^\circ)^2} = 1$$

which is the relation found empirically.

There is no way within thermodynamics to determine the activity of a single ion since we cannot vary the concentration of a single ion while keeping the amounts of the other ions constant. Since a_+ and a_- approach m_2 at infinite dilution for a uni-univalent electrolyte, a_+ must equal a_- at infinite dilution. However, at any nonzero concentration the difference between a_+ and a_- is unknown, although it may be negligibly small in dilute solution. Nevertheless, in a solution of any concentration, the *mean* activity of the ions can be determined. By the mean activity, $a_\pm$, we refer to the geometric mean, which for a uni-univalent electrolyte is defined by the equation

$$a_\pm = (a_+a_-)^{1/2} = a_2^{1/2} \tag{20-8}$$

We can also define an *activity coefficient*, γ_i, for each ion in an electrolyte solution. For each ion of a uni-univalent electrolyte

$$\gamma_+ = \frac{a_+}{m_+/m^\circ}$$

and (20-9)

$$\gamma_- = \frac{a_-}{m_-/m^\circ}$$

These individual-ion activity coefficients have the desired property of approaching one at infinite dilution, since each ratio $a_1/(m_1/m^\circ)$ approaches one. However, individual-ion activity coefficients, like individual-ion activities, cannot be determined experimentally. Therefore it is customary to deal with the mean activity coefficient, $\gamma_\pm$, which for a uni-univalent electrolyte can be related to measurable quantities as follows:

$$\begin{aligned}\gamma_\pm &= (\gamma_+\gamma_-)^{1/2}\\ &= \left(\frac{a_+}{m_+/m^\circ}\right)\left(\frac{a_-}{m_-/m^\circ}\right)^{1/2}\\ &= \frac{a_\pm}{m_2/m^\circ} = \frac{a_2^{1/2}}{m_2/m^\circ}\end{aligned} \tag{20-10}$$

From Equations 20-1, 20-8, and 20-10 we can see that

$$\lim_{m_2\to 0} \gamma_\pm = 1 \tag{20-11}$$

Multivalent Electrolytes

Symmetrical Salts. For salts in which anions and cations have the same valence, activities and related quantities are defined in exactly the same way as for uni-univalent electrolytes. For example, for $MgSO_4$ a finite limiting slope is obtained when the activity is plotted against the square of the molality ratio. Furthermore, m_+ equals m_-. Consequently, the treatment of symmetrical salts does not differ from that just described for uni-univalent electrolytes.

Unsymmetrical Salts. As an example of unsymmetrical salts, let us consider a salt such as $BaCl_2$, which dissociates into one cation and two anions. By analogy with the case of a uni-univalent electrolyte we can define the ion activities by the expression

$$\begin{aligned} a_2 &= (a_+)(a_-)(a_-) \\ &= (a_+)(a_-)^2 \end{aligned} \tag{20-12}$$

In this case, the mean ionic activity, $a_\pm$, also is the geometric mean of the individual-ion activities:

$$a_\pm = [(a_+)(a_-)^2]^{1/3} = a_2^{1/3} \tag{20-13}$$

It is desirable that the individual-ion activities approach the molality ratio of the ions in the limit of infinite dilution. That is,

$$\lim_{m_2 \to 0} \frac{a_+}{m_+/m^\circ} = \lim_{m_2 \to 0} \frac{a_+}{m_2/m^\circ} = 1$$

and (20-14)

$$\lim_{m_2 \to 0} \frac{a_-}{m_-/m^\circ} = \lim_{m_2 \to 0} \frac{a_-}{2m_2/m^\circ} = 1$$

It follows from Equations 20-13 and 20-14 that

$$\begin{aligned} \lim_{m_2 \to 0} a_\pm &= \lim_{m_2 \to 0} [(a_+)(a_-)^2]^{1/3} \\ &= [(m_2/m^\circ)(2m_2/m^\circ)^2]^{1/3} \\ &= 4^{1/3}(m_2/m^\circ) \end{aligned} \tag{20-15}$$

It is also desirable that the mean ionic activity coefficient, $\gamma_\pm$, approach unity in the limit of infinite dilution. We can achieve this result if, as in Equation 20-9, we define

$$\gamma_+ = \frac{a_+}{m_+/m^\circ}$$

and

$$\gamma_- = \frac{a_-}{m_-/m^\circ}$$

and we define

$$\gamma_{\pm} = [(\gamma_{+})(\gamma_{-})^2]^{1/3} \tag{20-16}$$

$$= \left[\left(\frac{a_{+}}{m_{+}/m^{\circ}}\right)\left(\frac{a_{-}}{m_{-}/m^{\circ}}\right)^2\right]^{1/3}$$

$$= \left[\frac{(a_{+})(a_{-})^2}{(m_2/m^{\circ})(2m_2/m^{\circ})^2}\right]^{1/3}$$

$$= \frac{a_{\pm}}{(4^{1/3})(m_2/m^{\circ})} \tag{20-17}$$

Then, from Equations 20-15 and 20-17,

$$\lim_{m_2 \to 0} \gamma_{\pm} = 1$$

For a uni-univalent electrolyte (Equation 20-10)

$$\gamma_{\pm} = \frac{a_{\pm}}{m_2/m^{\circ}}$$

To achieve a uniform definition of $\gamma_{\pm}$ for all electrolytes, it is convenient to define a *mean molality,* $m_{\pm}$ (for $BaCl_2$, for example), as

$$m_{\pm} = [(m_{+})(m_{-})^2]^{1/3}$$

$$= [(m_2)(2m_2)^2]^{1/3}$$

$$= 4^{1/3}m_2 \tag{20-18}$$

With this definition, the relationship for the mean activity coefficient

$$\gamma_{\pm} = \frac{a_{\pm}}{m_{\pm}/m^{\circ}} \tag{20-19}$$

holds for any electrolyte.

It follows from Equation 20-13 and Equation 20-15 that

$$\lim_{m_2 \to 0} \frac{a_2}{(4m_2/m^{\circ})^3} = 1 \tag{20-20}$$

Equation 20-20 is consistent with the empirical observation that a nonzero initial slope is obtained when the activity of a ternary electrolyte such as $BaCl_2$ is plotted against the cube of (m_2/m°). Since the activity in the standard state is equal to 1, by definition, the standard state of a ternary electrolyte is that hypothetical state of unit molality ratio with an activity one-fourth of the activity obtained by extrapolation of dilute solution behavior to m_2/m° equal to 1, as shown in Figure 20-3.

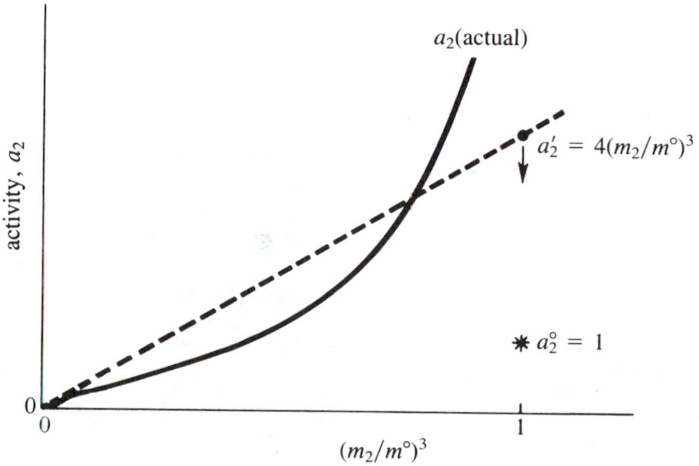

Figure 20-3. Establishment of the standard state for a ternary electrolyte.

General Case. If an electrolyte $A_{\nu_+}B_{\nu_-}$ dissociates into ν_+ positive ions of charge Z_+ and ν_- negative ions of charge Z_-, the general definitions for the activities and the activity coefficients are

$$a_2 = (a_+)^{\nu_+}(a_-)^{\nu_-} \tag{20-21}$$

$$a_\pm = (a_2)^{1/(\nu_+ + \nu_-)} = (a_2)^{1/\nu} \tag{20-22}$$

$$= [(a_+)^{\nu_+}(a_-)^{\nu_-}]^{1/\nu} \tag{20-23}$$

and

$$m_\pm = [(m_+)^{\nu_+}(m_-)^{\nu_-}]^{1/\nu}$$

$$= [(\nu_+ m_2)^{\nu_+}(\nu_- m_2)^{\nu_-}]^{1/\nu} \tag{20-24}$$

The appropriate limiting law that is consistent with experimental observation is

$$\lim_{m_2 \to 0} \frac{a_2}{(m_2/m^\circ)^\nu} = (\nu_+)^{\nu_+}(\nu_-)^{\nu_-} \tag{20-25}$$

and, as before,

$$\gamma_\pm = \frac{a_\pm}{m_\pm/m^\circ} \tag{20-26}$$

Table 20-1 summarizes the empirical expression of the limiting law and the definitions of the ionic activities, molality ratios, and activity coefficients for a few substances and for the general case of any electrolyte.

Table 20-1 Thermodynamic Functions for Dissolved Solutes

	Sucrose	NaCl	Na_2SO_4	$AlCl_3$	$MgSO_4$	$A_{\nu_+}B_{\nu_-}$
$\lim_{m_2 \to 0} \frac{a_2}{m_2/m^\circ}$	1	1	4	27	1	$\nu_+^{\nu_+}\nu_-^{\nu_-}$
$a_2 =$	a_{sucrose}	$(a_+)(a_-)$	$(a_+)^2(a_-)$	$(a_+)(a_-)^3$	$(a_+)(a_-)$	$(a_+)^{\nu_+}(a_-)^{\nu_-}$
$a_\pm =$		$[(a_+)(a_-)]^{1/2}$	$[(a_+)^2(a_-)]^{1/3}$	$[(a_+)(a_-)^3]^{1/4}$	$[(a_+)(a_-)]^{1/2}$	$[(a_+)^{\nu_+}(a_-)^{\nu_-}]^{1/(\nu_+ + \nu_-)}$
$m_\pm =$		m_2	$4^{1/3}m_2$	$27^{1/4}m_2$	m_2	$[(\nu_+)^{\nu_+}(\nu_-)^{\nu_-}]^{1/(\nu_+ + \nu_-)}m_2$
$\gamma_\pm =$		$\frac{a_\pm}{(m_\pm/m^\circ)}$	$\frac{a_\pm}{(m_\pm/m^\circ)}$	$\frac{a_\pm}{(m_\pm/m^\circ)}$	$\frac{a_\pm}{(m_\pm/m^\circ)}$	$\frac{a_\pm}{(m_\pm/m^\circ)}$

Mixed Electrolytes

In a solution of mixed electrolytes, the presence of common ions must be considered when calculating the mean molality. For example, in a solution in which $m_{\text{NaCl}} = 0.1$ and $m_{\text{MgCl}_2} = 0.2$, the mean molality, $m_{\pm}$, for NaCl is

$$\begin{aligned} m_{\pm\text{NaCl}} &= [(m_{\text{Na}^+})(m_{\text{Cl}^-})]^{1/2} \\ &= [(0.1)(0.5)]^{1/2} \\ &= 0.224 \text{ mol kg}^{-1} \end{aligned}$$

For $MgCl_2$

$$\begin{aligned} m_{\pm\text{MgCl}_2} &= [(m_{\text{Mg}^{2+}})(m_{\text{Cl}^-})^2]^{1/3} \\ &= [(0.2)(0.5)^2]^{1/3} \\ &= 0.368 \text{ mol kg}^{-1} \end{aligned}$$

Thus, when calculating the mean molality of an electrolyte in a mixture, we must use the total molality of each ion, regardless of the source of the ion.

Both on the basis of empirical data and on the grounds of electrostatic theory, it has been found convenient to introduce a quantity known as the *ionic strength* when considering the effects of several electrolytes on the activity of one of them.

The contribution of each ion to the ionic strength, I, is obtained by multiplying the molality of the ion by the square of its charge. One-half the sum of these contributions for all of the ions present is defined as the ionic strength. That is

$$I = (\tfrac{1}{2}) \sum m_i z_i^2 \tag{20-27}$$

The factor $\frac{1}{2}$ has been included so that the ionic strength will be equal to the molality for a uni-univalent electrolyte. Thus, for NaCl,

$$\begin{aligned} I &= (\tfrac{1}{2})[m_2(1)^2 + m_2(1)^2] \\ &= m_2 \end{aligned}$$

However, for $BaCl_2$

$$\begin{aligned} I &= (\tfrac{1}{2})[m_2(2)^2 + 2m_2(1)^2] \\ &= 3m_2 \end{aligned}$$

since $m_+ = m_2$ and $m_- = 2m_2$, in which m_2 is the molality of the electrolyte. Several examples, together with a general formulation, are shown in Table 20-2.

Table 20-2 Relationships between Ionic Strength and Molality

Salt	NaCl	Na_2SO_4	$AlCl_3$	$MgSO_4$	$A_{\nu_+}B_{\nu_-}$
I =	m_2	$3m_2$	$6m_2$	$4m_2$	$\frac{1}{2}\nu_-\left(\frac{\nu_-}{\nu_+}+1\right)z_-^2\, m_2$

20-2 DETERMINATION OF ACTIVITIES OF STRONG ELECTROLYTES

All of the methods used in the study of nonelectrolytes also can be applied *in principle* to the determination of activities of electrolyte solutes. However, in practice, several methods are difficult to adapt to electrolytes because it is impractical to obtain data for solutions sufficiently dilute to allow the necessary extrapolation to infinite dilution. For example, some data are available for the vapor pressures of the hydrogen halides in their aqueous solutions, but these measurements by themselves do not permit us to determine the activity of the solute because significant data cannot be obtained at concentrations below 4 molal.

Activity data for electrolytes usually are obtained by one or more of three independent experimental methods: measurement of the potentials of electrochemical cells, solubility measurement, and the measurement of colligative properties.

A great deal of information on activities of electrolytes also has been obtained by the isopiestic method, in which a comparison is made of the concentrations of two solutions with equal solvent vapor pressure. The principles of this method were discussed in Section 19-4.

Measurement of Cell Potentials

For the cell composed of a hydrogen electrode and a silver–silver chloride electrode immersed in a solution of HCl, represented by the notation

$$H_2(g),\ HCl(m_2),\ AgCl(s),\ Ag(s) \tag{20-28}$$

the convention that we have adopted (see Section 19-3) describes the cell reaction as

$$\tfrac{1}{2}H_2(g) + AgCl(s) = HCl(m_2) + Ag(s) \tag{20-29}$$

By this convention, the potential of the cell is defined as the potential of the electrode on the right, at which reduction occurs, minus the potential of the electrode on the left, at which oxidation occurs.

We know from Equation 8-108 that the free energy change of the reaction is related to the cell potential by

$$\Delta G = -n\mathcal{F}\mathcal{E}$$

and from Equation 17-22 that ΔG is related to the activities of reactants and products by

$$\Delta G = \Delta G^\circ + RT \ln \left(\frac{a_{\text{HCl}}\, a_{\text{Ag}}}{a_{\text{H}_2}^{1/2} a_{\text{AgCl}}} \right) \tag{20-30}$$

Substituting from Equation 8-108 into Equation 20-30, we obtain

$$-n\mathcal{F}\mathcal{E} = -n\mathcal{F}\mathcal{E}^\circ + RT \ln \left(\frac{a_{\text{HCl}}\, a_{\text{Ag}}}{a_{\text{H}_2}^{1/2} a_{\text{AgCl}}} \right) \tag{20-31}$$

or

$$\mathcal{E} = \mathcal{E}^\circ - \left(\frac{RT}{n\mathcal{F}} \right) \ln \left(\frac{a_{\text{HCl}}\, a_{\text{Ag}}}{a_{\text{H}_2}^{1/2} a_{\text{AgCl}}} \right) \tag{20-32}$$

If the pressure of hydrogen gas is maintained at the standard pressure of 1 bar, a pressure that is essentially equal to the fugacity, then the hydrogen can be considered to be in its standard state, with an activity equal to 1. Since pure solid Ag and pure solid AgCl are in their standard states, their activities also are equal to 1. Thus Equation 20-32 can be written

$$\mathcal{E} = \mathcal{E}^\circ - \left(\frac{RT}{n\mathcal{F}} \right) \ln a_{\text{HCl}} \tag{20-33}$$

Hence, the cell indicated by Equation 20-29 can be used to determine the activity of dissolved HCl.

To apply Equation 20-33 to experimental data, we must specify our choice of standard states, since the values of $\mathcal{E}^\circ$ and of a_{HCl} depend on this choice. We shall use the hypothetical unit molality ratio standard state obtained by extrapolation from the infinitely dilute solution. By convention, m° is taken equal to 1 mol kg^{-1}.

From Equations 20-5, 20-9, 20-10, and 20-19, we can write, for dissolved HCl,

$$\begin{aligned} a_{\text{HCl}} &= (a_{\text{H}^+})(a_{\text{Cl}^-}) \\ &= [(m_{\text{H}^+}/m^\circ)\, \gamma_{\text{H}^+}][(m_{\text{Cl}^-}/m^\circ)\, \gamma_{\text{Cl}^-}] \\ &= (m_\pm/m^\circ)^2(\gamma_\pm)^2 \end{aligned} \tag{20-34}$$

Substituting from Equation 20-34 into Equation 20-33, we have

$$\mathscr{E} = \mathscr{E}^\circ - \left(\frac{RT}{n\mathscr{F}}\right) \ln(m_\pm/m^\circ)^2(\gamma_\pm)^2$$

$$= \mathscr{E}^\circ - 2\left(\frac{RT}{n\mathscr{F}}\right) \ln(m_\pm/m^\circ)(\gamma_\pm)$$

$$= \mathscr{E}^\circ - 2\left(\frac{RT}{n\mathscr{F}}\right) \ln(m_\pm/m^\circ) - 2\left(\frac{RT}{n\mathscr{F}}\right) \ln \gamma_\pm \qquad (20\text{-}35)$$

We defined the limiting behavior of $\gamma_\pm$ (Equation 20-11) so that

$$\lim_{m_2 \to 0} \gamma_\pm = 1$$

Consequently

$$\lim_{m_2 \to 0} \ln \gamma_\pm = 0$$

and the third term on the right in Equation 20-35 becomes equal to zero in the limit of infinite dilution. However, the limits of both $\ln(m_\pm/m^\circ)$ and $\mathscr{E}$ as m_2 goes to zero are indeterminate, as is illustrated in Figure 20-4. Therefore, we rewrite Equation 20-35 in the form

$$\mathscr{E}' = \mathscr{E} + 2\left(\frac{RT}{n\mathscr{F}}\right) \ln(m_\pm/m^\circ)$$

$$= \mathscr{E}^\circ - 2\left(\frac{RT}{n\mathscr{F}}\right) \ln \gamma_\pm \qquad (20\text{-}36)$$

in which $\mathscr{E}'$ is a convenient notation for

$$\mathscr{E} + 2\left(\frac{RT}{n\mathscr{F}}\right) \ln(m_\pm/m^\circ)$$

The value of $\mathscr{E}'$ can be calculated from experimental data (see Figure 20-4), and the $\ln \gamma_\pm$ on the right of Equation 20-36 goes to zero at infinite dilution. Therefore, an extrapolation of $\mathscr{E}'$ to zero molality should yield a value of $\mathscr{E}^\circ$. That is,

$$\lim_{m_2 \to 0} \mathscr{E}' = \lim_{m_2 \to 0} \left[\mathscr{E} + 2\left(\frac{RT}{n\mathscr{F}}\right) \ln(m_\pm/m^\circ)\right]$$

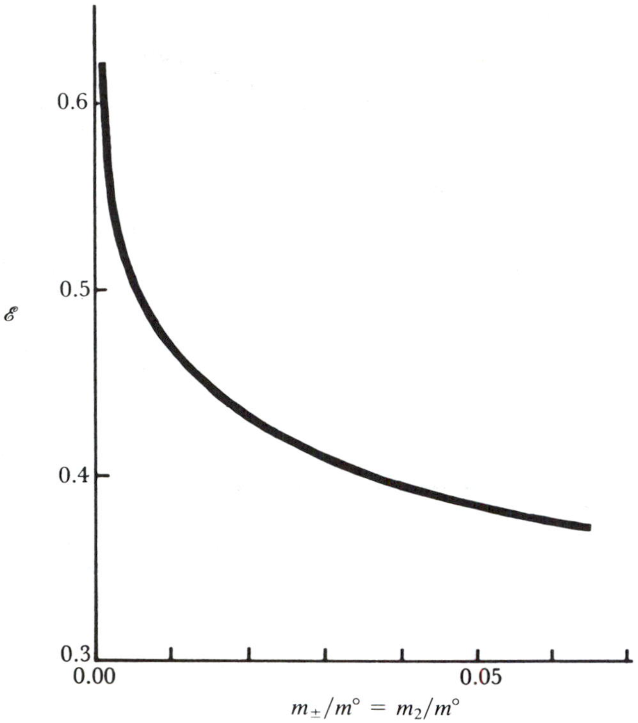

Figure 20-4. Potentials at 25°C of the cell: H_2, $HCl(m_2)$, AgCl, Ag.

$$= \lim_{m_2 \to 0} \left[\mathcal{E}^{\circ} - 2\left(\frac{RT}{n\mathcal{F}}\right) \ln \gamma_{\pm} \right]$$

$$= \mathcal{E}^{\circ} \tag{20-37}$$

If $\mathcal{E}'$ is plotted against the molality ratio, as in Figure 20-5, the values approach the vertical axis with a very steep slope that makes extrapolation impossible. If $\mathcal{E}'$ is plotted against the square root of the molality ratio, the extrapolation can be carried out to yield a precise value of $\mathcal{E}^{\circ}$, as shown in Figure 20-6. The choice of the square root was based both on experience with other data for electrolytes and on electrostatic theory.

Having obtained $\mathcal{E}^{\circ}$, we can calculate mean activity coefficients from a rearrangement of Equation 20-36 to the form

$$\ln \gamma_{\pm} = \left(\frac{n\mathcal{F}}{2RT}\right)(\mathcal{E}^{\circ} - \mathcal{E}') \tag{20-38}$$

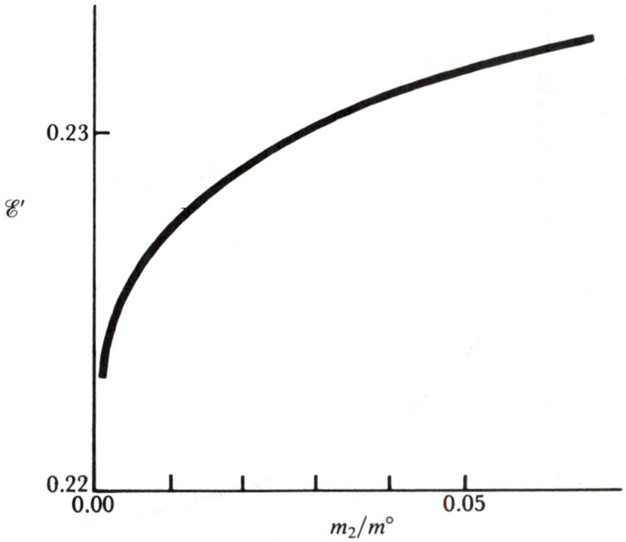

Figure 20-5. $\mathscr{E}'$ at 25°C as a function of the molality ratio for the cell: H_2, $HCl(m_2)$, AgCl, Ag.

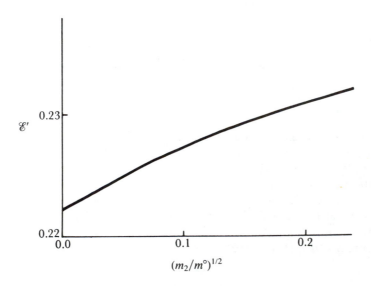

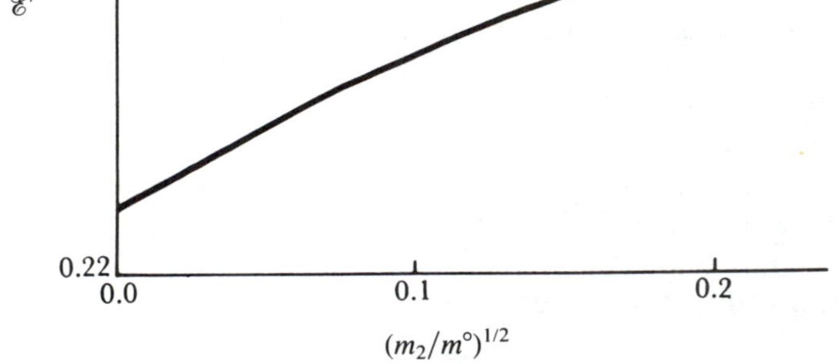

Figure 20-6. Appropriate axes for the extrapolation of $\mathscr{E}'$ at 25°C to determine $\mathscr{E}^\circ$ for the cell: H_2, $HCl(m_2)$, AgCl, Ag.

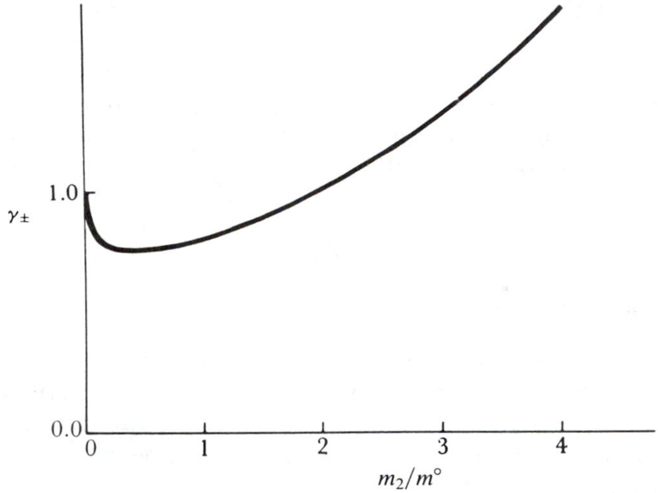

Figure 20-7. Mean activity coefficients of aqueous hydrochloric acid at 25°C. Based on data from H. S. Harned and B. B. Owen, *The Physical Chemistry of Electrolyte Solutions*, 3d ed., Reinhold, New York, 1958, p. 466.

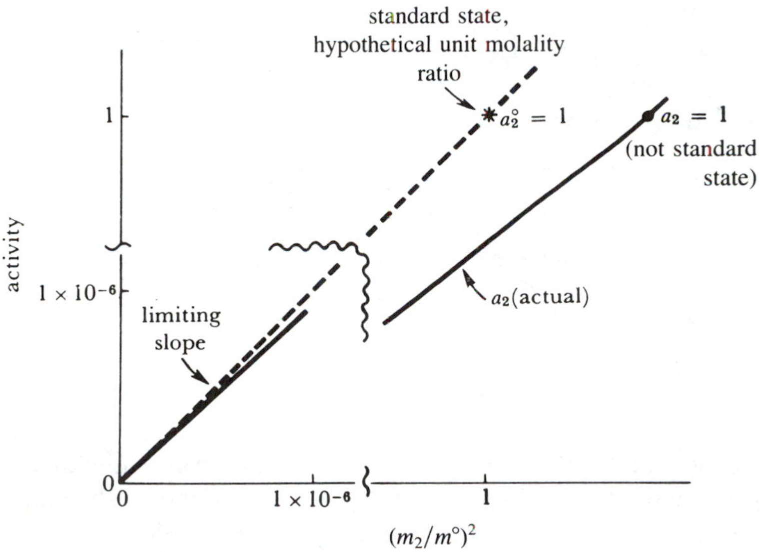

Figure 20-8. Activities of aqueous solutions of hydrochloric acid.

The mean activity coefficient, $\gamma_{\pm}$, for HCl is plotted against m_2/m° in Figure 20-7 as an example of the behavior of a uni-univalent electrolyte in aqueous solution. From these data [1] the activity, a_2, has been calculated, and it is illustrated as a function of the square of the molality in Figure 20-8; the dashed line indicates the limiting slope. The point on the dashed line corresponding to $(m_2/m^{\circ}) = 1$ is the activity of the hypothetical unit molality ratio standard state.

Solubility Measurements

As long as a pure solid, A, is in equilibrium with a dissolved solute, A′, the activity of the dissolved solute must be constant, since the activity of the solid, a_s, is constant at a fixed temperature and pressure. Thus any change in the solubility with the addition of other electrolytes must be due to changes in the activity coefficient, $\gamma_{\pm}$. For the reaction

$$\text{A(pure solid)} = \text{A(solute in solution)}$$

the equilibrium constant for a uni-univalent solute is

$$\begin{aligned} K &= \frac{a_2}{a_{2(s)}} \\ &= \frac{a_2}{1} \\ &= a_+ a_- \\ &= a_{\pm}^2 \\ &= (m_{\pm}/m^{\circ})^2\, \gamma_{\pm}^2 \end{aligned} \tag{20-39}$$

or, in logarithmic form,

$$\log K^{1/2} = \log(m_{\pm}/m^{\circ}) + \log \gamma_{\pm} \tag{20-40}$$

Since

$$\lim_{I \to 0} \gamma_{\pm} = 1$$

then

$$\lim_{I \to 0} \log(m_{\pm}/m^{\circ}) = \log K^{1/2} \tag{20-41}$$

To obtain a value of K by extrapolation, the appropriate functions to plot are the logarithm of the molality ratio against the square root of the ionic strength. Such a graph is shown in Figure 20-9 for solutions of AgCl in various aqueous electrolytes. From the extrapolated value for the constant K, the following

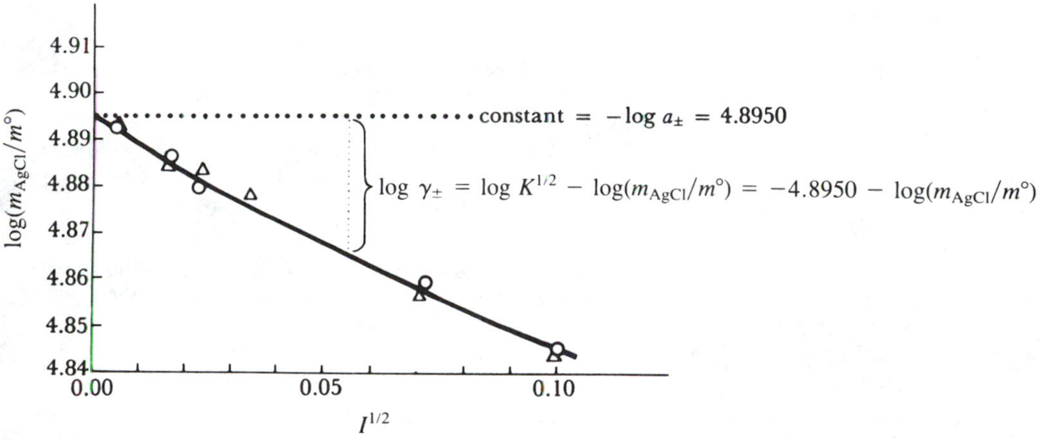

Figure 20-9. Activity coefficients of AgCl in aqueous solution obtained from variation of solubility with ionic strength. Added electrolyte: ○, $NaNO_3$; △, KNO_3. Based on data from S. Popoff and E. W. Neuman, *J. Phys. Chem.* **34**, 1853 (1930).

equation can be written for the mean activity coefficient of AgCl (since $m_\pm$ equals m_{AgCl}):

$$\log \gamma_\pm = -4.8950 - \log(m_{AgCl}/m°) \qquad (20\text{-}42)$$

Colligative Property Measurement

As we saw in Section 19-4, the activity coefficient of a nonelectrolyte solute can be calculated from the activity coefficient of the solvent, which, in turn, can be obtained from the measurement of colligative properties such as vapor pressure lowering, freezing point depression, or osmotic pressure. We used the Gibbs–Duhem equation in the form (Equation 19-33)

$$d \ln \gamma_2 = -\frac{X_1}{X_2}\, d \ln \gamma_1 \qquad (20\text{-}43)$$

The activity coefficients of solute and solvent are of comparable magnitudes in dilute solutions of nonelectrolytes, so that Equation 19-33 is a useful relationship. But the activity coefficients of an electrolyte solute differ substantially from unity even in very dilute solutions in which the activity coefficient of the solvent differs from unity by less than 1×10^{-3}. The data in the first three columns of Table 20-3 illustrate the situation. It can be seen that the calculation of the activity coefficient of solute from the activity coefficient of H_2O would be imprecise at best.

To deal with this problem, Bjerrum [2] suggested that the deviation of solvent behavior from Raoult's law be described by the *osmotic coefficient g*

Table 20-3 Activity Coefficients and Osmotic Coefficients[a]

$[KNO_3]/(\text{mol dm}^{-3})$	$\gamma(KNO_3)$	$\gamma(H_2O)$	$g(H_2O)$
0.01	0.8993	1.00001	0.9652
0.05	0.7941	1.00005	0.9252
0.1	0.7259	1.0002	0.8965
1.0	0.3839	1.0056	0.6891

[a]G. Scatchard, S. S. Prentice, and P. T. Jones, *J. Am. Chem. Soc.* **54**, 2690 (1932). The activity coefficient of the solute is based on the unit molarity ratio standard state, whereas the activity coefficient of the solvent is based on the unit mole fraction standard state.

rather than by the activity coefficient γ_1. The osmotic coefficient is defined by the relationships

$$\mu_1 = \mu_1^\circ + gRT \ln X_1 \qquad (20\text{-}44)$$

and

$$\lim_{X_1 \to 1} g = 1$$

The greater sensitivity, and hence usefulness, of g over γ_1, for solutions of electrolytes can be seen from the values in the third and fourth columns of Table 20-3.

Equating the expressions for μ_1 from Equation 17-1 and Equation 20-44, we obtain

$$\ln a_1 = g \ln X_1$$

or

$$g = \frac{\ln a_1}{\ln X_1} \qquad (20\text{-}45)$$

When the activity of the solvent is determined from a colligative property, then g can be calculated with Equation 20-45. If we use the osmotic pressure as an example, we can combine Equation 16-22 and Equation 17-1 to obtain the expression

$$\mu_1^\bullet(P_0) = \mu_1^\circ(P) + RT \ln a_1(P)$$

or

$$RT \ln a_1(P) = \mu_1^\bullet(P_0) - \mu_1^\circ(P) \qquad (20\text{-}46)$$

Each of the two terms on the right in Equation 20-46 refers to pure solvent at the same temperature, but at a pressure of P_0 for the first term and P for the second term. Thus, the right side of Equation 20-46 is also given by

$$\mu_1^\bullet(P_0) - \mu_1^\circ(P) = \int_P^{P_0} \left(\frac{\partial \mu_1^\bullet}{\partial P}\right) dP$$

$$= \int_P^{P_0} \bar{V}_1^\bullet \, dP \qquad (20\text{-}47)$$

Since liquids are relatively incompressible, $\bar{V}_1^\bullet$ can be assumed to be independent of pressure, and Equation 20-47 can be integrated to obtain

$$\mu_1^\bullet(P_0) - \mu_1^\circ(P) = \bar{V}_1^\bullet(P_0 - P)$$

If we again define $P - P_0$ as the osmotic pressure, Π, and substitute in Equation 20-46, we obtain

$$\ln a_1(P) = -\frac{\bar{V}_1^\bullet \Pi}{RT} \qquad (20\text{-}48)$$

As indicated, the activity of solvent is at pressure P, and the small correction to pressure P_0 can be obtained if desired from Equation 17-31. Substituting from Equation 20-48 into Equation 20-45, we find

$$g = \frac{-\bar{V}_1^\bullet \Pi / RT}{\ln X_1} \qquad (20\text{-}49)$$

For a solution of electrolytes,

$$X_1 = \frac{n_1}{n_1 + \sum_+ n_+ + \sum_- n_-}$$

$$= \frac{n_1}{n_1 + \nu n_2} \qquad (20\text{-}50)$$

From Equation 20-50

$$\ln X_1 = \ln \left(\frac{n_1}{n_1 + \nu n_2}\right)$$

$$= \ln \left(\frac{1}{1 + (\nu n_2/n_1)}\right)$$

$$= -\ln[1 + (\nu n_2/n_1)] \qquad (20\text{-}51)$$

For the dilute solutions for which the osmotic coefficient is most useful, the natural logarithm in Equation 20-51 can be expanded in a Taylor's series, and terms of higher powers can be neglected. The result is

$$\ln X_1 = \frac{-\nu n_2}{n_1} \qquad (20\text{-}52)$$

Substituting in Equation 20-49, we obtain

$$\Pi = (g\nu RT)\left(\frac{n_2}{n_1\bar{V}_1^\bullet}\right) \tag{20-53}$$

The product $n_1\bar{V}_1^\bullet$ is essentially equal to V, the volume of solution, for dilute solutions, so that Equation 20-53 reduces to

$$\Pi = g\nu c_2 RT \tag{20-54}$$

It can be observed that g is the ratio between the observed osmotic pressure and the osmotic pressure that would be observed for a completely dissociated electrolyte that follows Henry's law, hence the name, osmotic coefficient. A similar result can be obtained for the freezing point depression and the vapor pressure lowering.

Once values of g as a function of solution composition have been obtained, the Gibbs–Duhem equation can be used to relate the osmotic coefficient of the solvent to the activity coefficient of the solute. For this purpose the chemical potential of the solvent is expressed as in Equation 20-44, with the approximation given in Equation 20-52, so that

$$\mu_1 = \mu_1^\circ - RT\nu g\frac{n_2}{n_1} \tag{20-55}$$

If we describe the composition of the solution in terms of molalities, Equation 20-55 becomes

$$\mu_1 = \mu_1^\circ - RT\nu g m_2 M_1$$

and

$$d\mu_1 = -RT\nu M_1(g\,dm_2 + m_2\,dg) \tag{20-56}$$

The chemical potential of the solute is

$$\begin{aligned}\mu_2 &= \mu_2^\circ + RT\ln a_2\\ &= \mu_2^\circ + RT\ln(m_\pm/m^\circ)^\nu + RT\ln(\gamma_\pm)^\nu\\ &= \mu_2^\circ + \nu RT\ln(m_\pm/m^\circ) + \nu RT\ln\gamma_\pm\end{aligned} \tag{20-57}$$

and

$$\begin{aligned}d\mu_2 &= \nu RTd\ln m_\pm + \nu RTd\ln\gamma_\pm\\ &= \nu RTd\ln m_2 + \nu RTd\ln\gamma_\pm\end{aligned} \tag{20-58}$$

the replacement of $m_\pm$ by m_2 following from Equation 20-24 if $\nu_+ = \nu_- = 1$.

Using the Gibbs–Duhem equation in the form

$$d\mu_1 = -\left(\frac{n_2}{n_1}\right)d\mu_2$$

we obtain

$$d \ln \gamma_{\pm} = -\left(\frac{1-g}{m_2}\right)dm_2 + dg \tag{20-59}$$

which is *Bjerrum's equation*. If Equation 20-59 is integrated from the infinitely dilute solution to some finite but still dilute molality, the result is

$$\int_0^{\ln \gamma_{\pm}} d \ln \gamma_{\pm} = -\int_0^{m_2}\left(\frac{1-g}{m_2}\right)dm_2 + \int_1^g dg$$

or

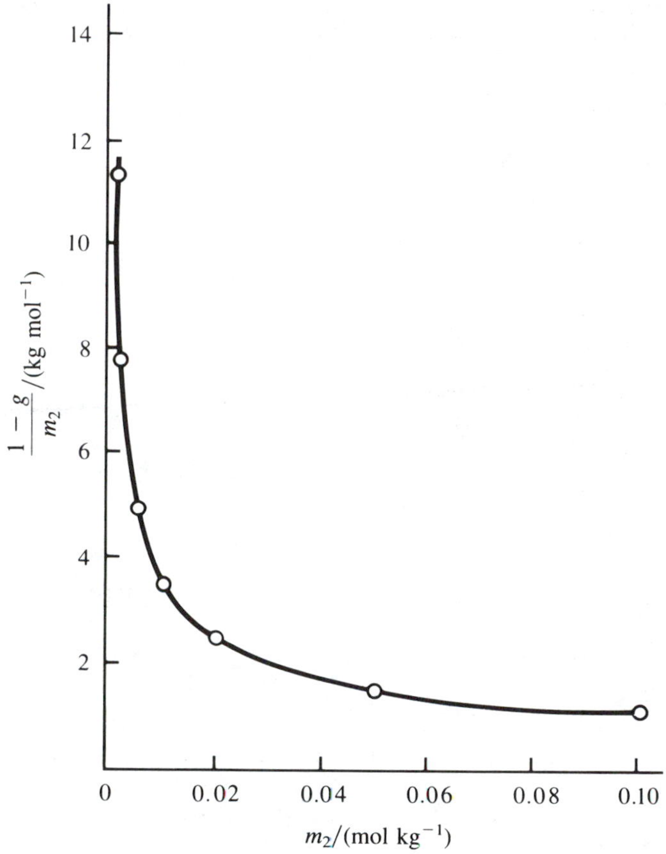

Figure 20-10. A plot of $(1 - g)/m_2$ against m_2 from the freezing point data on KNO_3 solutions. [G. Scatchard, S. S. Prentice, and P. T. Jones, *J. Am. Chem. Soc.* **54**, 2690 (1932).] See Equation 20-60.

$$\ln \gamma_{\pm} = -\int_0^{m_2} \left(\frac{1-g}{m_2}\right) dm_2 + (g-1) \qquad (20\text{-}60)$$

The integral in Equation 20-60 is usually evaluated graphically or numerically. Though both $1 - g$ and m_2 go to zero as $m_2 \rightarrow 0$, the ratio has a finite limit. Such a finite limit is not apparent from a plot of $(1 - g)/m_2$ against m_2 for electrolytes, as can be seen from Figure 20-10. But, if the integral in Equation 20-60 is transformed to

$$2\int_0^{(m_2)^{1/2}} \left(\frac{1-g}{m_2^{1/2}}\right) d(m_2^{1/2}) \qquad (20\text{-}61)$$

the finite limit can be seen clearly, as in Figure 20-11.

At concentrations beyond the region of validity of the expressions we have used for the osmotic coefficient, the activity coefficient of the solvent is sufficiently different from 1 that it can be used to calculate the activity coefficient of the solute.

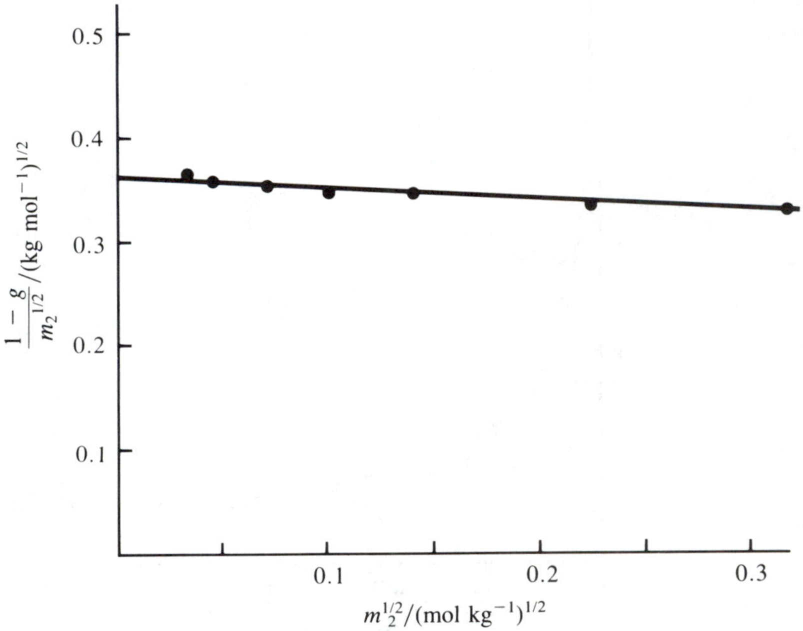

Figure 20-11. A plot of $(1 - g)/m_2^{1/2}$ against $m_2^{1/2}$ from the freezing point data on KNO_3 solutions. [G. Scatchard, S. S. Prentice, and P. T. Jones, *J. Am. Chem. Soc.* **54**, 2690 (1932).] See Equation 20-61.

20-3 ACTIVITY COEFFICIENTS OF SOME STRONG ELECTROLYTES

Experimental Values

With the experimental methods described, as well as with several others, the activities of numerous strong electrolytes of various valence types have been calculated. Many of these data have been assembled and examined critically by Harned and Owen [3].

The behavior of a few typical electrolytes is illustrated in Figure 20-12. By definition $\gamma_{\pm}$ is one at zero molality for all electrolytes. Furthermore, in every case $\gamma_{\pm}$ decreases rapidly with increasing molality at low values of m_2. However, the steepness of this initial drop varies with the valence type of the electrolyte. For a given valence type $\gamma_{\pm}$ is substantially independent of the chemical nature of the constituent ions, as long as m_2 is below about 0.01. At higher concentrations curves for $\gamma_{\pm}$ begin to separate widely and to exhibit marked specific ion effects.

Theoretical Correlation

No adequate theoretical model based on the atomic characteristics of the ions has been developed yet that is capable of accounting for the thermodynamic properties of aqueous solutions in wide ranges of concentration. However, for dilute solutions of completely ionized electrolytes, expressions have been derived [4] that predict exactly the limiting behavior of activity coefficients in an infinitely dilute solution, and that provide very useful equations for describing these quantities at small, finite concentrations. Although it is beyond the objectives of this text to consider the development of the Debye–

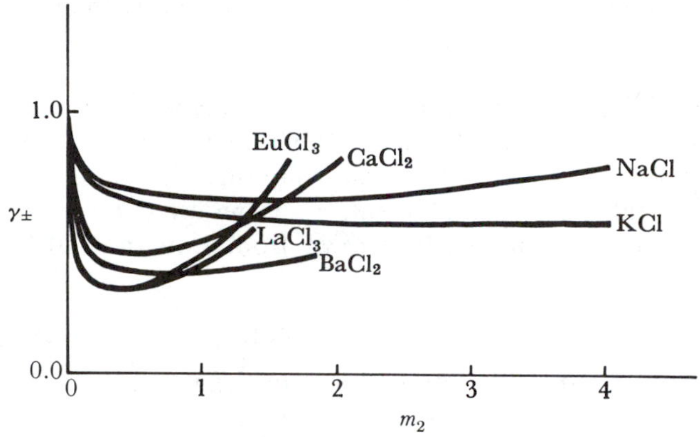

Figure 20-12. Mean activity coefficients at 25°C for some typical electrolytes in aqueous solution.

Table 20-4 Values of Constants in Debye–Hückel Equation for Activity Coefficients in Aqueous Solutions[a]

$t/°C$	A	$B \times 10^{-8}$
0	0.4883	0.3241
5	0.4921	0.3249
10	0.4960	0.3258
15	0.5002	0.3267
20	0.5046	0.3276
25	0.5091	0.3286
30	0.5139	0.3297
35	0.5189	0.3307
40	0.5241	0.3315
45	0.5295	0.3330
50	0.5357	0.3342
55	0.5410	0.3354
60	0.5470	0.3366

[a]From *The Physical Chemistry of Electrolytic Solutions*, 3d ed., by Harned and Owen. © 1958 by Litton Educational Publishing, Inc. Reprinted by permission of Van Nostrand Reinhold Company.

Hückel theory, it is desirable to present some of the final results because they are of great value in the treatment of experimental data.

According to the Debye–Hückel theory, in the limit of the infinitely dilute solution, individual-ion activity coefficients are given by the equation

$$\log \gamma_i = -Az_i^2 I^{1/2} \tag{20-62}$$

in which A is a constant for a given solvent at a specified temperature. Values of A for aqueous solutions are listed in Table 20-4. Equation 20-62 has been found useful up to $I^{1/2}$ near 0.1, that is, for solutions with ionic strengths as high as 0.01. The mean activity coefficient, $\gamma_\pm$, is required for comparison with experimental data, and it can be shown from Equation 20-62 that *for an electrolyte with two kinds of ions* [5]

$$\log \gamma_\pm = -A\,|z_+ z_-|\,I^{1/2} \tag{20-63}$$

in which z_+ and z_- are the charge (equal to the valence) of the cation and anion, respectively.

The symbol $|z_+ z_-|$ is used to indicate absolute value, without regard to the sign of the charges.

Table 20-5 Effective Diameters of Hydrated Ions in Aqueous Solutions [6]

$10^8 a_i/\text{cm}$	
	Inorganic Ions: Charge 1
9	H^+
6	Li^+
4–4.5	Na^+, $CdCl^+$, ClO_2^-, IO_3^-, HCO_3^-, $H_2PO_4^-$, HSO_3^-, $H_2AsO_4^-$, $Co(NH_3)_4(NO_2)_2^+$
3.5	OH^-, F^-, NCS^-, NCO^-, HS^-, ClO_3^-, ClO_4^-, BrO_3^-, IO_4^-, MnO_4^-
3	K^+, Cl^-, Br^-, I^-, CN^-, NO_2^-, NO_3^-
2.5	Rb^+, Cs^+, NH_4^+, Tl^+, Ag^+
	Inorganic Ions: Charge 2
8	Mg^{2+}, Be^{2+}
6	Ca^{2+}, Cu^{2+}, Zn^{2+}, Sn^{2+}, Mn^{2+}, Fe^{2+}, Ni^{2+}, Co^{2+}
5	Sr^{2+}, Ba^{2+}, Ra^{2+}, Cd^{2+}, Hg^{2+}, S^{2-}, $S_2O_4^{2-}$, WO_4^{2-}
4.5	Pb^{2+}, CO_3^{2-}, SO_3^{2-}, MoO_4^{2-}, $Co(NH_3)_5Cl^{2+}$, $Fe(CN)_5NO^{2-}$
4	Hg_2^{2+}, SO_4^{2-}, $S_2O_3^{2-}$, $S_2O_8^{2-}$, SeO_4^{2-}, CrO_4^{2-}, HPO_4^{2-}, $S_2O_6^{2-}$
	Inorganic Ions: Charge 3
9	Al^{3+}, Fe^{3+}, Cr^{3+}, Sc^{3+}, Y^{3+}, La^{3+}, In^{3+}, Ce^{3+}, Pr^{3+}, Nd^{3+}, Sm^{3+}
6	$Co(\text{ethylenediamine})_3^{3+}$
4	PO_4^{3-}, $Fe(CN)_6^{3-}$, $Cr(NH_3)_6^{3+}$, $Co(NH_3)_6^{3+}$, $Co(NH_3)_5H_2O^{3+}$
	Inorganic Ions: Charge 4
11	Th^{4+}, Zr^{4+}, Ce^{4+}, Sn^{4+}
6	$Co(S_2O_3)(CN)_5^{4-}$
5	$Fe(CN)_6^{4-}$
	Inorganic Ions: Charge 5
9	$Co(SO_3)_2(CN)_4^{5-}$
	Organic Ions: Charge 1
8	$(C_6H_5)_2CHCOO^-$, $(C_3H_7)_4N^+$
7	$[OC_6H_2(NO_2)_3]^-$, $(C_3H_7)_3NH^+$, $CH_3OC_6H_4COO^-$
6	$C_6H_5COO^-$, $C_6H_4OHCOO^-$, $C_6H_4ClCOO^-$, $C_6H_5CH_2COO^-$, $CH_2{=}CHCH_2COO^-$, $(CH_3)_2CHCH_2COO^-$, $(C_2H_5)_4N^+$, $(C_3H_7)_2NH_2^+$
5	$CHCl_2COO^-$, CCl_3COO^-, $(C_2H_5)_3NH^+$, $(C_3H_7)NH_3^+$
4.5	CH_3COO^-, CH_2ClCOO^-, $(CH_3)_4N^+$, $(C_2H_5)_2NH_2^+$, $NH_2CH_2COO^-$
4	$NH_3^+CH_2COOH$, $(CH_3)_3NH^+$, $C_2H_5NH_3^+$
3.5	$HCOO^-$, $H_2\text{-citrate}^-$, $CH_3NH_3^+$, $(CH_3)_2NH_2^+$
	Organic Ions: Charge 2
7	$OOC(CH_2)_5COO^{2-}$, $OOC(CH_2)_6COO^{2-}$, Congo red anion^{2-}
6	$C_6H_4(COO)_2^{2-}$, $H_2C(CH_2COO)_2^{2-}$, $(CH_2CH_2COO)_2^{2-}$
5	$H_2C(COO)_2^{2-}$, $(CH_2COO)_2^{2-}$, $(CHOHCOO)_2^{2-}$
4.5	$(COO)_2^{2-}$, H-citrate^{2-}
	Organic Ions: Charge 3
5	Citrate^{3-}

Table 20-6 Activity Coefficients of Individual Ions in Water at 25°C [6]

Parameter	Volume ionic strength,[a] I							
$10^8 a_i$/cm	0.005	0.001	0.0025	0.005	0.01	0.025	0.05	0.1
			Ion Charge 1					
9	0.975	0.967	0.950	0.933	0.914	0.88	0.86	0.83
8	0.975	0.966	0.949	0.931	0.912	0.880	0.85	0.82
7	0.975	0.965	0.948	0.930	0.909	0.875	0.845	0.81
6	0.975	0.965	0.948	0.929	0.907	0.87	0.835	0.80
5	0.975	0.964	0.947	0.928	0.904	0.865	0.83	0.79
4.5	0.975	0.964	0.947	0.928	0.902	0.86	0.82	0.775
4	0.975	0.964	0.947	0.927	0.901	0.855	0.815	0.77
3.5	0.975	0.964	0.946	0.926	0.900	0.855	0.81	0.76
3	0.975	0.964	0.945	0.925	0.899	0.85	0.805	0.755
2.5	0.975	0.964	0.945	0.924	0.898	0.85	0.80	0.75
			Ion Charge 2					
8	0.906	0.872	0.813	0.755	0.69	0.595	0.52	0.45
7	0.906	0.872	0.812	0.755	0.685	0.58	0.50	0.425
6	0.905	0.870	0.809	0.749	0.675	0.57	0.485	0.405
5	0.903	0.868	0.805	0.744	0.67	0.555	0.465	0.38
4.5	0.903	0.868	0.805	0.742	0.665	0.55	0.455	0.37
4	0.903	0.867	0.803	0.740	0.660	0.545	0.445	0.355
			Ion Charge 3					
9	0.802	0.738	0.632	0.54	0.445	0.325	0.245	0.18
6	0.798	0.731	0.620	0.52	0.415	0.28	0.195	0.13
5	0.796	0.728	0.616	0.51	0.405	0.27	0.18	0.115
4	0.796	0.725	0.612	0.505	0.395	0.25	0.16	0.095
			Ion Charge 4					
11	0.678	0.588	0.455	0.35	0.255	0.155	0.10	0.065
6	0.670	0.575	0.43	0.315	0.21	0.105	0.055	0.027
5	0.668	0.57	0.425	0.31	0.20	0.10	0.048	0.021
			Ion Charge 5					
9	0.542	0.43	0.28	0.18	0.105	0.045	0.020	0.009

[a]Concentrations expressed in mol L^{-1}.

At ionic strengths near 0.01 it is convenient to use the more complete form of the Debye–Hückel expression:

$$\log \gamma_i = \frac{-Az_i^2 I^{1/2}}{1 + Ba_i I^{1/2}} \tag{20-64}$$

in which A has the same significance as in Equation 20-62, B is a constant for a given solvent at a specified temperature (Table 20-4), and a_i may be thought of as the "effective diameter" of the ion *in the solution*. Since no independent method is available for evaluating a_i, this quantity is an empirical parameter, but the a_i's obtained are of a magnitude expected for ion sizes.

In connection with the use of Equation 20-64, Kielland [6] has assembled a series of effective diameters for a large number of ions, grouped according to charge. These values of a_i are listed in Table 20-5. With these values of a_i, he has calculated individual-ion activity coefficients from Equation 20-64, which are summarized in Table 20-6. Although the activity coefficient of an individual ion cannot be determined or perhaps even defined accurately [7], these tables provide very useful semiempirical information for estimating γ's for use in calculations of free energy changes in solutions of ionic strength below 0.1.

For solutions above $I = 0.1$, various extensions of the Debye–Hückel theory have been proposed. Attempts also have been made to superimpose terms, taking account of ion–solvent interactions. Some of these efforts are described in a monograph by Robinson and Stokes [8]. Other descriptions of treatments beyond the Debye–Hückel theory are discussed in the references [9].

EXERCISES

1. Prove that the following equation is valid for an electrolyte that dissociates into ν particles:

$$\left(\frac{\partial \ln \gamma_{\pm}}{\partial T}\right)_{P,m_2} = -\frac{L_2}{\nu RT^2} \tag{20-65}$$

2. For the equilibrium between a pure solute and its saturated solution,

$$\text{solute}_{(\text{pure})} = \text{solute}_{(\text{satd soln})}$$

the equilibrium constant, K, is given by

$$K = \frac{a_{2(\text{satd})}}{a_{2(\text{pure})}} = \frac{a_{2(\text{satd})}}{1} \tag{20-66}$$

a. Show that

$$\left(\frac{\partial \ln a_2}{\partial T}\right)_{\text{satd}} = -\frac{L_2}{RT^2} \tag{20-67}$$

b. For the general case of an electrolyte that dissociates into ν particles show that

$$\nu\left[\left(\frac{\partial \ln \gamma_{\pm}}{\partial T}\right)_{\text{satd}} + \left(\frac{\partial \ln m_{\pm}}{\partial T}\right)_{\text{satd}}\right] = -\frac{L_2}{RT^2} \tag{20-68}$$

c. Considering $\ln \gamma_{\pm}$ as a function of temperature and molality (pressure being maintained constant) show that

$$\left(\frac{\partial \ln \gamma_\pm}{\partial T}\right)_{\text{satd}} = \left(\frac{\partial \ln \gamma_\pm}{\partial T}\right)_{m_2} + \left(\frac{\partial \ln \gamma_\pm}{\partial m_2}\right)_T \left(\frac{\partial m_2}{\partial T}\right)_{\text{satd}} \quad (20\text{-}69)$$

d. Derive the equation

$$\left(\frac{\partial \ln \gamma_\pm}{\partial T}\right)_{m_2} + \left(\frac{\partial m_2}{\partial T}\right)_{\text{satd}} \left[\left(\frac{\partial \ln \gamma_\pm}{\partial m_2}\right)_T + \left(\frac{1}{m_2}\right)_{\text{satd}}\right] = -\frac{L_2}{\nu RT^2} \quad (20\text{-}70)$$

e. Using Equation 20-65, derive the following expression:

$$\Delta H_{\text{soln}} = \nu RT^2 \left(\frac{\partial m_2}{\partial T}\right)_{\text{satd}} \left[\left(\frac{\partial \ln \gamma_\pm}{\partial m_2}\right)_T + \left(\frac{1}{m_2}\right)_{\text{satd}}\right] \quad (20\text{-}71)$$

in which ΔH_{soln} is the heat absorbed per mole of solute dissolved in the nearly saturated solution.

f. To what does Equation 20-71 reduce if the solute is a nonelectrolyte?

g. To what does Equation 20-71 reduce if the solution is ideal?

3. Calculate the electromotive force of the cell pair

$$H_2(g),\ HCl(m_2 = 4),\ AgCl,\ Ag{-}Ag,\ AgCl,\ HCl(m_2 = 10),\ H_2(g)$$

from vapor pressure data (Lewis and Randall, *Thermodynamics*, McGraw-Hill, New York, 1923, p. 330). $p_{HCl} = 0.2395 \times 10^{-4}$ atm for the 4-molal solution, and $p_{HCl} = 55.26 \times 10^{-4}$ atm for the 10-molal solution.

4. The electromotive force at 25°C of the cell

$$H_2(g),\ HCl(m_2 = 0.002951),\ AgCl,\ Ag$$

is 0.52393 V when the apparent barometric height (as read on a brass scale) is 75.10 cm Hg at 23.8°C and when the hydrogen is bubbled to the atmosphere through a column of solution 0.68 cm high. Calculate the partial pressure of the hydrogen, and the electromotive force of the cell when the partial pressure of hydrogen is 1 atm.

5. Unpublished data of T. F. Young and N. Anderson on the potentials at 25°C of the cell

$$H_2,\ HCl(m_2),\ AgCl,\ Ag$$

are given in Table 20-7.

a. Plot $\mathscr{E}'$ against an appropriate composition variable, draw a smooth curve, and determine $\mathscr{E}^\circ$ by extrapolation.

b. On the graph in (a) show the significance of the activity coefficient.

c. Draw the asymptote predicted by the Debye–Hückel limiting law for the curve in (a).

Table 20-7

$\sqrt{m_\pm}$/(mol kg^{-1})$^{1/2}$	$m_\pm$/(mol kg^{-1})[a]	$\mathscr{E}$/V	$\log(m_\pm/m^\circ)$	$\mathscr{E}'$/V
0.054322	0.0029509	0.52456	−2.53005	0.22524
0.044115	0.0019461	0.54541	−2.71083	0.22471
0.035168	0.0012368	0.56813	−2.90770	0.22413
0.029908	0.0008945	0.58464	−3.04842	0.22399
0.027024	0.0007303	0.59484	−3.13650	0.22378
0.020162	0.0004065	0.62451	−3.39094	0.22334
0.015033	0.00022599	0.65437	−3.64591	0.22303
0.011631	0.00013528	0.68065	−3.86877	0.22296
0.009704	0.00009417	0.69914	−4.02606	0.22283
0.007836	0.00006140	0.72096	−4.21185	0.22267
0.007343	0.00005392	0.72759	−4.26827	0.22263
0.005376	0.000028901	0.75955	−4.53909	0.22255

[a]In the most dilute solutions studied, the molality of chloride ion was considerably greater than the molality of the hydrogen ion. Therefore the mean molality, $m_\pm$, is tabulated, rather than the molality of either ion.

d. Determine the mean activity coefficient of the ions of HCl at $m_\pm = 0.001$, 0.01, and 0.1, respectively. Compare these values with those computed from the Debye–Hückel limiting law.

e. What error is introduced into the calculated activity coefficients by an error of 0.00010 V in $\mathscr{E}$ or $\mathscr{E}^\circ$?

6. Table 20-8 lists S. Popoff and E. W. Neuman's values [*J. Phys. Chem.* **34**, 1853 (1930)] of the solubility of silver chloride in water containing "solvent" electrolytes at the concentrations indicated. According to the same authors, the solubility of silver chloride in pure water is 1.278×10^{-5} mole liter^{-1}.

   a. Using distinctive symbols to represent each of the four series of data, plot (with reference to a single pair of axes) the solubility of silver chloride versus the concentration of each solvent electrolyte. On the same graph plot the solubility of silver chloride against three times the concentration of barium nitrate.

   b. Show that these data are in accordance with the ionic-strength principle.

   c. Verify several of the tabulated values of the total ionic strength; then plot the logarithm of the reciprocal of the solubility versus the square root of the ionic strength.

   d. Draw a line representing the Debye–Hückel limiting law for comparison with the data.

   e. Determine the activity of silver chloride in a solution containing only silver chloride and in solutions in which the ionic strength is 0.001 and 0.01, respectively.

Table 20-8

Concentration of solvent electrolyte/ (mole liter^{-1})	Concentration of AgCl/ (mole liter^{-1})	I	$I^{1/2}$	$-\log[(AgCl)/C°]$
KNO_3				
0.000 012 80	1.280×10^{-5}	0.000 025 6	0.005 06	4.892 8
0.000 260 9	1.301	0.000 273 9	0.016 55	4.885 7
0.000 509 0	1.311	0.000 522 1	0.022 85	4.882 4
0.001 005	1.325	0.001 018	0.031 91	4.877 8
0.004 972	1.385	0.004 986	0.070 61	4.858 6
0.009 931	1.427	0.009 945	0.099 72	4.845 6
$NaNO_3$				
0.000 012 81	1.281	0.000 025 0	0.005 06	4.892 5
0.000 264 3	1.300	0.000 277 3	0.016 65	4.886 1
0.000 515 7	1.315	0.000 528 9	0.023 00	4.881 1
0.005 039	1.384	0.005 053	0.071 08	4.858 9
0.010 076	1.428	0.010 090	0.100 45	4.845 3
HNO_3				
0.000 012 8	1.280	0.000 025 6	0.005 06	4.892 8
0.000 723 3	1.318	0.000 736 5	0.027 14	4.880 1
0.002 864	1.352	0.002 877 5	0.053 64	4.869 0
0.005 695	1.387	0.005 709	0.075 56	4.857 9
0.009 009	1.422	0.009 023	0.094 99	4.847 1
$Ba(NO_3)_2$				
0.000 006 40	1.280	0.000 032 0	0.005 66	4.892 8
0.000 036 15	1.291	0.000 121 4	0.011 02	4.889 1
0.001 211 08	1.309	0.000 646 3	0.025 42	4.883 1
0.000 706 4	1.339	0.002 133	0.046 18	4.873 2
0.001 499	1.372	0.004 511	0.067 16	4.862 7
0.002 192	1.394	0.006 590	0.081 18	4.855 7
0.003 083	1.421	0.009 263	0.096 24	4.847 4

f. Calculate the solubility product of silver chloride. What is the activity of silver chloride in any of the saturated solutions?

7. From the vapor pressure data in the International Critical Tables, Vol. III, p. 297, calculate the activity of the water in 1.0-, 2.0-, 2.8-, and 4.0-molal NaCl solutions at 25°C.

8. Calculate the emf of each of the following cells at 25°C. Use approximate values of the activity coefficient as calculated from the Debye–Hückel limiting law.

   a. H_2, $HCl(m_2 = 0.0001)$, Cl_2—Cl_2, $HCl(m_2 = 0.001)$, H_2.

   b. Mg, $MgSO_4(m_2 = 0.001)$, Hg_2SO_4, Hg—Hg, Hg_2SO_4, $MgSO_4(m_2 = 0.0001)$, Mg.

9. Table 20-9 contains data of H. S. Harned and L. F. Nims [*J. Am. Chem. Soc.* **54**, 423 (1932)] for the cell

Table 20-9

t/°C	$\mathscr{E}$/V	$(\mathscr{E}/T)$/(V/K)
15	0.18265	0.00063398
20	0.18663	0.00063675
25	0.19044	0.00063885
30	0.19407	0.00064028
35	0.19755	0.00064119

Ag, AgCl, NaCl(m_2 = 4), Na(amalgam)—Na(amalgam), NaCl(m_2 = 0.1), AgCl, Ag.

a. Write the cell reaction.

b. Calculate $\Delta\bar{H}_2$ or $\Delta\bar{L}_2$ for the reaction in (a) at 25°C.

c. Compare the result in (b) with that which you would obtain from the direct calorimetric data of Table 20-10.

Table 20-10

m_2/(mol kg^{-1})	$\bar{L}_2$/(cal mol^{-1}) at 25°C
0.0100	56
0.0400	86
0.0900	95
0.1600	86
0.2500	63
0.4900	−14
0.6400	−65
1.0000	−186
1.9600	−452
2.5600	−561
3.2400	−643
4.0000	−688
4.8400	−679
5.7600	−605
6.0025	−573

d. How precise is the result in (b) if $\mathscr{E}$ can be measured to ±0.00010 V?

e. Obtain activity coefficients for the solutions of NaCl in this cell from Harned and Owen, *Physical Chemistry of Electrolytic Solutions*, 3d ed., Reinhold, New York, 1958, p. 731. Calculate the emf of the cell at 25°C. Compare your result with the value listed in Table 20-9.

10. The solubility (moles per kilogram of H_2O) of cupric iodate, $Cu(IO_3)_2$, in aqueous solutions of KCl at 25°C, as determined by R. M. Keefer [*J. Am. Chem. Soc.* **70**, 476 (1948)], is given in Table 20-11.

Table 20-11

m_{KCl}/(mol kg^{-1})	$m_{(satd)Cu(IO_3)_2}$/(mol kg^{-1})
0.00000	3.245×10^{-3}
0.00501	3.398
0.01002	3.517
0.02005	3.730
0.03511	3.975
0.05017	4.166
0.07529	4.453
0.1005	4.694

a. Plot the logarithm of the reciprocal of the solubility versus $I^{1/2}$ and extrapolate to infinite dilution.

b. Using the Debye–Hückel limiting law to evaluate activity coefficients, show that

$$\log (m_{Cu^{2+}})(m_{IO_3^-})^2 = \log K + 3.051\, I^{1/2} \tag{20-72}$$

c. Compute log K. Keefer reports a value of −7.1353.

d. Draw a line representing the limiting law on the graph in (a).

e. Keefer has found that the equation

$$\log (m_{Cu^{2+}})(m_{IO_3^-})^2 = -7.1353 + \frac{3.036\, I^{1/2}}{1 + 1.08\, I^{1/2}} \tag{20-73}$$

represents the solubility data given in the table with great precision. (The constant 3.036 is slightly different from that of 3.051 used in Equation 20-72 because it is based on older values of the electronic charge.) For example, at the highest concentration of KCl, the calculated solubility of $Cu(IO_3)_2$ is 4.697×10^3 mole (kg H_2O)$^{-1}$. Verify this calculation.

REFERENCES

1. H. S. Harned and B. B. Owen, *The Physical Chemistry of Electrolytic Solutions*, 3d ed., Reinhold, New York, 1958, pp. 430–431.
2. N. Bjerrum, *Z. Electrochem.* **24**, 321 (1918); *Z. phys. Chem.* **104**, 406 (1932).
3. Harned and Owen, *Chemistry of Electrolytic Solutions.*
4. P. Debye and E. Hückel, *Physik. Z.* **24**, 185 (1923).

5. For the general case of any number of ions see Harned and Owen, *Chemistry of Electrolyte Solutions*, p. 62.

6. J. Kielland, *J. Am. Chem. Soc.* **59**, 1675 (1937).

7. E. A. Guggenheim, *J. Phys. Chem.* **33**, 842 (1929); *J. Phys. Chem.* **34**, 1541 (1930).

8. R. A. Robinson and R. H. Stokes, *Electrolyte Solutions*, 2d ed., Academic Press, New York, 1959, pp. 238–253.

9. D. A. McQuarrie, *Statistical Mechanics*, Harper and Row, New York, 1976, Chapter 15. K. S. Pitzer, *Accounts Chem. Res.* **10**, 371 (1977).

CHAPTER 21

Free Energy Changes for Processes Involving Solutions

Deviations from ideality in real solutions have been discussed in some detail to provide an experimental and theoretical basis for precise calculations of changes in free energy in transformations involving solutions. Therefore, it is appropriate to continue our discussions of the principles of chemical thermodynamics with a consideration of some typical calculations of free energy changes in real solutions.

21-1 ACTIVITY COEFFICIENTS OF WEAK ELECTROLYTES [1]

Let us consider a typical weak electrolyte, such as acetic acid, whose ionization can be represented by the equation

$$HC_2H_3O_2 = H^+ + C_2H_3O_2^- \tag{21-1}$$

In discussions in Chapter 20 of the activity coefficient,

$$\gamma_2 = \frac{a_2}{(m_2/m^\circ)}$$

for strong electrolytes, we always used the stoichiometric (or total) molality for m_2 and disregarded the possibility of incomplete dissociation. On the same basis for acetic acid we could define

$$\gamma_+ = \frac{a_+}{(m_s/m^\circ)}$$

$$\gamma_- = \frac{a_-}{(m_s/m^\circ)}$$

and

$$\gamma_{\pm} = \left[\frac{(a_+)(a_-)}{(m_s/m^\circ)^2}\right]^{1/2} \tag{21-2}$$

in which m_s represents the stoichiometric (or total) molality of acetic acid.

Since it is possible to measure (or closely approximate) the ionic concentration of a weak electrolyte, it is more convenient to define ionic activity coefficients, f_i, based on the *actual* ionic concentrations, m_+ or m_-. Thus

$$f_+ = \frac{a_+}{(m_+/m^\circ)}$$

$$f_- = \frac{a_-}{(m_-/m^\circ)}$$

and

$$f_{\pm} = \left[\frac{(a_+)(a_-)}{(m_+)(m_-)}\right]^{1/2}\left(\frac{1}{m^\circ}\right) \tag{21-3}$$

Similarly, for the undissociated species of molality m_u

$$f_u = \frac{a_u}{(m_u/m^\circ)} \tag{21-4}$$

Since the degree of dissociation, α, of a uni-univalent weak electrolyte such as acetic acid is given by the equation

$$\alpha = \frac{m_+}{m_s} = \frac{m_-}{m_s} \tag{21-5}$$

it follows that

$$\gamma_+ = \alpha f_+ \qquad \text{and} \qquad \gamma_- = \alpha f_- \tag{21-6}$$

21-2 DETERMINATION OF EQUILIBRIUM CONSTANTS FOR DISSOCIATION OF WEAK ELECTROLYTES

Three experimental methods have been developed that are capable of determining dissociation constants with a precision of the order of tenths of 1%. Each of these—the electromotive force method [2], the conductance method [3], and the optical method [4]—provides data that can be treated approximately, assuming that the solutions obey Henry's law or, more exactly, on the basis of the methods developed in Chapter 20. We will apply the more exact procedures. Since the optical method can be used only if the acid and conjugate base show substantial differences in absorption of visible or ultraviolet light, we shall limit our discussion to the two electrical methods.

From emf Measurements

It is possible to select a cell containing a weak acid in solution, whose potential depends on the ion concentrations in the solution and hence on the dissociation constant of the acid. As an example, we will consider acetic acid in a cell containing a hydrogen electrode and a silver–silver chloride electrode:

$$H_2(g, P = 1\ \text{bar});\ HC_2H_3O_2(m_2),\ NaC_2H_3O_2(m_3),\ NaCl(m_4);\ AgCl(s),\ Ag(s) \tag{21-7}$$

Since the reaction that occurs in this cell is (Equation 20-29)

$$\tfrac{1}{2}H_2(g) + AgCl(s) = Ag(s) + HCl(aq)$$

the electromotive force must be given by the expression (Equations 20-33 and 20-34)

$$\mathscr{E} = \mathscr{E}^\circ - \frac{RT}{\mathscr{F}} \ln a_{HCl} = \mathscr{E}^\circ - \frac{RT}{\mathscr{F}} \ln[(m_{H^+} m_{Cl^-} \gamma_{H^+} \gamma_{Cl^-})/(m^\circ)^2]$$

Since the molality, m_{H^+}, depends on the acetic acid equilibrium, which we can indicate in a simplified notation by the equation

$$HAc = H^+ + Ac^- \tag{21-8}$$

we can introduce the dissociation constant, K, for acetic acid into the equation for the emf. For acetic acid, K is given by

$$K = \frac{m_{H^+} m_{Ac^-}}{m_{HAc}} \frac{f_{H^+} f_{Ac^-}}{f_{HAc}} \left(\frac{1}{m^\circ}\right) \tag{21-9}$$

from which m_{H^+} can be expressed in terms of the other variables and substituted into Equation 20-34. Thus we obtain

$$\mathscr{E} = \mathscr{E}^\circ - \frac{RT}{\mathscr{F}} \ln \left(K \frac{m_{HAc}}{m_{Ac^-}} \frac{m_{Cl^-}}{m^\circ} \frac{\gamma_{H^+} \gamma_{Cl^-} f_{HAc}}{f_{H^+} f_{Ac^-}} \right) \tag{21-10}$$

We normally use f_{H^+} in connection with molality of the hydrogen ion in equilibrium with the undissociated acetic acid. The electromotive cell (Equation 21-7) measures only the HCl in the solution (see Equations 20-32 to 20-34) as $m_{H^+}\gamma_{H^+} m_{Cl^-}\gamma_{Cl^-}$. For the system in Equation 21-7, these different expressions of hydrogen ion molality are equal, so

$$f_{H^+} = \gamma_{H^+} \tag{21-11}$$

and

$$\mathscr{E} = \mathscr{E}^\circ - \frac{RT}{\mathscr{F}} \ln \left(K \frac{m_{HAc}}{m_{Ac^-}} \frac{m_{Cl^-}}{m^\circ} \frac{\gamma_{Cl^-} f_{HAc}}{f_{Ac^-}} \right) \tag{21-12}$$

This equation can be rearranged to the form

$$\mathscr{E} - \mathscr{E}^\circ + \frac{RT}{\mathscr{F}} \ln \frac{m_{HAc} m_{Cl^-}}{m_{Ac^-} m^\circ} = -\frac{RT}{\mathscr{F}} \ln K - \frac{RT}{\mathscr{F}} \ln \frac{\gamma_{Cl^-} f_{HAc}}{f_{Ac^-}} \equiv \frac{RT}{\mathscr{F}} \ln K' \tag{21-13}$$

in which

$$K' \equiv K \frac{\gamma_{Cl^-} f_{HAc}}{f_{Ac^-}} \tag{21-14}$$

All the terms on the left in Equation 21-13 are known from previous experiments (see Section 20-2 for the determination of $\mathscr{E}^\circ$) or can be calculated from the composition of the solution in the cell. Thus

$$m_{Cl^-} = m_4 \tag{21-15}$$

$$m_{HAc} = m_2 - m_{H^+} \tag{21-16}$$

and

$$m_{Ac^-} = m_3 + m_{H^+} \tag{21-17}$$

Generally, $m_{H^+} \ll m_2$ or m_3, so it can be estimated from Equation 21-9 by inserting an approximate value of K and neglecting the activity coefficients. Thus it is possible to obtain tentative values of $-(RT/\mathscr{F}) \ln K'$ and hence of K' at various concentrations of acetic acid, sodium acetate, and sodium chloride, respectively. The ionic strength, I, can be estimated as

$$I = m_3 + m_4 + m_{H^+} \tag{21-18}$$

It can be seen from the limiting behavior of activity coefficients (Equation 20-11) that

$$\lim_{I \to 0} \frac{\gamma_{Cl^-} f_{HAc}}{f_{Ac^-}} = 1$$

Thus if

$$-\frac{RT}{\mathscr{F}} \ln K'$$

or K' is plotted against some function of the ionic strength and extrapolated to $I = 0$, the limiting form of Equation 21-13 is

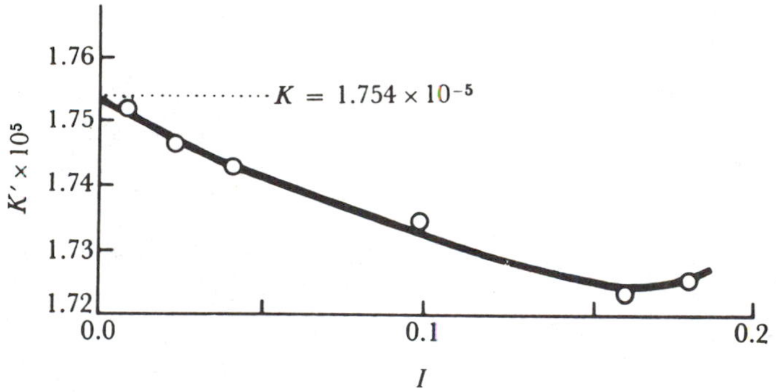

Figure 21-1. Extrapolation of K' values in the determination of the ionization constant of acetic acid at 25°C. Based on data from H. S. Harned and R. W. Ehlers, *J. Am. Chem. Soc.* **54**, 1350 (1932).

$$\lim_{I \to 0} \frac{RT}{\mathscr{F}} \ln K' = \frac{RT}{\mathscr{F}} \ln K$$

and

$$\lim_{I \to 0} K' = K \tag{21-19}$$

It is found that the ionic strength to the first power as the abscissa yields a meaningful extrapolation.

A typical extrapolation of the data for acetic acid is illustrated in Figure 21-1. At 25°C the value of 1.754×10^{-5} has been found for K by this method.

If the equilibrium constant is not already known fairly well, the K determined by this procedure can be looked upon as a first approximation. It then can be used to estimate m_{H^+} for substitution into Equations 21-16, 21-17, and 21-18, and a second extrapolation can be carried out. In this way a second value of K is obtained. The process then can be repeated until successive estimates of K agree within the precision of the experimental data [5].

From Conductance Measurements

Like the electromotive force method, conductance measurements have been used for the estimation of dissociation constants of weak electrolytes. If we use acetic acid as an example, we find that the equivalent conductance, Λ, shows a strong dependence on concentration, as illustrated in Figure 21-2. The rapid decline in Λ with increasing concentration is due largely to a decrease in the fraction of dissociated molecules.

In the approximate treatment of the conductance of weak electrolytes the decrease in Λ is treated as due only to changes in the degree of dissociation, α.

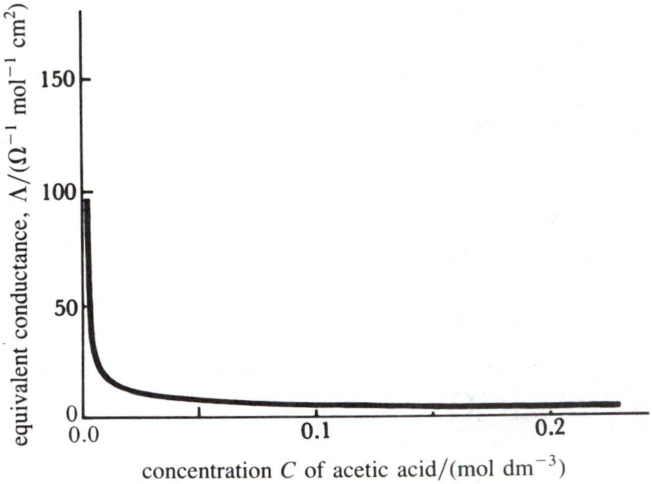

Figure 21-2. Equivalent conductance of aqueous solutions of acetic acid at 25°C. Based on data from D. A. MacInnes and T. Shedlovsky, *J. Am. Chem. Soc.* **54**, 1429 (1932).

On this basis, it can be shown that an apparent degree of dissociation, α', can be obtained from

$$\alpha' = \frac{\Lambda}{\Lambda_0} \tag{21-20}$$

in which Λ_0 is the equivalent conductance of the weak electrolyte at infinite dilution. Hence the apparent dissociation constant, K', is obtainable from the expression

$$K' = \frac{C'_{\mathrm{H^+}}C'_{\mathrm{Ac^-}}}{C'_{\mathrm{HAc}}C^\circ} = \frac{(\alpha' C)(\alpha' C)}{(1-\alpha')C\,C^\circ}$$

$$= \frac{(\alpha')^2 C}{(1-\alpha')C^\circ}$$

$$= \frac{(\Lambda/\Lambda_0)^2 C}{[1-(\Lambda/\Lambda_0)]C^\circ} \tag{21-21}$$

in which C is the total (stoichiometric) concentration of acetic acid in moles per liter. Generally, Λ_0 is evaluated from data at infinite dilution for strong electrolytes. Thus for acetic acid, Λ_0 is obtained as follows:

$$\Lambda_0(\mathrm{H^+ + Ac^-}) = \Lambda_0(\mathrm{H^+ + Cl^-}) + \Lambda_0(\mathrm{Na^+ + Ac^-}) - \Lambda_0(\mathrm{Na^+ + Cl^-}) \tag{21-22}$$

However, for more precise calculations it is necessary to consider that the mobility (hence the conductance) of ions changes with concentration, even when dissociation is complete, because of interionic forces. Thus Equation

21-20 is oversimplified in its use of Λ_0 to evaluate α, since at any finite concentration the equivalent conductances of the H^+ and Ac^- ions, even when dissociation is complete, do not equal Λ_0.

To allow for the change in mobility due to changes in ion concentrations MacInnes and Shedlovsky [6] proposed the use of a quantity, Λ_e, in place of Λ_0. The quantity Λ_e is the sum of the equivalent conductances of the H^+ and Ac^- ions at the concentration C_i at which they exist in the acetic acid solution. For example, for acetic acid, Λ_e is obtained from the equivalent conductances of HCl, NaAc, and NaCl at a concentration, C_i, equal to that of the ions in the solution of acetic acid. Thus, since

$$\Lambda_{\mathrm{HCl}} = 426.04 - 156.70\sqrt{C} + 165.5C(1 - 0.2274\sqrt{C}) \qquad (21\text{-}23)$$

$$\Lambda_{\mathrm{NaAc}} = 90.97 - 80.48\sqrt{C} + 90.0C(1 - 0.2274\sqrt{C}) \qquad (21\text{-}24)$$

and

$$\Lambda_{\mathrm{NaCl}} = 126.42 - 88.53\sqrt{C} + 89.5C(1 - 0.2274\sqrt{C}) \qquad (21\text{-}25)$$

the effective equivalent conductance, Λ_e, of completely dissociated acetic acid is given by

$$\Lambda_e = \Lambda_{\mathrm{HCl}} + \Lambda_{\mathrm{NaAc}} - \Lambda_{\mathrm{NaCl}} = 390.59 - 148.65\sqrt{C_i} + 166.0C_i(1 - 0.2274\sqrt{C_i}) \qquad (21\text{-}26)$$

Assuming that the degree of dissociation at the stoichiometric molar concentration, C, is given by the expression

$$\alpha'' = \frac{\Lambda}{\Lambda_e} \qquad (21\text{-}27)$$

we obtain a better approximation for the dissociation constant than Equation 21-21:

$$K'' = \frac{C''_{\mathrm{H}^+}C''_{\mathrm{Ac}^-}}{C''_{\mathrm{HAc}}C^\circ} = \frac{(\alpha''C)^2}{(1 - \alpha'')C\,C^\circ} = \frac{(\Lambda/\Lambda_e)^2C}{[1 - (\Lambda/\Lambda_e)]C^\circ} \qquad (21\text{-}28)$$

Now if we insert appropriate activity coefficients, we obtain a third approximation for the dissociation constant:

$$K''' = \frac{C''_{\mathrm{H}^+}C''_{\mathrm{Ac}^-}}{C''_{\mathrm{HAc}}C^\circ}\,\frac{f_{\mathrm{H}^+}f_{\mathrm{Ac}^-}}{f_{\mathrm{HAc}}} \qquad (21\text{-}29)$$

This equation can be converted into logarithmic form to give

$$\log K''' = \log K'' + \log\left(\frac{f_\pm^2}{f_u}\right) \qquad (21\text{-}30)$$

in which

$$f_{\pm}^2 = f_{H^+} f_{Ac^-} \tag{21-31}$$

and

$$f_u = f_{HAc} \tag{21-32}$$

To evaluate log K'' it is necessary to know Λ_e and therefore C_i. Yet to know C_i we must have a value for α, which depends on a knowledge of Λ_e. In practice, this impasse is overcome by a method of successive approximations. To begin, we take $\Lambda_e = \Lambda_0$ and make a first approximation for α'' from Equation 21-27. With this value of α'' we can calculate a tentative C_i, which can be inserted into Equation 21-26 to give a tentative value of Λ_e. From Equation 21-26 and Equation 21-27 a new value of α'' is obtained, which leads to a revised value for C_i and subsequently for Λ_e. This method is continued until successive calculations give substantially the same value of α''. Thus for a 0.02000-molar solution of acetic acid, with an equivalent conductance of 11.563 ohms^{-1} mol^{-1} cm^2, a first approximation for α is

$$\alpha' = \frac{11.563}{390.59} = 0.029604$$

since $\Lambda_0 = 390.59$ ohms^{-1} mol^{-1} cm^2. Therefore

$$C_i = \alpha' C = 0.029604(0.02000) = 0.00059208 \text{ mol L}^{-1}$$

and

$$\sqrt{C_i} = 0.024333$$

Substitution of this value of $\sqrt{C_i}$ into Equation 21-26 yields a value of

$$\Lambda_e = 387.07 \text{ ohms}^{-1} \text{ mol}^{-1} \text{ cm}^2$$

Coupling this value with 11.563 for Λ we obtain

$$\alpha'' = \frac{11.563}{387.07} = 0.029873$$

$$C_i = 0.00059746 \text{ mol L}^{-1}$$

$$\sqrt{C_i} = 0.024443$$

and

$$\Lambda_e = 387.06 \text{ ohms}^{-1} \text{ mol}^{-1} \text{ cm}^2$$

A third calculation of α'' gives 0.029874, which is substantially the same as the result of the second approximation; hence, it can be used in Equation 21-28. Once a value of log K'' is obtained, the value of log K can be determined by an extrapolation procedure. From Equation 20-11

$$\lim_{I \to 0} f_{\pm}^2 = 1$$

and

$$\lim_{I \to 0} \log f_{\pm}^2 = 0$$

From Equation 17-4

$$\lim_{I \to 0} f_u = \lim_{m_2 \to 0} f_u = 1$$

and

$$\lim_{I \to 0} \log f_u = 0$$

Thus the limiting form of Equation 21-30 is

$$\lim_{I \to 0} \log K'' = \lim_{I \to 0} \log K''' = \log K \tag{21-33}$$

in which K is the thermodynamic dissociation constant.

The best functions to use in the extrapolation can be determined from the dependence of $f_{\pm}^2$ and f_u on the ionic strength. Theoretically, little is known about the dependence of f_u on concentration [7], but from the Debye–Hückel theory we should expect $\log f_{\pm}^2$ to depend on $\sqrt{I}$, the dependence approaching linearity with increasing dilution. (The Debye–Hückel theory gives the actual ionic activity coefficient, f_i, rather than the stoichiometric activity coefficient. For strong electrolytes it customarily is assumed that dissociation is complete, hence γ_i has been used in Equations 20-62 through 20-64.)

The data for acetic acid, when $\log K''$ is plotted against the square root of the ionic strength (Figure 21-3), provide a meaningful value for K by extrapolation. MacInnes and Shedlovsky report a value for K of 1.753×10^{-5} at 25°C.

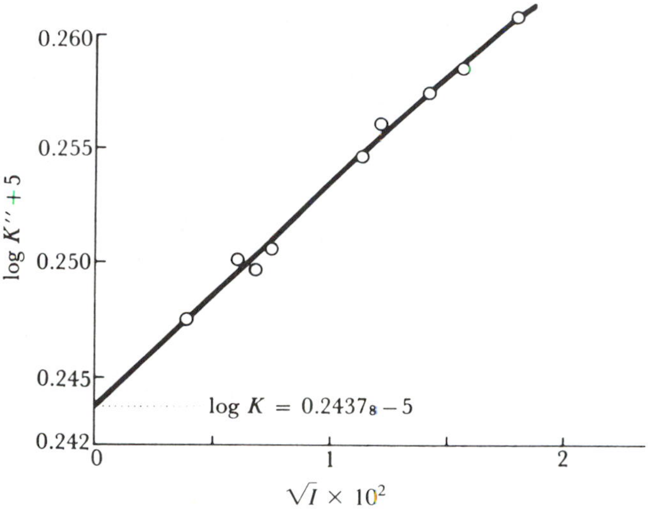

Figure 21-3. Extrapolation of ionization constants of acetic acid.

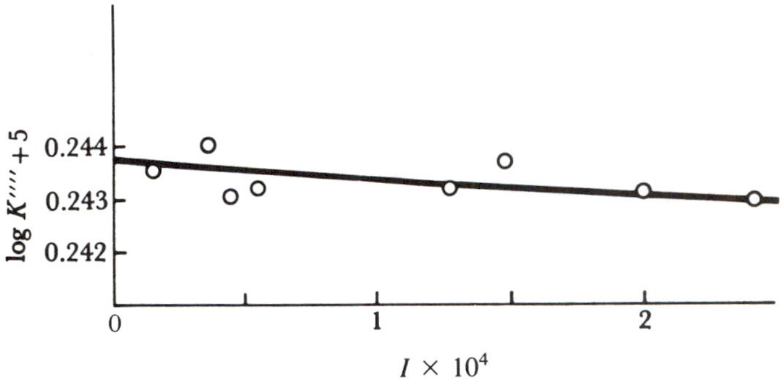

Figure 21-4. Alternative method of extrapolation of ionization constants of acetic acid.

An alternative method of extrapolation, in which the slope is reduced almost to zero, can be carried out by the following modification of Equation 21-30. Separating the activity coefficients, we obtain

$$\log K''' = \log K'' + \log f_{\pm}^2 - \log f_u \qquad (21\text{-}34)$$

We know from experiment that $\log f_u$ is a linear function of I. The value of $\log f_{\pm}^2$ is described quite well by the Debye–Hückel limiting law in very dilute solution. Thus we can substitute the expression

$$\log f_{\pm}^2 = 2 \log f_{\pm} = 2[-0.509\,|z_+z_-|\,\sqrt{I}] = -1.018\sqrt{I} \qquad (21\text{-}35)$$

into Equation 21-34 and rearrange the resultant equation into the relationship

$$\log K''' + \log f_u = \log K'' - 1.018\,\sqrt{I} \equiv \log K'''' \qquad (21\text{-}36)$$

Since both K'' and $\sqrt{I}$ can be calculated from experimental data, $\log K''''$ can be determined. A plot of $\log K''''$ against I gives a curve with small slope, such as illustrated in Figure 21-4. The determination of the intercept in this graph is easier than it is in Figure 21-3 for uni-univalent ions, and the improvement is greater when the dissociation process involves polyvalent ions.

An alternative procedure uses the Fuoss conductance–concentration function to relate the measured conductance to the ionic concentrations at equilibrium [8].

21-3 SOME TYPICAL CALCULATIONS FOR $\Delta G f°$

Standard Free Energy of Formation of Aqueous Solute: HCl

We have discussed in some detail the various methods that can be used to obtain the standard free energy of formation of a pure gaseous compound such as HCl(g). Since many of its reactions are carried out in water solution, it also is desirable to know $\Delta Gf°$ for HCl(aq).

Our problem is to find $\Delta G°$ for the reaction

$$HCl(g, a = f = 1) = HCl(aq, a_2 = 1) \tag{21-37}$$

since to this $\Delta G°$ we always can add $\Delta Gf°$ of HCl(g). Although in Equation 21-37 a_{HCl} is 1 on both sides, the standard states are not the same for the gaseous and aqueous phases; hence, the chemical potentials are not equal. To obtain $\Delta G°$ for Reaction 21-37 we can break up the reaction into a set of transformations for which we can find values of ΔG.

We can write the following three equations, whose sum is equivalent to Reaction 21-37:

1. $HCl(g, a = f = p = 1) = HCl(g, p = 0.2395 \times 10^{-4}\ \text{atm}),$

$$\Delta G_{298} = RT \ln (0.2395 \times 10^{-4}) = -26.401\ \text{kJ mol}^{-1} \tag{21-38}$$

in which 0.2395×10^{-4} atm is the equilibrium partial pressure of HCl(g) [9] in equilibrium with a 4-molal solution of HCl.

2. $HCl(g, p = 0.2395 \times 10^{-4}\ \text{atm}) = HCl(aq, m_2 = 4),$

$$\Delta G_{298} = 0\ \text{(equilibrium reaction)} \tag{21-39}$$

3. $HCl(aq, m_2 = 4, a_2' = 49.66) = HCl(aq, a_2 = 1),$

$$\Delta G_{298} = RT \ln \frac{a_2}{a_2'} = RT \ln \frac{1}{49.66} = -9.680\ \text{kJ mol}^{-1} \tag{21-40}$$

The activity of HCl in a 4-molal solution required for the ΔG in Equation 21-40 was calculated from the mean activity coefficient, 1.762, taken from tables of Harned and Owen [10], as follows:

$$a_2 = m_\pm^2 \gamma_\pm^2 = (4)^2(1.762)^2 = 49.66 \tag{21-41}$$

Now we can obtain the standard free energy change for Equation 21-37, since the sum of Equations 21-38 through 21-40 yields

$$HCl(g, a = 1) = HCl(aq, a_2 = 1), \qquad \Delta G_{298}^\circ = -36.081\ \text{kJ mol}^{-1} \tag{21-42}$$

Having obtained the standard free energy change accompanying the transfer of HCl from the gaseous to the aqueous state, we can add it to the standard free energy of formation of gaseous HCl [11],

$$\tfrac{1}{2}H_2(g) + \tfrac{1}{2}Cl_2(g) = HCl(g), \qquad \Delta Gf_{298}^\circ = -95.299\ \text{kJ mol}^{-1} \tag{21-43}$$

and obtain the standard free energy of formation of aqueous HCl:

$$\tfrac{1}{2}H_2(g) + \tfrac{1}{2}Cl_2(g) = HCl(aq), \qquad \Delta Gf_{298}^\circ = -131.386\ \text{kJ mol}^{-1} \tag{21-44}$$

Standard Free Energy of Formation of Individual Ions: HCl

Since it has been shown that the free energy of formation of an individual ion has no operational meaning [12], there is no way to determine such a quantity.

However, for the purposes of tabulation and calculation, it is possible to separate $\Delta Gf°$ *arbitrarily* into two or more parts, corresponding to the number of ions formed, in a way analogous to that used in tables of standard electrode potentials. In both cases the standard free energy of formation of aqueous H^+ is defined to be zero at every temperature:

$$\tfrac{1}{2}H_2(g) = H^+(aq) + e^-, \qquad \Delta Gf° = 0 \qquad (21\text{-}45)$$

With this definition it is possible to calculate the standard free energies of formation of other ions. For example, for Cl^- ion we proceed by adding appropriate equations to Equation 21-44. For the reaction

$$HCl(aq, a_2 = 1) = H^+(aq, a_+ = 1) + Cl^-(aq, a_- = 1) \qquad (21\text{-}46)$$

$\Delta G°$ is equal to zero because our definition of the individual ion activities (Equation 20-5) is

$$a_2 = (a_+)(a_-)$$

If we add Equation 21-46 to Equation 21-44 and then subtract Equation 21-45:

$$\tfrac{1}{2}H_2(g) + \tfrac{1}{2}Cl_2(g) = HCl(aq), \qquad \Delta Gf°_{298} = -131.386 \text{ kJ mol}^{-1}$$

$$HCl(aq) = H^+(aq) + Cl^-(aq), \qquad \Delta G° = 0$$

$$H^+(aq) + e^- = \tfrac{1}{2}H_2(g), \qquad \Delta G° = 0$$

we obtain the standard free energy of formation of Cl^- ion:

$$\tfrac{1}{2}Cl_2(g) + e^- = Cl^-(aq), \qquad \Delta Gf°_{298} = -131.386 \text{ kJ mol}^{-1} \qquad (21\text{-}47)$$

This $\Delta Gf°$ corresponds to the value that can be calculated from the standard electrode potential.

Standard Free Energy of Formation of Solid Solute in Aqueous Solution

Solute Very Soluble: NaCl. Since the standard free energy of formation of NaCl(s) is available [11], the $\Delta Gf°_{298}$ for NaCl(aq) can be obtained by a summation of the following processes:

$$Na(s) + \tfrac{1}{2}Cl_2(g) = NaCl(s),$$

$$\Delta Gf°_{298} = -384.138 \text{ kJ mol}^{-1} \qquad (21\text{-}48)$$

$$NaCl(s) = NaCl(aq, \text{satd}, m_2 = 6.12),$$

$$\Delta G = 0 \text{ (equilibrium)} \qquad (21\text{-}49)$$

$NaCl(m_2 = 6.12, a_2' = 38.42) = NaCl(a_2 = 1),$

$$\Delta G = RT \ln \frac{1}{a_2'} = -9.044 \text{ kJ mol}^{-1} \quad (21\text{-}50)$$

$NaCl(a_2 = 1) = Na^+(a_+ = 1) + Cl^-(a_- = 1),$

$$\Delta G° = 0 \quad (21\text{-}51)$$

$Na(s) + \frac{1}{2}Cl_2(g) = Na^+(a_+ = 1) + Cl^-(a_- = 1),$

$$\Delta Gf°_{298} = -393.182 \text{ kJ mol}^{-1} \quad (21\text{-}52)$$

The value of a_2' in Equation 21-50 is obtained as follows:

$$a_2' = (a_+)(a_-) = \frac{(m_\pm)^2(\gamma_\pm)^2}{(m°)^2} = \frac{(6.12)^2(1.013)^2}{(1.00)^2} = 38.42 \quad (21\text{-}53)$$

From the standard free energy of formation of the aqueous ions we also can obtain that for the Na^+ ion alone by subtracting Equation 21-47 from Equation 21-52. Thus we obtain

$$Na(s) = Na^+(aq) + e^-, \qquad \Delta Gf°_{298} = -261.796 \text{ kJ mol}^{-1} \quad (21\text{-}54)$$

Slightly Soluble Solute: AgCl. For AgCl we can add the following equations:

$Ag(s) + \frac{1}{2}Cl_2(g) = AgCl(s),$

$$\Delta Gf°_{298} = -109.789 \text{ kJ mol}^{-1} \text{ [11]} \quad (21\text{-}55)$$

$AgCl(s) = AgCl(aq, satd),$

$$\Delta G = 0 \text{ (equilibrium)} \quad (21\text{-}56)$$

$AgCl(aq, satd, a_2' = a_+'a_-') = Ag^+(a_+ = 1) + Cl^-(a_- = 1),$

$$\Delta G = RT \ln \frac{(a_+a_-)}{(a_+'a_-')_{\text{satd soln}}} = -RT \ln K_{sp} = 55.669 \text{ kJ mol}^{-1} \quad (21\text{-}57)$$

$Ag(s) + \frac{1}{2}Cl_2(g) = Ag^+(a_+ = 1) + Cl^-(a_- = 1),$

$$\Delta Gf°_{298} = -54.120 \text{ kJ mol}^{-1} \quad (21\text{-}58)$$

We can calculate a value for the Ag^+ ion by subtracting Equation 21-47 from Equation 21-58. Thus we obtain

$$Ag(s) = Ag^+(a_+ = 1) + e^-, \qquad \Delta Gf°_{298} = 77.266 \text{ kJ mol}^{-1} \quad (21\text{-}59)$$

Standard Free Energy of Formation of Ion of Weak Electrolyte

As part of a program to determine the free energy changes in the reactions by which glucose is oxidized in a living cell, Borsook and Schott [13] calculated the free energy of formation at 25°C of the first anion of succinic acid, $C_4H_5O_4^-$. The solubility of succinic acid in water at 25°C is 0.715 mole (kg H_2O)$^{-1}$. In such a solution the acid is 1.12% ionized ($\alpha = 0.0112$) and the undissociated portion has an activity coefficient of 0.87. Knowing the first dissociation constant of succinic acid, 6.4×10^{-5}, we can calculate ΔGf° of the $C_4H_5O_4^-$ ion by adding the following equations:

$$3H_2(g) + 4C(\text{graphite}) + 2O_2(g) = C_4H_6O_4(s),$$
$$\Delta Gf^\circ_{298} = -748.100 \text{ kJ mol}^{-1} \quad (21\text{-}60)$$

$$C_4H_6O_4(s) = C_4H_6O_4(\text{aq, satd}),$$
$$\Delta G = 0 \text{ (equilibrium)} \quad (21\text{-}61)$$

$$C_4H_6O_4(\text{aq, satd}, a_2') = C_4H_6O_4(a_2 = 1),$$
$$\Delta G = RT \ln (a_2/a_2') = 1.21 \text{ kJ mol}^{-1} \quad (21\text{-}62)$$

$$C_4H_6O_4(a_2 = 1) = H^+(a_+ = 1) + C_4H_5O_4^-(a_- = 1),$$
$$\Delta G^\circ = -RT \ln K = 24.0 \text{ kJ mol}^{-1} \quad (21\text{-}63)$$

$$3H_2(g) + 4C(\text{graphite}) + 2O_2(g) = H^+(a_+ = 1) + C_4H_5O_4^-(a_- = 1),$$
$$\Delta Gf^\circ_{298} = -722.9 \text{ kJ mol}^{-1} \quad (21\text{-}64)$$

For the free energy change in Equation 21-62, a_2' of the undissociated species of succinic acid in the saturated solution is obtained as follows:

$$a_2' = m_u \gamma_u = m_{\text{stoichiometric}} (1 - \alpha) \gamma_u = (0.715)(1 - 0.0112)(0.87) \quad (21\text{-}65)$$

Standard Free Energy of Formation of Moderately Strong Electrolyte

Moderately strong electrolytes, such as aqueous HNO_3, generally have been treated thermodynamically as completely dissociated substances. Thus for HNO_3(aq) the value for ΔGf° of -111.25 kJ mol^{-1} listed in [11] refers to the reaction

$$\tfrac{1}{2}H_2(g) + \tfrac{1}{2}N_2(g) + \tfrac{3}{2}O_2(g) = H^+(\text{aq}, a_+^\circ = 1) + NO_3^-(\text{aq}, a_-^\circ = 1) \quad (21\text{-}66)$$

The activity of the nitric acid is defined by the equation

$$a_{HNO_3} = a_{H^+} a_{NO_3^-} = m_s^2 \gamma_\pm^2 \quad (21\text{-}67)$$

in which m_s is the stoichiometric (or total) molality of the acid.

In recent years optical and nuclear magnetic resonance methods applicable to moderately strong electrolytes have been made increasingly precise [14]. By these methods it has proved feasible to determine dissociation constants and hence concentrations of the undissociated species. Thus for HNO_3 in aqueous solution [14] at 25°C, K is 24. However, in defining this equilibrium constant we have changed the standard state for aqueous nitric acid, and the activity of the undissociated species is given by the equation

$$a'_{HNO_3} = m_u f_u = a_u \qquad (21\text{-}68)$$

in which the subscript u refers to the undissociated species. [The standard states of the ions are unchanged despite the change in the standard state of the undissociated acid. In the limit of infinitely dilute solutions, the solute is completely dissociated; hence, $\gamma_\pm$ (Equation 21-2) becomes equal to $f_\pm$ (Equation 21-3). Thus the limiting law is the same for both definitions of ion activity coefficient.] Therefore Equation 21-67 is no longer applicable, and in its place we have

$$\frac{a_{H^+} a_{NO_3^-}}{a_u} = K = 24 \qquad (21\text{-}69)$$

With the preceding considerations clearly in mind we are able to calculate $\Delta Gf°$ of undissociated HNO_3. For this purpose we can add the following equation to Equation 21-66:

$$H^+(aq, a_+^\circ = 1) + NO_3^-(aq, a_-^\circ = 1) = HNO_3(aq, a_u^\circ = 1) \qquad (21\text{-}70)$$

For this reaction

$$\Delta G° = -RT \ln \frac{1}{K} = 7.9 \text{ kJ mol}^{-1} \qquad (21\text{-}71)$$

Hence the standard free energy of formation of molecular, undissociated, aqueous HNO_3 is -103.4 kJ mol^{-1}, which is the sum of the $\Delta G°$'s for Reactions 21-66 and 21-70:

$$\tfrac{1}{2}H_2(g) + \tfrac{1}{2}N_2(g) + \tfrac{3}{2}O_2(g) = HNO_3(aq, a_u = 1),$$

$$\Delta Gf^\circ_{298} = -103.4 \text{ kJ mol}^{-1} \qquad (21\text{-}72)$$

Effect of Salt Concentration on Geological Equilibrium Involving Water

In Section 14-3 we discussed the gypsum–anhydrite equilibrium (Equation 14-10)

$$CaSO_4 \cdot 2H_2O(s) = CaSO_4(s) + 2H_2O(l)$$

on the assumption that the liquid phase is pure water, and that ΔG for the reaction is dependent only on T, P_S, and P_F (Equation 14-11). If there is dissolved salt in the water, as is likely in a rock formation, the chemical potential and activity of the water (as shown in Chapter 20) depend on the salt concentration, as does ΔG for Equation 14-10. The equation for ΔG would be a modified form of Equation 14-11, with a term taking into account possible variation in the activity of water, as follows (with P_S and P_F much greater than one):

$$\Delta G(P_F, P_S, T) = \Delta G(P = 1, T, a_{H_2O} = 1) + P_S(\Delta V_S) + P_F(\Delta V_F) + 2\,RT \ln X_{H_2O}\gamma_{H_2O} \quad (21\text{-}73)$$

The activity coefficient of water in NaCl solutions of varying concentration can be calculated from data such as that given in Exercise 7, Chapter 20. From the resulting values of the activity coefficients, the effect of NaCl concentration on the equilibrium temperature for Equation 14-11 can be determined. The results of some calculations for a constant pressure of 1 atm are shown in Figure 21-5 [15].

General Comments

The preceding problems illustrate some of the methods that can be used to combine data for the free energies of pure phases with information on the behavior of constituents of a solution to calculate fundamental thermodynamic properties for the compound in solution, and to apply them to chemical reactions. The solutions to the problems illustrated, together with some of the exercises at the end of this chapter, should help students apply the same principles to particular problems in which they are interested.

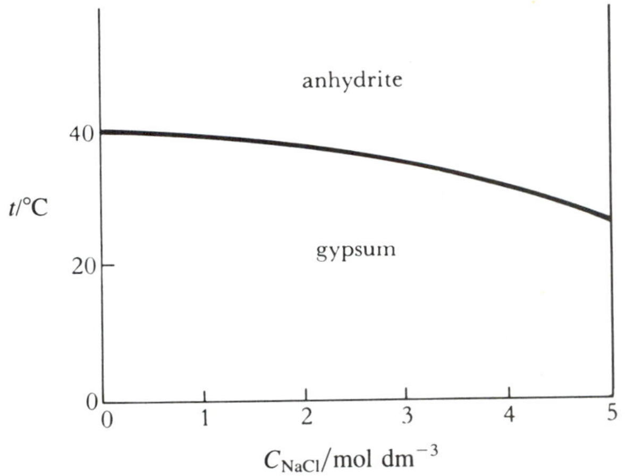

Figure 21-5. Effect of NaCl concentration on the equilibrium temperature of the anhydrite-gypsum reaction at 1 atm. From R. Kern and A. Weisbrod, *Thermodynamics for Geologists*, Freeman, Cooper, San Francisco, 1967, p. 277.

21-4 ENTROPIES OF IONS

In dealing with solutions we encounter the same difficulties in connection with free energy data that led to the development of the third law for pure substances. It frequently may be necessary to obtain values of ΔG for a process in solution for which only thermal data are available. If absolute entropy data also could be obtained for solutions, then it would be possible to calculate ΔG° from a calorimetric determination of ΔH° and the equation (8-27)

$$\Delta G^\circ = \Delta H^\circ - T\Delta S^\circ$$

For aqueous solutions of electrolytes, a concise method of tabulating such entropy data is in terms of the individual ions, since entropies for these species can be combined to give information on a wide variety of salts. The initial assembling of the ionic entropies generally is carried out by a reverse application of Equation 8-27; that is, $\Delta_S f^\circ$ of a salt is calculated from known values of $\Delta_G f^\circ$ and $\Delta_H f^\circ$ for that salt. After a suitable convention has been adopted, the entropy of formation of the cation and anion together then can be separated into the entropies for the individual ions.

The Entropy of an Aqueous Solution of a Salt

From Equation 21-52 we have

$$\mathrm{Na(s)} + \tfrac{1}{2}\mathrm{Cl_2(g)} = \mathrm{Na^+}(a_+ = 1) + \mathrm{Cl^-}(a_- = 1),$$

$$\Delta_G f^\circ_{298} = -393.188 \text{ kJ mol}^{-1}$$

The heat of formation of NaCl(s) is -411.153 kJ mol^{-1} [11]. Hence, we may write

$$\mathrm{Na(s)} + \tfrac{1}{2}\mathrm{Cl_2(g)} = \mathrm{NaCl(s)}, \qquad \Delta_H f^\circ_{298} = -411.153 \text{ kJ mol}^{-1} \tag{21-74}$$

To this we need to add the enthalpy change for the reaction

$$\mathrm{NaCl(s)} = \mathrm{Na^+}(a_+ = 1) + \mathrm{Cl^-}(a_- = 1) \tag{21-75}$$

Since the enthalpy of the dissolved sodium chloride in its standard state is that of the infinitely dilute solution, ΔH for the reaction in Equation 21-75 is

$$\Delta H = \bar{H}_2^\infty - H_{2(s)} = \bar{H}_2^\circ - H_{2(s)}$$

which, according to Equation 18-12, is $-L_{2(s)}$, the relative partial molar enthalpy of solid sodium chloride ($-3{,}861$ kJ mol^{-1}) [16]. Thus for Equation 21-52

$$\left.\begin{aligned} \Delta G^\circ &= -393.188 \text{ kJ mol}^{-1} \\ \Delta H^\circ &= -407.292 \text{ kJ mol}^{-1} \\ \text{and} \qquad \Delta S^\circ &= -47.31 \text{ J mol}^{-1}\text{ K}^{-1} \end{aligned}\right\} \tag{21-76}$$

Calculation of Entropy of Formation of Individual Ions

As in the case of free energy changes, we also can divide the entropy change for a reaction (such as 21-52) into two parts and assign one portion to each ion. Since *actual* values of individual-ion entropies cannot be determined, we must establish some convention for apportioning the entropy among the constituent ions.

In treating the free energies of individual ions we adopted the convention that ΔGf° of the hydrogen ion equals zero for all temperatures; that is (Equation 21-45),

$$\tfrac{1}{2}\mathrm{H_2(g)} = \mathrm{H^+(aq)} + \mathrm{e^-}, \qquad \Delta Gf^\circ = 0$$

We also have shown previously (Equation 8-43) that

$$\left(\frac{\partial\, \Delta G}{\partial T}\right)_P = -\Delta S$$

If Equation 21-45 is valid at all temperatures, it follows that the entropy change in the formation of hydrogen ion from gaseous hydrogen must be zero; that is,

$$\tfrac{1}{2}\mathrm{H_2(g)} = \mathrm{H^+(aq)} + \mathrm{e^-}, \qquad \Delta Sf^\circ = -\left(\frac{\partial\, \Delta Gf^\circ}{\partial T}\right)_P = 0 \tag{21-77}$$

Therefore a consistent convention would set the standard entropy of aqueous H^+ ion equal to $\frac{1}{2}S^\circ_{\mathrm{H_2(g)}} - S^\circ_{\mathrm{e}}$ or 65.342 J mol^{-1} K^{-1} $- S^\circ_{\mathrm{e}}$ [11].

Historically, the usefulness of ionic entropies first was emphasized by Latimer and Buffington [17], who established the convention of setting the standard entropy of hydrogen ion equal to zero; that is,

$$\bar{S}^\circ_{\mathrm{H^+}} \equiv 0 \tag{21-78}$$

Therefore, to maintain the validity of Equation 21-77, we should assign a value of 65.342 J mol^{-1} K^{-1}, ($\frac{1}{2}\bar{S}^\circ_{\mathrm{H_2}}$), to $\bar{S}^\circ$ for a mole of electrons. In practice, half-reactions are combined to calculate ΔS° for an overall reaction in which there is no net gain or loss of electrons; hence, any value assumed for $\bar{S}^\circ$ of the electron will cancel out.

Having chosen a value for $\bar{S}^\circ_{\mathrm{H^+}}$, we can proceed to obtain $\bar{S}^\circ_{298}$ for Cl^- ion from any one of several reactions (for example, Equation 21-44) for the formation of aqueous H^+ and Cl^-. Using values of -131.386 kJ mol^{-1} and -167.159 kJ mol^{-1} for ΔG° and ΔH°, respectively [12], we can calculate -119.98 J mol^{-1} K^{-1} for ΔS° at 298.15 K. If we adopt the convention stated in Equation 21-78, and if S°_{298} for $Cl_2(g)$ is taken as 223.066 J mol^{-1} K^{-1}, it follows that for Reaction 21-44

$$\Delta Sf^\circ_{298} = \bar{S}^\circ_{\mathrm{H^+}} + \bar{S}^\circ_{\mathrm{Cl^-}} - \tfrac{1}{2}\bar{S}^\circ_{\mathrm{H_2}} - \tfrac{1}{2}\bar{S}^\circ_{\mathrm{Cl_2}} = -120.0 \text{ J mol}^{-1}\text{ K}^{-1} \tag{21-79}$$

Hence

Table 21-1 Entropies of Aqueous Ions at 298.15 K [11]

Ion	$\bar{s}°$/J mol^{-1} K^{-1}	Ion	$\bar{s}°$/J mol^{-1} K^{-1}
H^+	(0.00)	OH^-	−10.75
Li^+	13.4	F^-	−13.8
Na^+	59.0	Cl^-	56.5
K^+	102.5	Br^-	82.4
Rb^+	121.50	I^-	111.3
Cs^+	133.05	ClO^-	42.
NH_4^+	113.4	ClO_2^-	101.3
Ag^+	72.68	ClO_3^-	162.3
$Ag(NH_3)_2^+$	245.2	ClO_4^-	182.0
Tl^+	125.5	BrO_3^-	161.71
Mg^{2+}	−138.1	IO_3^-	118.4
Ca^{2+}	− 53.1	HS^-	62.8
Sr^{2+}	− 32.6	HSO_3^-	139.7
Ba^{2+}	9.6	SO_3^{2-}	−29.
Fe^{2+}	−137.7	HSO_4^-	131.8
Cu^{2+}	− 99.6	SO_4^{2-}	20.1
Zn^{2+}	−112.1	NO_2^-	123.0
Cd^{2+}	− 73.2	NO_3^-	146.4
Hg_2^{2+}	84.5	$H_2PO_4^-$	90.4
Sn^{2+}	− 17.[a]	HPO_4^{2-}	−33.5
Pb^{2+}	10.5	PO_4^{3-}	−222.
Al^{3+}	−321.7	HCO_3^-	91.2
Fe^{3+}	−315.9	CO_3^{2-}	−56.9
		$C_2O_4^{2-}$	45.6
		CN^-	94.1
		MnO_4^-	191.2
		CrO_4^{2-}	50.21
		$H_2AsO_4^-$	117.

[a] In 3-*M* $NaClO_4$. For entropies of aqueous ions at high temperatures see C. M. Criss and J. W. Cobble, *J. Am. Chem. Soc.* **86**, 5385 (1964).

$$\bar{s}^{\circ}_{Cl^-} = 56.89 \text{ J mol}^{-1} \text{ K}^{-1} \tag{21-80}$$

Having a value for $\bar{s}^{\circ}_{Cl^-}$, we can proceed to obtain the entropy of formation of Na^+(aq) from $\Delta S°$ for Reaction 21-52:

$$\Delta sf^{\circ}_{298} = \bar{s}^{\circ}_{Na^+} + \bar{s}^{\circ}_{Cl^-} - \bar{s}^{\circ}_{Na} - \tfrac{1}{2}\bar{s}^{\circ}_{Cl_2} = -47.31 \text{ J mol}^{-1} \text{ K}^{-1} \tag{21-81}$$

Consequently, with $\bar{s}^{\circ}_{Na} = 51.21$ J mol^{-1} K^{-1} [11],

$$\bar{s}^{\circ}_{Na^+} = 58.54 \text{ J mol}^{-1} \text{ K}^{-1} \tag{21-82}$$

By procedures analogous to those described in the preceding two examples, we can obtain entropies for many aqueous ions. A list of such values is assembled in Table 21-1.

Utilization of Ion Entropies in Thermodynamic Calculations

With tables of ion entropies available, it is possible to estimate a free energy change that cannot be obtained at all, or not with any reasonable precision, by direct experiment. For example, an adequate calcium electrode has yet to be prepared; nevertheless, it is possible to calculate the electrode potential or, more specifically, the free energy change in the reaction

$$Ca(s) + 2H^+(aq) = Ca^{2+}(aq) + H_2(g) \tag{21-83}$$

from the data in Table 21-1 plus a knowledge of $\Delta H°$ of this reaction. Thus

$$\Delta S^\circ_{298} = \bar{S}^\circ_{Ca^{2+}} + S^\circ_{H_2} - S^\circ_{Ca} - 2\bar{S}^\circ_{H^+}$$
$$= -53.1 + 130.684 - 41.42 - 0 = 36.1 \text{ J mol}^{-1}\text{ K}^{-1} \tag{21-84}$$

Since $\Delta H°$ is -542.83 kJ mol^{-1}, we find a value of $\Delta G°$ of -542.07 kJ mol^{-1}. Hence $\mathscr{E}°$ for Reaction 21-83 is 2.8090 V.

EXERCISES

1. The free energy of formation of $NH_4^+(aq)$ can be obtained by the following procedure:

   a. $\Delta_{G}f°$ of $NH_3(g)$ is -16.45 kJ mol^{-1} at 298.15 K. That is,

   $$\tfrac{1}{2}N_2(g) + \tfrac{3}{2}H_2(g) = NH_3(g, a = f = 1), \qquad \Delta_{G}f^\circ_{298} = -16.45 \text{ kJ mol}^{-1} \tag{21-85}$$

   b. A graph of the partial pressure of $NH_3(g)$ against the molality of undissociated ammonia dissolved in water has a limiting slope, p/m_2, of 0.01764 atm (mol kg^{-1})$^{-1}$ as m_2 approaches zero. If the fugacity is assumed to be equal to the partial pressure, show that the Henry's-law constant, k_2'' (Equation 16-10), is 0.01764 atm. For this calculation, the quantity of ammonia that is dissociated in solution is negligible (that is, the limiting slope is reached at relatively high values of m_2, where the fraction of NH_4^+ ions is very small).

   c. Show that ΔG°_{298} must be -10.009 kJ mol^{-1} for the reaction

   $$NH_3(g, a = f = 1) = NH_3(aq, a_2 = 1) \tag{21-86}$$

   Keep in mind that although a and a_2 are both 1, they have not been defined on the basis of the same standard state.

   d. The equilibrium constant for the reaction

   $$NH_3(aq) + H_2O(l) = NH_4^+(aq) + OH^-(aq) \tag{21-87}$$

   can be taken as 1.8×10^{-5} and the standard free energy of formation of $H_2O(l)$ at 298.15 K as -237.129 kJ mol^{-1}. Show that $\Delta_{G}f^\circ_{298}$ for $NH_4^+ + OH^-$ ions must be -236.51 kJ mol^{-1}.

e. If ΔGf°_{298} for OH^- is -157.244 kJ mol^{-1} show that the corresponding quantity for NH_4^+ is -79.27 kJ mol^{-1}.

2. Given the following information on CO_2 and its aqueous solutions, and using standard sources of reference for any other necessary information, calculate the standard free energy of formation at 298.15 K for the CO_3^{2-} ion.

   Henry's-law constant for solubility in H_2O = 29.5 for p in atmospheres and m_2 in moles (kg H_2O)$^{-1}$.
   Ionization constants of H_2CO_3: $K_1 = 3.5 \times 10^{-7}$
   $K_2 = 3.7 \times 10^{-11}$

3. The solubility of pure solid glycine at 25°C in water is 3.33 moles (kg H_2O)$^{-1}$. The activity coefficient of glycine in such a saturated solution is 0.729. Data on the relative partial molar enthalpies of glycine in aqueous solution are tabulated in Exercise 17, Chapter 18. Given ΔGf° and ΔHf° for solid glycine, complete the following table:

	Glycine(s)	Glycine(aq)
ΔGf°	-370.54 kJ mol^{-1}	
ΔHf°	-528.31 kJ mol^{-1}	
ΔSf°		

Correct answers can be found in an article by F. T. Gucker, Jr., H. B. Pickard, and W. L. Ford, *J. Am. Chem. Soc.* **62**, 2698 (1940).

4. The solubility of α-D-glucose in aqueous 80% ethanol at 20°C is 20 g L^{-1}; that of β-D-glucose, 49 g L^{-1}. [C. S. Hudson and E. Yanovsky, *J. Am. Chem. Soc.* **39**, 1013 (1917)]. If an excess of solid α-D-glucose is allowed to remain in contact with its solution for sufficient time, some β-D-glucose is formed, and the total quantity of dissolved glucose increases to 45 g L^{-1}. If excess solid β-D-glucose remains in contact with its solution, some α-D-glucose is formed in the solution, and the total concentration rises to a limit, which we will refer to as T_β.

   a. Assuming that the activity of each sugar is proportional to its concentration and that neither substance has an appreciable effect on the chemical potential of the other, determine ΔG° for: (i) α-D-glucose(s) = β-D-glucose(s); (ii) α-D-glucose(s) = α-D-glucose(solute); (iii) α-D-glucose(solute) = β-D-glucose(solute).

   b. Compute the concentration T_β. Is such a solution stable? Explain.

5. The dissociation constant of acetic acid is 1.754×10^{-5}. Calculate the degree of dissociation of 0.01-molar acid in the presence of 0.01-molar NaCl. Use the Debye–Hückel limiting law to calculate the activity coefficients of the ions. Take the activity coefficient of the undissociated acid as unity. In your calculation neglect the concentration of H^+ in comparison with the concentration of Na^+.

6. When ΔG° is calculated from an equilibrium constant, K, what error would result at 25°C from an error of a factor of 2 in K?

7. a. What will happen to the degree of dissociation of 0.02-molar acetic acid if NaCl is added to the solution? Explain.

   b. What will happen to the degree of hydrolysis of 0.02-molar sodium acetate if NaCl is added? Explain.

8. Two forms of solid A exist. For the transition A′ = A″, $\Delta G^\circ = -1000$ J mol^{-1}. A′ and A″ produce the same dissolved solute. Which is the more soluble?

9. Henry's law is obeyed by a solute in a certain temperature range. Prove that $\bar{L}_2$, $\bar{L}_1$, and the integral heat of dilution are zero within this range. Do not assume that the Henry's-law constant is independent of temperature; generally, it is not.

10. The average value of $\bar{L}_2$ between 0°C and 25°C of NaCl in 0.01-molar solution is about 188 J mol^{-1}. According to the Debye–Hückel limiting law, $\gamma_\pm$ of NaCl in this solution at 25°C is 0.89.

    a. Calculate $\gamma_\pm$ at 0°C from the thermodynamic relationship for the temperature coefficient of log $\gamma_\pm$ (Equation 20-65).

    b. Calculate $\gamma_\pm$ at 0°C from the Debye–Hückel limiting law at that temperature. Compare the result with that obtained in (a).

11. The activity coefficient of $CdCl_2$ in 6.62-molal solution is 0.025. The emf of the cell

$$\text{Cd, CdCl}_2(m_2 = 6.62), \text{Cl}_2$$

is 1.8111 V at 25°C. The 6.62-molal solution is a saturated one; it can exist in equilibrium with solid $CdCl_2 \cdot \frac{5}{2}H_2O$ and water vapor at a pressure of 16.5 mm Hg. Calculate ΔG° for the reaction

$$\text{Cd(s)} + \text{Cl}_2\text{(g)} + \tfrac{5}{2}\text{H}_2\text{O(l)} \rightarrow \text{CdCl}_2 \cdot \tfrac{5}{2}\text{H}_2\text{O(s)}$$

For the solution of Exercises 12 through 18 the Debye–Hückel limiting law is sufficiently accurate, and the difference between molality and molarity can be neglected. The temperature to be used is 25°C.

12. Calculate the emf of the cell pair

$$\text{H}_2\text{; HCl(0.001), KNO}_3\text{(0.009); AgCl, Ag—Ag, AgCl; HCl(0.01); H}_2$$

13. Calculate the emf of the cell

$$\text{H}_2\text{; HCl(0.001), KCl(0.009); AgCl, Ag}$$

using the fact that the cell

$$\text{H}_2\text{; HCl(0.001), AgCl, Ag}$$

has an emf of 0.46395 V.

14. Calculate the emf of the cell pair

$$\text{Ag, AgCl; KCl(0.01), TlCl(satd), Tl—Tl, TlCl(satd), KCl(0.001); AgCl, Ag}$$

15. MacInnes and Shedlovsky have made conductance measurements that indicate that 0.02-molar acetic acid is 2.987% ionized. Assuming that the activity coefficient of the undissociated acid is 1:

a. Compute the ionization constant of acetic acid.

b. Using this constant, calculate the degree of dissociation of 0.01-molar acetic acid.

16. Two solutions contain only hydrochloric acid, acetic acid, and water. In the first, the concentration of acetate ion is 0.0004-molar; in the second, 0.0001-molar. The total ionic strength in each solution is 0.01-molar. Compute the ratio of the activities of acetic acid in the two solutions.

17. L. E. Strong et al. [8] have determined the ionization constant of benzoic acid in H_2O as a function of temperature by conductance methods. Their data are listed below, with the pK_a based on a hypothetical standard state of 1 mol dm^{-3}. Temperature was controlled to ± 0.002°C.

t/°C	pK_a
5	4.2245
10	4.2148
15	4.2076
20	4.2034
25	4.1998
30	4.2013
35	4.2061
40	4.2115
45	4.2190
50	4.2287
55	4.2400
60	4.2525
65	4.2665
70	4.2821
75	4.2991
80	4.3170

To change pK_a to a hypothetical standard state of 1 mol kg^{-1}, it is necessary to add to each value log $[\rho_{H_2O}/\text{g dm}^{-3}]$, where $\rho = (999.83952 + 16.945176t - 7.9870401 \times 10^{-3}t^2 - 46.170461 \times 10^{-6}t^3 + 105.56302 \times 10^{-9}t^4 - 280.54243 \times 10^{-12}t^5)/(1 + 0.01687985t)$. Calculate pK_a on the molal scale and $\Delta G°$ for each temperature. Use the method of numerical differentiation described in Chapter 2 to obtain values of $\Delta H°$, ΔC_p°, and $\Delta S°$. Also use the difference table method suggested by D. J. G. Ives and P. G. N. Mosely, *J. Chem. Soc. Faraday Trans. I* **72**, 1132 (1976), and compare the results with those obtained with numerical differentiation.

18. Two solutions contain only sodium chloride, acetic acid, and water. In the first solution the concentration of acetate ion is 0.0004-molar; in the second it is 0.0001-molar. The total ionic strength of each solution is 0.01. Compute

the ratio of the activities of acetic acid in the two solutions. What can be said about the relative partial vapor pressures of the monomeric form of acetic acid above the solutions? Of acetic acid dimer?

19. The solubility in 0.0005-molar KNO_3 of hexaamminecobalt(III) hexacyanoferrate(III), $[Co(NH_3)_6][Fe(CN)_6]$, was found by V. K. La Mer, C. V. King, and C. F. Mason [*J. Am. Chem. Soc.* **49**, 363 (1927)] to be 3.251×10^{-5} mol L^{-1}.

    a. In what concentration of $MgSO_4$ is its solubility the same?

    b. What is its solubility in pure water? Since the ionic strength is not known, the problem can be solved by a method of successive approximations (or by a graphical method).

    c. Calculate its solubility in 0.0025-molar $MgSO_4$. How much error is introduced by neglecting the contribution of the complex salt itself to the ionic strength?

    d. The total molarity (C of the cobalt complex plus C of NaCl) of a solution containing NaCl and saturated with the complex salt is 0.0049. Calculate the molarities of each of the solutes.

20. The solubility of $SrSO_4$ in water is 0.00087 molar. Calculate $\Delta G°$ for

$$SrSO_4(s) = Sr^{2+}(aq) + SO_4^{2-}(aq)$$

21. According to recent measurements, a saturated solution of AgCl in water at 25°C contains Ag^+ and Cl^- at the concentration 1.338×10^{-5} mole $(kg\ H_2O)^{-1}$. It also contains dissolved undissociated AgCl, whose dissociation constant is 0.49×10^{-3}.

    a. Determine the mean activity coefficient of the Ag^+ and Cl^- ions from the Debye–Hückel limiting law.

    b. Calculate the concentration of undissociated AgCl.

    c. Calculate $\Delta G°$ for the reaction

$$AgCl(s) = AgCl(aq, \text{undissociated})$$

    d. Compute the mean stoichiometric activity coefficient of the Ag^+ and Cl^- ions. Why must it be less than the activity coefficient calculated from the Debye–Hückel theory?

REFERENCES

1. T. F. Young and A. C. Jones, *Ann. Rev. Phys. Chem.* **3**, 275 (1952).
2. H. S. Harned and R. W. Ehlers, *J. Am. Chem. Soc.* **54**, 1350 (1932).
3. D. A. MacInnes and T. Shedlovsky, *J. Am. Chem. Soc.* **54**, 1429 (1932).
4. H. von Halban and G. Kortum, *Z. physik. Chem.* **170**, 212, 351 (1934).
5. For the effect of dimerization of acetic acid molecules on K, see A. Katchalsky, J. Eisenberg, and S. Lifson, *J. Am. Chem. Soc.* **73**, 5889 (1951).
6. MacInnes and Shedlovsky, *J. Am. Chem. Soc.* **54**, 1429 (1932).

7. See M. Randall and C. F. Failey, *Chem. Rev.* **4**, 291 (1927).

8. L. E. Strong, T. Kinney, and P. Fischer, *J. Solution Chem.* **8**, 329 (1979). R. M. Fuoss, *J. Phys. Chem.* **80**, 2091 (1976).

9. S. J. Bates and H. D. Kirschman, *J. Am. Chem. Soc.* **41**, 1991 (1919); see also G. N. Lewis and M. Randall, *Thermodynamics*, McGraw-Hill, New York, 1923, p. 330.

10. H. S. Harned and B. B. Owen, *The Physical Chemistry of Electrolytic Solutions*, 3d ed., Reinhold, New York, 1958, p. 466.

11. Wagman, et al., "The NBS Tables of Chemical Thermodynamic Properties," *J. Phys. Chem. Reference Data* **11** (1982), Supplement No. 2.

12. E. A. Guggenheim, *J. Phys. Chem.* **33**, 842 (1929); *J. Phys. Chem.* **34**, 1541 (1930).

13. H. Borsook and H. F. Schott, *J. Biol. Chem.* **92**, 535 (1931).

14. For a general review, see T. F. Young, L. F. Maranville, and H. M. Smith, in W. J. Hamer (ed.), *The Structure of Electrolytic Solutions*, Wiley, New York, 1959, pp. 35–63.

15. R. Kern and A. Weisbrod, *Thermodynamics for Geologists*, Freeman, Cooper and Co., San Francisco, 1967, p. 277.

16. T. F. Young and O. G. Vogel, *J. Am. Chem. Soc.* **54**, 3030 (1932).

17. W. M. Latimer and R. M. Buffington, *J. Am. Chem. Soc.* **48**, 2297 (1926).

CHAPTER 22

Systems Subject to a Gravitational Field

In most circumstances of interest to chemists, the dominant experimental variables are temperature, pressure, and composition, and our attention has been concentrated on the dependence of a transformation on these factors. There are occasions, however, when a transformation takes place in a field: gravitational, electrical, or magnetic. It behooves us, therefore, to see how we can approach such problems. Since a gravitational field is the most familiar in common experience, we shall focus particularly on some representative problems in this area.

22-1 DEPENDENCE OF FREE ENERGY ON FIELD

In our exposition of the properties of the free energy G (Chapter 8), we examined systems with constraints on them in addition to the ambient pressure. We found that changes in free energy are related to the maximum work obtainable from an isothermal transformation. In particular, for a reversible transformation at constant pressure and temperature,

$$dG_{T,P} = DW_{\text{net}} \tag{8-103}$$

where DW is the net useful (non-PdV) reversible work associated with the change in free energy.

Of course, the equality in Equation 8-103 is symmetric; that is, the equation may be read in the mirror-image direction: If we do reversible (non-PdV) work DW_{net} on a system at constant pressure and temperature, we increase its free energy by the amount $dG_{T,P}$. For example, if we reversibly change the position, x, of a body in the gravitational field of the earth [Figure 22-1(a)], we do an amount of work given by

$$DW_{\text{net}} = mgdx \tag{22-1}$$

where m is the mass of the body, g is the gravitational acceleration, and x is positive in the upward direction. It follows then from Equation 8-103 that

$$\left(\frac{\partial G}{\partial x}\right)_{T,P} = mg = \text{force exerted on body to move it against gravitational field} \tag{22-2}$$

If we consider lowering a body down a shaft [Figure 22-1(b)], it is convenient to change our convention regarding the positive direction of x to downward. Hence

$$\left(\frac{\partial G}{\partial x}\right)_{T,P} = -mg \tag{22-3}$$

More generally, for constraints other than gravity, we can also state that

$$DW_{\text{net in field}} = (\text{force exerted against field})dx \tag{22-4}$$

Consequently, it follows that

$$\left(\frac{\partial G}{\partial x}\right)_{T,P} = \text{force exerted against field} \equiv \mathscr{F} \tag{22-5}$$

Turning back to a system where fields are absent, we found in Chapter 8 that since the free energy G is a function of pressure and temperature,

$$G = f(T,P)$$

we can write for the total differential

$$dG = \left(\frac{\partial G}{\partial T}\right)_P dT + \left(\frac{\partial G}{\partial P}\right)_T dP \tag{8-36}$$

Subsequently, when we examined systems in which composition, as well as T and P, can be varied (but fields are still absent, or constant), we found (Chapter 12) that

$$G = f(T, P, n_1, n_2, \ldots, n_i) \tag{12-1}$$

where $n_1, n_2, \ldots$ are the moles of the respective components. So the total differential now becomes

$$dG = \left(\frac{\partial G}{\partial T}\right)_{P,n_i} dT + \left(\frac{\partial G}{\partial P}\right)_{T,n_i} dP + \sum \left(\frac{\partial G}{\partial n_i}\right)_{T,P,n_j} dn_i \tag{12-2}$$

Now let us remove the constraint of a fixed field. To be concrete, let us move some unit of material from one position in the gravitational field of the earth to another. Under these circumstances, the free energy G also depends on x, the position in the field, so we may write for the most general circumstances

$$G = f(T, P, n_1, n_2, \ldots, n_i, x) \tag{22-6}$$

Consequently it follows that the total differential should be expressed as

$$dG = \left(\frac{\partial G}{\partial T}\right)_{P,n_i,x} dT + \left(\frac{\partial G}{\partial P}\right)_{T,n_i,x} dP + \sum \left(\frac{\partial G}{\partial n_i}\right)_{T,P,n_j,x} dn_i + \left(\frac{\partial G}{\partial x}\right)_{T,P,n_i} dx \tag{22-7}$$

For the field of the earth's gravitation, the partial derivative in the last term of Equation 22-7 can be replaced by mg (see Equation 22-2). Furthermore, in view of the relation in Equation 12-4, the second term on the right-hand side can be replaced by

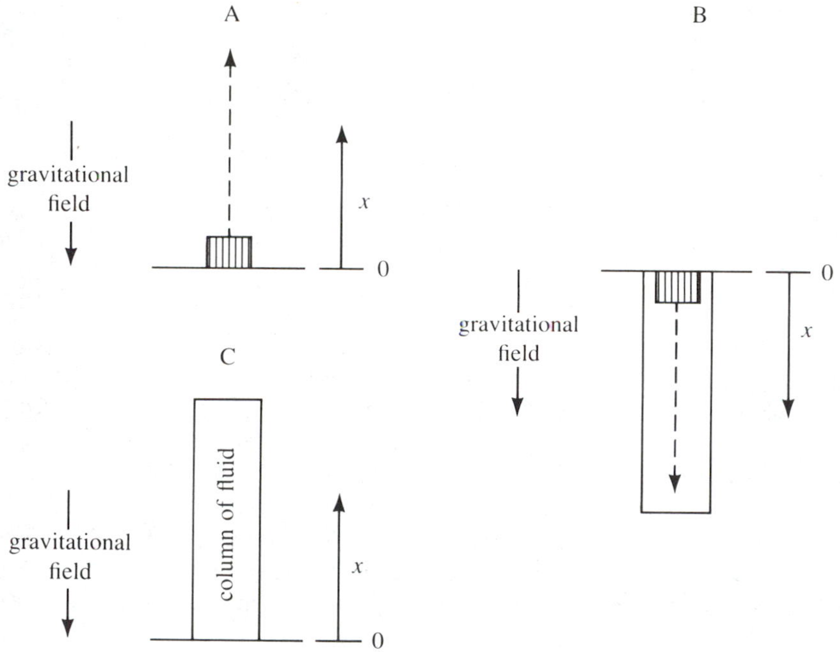

Figure 22-1. Reversible processes in a gravitational field.

$$\left(\frac{\partial G}{\partial P}\right)_{T,n_i,x} dP = VdP \tag{22-8}$$

22-2 SYSTEM IN A GRAVITATIONAL FIELD

Let us now analyze some specific systems. First, we examine a column of pure fluid perpendicular to the surface of the earth [Figure 22-1(c)] and at equilibrium. In this case it can be shown by the following argument that the pressure within the fluid varies with position in the fluid.

The column of pure fluid [Figure 22-1(c)] is at a constant temperature, and the pressure on it is constant. Thus for the column of fluid

$$dG = 0 \tag{22-9}$$

for any transfer of fluid from one level to another.

Let us now analyze the contributions to dG if we take one mole of the pure fluid in the column at position x and move it to the position $x + dx$. At each level within the column the pressure is different (since the weight of fluid above it is different), although it remains fixed at each level. Hence, since the unit of pure fluid is being moved from one position to another, Equation 22-7 can be written as

$$dG = d\textsc{g} = \left(\frac{\partial \textsc{g}}{\partial P}\right)_{T,n_i,x} dP + \left(\frac{\partial \textsc{g}}{\partial x}\right)_{T,P,n_i} dx$$

$$= \textsc{v}dP + Mgdx = 0 \tag{22-10}$$

where $\textsc{g}$ and $\textsc{v}$ denote values of the respective properties per mole and M is the molar mass. From this equation we conclude that

$$\frac{\partial P}{\partial x} = -\frac{M}{\textsc{v}}g = -\rho g \tag{22-11}$$

where ρ is the density of the fluid. Thus the pressure in the fluid is a function of x, and for the column rising from the surface of the earth [Figure 22-1(c)], the pressure decreases as the distance, x, above the surface increases.

If the pure fluid were in the shaft extending below the surface [Figure 22-1(b)], our analysis would correspond in every detail to that in the column above the surface except that the negative sign in Equation 22-11 would be replaced by a positive sign, because x is positive in the direction of the gravitational field. Thus the pressure within the fluid would *increase* as the depth, x, down the shaft increases.

Let us examine now a column of a *solution* in the shaft of Figure 22-1(b). For simplicity, we shall assume there is only one dissolved solute in a single-component solvent. If equilibrium has been attained, we find that the molality, m, of solute varies with the depth, and we can derive an analytic expression for this dependence of molality on depth.

Consider the transfer of one mole of solute from one position, x, in the column at equilibrium, to another position, $x + dx$. The transfer of the solute, in a column of very large cross section, does not change the molality at any position. The fluid in the shaft is at equilibrium, its temperature is invariant, and the pressure on it is fixed; hence, $dG = 0$. A mole of solute in solution has a molar free energy $\bar{G}_2$. If the solute is moved from one position, x, to another, $x + dx$, it could undergo a change in free energy $d\bar{G}_2$ due to the difference in pressure in the fluid, because of the change of position in the gravitational field, and as a result of any change in molality of solute at different levels. Since, in our thought experiment, this transposition is the only change being made in the system, we can write [1] in place of Equation 22-7, using ln m in place of n_i,

$$dG = d\bar{G}_2 = \left(\frac{\partial \bar{G}_2}{\partial \ln m}\right)_{P,x} d \ln m + \left(\frac{\partial \bar{G}_2}{\partial P}\right)_{\ln m,x} dP + \left(\frac{\partial \bar{G}_2}{\partial x}\right)_{\ln m,P} dx$$

$$= 0 \qquad (22\text{-}12)$$

From Equation 16-11 we can obtain the following relation;

$$\bar{G}_2 = RT \ln (m/m^\circ) + \bar{G}_2^\circ; \qquad \frac{\partial \bar{G}_2}{\partial \ln m} = RT \qquad (22\text{-}13)$$

From an equation analogous to Equation 12-17, it follows (see Exercise 4 of Chapter 12) that

$$\left(\frac{\partial \bar{G}_2}{\partial P}\right)_{\ln m,x} = \bar{V}_2 = M_2\bar{v}_2 \qquad (22\text{-}14)$$

where $\bar{V}_2$ is the partial molar volume (and $\bar{v}_2$, the partial specific volume) of the solute and M_2 its molar mass. The dependence of free energy G on the gravitational field is expressed in Equation 22-3, which can be converted to

$$\frac{\partial \bar{G}_2}{\partial x} = -M_2 g \qquad (22\text{-}15)$$

since M_2 and g are constants in the situation being analyzed. Recognizing that at equilibrium $dG = 0$, and substituting Equations 22-13 to 22-15 into Equation 22-12, we find

$$RTd \ln m = M_2 g dx - M_2\bar{v}_2 dP$$

$$= M_2 g dx - M_2\bar{v}_2 \rho g dx \qquad \text{(see Equation 22-11)} \qquad (22\text{-}16)$$

The integrated form of this equation is

$$RT \ln \frac{m_{\text{at depth } d}}{m_{\text{at surface}}} = M_2 g\,(1 - \bar{v}_2\rho)d \tag{22-17}$$

where d is the depth below the surface.

Thus whether an increase or decrease in molality occurs at depth d, in comparison to the surface, is determined by the factor $(1 - \bar{v}_2\rho)$. If $\bar{v}_2\rho > 1$, the molality of solute will decrease with increasing depth. Contrariwise, if $\bar{v}_2\rho < 1$, the molality of solute will be greater at greater depths.

Let us illustrate this phenomenon with a practical example, the variation of oxygen and of nitrogen equilibrium solubilities with depth in the ocean [2]. For seawater, the density ρ depends on temperature and salinity, and could vary from 1.025 to 1.035 g cm^{-3}. For the lower limit and water temperature near 25°C, Equation 22-17 becomes (if d is expressed in meters)

$$\log_{10} \frac{m_d}{m_{\text{surface}}} = 3.2 \times 10^{-7}\, d \tag{22-18}$$

for dissolved oxygen ($\bar{v}_2 = 0.97$ cm^3/g). Thus, for example, at a depth of 1000 meters, the solubility of oxygen increases by a factor of 1.0007. If the density of the seawater is as high as 1.035, then the solubility of oxygen at 1000 meters depth *decreases* by a similar factor.

On the other hand, the situation with nitrogen is markedly different. Here, variations in salinity and temperature have little effect on the factor $(1 - \bar{v}_2\rho)$ because the $\bar{v}_2$ of nitrogen, 1.43 $cm^3\ g^{-1}$, is relatively so large. Thus for nitrogen, Equation 22-17 becomes

$$\log_{10} \frac{m_d}{m_{\text{surface}}} = -2.4 \times 10^{-5}\, d \tag{22-19}$$

At a depth of 1000 meters, the solubility decreases by 5–6%. In contrast to oxygen, the (equilibrium) solubility of nitrogen always decreases progressively with depth.

22-3 SYSTEM IN A CENTRIFUGAL FIELD

Near the surface of the earth, the gravitational acceleration, g, is essentially constant. For contrast, let us turn our attention next to a centrifugal field, where the acceleration is very sensitive to the distance from the center of rotation.

The centrifugal force at a distance, r, from the axis of rotation (Figure 22-2) is

$$\mathscr{F}_{\text{centrifugal}} = m\omega^2 r \tag{22-20}$$

where m is the mass of the entity being centrifuged and ω is the angular velocity. Thus, by an analysis similar to that presented for Equation 22-5, we find that

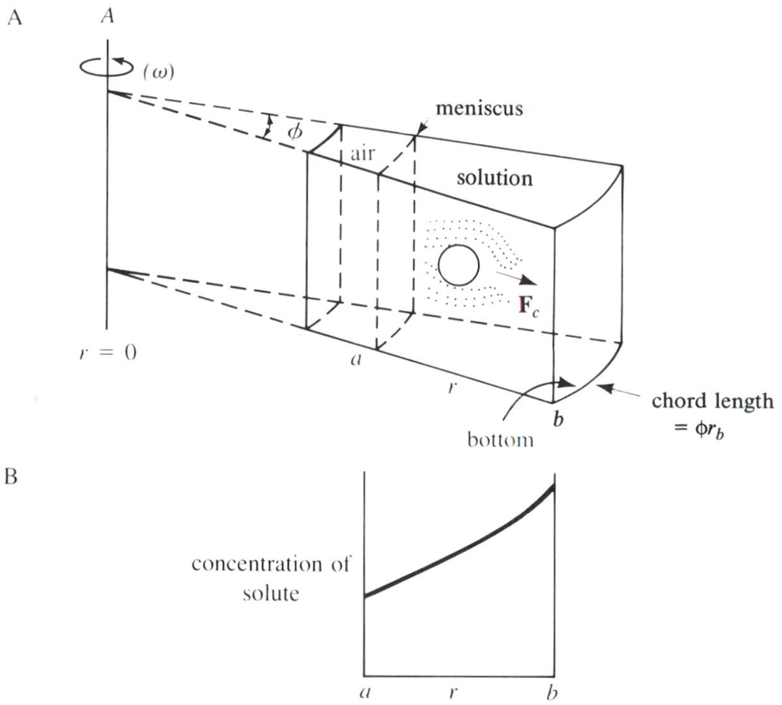

Figure 22-2. *a.* Schematic diagram of a sedimentation apparatus for determination of molecular weight of a solute molecule. The sector-shaped container is actually mounted in a rotor that spins about an axis of rotation *A*, at an angular velocity ω. $\mathbf{F}_c$ shows the direction of the centrifugal field. The distance from the axis of rotation to any position in the cell is *r*. The field $\mathbf{F}_c$ at any point is directed along the axis *r*. At any position, *r*, the chord length is ϕr, which increases with increasing *r*. The height of the cell, a, is constant.
b. Concentration distribution of solute in cell at sedimentation equilibrium.

$$\left(\frac{\partial G}{\partial r}\right)_{P,T,n_i} = -m\omega^2 r \qquad (22\text{-}21)$$

since m and ω are constant. This relation is analogous to that in the gravitational field, with the angular acceleration $\omega^2 r$ replacing the gravitational acceleration g. The negative sign on the right-hand side arises from the fact that the free energy decreases as r increases; that is, work would have to be done on the sedimenting particle to bring it back from a longer axial position to an initial shorter one.

If we now examine a cell (Figure 22-2) containing a pure liquid (rather than a solution as shown), then again (as in the case of a gravitational field) we can show that the pressure in the pure fluid varies with position in the centrifugal field. Starting with Equation 22-7, we arrive at

$$vdP - M\omega^2 r\,dr = 0 \tag{22-22}$$

in place of Equation 22-10 for the gravitational field. From this we conclude that

$$\frac{\partial P}{\partial r} = \frac{M}{v}\,\omega^2 r = \rho\omega^2 r \tag{22-23}$$

Thus the ambient pressure within the fluid in the cell in Figure 22-2 increases with increasing distance from the axis of rotation.

Finally, we consider the behavior of a solute in a solution in the cell subjected to the centrifugal field. At a suitable angular velocity, the tendency of the solute to sediment toward the bottom of the cell is countered by its tendency to diffuse backward toward the meniscus, owing to the depletion of its concentration at that end, and the increasing concentration toward the bottom of the cell. At some time, a sedimentation equilibrium is attained. A typical concentration distribution is depicted in Figure 22-2(b). What is a quantitative analytical expression for this curve?

As in the gravitational field, we consider a transfer of one mole of any single solute i from one position r in the cell in the centrifugal field to a second position $r + dr$. For this transfer at equilibrium, $dG = 0$, so we write in place of Equation 22-12

$$0 = d\bar{G}_i = \left(\frac{\partial \bar{G}_i}{\partial P}\right)_{c_i,c_k,r} dP + \left(\frac{\partial \bar{G}_i}{\partial c_i}\right)_{P,c_k,r} dc_i + \sum_{k\neq i}\left(\frac{\partial \bar{G}_i}{\partial c_k}\right)_{P,c_i,r} dc_k + \left(\frac{\partial \bar{G}_i}{\partial r}\right)_{P,c_i,c_k} dr \tag{22-24}$$

We distinguish between c_i, the concentration of solute whose distribution we are focusing upon, and the c_k's of other solutes because this type of system is of frequent practical interest. The ultracentrifuge is used widely to determine the molecular weight of a macromolecule from its concentration distribution at equilibrium. The large molecule, natural or synthetic, which may be designated by i, is frequently dissolved in an aqueous solution containing other solute species, k, to buffer the solution or to provide an appropriate ionic strength.

For the individual terms and factors on the right-hand side of Equation 22-24 we may insert the following substitutions:

$$\frac{\partial \bar{G}_i}{\partial P} = \bar{V}_i = M_i \bar{v}_i \tag{22-25}$$

$$dP = \rho\omega^2 r\,dr \tag{22-26}$$

$$\bar{G}_i = RT \ln a_i + \bar{G}_i^\circ \tag{22-27}$$

$$\frac{\partial \bar{G}_i}{\partial c_i} = RT\left[\frac{\partial \ln c_i}{\partial c_i} + \frac{\partial \ln \gamma_i}{\partial c_i}\right] \qquad (22\text{-}28)$$

$$\left(\frac{\partial \bar{G}_i}{\partial c_k}\right)_{c_i} = RT\,\frac{\partial \ln a_i}{\partial c_k} = RT\,\frac{\partial \ln \gamma_i}{\partial c_k} \qquad (22\text{-}29)$$

$$\frac{\partial \bar{G}_i}{\partial r} = -M_i\omega^2 r \qquad (22\text{-}30)$$

Although we shall carry along the term in Equation 22-28 for the variation of $\ln \gamma_i$ with c_i—for in practice the macromolecule concentration may cover a wide range from meniscus to bottom of the cell (Figure 22-2)—we shall assume that the change in $\ln \gamma_i$ of the macromolecule with change in concentration of other solutes, c_k, in the solution, is negligible to a good approximation. Within these specifications, Equation 22-24 can be reduced to

$$0 = M_i\bar{v}_i\rho\omega^2 r\,dr + RT\left[\frac{1}{c_i} + \frac{\partial \ln \gamma_i}{\partial c_i}\right]dc_i - M_i\omega^2 r\,dr \qquad (22\text{-}31)$$

which in turn can be converted into

$$\frac{1}{r}\,\frac{1}{c_i}\,\frac{\partial c_i}{\partial r} = \frac{\omega^2}{RT}\,\frac{M_i(1-\bar{v}_i\rho)}{\left[1 + c_i\dfrac{\partial \ln \gamma_i}{\partial c_i}\right]} \qquad (22\text{-}32)$$

An alternative form is

$$\frac{\partial \ln c_i}{\partial (r^2)} = \frac{\omega^2}{2RT}\,\frac{M_i(1-\bar{v}_i\rho)}{\left[1 + c_i\dfrac{\partial \ln \gamma_i}{\partial c_i}\right]} \qquad (22\text{-}33)$$

This equation suggests that a convenient graphical representation of the concentration distribution of species i would be one plotting $\ln c_i$ versus r^2. Three representative possible curves are illustrated in Figure 22-3.

If the solute i is monodisperse—that is, if there is no dissociation or aggregation of the (macro)molecules and each one has exactly the same molecular weight at every position in the cell—then M_i is a constant. If, furthermore, these solute molecules do not interact with each other—that is, if they behave ideally—the term $\partial \ln \gamma_i/\partial c_i = 0$. Under these circumstances, $\ln c_i$ varies linearly with r^2, as shown in line A of Figure 22-3. If the molecular

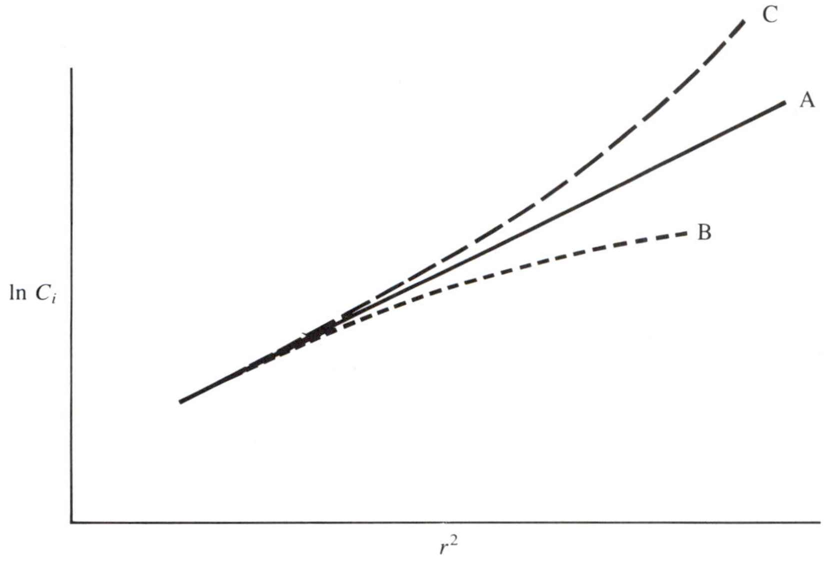

Figure 22-3. Concentration distribution of solute in solution at sedimentation equilibrium. A, ideal behavior, monodisperse solute; B, nonideality; C, polydisperse system.

weight of species i is unknown, it can be determined from the slope of line A, because Equation 22-33 becomes

$$\frac{\partial \ln c_i}{\partial(r^2)} = \frac{\omega^2}{2RT} M_i(1 - \bar{v}_i\rho) \tag{22-34}$$

Equilibrium ultracentrifugation has played a crucial role in establishing the molecular weights of protein molecules on an *ab initio* basis [3,4], that is, without requiring calibration with macromolecules of known molecular weight.

Should the macromolecules interact with each other, then $\partial \ln \gamma_i/\partial c_i$ does not vanish. In actual experience, its value is almost always positive, largely because of excluded volume effects. As is apparent from Equation 22-33, $c_i[\partial \ln \gamma_i/\partial c_i]$ will then increase in magnitude as c_i increases. Thus the curvature shown in curve B of Figure 22-3 is typical of nonideal behavior.

It is also possible to observe upward curvature in a plot of $\ln c_i$ versus r^2, as in curve C of Figure 22-3. This occurs when the macromolecules are polydisperse, that is, when they possess a range of molecular weights. Common sense tells us, in this case correctly, that the heavier species in the class i will congregate toward the bottom of the cell. Since the slope depends on M_i, curve C will become steeper as we move toward the bottom of the cell. A mathematical analysis of equilibrium behavior in a polydisperse system leads to the conclusion that from the slope at any point on curve C, we can obtain a weight average molecular weight at that local concentration of solute i.

EXERCISES

1. Suppose a shaft [Figure 22-1(b)] is filled with a column of an ideal gas at a uniform temperature T. Show that the variation of pressure, P, with depth, d, from the top of the column is given by the equation

$$P = P_0 e^{(Mg/RT)d}$$

where P_0 is the pressure at the surface (that is, when $d = 0$).

2. The partial specific volume, $\bar{v}_2$, for hydrogen gas dissolved in water is exceptionally large: 13.0 mL gm^{-1}. Find an equation for m_d/m_{surface} for dissolved H_2.

3. In one of his experiments [*Ann. Chim. Phys.* **18**, 55–63 (1909)], Perrin counted a total of 13,000 particles of gamboge and found average concentrations proportional to the following numbers at the given heights above the bottom of the vessel.

$h/10^{-6}$ m	Relative concentration
5	100
35	47
65	22.6
95	12

The experiment was carried out at 298 K, and the density of the particles was 0.2067 kg dm^{-3} greater than that of the water in which they were suspended. The mean radius of the particles was 2.12×10^{-7} m. Calculate Avogadro's constant from these data.

4. K. O. Pederson, a colleague of Svedberg's, studied the sedimentation equilibrium of lactoglobulin, a protein from milk, in the analytical ultracentrifuge [*Biochem. J.*, **30**, 967 (1936)]. By graphical integration of the concentration gradient curve obtained with a Schlieren optical system, he calculated the concentration of protein as a function of position in the cell:

$x/10^{-3}$ m	$c'/(\text{g dm}^{-3})$
49.0	1.30
49.5	1.46
50.0	1.64
50.5	1.84
51.0	2.06
51.5	2.31

The partial specific volume of the protein was found to be 0.7514 $dm^3\ kg^{-1}$, and the density of the solution was 1.034 kg dm^{-3}. The rotor was turning at a rate of 182.8 rps. Calculate the molar mass.

REFERENCES

1. In the absence of a field, $\bar{G}$ and μ are identical. However, in the presence of a gravitational (or other) field, that identity no longer is valid because of historical reasons. As defined by Gibbs (J. W. Gibbs, *The Collected Works of J. Willard Gibbs*, Longmans, Green and Co., New York, 1928, Vol. 1, pp. 144–150), the chemical potential μ is not a function of position x in a field. On the other hand, as used by G. N. Lewis (G. N. Lewis and M. Randall, *Thermodynamics*, McGraw-Hill, New York, 1923, pp. 242–244), the partial molar free energy, $\bar{G}$, includes the energy associated with position, x, in a field. For this reason we have expressed our derivations in terms of $\bar{G}$. In a gravitational field (over distances for which g is essentially constant), $\bar{G}$ and μ are related by the equation

$$\bar{G} = \mu + Mgx$$

We also find some thermodynamicists who define a "total chemical potential," μ_{total}, or "gravitochemical potential," as the sum of $\mu + Mgx$; hence, they are essentially using $\bar{G}$.

2. I. M. Klotz, *Limnology and Oceanography* **8**, 149–151 (1963).
3. H. K. Schachman, *Ultracentrifugation in Biochemistry*, Academic Press, New York, 1959.
4. H. Kim, R. C. Deonier, and J. W. Williams, *Chem. Rev.* **77**, 659 (1977).

CHAPTER 23

Estimation of Thermodynamic Quantities

In this chapter we shall consider some semiempirical methods of estimation of thermodynamic quantities associated with chemical transformations.

23-1 APPROXIMATE METHODS

Precise data are available for relatively few compounds. However, in many situations it is desirable to have some idea of the feasibility or impossibility of a given chemical transformation even though the necessary thermodynamic data are not available. Several groups of investigators [1] have proposed empirical methods of correlation that allow us to estimate the thermodynamic properties required to calculate free energies and equilibrium constants. We will consider in some detail only one of these procedures—that of Andersen, Beyer, and Watson—to illustrate the type of approach used in these approximation methods.

Group Contribution Method of Andersen, Beyer, and Watson

Like several other systems, this method is based on the assumption that a given thermodynamic property, such as entropy, of an organic substance can be resolved into contributions from each of the constituent groups in the molecule. With tables of such group contributions assembled from available experimental data, we can estimate the thermodynamic properties of any molecule by adding the contributions of the constituent groups.

Generally, there are several alternative methods of associating groups into a specified molecule. In the Andersen–Beyer–Watson approach, the thermodynamic properties in *the ideal gaseous state* are estimated by considering a given compound as built up from a base group (such as one of those listed in Table 23-1), which has been modified by appropriate substitutions to yield the desired molecule. Thus aliphatic hydrocarbons can be built up from methane by repeated substitutions of methyl groups for hydrogen atoms. Similarly, any amide can be viewed as a derivative of formamide and any primary amine as a derivative of methylamine.

The thermodynamic quantities for large, complex molecules are obtained by adding the contributions of the appropriate substitution group to the value for the base group. Table 23-2 gives the contributions for the primary substitution of a CH_3 group on a single carbon atom in each of the nine base groups listed in Table 23-1. For the cyclic base groups—cyclopentane, benzene, and naphthalene—several carbon atoms are available for successive primary substitutions (no more than one on each carbon atom), and the magnitude of the contribution depends on the number and position of the added methyl groups as well as on the type of base ring. Benzene and naphthalene may be considered together, but cyclopentane forms a category of its own. In the latter ring system, *ortho* refers to adjacent carbon atoms on the nucleus and *meta* refers to carbon atoms separated by at least one carbon within the ring. In the naphthalenes formed by enlargement of the cyclopentane ring, we also may refer to a *para* substitution when the second replacement is on a carbon atom in the ring separated by at least two carbon atoms from the atom on which the first substitution was made.

Once the building up process has started, primary substitution on any methyl group in the molecule is assumed to have the same value as a primary substitution in the base group.

A second substitution of a methyl group for a hydrogen on a single carbon atom of a base group is called a secondary substitution. These secondary replacements have to be treated in more detail because the changes in thermodynamic properties depend on the nature of the carbon atom on which the replacement is being made and on the nature of the adjacent carbon atom. For this reason these carbon atoms are characterized by "type numbers," as shown in Table 23-3. The thermodynamic changes associated with secondary methyl substitutions then can be tabulated as in Table 23-4. The number in Column A is the type number of the carbon atom on which the second methyl substitution is made, and that in Column B is the highest type number of an adjacent carbon atom, each number referring to the status of the carbon atom before the substitution is made.

In connection with the secondary methyl substitutions in Table 23-4, it is necessary to introduce two special categories for use in estimating changes in thermodynamic properties for esters and ethers. The last entry in Table 23-4 gives the changes accompanying the replacement of the H atom in the OH of a carboxyl group by a CH_3 group to form a methyl ester. Similarly, the next to the last entry refers to the replacement of an H atom on the methyl group of an OCH_3 group in either an ether or an ester to form the corresponding ethyl ether or ester.

Table 23-1 Base-Group Properties

Group	$\Delta Hf^{\circ}_{298.1}(g)$/kJ mol^{-1}	$s^{\circ}_{298.1}(g)$/J mol^{-1} K^{-1}	Heat Capacity Constants[a] for Ideal Gas at T/K		
			a	$b(10^3)$	$c(10^6)$
Methane	−74.9	186.2	14.31	74.68	−17.41
Cyclopentane	−89.5	295.8	10.96	345.89	−103.43
Benzene	75.7	269.4	0.96	325.64	−113.64
Naphthalene	148.1	337.6	13.18	457.73	−145.56
Methylamine	−29.7	241.4	16.82	128.53	−36.40
Dimethylamine	−32.6	272.8	16.40	202.13	−58.95
Trimethylamine	−45.6		16.44	275.52	−81.50
Dimethyl ether	−192.5	266.5	26.86	165.85	−47.91
Formamide	−207.1		27.24	105.35	−31.25

[a]See Equation 4-35 and Table 4-1.

Table 23-2 Contributions of Primary CH_3 Substitution Groups Replacing Hydrogen on Carbon

Base Group	$\Delta(\Delta Hf^{\circ}_{298.1}(g))$/ kJ mol^{-1}	$\Delta s^{\circ}_{298.1}(g)$/ J mol^{-1} K^{-1}	Heat Capacity Constants for Ideal Gas at T/K		
			Δa	$\Delta b(10^3)$	$\Delta c(10^6)$
Methane	−9.2	43.5	−8.54	100.42	−40.46
Cyclopentane					
Enlargement of ring	−38.9	2.9	−4.35	80.75	−24.23
First substitution	−21.8	48.1	−0.29	77.70	−24.14
Second substitution:					
ortho	−51.0				
meta	−35.1		−1.00	69.29	−21.13
para	−29.7				
Third substitution	−29.3				
Benzene and naphthalene					
First substitution	−18.8	50.2	1.51	73.85	−24.60
Second substitution:					
ortho	−26.4	33.9	21.76	25.19	4.94
meta	−27.2	38.5	7.20	59.33	−15.73
para	−33.5	32.6	5.36	60.96	−16.65
Third substitution (*sym*)		33.5	2.38	69.08	−21.71
Methylamine	−23.8				
Dimethylamine	−25.4		−0.42	73.30	−22.38
Trimethylamine	−17.2				
Formamide					
Substitution on C atom	−37.7		25.56	− 7.32	19.87

Table 23-3 Type Numbers of Different Carbon Atoms

Type Number	Nature
1	—CH_3
2	—CH_2—
3	—CH< (three bonds)
4	>C< (four bonds)
5	C in benzene or naphthalene ring

Table 23-4 Secondary Methyl Substitutions Replacing Hydrogen

A	B	$\Delta(\Delta Hf^\circ_{298.1}(g))$/ kJ mol^{-1}	$\Delta S^\circ_{298.1}(g)$/ J mol^{-1} K^{-1}	Heat Capacity Constants for Ideal Gas at T/K: Δa	$\Delta b(10^3)$	$\Delta c(10^6)$
1	1	−18.8	41.0	−4.06	95.65	−36.61
1	2	−21.8	38.5	4.64	77.28	−28.66
1	3	−23.0	39.7	4.18	83.18	−33.60
1	4	−20.9	46.0	5.82	71.63	−24.60
1	5	−25.5	41.8	0.42	71.88	−21.76
2	1	−27.6	24.3	7.91	73.64	−25.98
2	2	−28.5	29.3	6.36	83.47	−35.86
2	3	−28.5	26.4	4.23	82.38	−32.76
2	4	−21.3	25.1	10.54	67.40	−24.60
2	5	−24.3	11.3	0.04	72.89	−22.30
3	1	−33.9	11.3	−4.02	114.93	−51.80
3	2	−33.5	20.1	−4.98	120.37	−53.18
3	3	−28.9	24.3	−13.68	129.54	−58.83
3	4	−23.8	7.1	−0.59	120.80	−42.97
3	5	−38.5	5.4	1.76	67.78	−19.58
1	—O—in ester or ether	−29.3	60.2	−0.04	73.55	−22.30
Substitution of H of an OH group to form an ester		+39.7	69.9	1.84	69.58	−20.71

Table 23-5 Multiple-Bond Contributions Replacing Single Bonds

A	Type of Bond	B	$\Delta(\Delta Hf^\circ_{298.1}(g))$/ kJ mol^{-1}	$\Delta s^\circ_{298.1}(g)$/ J mol^{-1} K^{-1}	Heat Capacity Constants for Ideal Gas at T/K: Δa	$\Delta b(10^3)$	$\Delta c(10^6)$
1	=	1	137.2	−8.8	5.56	−53.09	+19.96
1	=	2	125.5	3.3	6.53	−62.22	+23.30
1	=	3	118.0	9.2	2.64	−98.95	+54.81
2	=	2	117.2	−3.8	1.67	−78.95	+41.38
2	=	2 cis	118.8	−2.5	1.67	−78.95	+41.38
2	=	2 trans	115.1	−5.0	1.67	−78.95	+41.38
2	=	3	111.7	6.7	2.64	−98.95	+54.81
3	=	3	106.7		−19.37	−74.64	+49.71
Additional correction for each pair of conjugated double bonds			−15.9	−43.5	Approximately zero		
1	≡	1	311.3	−28.5	23.35	−130.50	+46.82
1	≡	2	289.1	−32.6	26.86	−152.34	+60.79
2	≡	2	272.4	−26.4	19.50	−151.04	+63.93
Correction for double bond adjacent to aromatic ring			−21.3	−18.0	Approximately zero		

The effect of introducing multiple bonds in a molecule is treated separately. The appropriate corrections have been assembled in Table 23-5 and require no special comments, except perhaps to emphasize the *additional* contribution that must be introduced every time a pair of conjugated double bonds is formed by any of the preceding substitutions in this table.

Finally, we will consider the changes in properties accompanying the introduction of various functional groups in place of one or two of the methyl groups on a given carbon atom. These are listed in Table 23-6. Data from Table 23-1 and Table 23-2 give the contributions for the appropriate methyl-substituted base groups. Observe particularly that the =O structure requires replacement of two methyl groups.

The procedure followed in the use of the tables of Andersen, Beyer, and Watson has been described for the estimation of standard entropies. These tables also include columns of base structure and group contributions for estimating $\Delta Hf^\circ_{298.1}$, the standard enthalpy of formation of a compound, as well as *a, b*, and *c*, the constants in the heat capacity equations described in

Table 23-6 Substitution-Group Contributions Replacing CH_3 Group

Group	$\Delta(\Delta Hf^\circ_{298.1}(g))$/ kJ mol^{-1}	$\Delta s^\circ_{298.1}(g)$/ J mol^{-1} K^{-1}	Heat Capacity Constants for Ideal Gas at T/K Δa	$\Delta b(10^3)$	$\Delta c(10^6)$
—OH (aliphatic, *meta, para*)	−136.8	10.9	13.26	−62.17	23.39
—OH *ortho*	−199.6				
$—NO_2$	5.0	8.4	26.4	−81.71	43.35
—CN	163.2	54.8	15.23	−58.24	18.95
—Cl	0 for first Cl on a carbon; 18.8 for each additional Cl	0	9.16	−78.87	26.19
—Br	41.8	12.6[a]	11.76	−81.21	26.48
—F	−146.4	−4.2[a]	9.37	−98.78	49.33
—I	103.8	20.9[a]	11.42	−72.68	17.11
=O aldehyde	− 54.0	−51.5	15.10	−232.71	95.06
=O ketone	− 55.2	−10.0	21.00	−276.48	126.40
—COOH	−364.0	64.4	35.56	−63.05	33.22
—SH	66.1	21.8	17.03	−104.43	51.76
$—C_6H_5$	135.1	90.8	−3.31	224.39	80.37
$—NH_2$	51.5	−19.7	5.27	−30.63	9.33

[a]Add 4.2 to the calculated entropy contributions of halides for methyl derivatives; for example, methyl chloride = 185.8(base) + 43.5(primary CH_3) − 0.0(Cl substitution) + 4.2.

Equation 4-35. Thus it is possible to estimate $\Delta G^\circ_{298.1}$ by appropriate summations of group contributions to $\Delta H^\circ_{298.1}$ and $\Delta S^\circ_{298.1}$. Then, if information is required at some other temperature, the constants of the heat capacity equations can be inserted into the appropriate equations for ΔG° as a function of temperature and ΔG° can be evaluated at any desired temperature (see Equation 8-80).

Typical Problems in Estimating Entropies

The use of Tables 23-1 through 23-6 will be illustrated by two examples, using the Andersen–Beyer–Watson method.

Example 1. Estimate the entropy, $s^\circ_{298.1}$, of *trans*-2-pentene(g).

	Contribution
Base group, H—C(H)(H)—H	186.2
Primary CH_3 substitution → CH_3—CH_3	43.5
Secondary CH_3 substitutions → CH_3—CH_2—CH_2—CH_2—CH_3:	

Type numbers: Carbon A	Carbon B	$\Delta s^\circ_{298.1}$
1	1	41.0
1	2	38.5
1	2	38.5
Introduction of double bond at 2-position:		
2	2 *trans*	−5.0
Summation of group contributions		342.7 J K^{-1} mol^{-1}
Experimental value (NBS tables)		342.29 J K^{-1} mol^{-1}

Example 2. Estimate the entropy, $s^\circ_{298.1}$, of acetaldehyde(g).

Base group, H—C(H)(H)—H	186.2
Primary CH_3 substitution → CH_3—CH_3	43.5
Secondary CH_3 substitutions → CH_3—CH(CH_3)—CH_3:	

Type numbers: Carbon A	Carbon B	$\Delta s^\circ_{298.1}$
1	1	41.0
2	1	24.3

Substitution of =O replacing 2 —CH_3 groups → CH_3—C(H)=O	−51.5
Summation of group contributions	243.5 J K^{-1} mol^{-1}
Experimental value (NBS tables)	250.2 J K^{-1} mol^{-1}

These examples illustrate the procedure used in the Andersen–Beyer–Watson method. The first example shows unusually good agreement; the second, unusually poor. Generally, it is preferable to consider the group substitutions in the same order as has been used in the presentation of the tables. Experience shows that when more than one base group is possible, the one with the largest entropy should be chosen. The best agreement with experimental values, when they are known, has been obtained by using the minimum number of substitutions necessary to construct the molecule. For cases in which several alternate routes with the minimum number of substitutions are possible the average of the different results should be used.

Other Methods

Although the tables presented by Parks and Huffman [2] are based on older data, they are often more convenient to use, since they are simpler and because they have been worked out for the liquid and solid states as well as for the gaseous phase. A complete survey and analysis of methods of estimating thermodynamic properties is available in Janz's monograph [3]. Thermodynamicists should have a general acquaintance with more than one method of estimating entropies so they can choose the best method for a particular application.

Accuracy of the Approximate Methods

Free energy changes and equilibrium constants calculated from the enthalpy and entropy values estimated by the group-contribution method generally are reliable only to the order of magnitude. For example, Andersen, Beyer, and Watson have found that their estimated enthalpies and entropies usually differ from experimental values [4] by less than 16.7 kJ mol^{-1} and 8.4 J mol^{-1} K^{-1}, respectively. If errors of this magnitude occurred cumulatively, the free energy change would be incorrect by approximately 19.2 kJ mol^{-1} near 25°C. Such an error in the free energy corresponds to an uncertainty of several powers of 10 in an equilibrium constant. With few exceptions such an error is an upper limit. Nevertheless, it must be emphasized that approximate methods of calculating these thermodynamic properties are reliable for estimating the feasibility of a projected reaction, but are not adequate for calculating equilibrium compositions to better than the order of magnitude.

EXERCISES

1. a. Estimate $S^{\circ}_{298.15}$ for *n*-heptane (gas) by the group-contribution method of Andersen, Beyer, and Watson. Compare with the result obtainable from the information in Exercise 11-15.

b. Estimate $s^\circ_{298.15}$ for liquid *n*-heptane from the rules of Parks and Huffman. Compare with the result obtained in (a).

2. a. Using the group-contribution method of Andersen, Beyer, and Watson, estimate $s^\circ_{298.15}$ for 1,2-dibromoethane(g).

   b. Calculate the entropy change when gaseous 1,2-dibromoethane is expanded from 1 atm to its vapor pressure in equilibrium with the liquid phase at 298.15 K. Neglect any deviations of the gas from ideal behavior. Appropriate data on vapor pressures have been assembled conveniently by D. R. Stull, *Ind. Eng. Chem.* **39**, 517 (1947).

   c. Using the data in Stull's publication, calculate the heat of vaporization of 1,2-dibromoethane at 298.15 K.

   d. Calculate the entropy, $s^\circ_{298.15}$, for liquid 1,2-dibromoethane.

   e. Compare the estimate obtained in (d) with that obtainable from the rules of G. S. Parks and H. M. Huffman in *The Free Energies of Some Organic Compounds*, Reinhold, New York, 1932.

   f. Compare the estimates of (d) and (e) with the value found by K. S. Pitzer, *J. Am. Chem. Soc.* **62**, 331 (1940).

3. Recent precision measurements of heats of formation and entropies are probably accurate to perhaps 250 J mol^{-1} and 0.8 J mol^{-1} K^{-1}, respectively. Show that either one of these uncertainties corresponds to a change of 10% in an equilibrium constant at 25°C.

4. It has been suggested [G. J. Janz, *Can. J. Res.* **25B**, 331 (1947)] that 1,4-dicyano-2-butene might be prepared in the vapor phase from the reaction of cyanogen with butadiene.

   a. Estimate ΔHf° and s° for the dicyanobutene at 25°C by the group-contribution method.

   b. With the aid of the following additional information, calculate the equilibrium constant for the suggested reaction:

Substance	$\Delta Hf^\circ_{298.15}$/J mol^{-1}	$s^\circ_{298.15}$/J mol^{-1} K^{-1}
Butadiene(g)	111,914	277.90
Cyanogen(g)	300,495	241.17

5. a. Estimate ΔHf° and s° for benzonitrile, $C_6H_5CN(g)$, at 750°C by the group-contribution method using toluene as the parent compound.

   b. Combining the result of (a) with published tables, estimate ΔG° at 750°C for the reaction

$$C_6H_6(g) + (CN)_2(g) = C_6H_5CN(g) + HCN(g)$$

An estimate of -77.0 kJ mol^{-1} has been reported by G. J. Janz, *J. Am. Chem. Soc.* **74**, 4529 (1952).

REFERENCES

1. G. S. Parks and H. M. Huffman, *The Free Energies of Some Organic Compounds*, Reinhold, New York, 1932; J. W. Andersen, G. H. Beyer, and K. M. Watson, *Natl. Petrol. News, Tech. Sec.* **36**, R476 (July 5, 1944); D. W. Van Krevelen and H. A. G. Chermin, *Chem. Eng. Sci.* **1**, 66 (1951); S. W. Benson and J. H. Buss, *J. Chem. Phys.* **29**, 546 (1958).
2. G. S. Parks and H. M. Huffman, *The Free Energies of Some Organic Compounds*, Reinhold, New York, 1932.
3. G. J. Janz, *Estimation of Thermodynamic Properties of Organic Compounds*, rev. ed., Academic Press, New York, 1967.
4. D. D. Wagman et al., "The NBS Tables of Chemical Thermodynamic Properties," *J. Phys. Chem. Reference Data* **11**, Supplement 2 (1982), p. 2–12.

CHAPTER 24

Concluding Remarks

With the discussion of the estimation of thermodynamic quantities, we conclude our technical considerations of the theory and methods of chemical thermodynamics. We have achieved our primary objective, in that we have established the principles and procedures by which the thermodynamic properties associated with a given transformation can be determined; and we have learned how these quantities can be used to judge the feasibility of change of state.

However, in emphasizing these aspects of the subject, we have neglected numerous broad fields within the realm of thermodynamics. Even within the areas to which we have limited ourselves, we have omitted any discussion of surface reactions [1], and have paid only superficial attention to problems of phase equilibria [2] and to electrochemical processes [3]. We also could have examined some of the topics of more theoretical interest, such as relativity and cosmology [4]. Similarly we could have considered chemical and phase equilibria at high temperature and pressure [5].

The point of view adopted toward thermodynamics in this book is the classical or phenomenological one. This approach is in a sense the most general but also the least illuminating in molecular insight. The three basic principles of phenomenological thermodynamics are extracted as postulates from general experience and no attempt is made to deduce them from equations describing the mechanical behavior of material bodies. Since it is independent of the laws governing the behavior of material bodies, classical thermodynamics cannot be used to derive any of these laws. Generally, thermodynamic reasoning leads to relationships between certain physical quantities, but classical thermodynamics does not allow us to calculate actual values of any of the quantities appearing in these relationships.

The phenomenological approach was inaugurated more than a century ago and reached its fruition in theoretical formulation near the turn of the century. Since then the major extension has been toward an analysis of nonequilibrium, nonisothermal processes. With the aid of additional phenomenological postulates, such as linear relationships between certain rates and appropriate forces, plus the Onsager reciprocity relationships, a conceptual system has been developed that is capable of analyzing a broad class of irreversible processes [6]. More recently, the laws of classical thermodynamics have been recast in the form of a Euclidean metric geometry whereby its formulas can be read from simple diagrams. It has been suggested that the relationship between the geometric representation of thermodynamics and the differential equations of Gibbs is analogous to the relationship between the matrix mechanics of Heisenberg and the wave mechanics of Schrodinger [7].

However, parallel with the phenomenological development there has arisen an alternative point of view toward thermodynamics—a statistical-mechanical approach. Its philosophy is more axiomatic and deductive than phenomenological. The kinetic theory of gases naturally led to attempts to derive equations describing the behavior of matter in bulk from the laws of mechanics (first classical, then quantum) applied to molecular particles. Since the number of molecules is so great, a detailed treatment of the mechanical problem presents insurmountable mathematical difficulties, and statistical methods are used to derive average properties of the assembly of molecules and of the system as a whole.

In the field of thermodynamics, statistical mechanics has provided a molecular model, which leads to a more concrete visualization of some of the abstract concepts (such as entropy) of classical thermodynamics. In addition, it has developed means for the analysis of microscopic fluctuation phenomena, such as Brownian motion and the density fluctuations that are the basis of light scattering. Furthermore, it has extended the range of thermodynamic reasoning to new kinds of experimental data such as spectroscopic properties of matter, and has been fundamental to the building of a bridge between the thermodynamics and the kinetics of chemical reactions. For these reasons a knowledge of statistical thermodynamics is essential as a companion to phenomenological thermodynamics for the effective solution of many current problems and for the formulation of stimulating new questions [8].

Although we have indicated some applications of thermodynamics to biological systems, more extensive discussions are available [9]. The study of equilibrium involving multiple reactions in multiphase systems and the estimation of thermodynamic properties are now easier as a result of the development of computers and appropriate algorithms [10].

REFERENCES

1. G. D. Holsey and C. M. Greenlief, *Ann. Rev. Phys. Chem.,* **21**, 129 (1970).
2. A. Reisman, *Phase Equilibria, Basic Principles, Applications, and Experimental Techniques*, Academic Press, New York, 1970.

3. B. Conway, J. O'M. Bockris, and E. Yeager, Eds., "Comprehensive Treatise of Electrochemistry," Vol. 5, *Thermodynamics and Transport Properties of Aqueous and Molten Electrolytes*, Plenum Press, New York, 1983.

4. B. Gal-Or, Ed., *Modern Developments in Thermodynamics*, John Wiley and Sons, New York, 1974.

5. O. Kubaschewski, P. J. Spencer, and W. A. Dench, in *Chemical Thermodynamics*, Vol. 1, The Chemical Society, London, 1973, p. 317. C. L. Young, in *Chemical Thermodynamics*, Vol. 2, The Chemical Society, London, 1978, p. 71.

6. I. Prigogine, *Introduction to Thermodynamics of Irreversible Processes*, Charles C. Thomas, Springfield, Illinois, 1955. A. Katchalsky and P. F. Curran, *Nonequilibrium Thermodynamics in Biophysics*, Harvard University Press, Cambridge, 1965.

7. F. Weinhold, "Thermodynamics and Geometry," *Physics Today*, March 1976.

8. H. L. Friedman, *A Course in Statistical Mechanics*, Prentice-Hall, Englewood Cliffs, N.J., 1985.

9. J. T. Edsall and H. Gutfreund, *Biothermodynamics*, John Wiley and Sons, New York, 1983. I. M. Klotz, *Introduction to Biomolecular Energetics*, Academic Press, New York, 1986.

10. F. Van Zeggern and S. H. Storey, *The Computation of Chemical Equilibria*, Cambridge University Press, Cambridge, 1970. W. R. Smith and R. W. Missen, *Chemical Reaction Equilibrium Analysis: Theory and Algorithms*, Wiley-Interscience, New York, 1982. W. J. Lyman, W. F. Reehl, and D. H. Rosenblatt, Eds., *Handbook of Chemical Property Estimation Methods*, McGraw-Hill, New York, 1982.

Index